P9-CBL-189

The Rocket Team

BY THE AUTHORS

HISTORY OF ROCKETRY & SPACE TRAVEL
Third Revised Edition
(Frederick I. Ordway III and Wernher von Braun)

DIVIDENDS FROM SPACE
(Frederick I. Ordway III, Mitchell R. Sharpe,
and Carsbie Adams)

PICTORIAL GUIDE TO PLANET EARTH
(Frederick I. Ordway III)

IT IS I, SEAGULL
the story of Valentina Tereshkova,
the first woman astronaut
(Mitchell R. Sharpe)

The Rocket Team

Frederick I. Ordway III
Mitchell R. Sharpe

FOREWORD BY WERNHER VON BRAUN

Thomas Y. Crowell, Publishers
Established 1834
New York

 For information address Thomas Y. Crowell, Publishers, 521 Fifth Avenue, New York, N.Y. 10017.

Library of Congress Cataloging in Publication Data
Ordway, Frederick Ira
The Rocket Team.
1. Project Saturn. I. Sharpe, Mitchell R., joint author.
II. Title.
TL789.8.U6S36 629.47′52′09 78–3313
ISBN 0–690–01656–5

---

79 80 81 82 83 10 9 8 7 6 5 4 3 2

Contents

	Foreword by Wernher von Braun	xi
	Preface	xiii
1	"Take Us to Ike!"	1
2	Countdown at Raketenflugplatz, Berlin	12
3	From Berlin to the Baltic	21
4	The Flowering of Peenemünde	30
5	Producing the Missiles	60
6	Intelligence and Reconnaissance	91
7	Attacking the Missile Installations	111
8	Events in Poland and Sweden	130
9	A4 to V-2: the Final Months	160
10	V-Weapon Offensive	174
11	Defense Against the V-Weapons	206
12	Effectiveness of the V-Weapon Campaign	239
13	Exodus from Peenemünde	254
14	Roundup in Bavaria	271
15	Backfire at Cuxhaven	294
16	The New Beginning	310
17	V-2s on the Steppes	318
18	V-2s in the Desert	344

19 Thrust into Space 363
Epilogue 401
Appendix A: Production Quantities of A4s 405
Appendix B: Addenda on Intelligence 409
Bibliography 414
Index 451

List of Illustrations

Between pages 78 and 79

Flames in Nazi Headquarters at Reutte, April 30, 1945 (*courtesy of the US Army*)

At Reutte, Austria, April 1945 (*Ordway*)

Set from *Frau im Mond* (*courtesy of the Smithsonian Institution*)

Following successful test of *Kegelduse* rocket, August 5, 1930 (*Ordway*)

Püllenberg with model of his proposed postal rocket (*courtesy of Albert Püllenberg*)

G. Edward Pendray and Klaus Riedel at Raketenflugplatz, April, 1931 (*Sharpe*)

Max Valier and Paul Heylandt (*Ordway*)

Hitler visits Kummersdorf in 1939 (*courtesy of Walter B. Dornberger*)

Von Braun and Bahr sailing in the Baltic Sea, about 1942 (*Sharpe*)

The *Wasserfall* missile (*courtesy of Wernher von Braun*)

Preparing the A3 rocket for launching, November 29, 1937 (*courtesy of the Deutsches Museum, Munich*)

Dornberger describes V-2 launching (*courtesy of Walter B. Dornberger*)

Peenemünde West in 1942 (*Sharpe*)

Housekeeping rules at Peenemünde (*courtesy of Wernher von Braun*)

Russian prisoners of war at Peenemünde (*courtesy of Wernher von Braun*)

A4 propulsion unit at Niedersachswerfen (*Ordway*)

Col. T. R. B. Sanders with stacked A4 combustion chambers (*Ordway*)

Entrance to *Rebstock*, 1944 (*Ordway*)

Between pages 174 and 175

Assembly of A4 ground equipment in *Rebstock*, 1944 (*Ordway*)

A4 engine undergoing test firing near Lehesten (*Ordway*)

Prof. Frederick A. Lindemann, Viscount Cherwell (*courtesy of the Imperial War Museum*)

Duncan Sandys, Churchill's son-in-law (*courtesy of the Imperial War Museum*)

Flight Officer Constance Babington Smith in her RAF office at Medmenham (*courtesy of Constance Babington Smith*)

Test Stand 7 experimental area (*courtesy of the Royal Air Force*)

Fi103 (V-1) on its launch ramp at Peenemünde (*courtesy of the Imperial War Museum; Constance Babington Smith*)

Section of Peenemünde experimental R & D establishment (*courtesy of the Imperial War Museum*)

Section of Peenemünde where numerous huts were later destroyed (*courtesy of the Imperial War Museum*)

Section of the plant under attack (*courtesy of the Imperial War Museum*)

Test Stand 7 experimental area after bombardment (*courtesy of the Imperial War Museum*)

V-1 launching platform (*courtesy of the Imperial War Museum*)

RAF reconnaissance photograph, 1944 (*courtesy of the Imperial War Museum*)

Landscape created by Allied bombing at Wizernes (*courtesy of the Imperial War Museum*)

Missile installation at Watten after RAF attacks (*Ordway*)

S. G. Culliford in his Dakota DC-3 in Italy (*courtesy of S. G. Culliford*)

RAF reconnaissance photograph of Blizna, Poland (*courtesy of the Imperial War Museum*)

Between pages 270 and 271

Heidelager camp at Blizna, Poland (*courtesy of Col. T. R. B. Sanders*)

British team looks at recovered German V-2 combustion chamber (*courtesy of Col. T. R. B. Sanders*)

Meilerwagen V-2 rocket trailers in the Haagsche Bosch (*courtesy of the Air Ministry; Constance Babington Smith*)

Explosion of German V-2 at Lange Nieux, November 27, 1944 (*courtesy of the US Army*)

Scene of balloon defense site where flying bomb was brought down (*courtesy of the Imperial War Museum*)

SS Gen. Kammler arriving at the V-1 site at Saleux, August 10, 1944 (*courtesy of Col. Max Wachtel*)

Air Chief Marshal Sir Roderic Hill (*courtesy of the Imperial War Museum*)

A V-1 in flight over England (*courtesy of the Imperial War Museum*)

This V-1 landed almost intact in the English countryside (*courtesy of the Imperial War Museum*)

V-1 just before impacting on London (*courtesy of the Imperial War Museum*)

V-2 rockets, 1945 (*courtesy of the Imperial War Museum*)

Entrance to mine at Dörnten (*courtesy of Dieter K. Huzel*)

Hitler inspects bomb damage, 1944 (*courtesy of the Imperial War Museum*)

V-2 removed from the Mittelwerk, May 31, 1945 (*courtesy of the Imperial War Museum*)

Winston Churchill talking with General Eisenhower in Germany, March 25, 1945 (*courtesy of the Imperial War Museum*)

Dependents' housing at Camp Overcast, 1945 (*Sharpe*)

Between pages 366 and 367

Hans Hüter and von Braun about to board C-47 at Munich, July, 1945 (*courtesy of Georg von Tiesenhausen*)

Operation Backfire, October, 1945 (*courtesy of Lt. Col. W. S. J. Carter*)

At Operation Backfire (*courtesy of Lt. Col. W. S. J. Carter*)

Five Russian officers show up at Operation Backfire (*courtesy of the Imperial War Museum*)

White Sands Proving Ground, New Mexico, 1945 (*Sharpe*)

Von Braun group at White Sands, 1946 (*courtesy of the US Army*)

Revealing the Explorer Satellite in 1957 (*courtesy of the US Army*)

Press Conference following the launch of Explorer I Satellite, January 31, 1958 (*courtesy of the US Army*)

Maj. Gen. Toftoy, von Braun and various missiles (*courtesy of the US Army*)

Tense moment at Cape Canaveral (*courtesy of the US Army*)

Just before a heated discussion over the Moon (*courtesy of NASA*)

Von Braun's "hardware" (*courtesy of NASA*)

Second Saturn Apollo space vehicle on its way to the launch pad (*courtesy of NASA*)

Foreword

Wernher von Braun read an essentially complete copy of the manuscript of The Rocket Team *some nine months before he died on June 16, 1977. This is his appraisal of it.*

This book is more than the story of events associated with Peenemünde, of the Allied response to German wartime missile developments, and of post-World War II rocketry and space flight. It is also the story of thousands of dedicated men and women from the old world and the new with whom, for some four decades, I have proudly served.

The authors have done an excellent job of telling the story of our team. They worked for many years gathering the material for this book, principally in the United States and Europe—though one of them ranged as far as New Zealand. Obviously, this book, as big as it is, is not the full story; no single volume could include every name and every interesting event, considering the length of the period and the size of the team. Nor can the reader expect that with so many people questioned, there will not be contradictions or differences in memory and judgment. Because of this, the authors constantly had to exercise their own judgment as they sifted through the tremendous mass of original source material that they had accumulated.

If I had been writing this book myself, it is possible that I would have told some things differently, but I have not immersed myself in the sources as they have. However, having said this, it is only fair to add that I think the authors have produced as complete and as accurate an account as we are likely ever to see, and I commend it to my fellow team members as well as to the general public.

I would like to point out, too, that it is a book essentially without end.

The efforts of our team did not cease with the landing of twelve men on the Moon and the launching of the Skylab embryonic space station with its three crews, the last of which set a record of eighty-four science-filled days in orbit around our planet Earth. Some members are at this moment helping develop the space shuttle and its key payload, the Spacelab, which together will become the major elements in tomorrow's reusable space transportation system and its versatile array of equally reusable laboratory equipment. The first shuttle orbiter vehicle, incidentally, was rolled out of its assembly plant in California in mid-September 1976 just as I was preparing this Foreword. It was named Enterprise by President Gerald R. Ford. While the shuttle program is advancing on schedule, still other team members are looking ahead to space manufacturing facilities and even permanent colonies of hundreds and perhaps thousands of people who will live and work in large space stations and space bases high above the Earth. And some are investigating ways that the bounteous energy of the Sun may be utilized here on the ground by first capturing it in arrays of solar collector stations up in orbit.

While the members of this magnificent team changed with time, the fundamental characteristics of the team itself never did. It always has been characterized by enthusiasm, professionalism, skill, imagination, a sense of perfectionism, and dedication to rocketry and space exploration. How can the story of such people and of the exciting programs with which they are involved ever end?

Wernher von Braun

Preface

The Rocket Team spans a couple of generations, from the late 1920s to the early 1970s. These years witnessed the Great Depression, World War II, the opening and maturing of the Atomic Age, the Cold War, Vietnam, the awakening of the environmental ethic, and humanity's early probings into space.

The men whose story we tell lived through this period of turmoil, change, and challenge. They started out as enthusiastic amateurs, cheering when their homemade rockets soared briefly into the air. Spurred by the fervor of preparations for war, and later of war itself, they organized themselves into one of history's most sophisticated and intensely supported research and development teams. The results of their work are fundamental and far-reaching.

During the final year of World War II, this team ushered in a new era of missile-based weaponry that has dominated military tactics and strategy ever since. Transferred to the United States, their former enemy, the nucleus of this team helped develop and guide the technology that carried man to the Moon two and a half decades later.

Both of us have known and worked for years with many members of this team whose exploits and activities we describe. This association has given us a close, and perhaps even special, insight into their motives, aspirations, and dreams. At least, such is our hope.

It took us about ten years of planning, research, and writing before the job was over. Extensive travel was partly responsible for this long period of time. In order to conduct some two hundred direct interviews, we had to go to many parts of the United States as well as to Germany, the United Kingdom, Belgium, the Netherlands, Poland, the Soviet Union, France,

Canada, Mexico, and New Zealand. These interviews, most of which were taped, are being stored for later use by historians of science and technology.

Another cause of delay in completing the manuscript was the security classification of much of the documentation. On the Allied side, for example, memos, reports, and papers relating to V-weapon intelligence, surveillance, reconnaissance, and countermeasures were often found still to be classified at the secret or even top-secret level. Lengthy, and sometimes rather involved, procedures had to be undertaken to secure downgrading. This resulted in the fact that ninety percent of the manuscript was completed several years before the last ten percent could be written. We are deeply grateful to many research, development, archival, and other organizations, including elements of the military services, for their collaboration in the documentation acquisition phase of our work.

We are equally grateful to many others who helped make this book possible. During the years of research, we were assisted by David L. Christensen, Ruth Heimburg, Elfriede J. Paetz, and Lily von Saurma (now von Maldeghem). Deserving of our deepest appreciation also is Mrs. Hazel Toftoy, widow of the late Major General Holger N. Toftoy, a part of whose story is told herein. She freely made available to us not only the files and photographs of her husband, but many delightful and informative memories of events that unfold in this book.

Dr. Wernher von Braun, former technical director of the German rocket and missile research and development center at Peenemünde and later director of NASA's George C. Marshall Space Flight Center, not only reviewed virtually the entire manuscript before he died in mid-1977, but throughout its preparation gave untiring aid, encouragement, and guidance. To his former military commander in Germany, Dr. Walter Dornberger, and to his mentor, Professor Hermann Oberth—both pioneers of rocketry and astronautics—we are also indebted.

Others, one way or another, involved in German prewar and wartime missilery to whom we extend our sincere appreciation are Albert Speer, Eberhard Rees, Arthur Rudolph, Ernst Stuhlinger, Rudolf Hermann, Walter Häussermann, Hermann Weidner, Krafft Ehricke, Ernst Steinhoff, Konrad and Ingeborg Dannenberg, Erich Neubert, Helmut J. Horn, Karl Heimburg, Hans H. Hüter, Walter Wiesman, Werner K. Rosinski, Hermann Ludewig, Ruth Hermann, Albert E. Schuler, Hans Maus, Otto Hirschler, Heinrich A. Schulze, Gerhard Drawe, Willibald Prasthofer, Otto F. Cerny, Ernst Klauss, Fridtjof Speer, Theodor Vowe, Gerhard Reisig, Hans H. Hosenthien, Werner K. Gengelbach, Friedrich Dohm, Dieter Grau, Georg von Tiesenhausen, Helmut Hölzer, Carl H. Mandel, Bernhard R. Tessmann, Robert Paetz, Alfred J. Finzel, Hermann W. Kröger, Erich Görner, Wilhelm A. Schulze, William A.

Mrazek, Hans W. Milde, Max E. Nowak, Otto A. Hoberg, Ernst D. Geissler, Leopold A. Hein, Ruth Hüter, Hans Billmeyer, Albin Wittmann, Wilhelm Angele, Werner Sieber, Erich Kaschig, Karl Reilmann, Werner Voss, Hans Palaoro, Friedrich Duerr, Hans Grüne, Theodor Poppel, Walter Jacobi, Willi Heybey, Heinz Hilton, Hans Paul, Heinz Kampmeier, Eric Wormser, Albert Zeiler, Kurt Debus, Emile Hellebrand, Gerhard Heller, Martin Schilling, and Ruth von Saurma. The order in which these names appear in no way reflects the relative degree of assistance given, but simply the sequence in which they were interviewed.

Among the interviews conducted in Germany were Professor Hermann Oberth, General Leo Zanssen, Kurt Kettler, Paul H. Figge, Lieutenant Colonel Wilhelm Zippelius, Max Meyer, Rudolf and Dorette Schlidt, Heinz A. Millinger, Hannes Lührsen, Eberhard Spohn, Klaus H. Scheufelen, Eduard M. Fischel, Herbert Axster, Hanns Weidinger, Oscar Scholze, Walter Schuran, Hannelore Ranft, and Ernst Klee. The late Dipl-Ing Klee was especially helpful in making available invaluable documentation at the Deutsches Museum in Munich, with which he was affiliated. Boris Kit, Krafft Ehricke, A. F. Staats, E. Roth-Oberth, and Werner Büdeler offered their assistance in the conduct of selected German interviews. We are indebted to M. Regel and Dr. Haupt at Bundesarchiv in Coblenz for their cooperation with the research. We also acknowledge the Berlin Document Center, Mission of the United States of America, Georg Borittscheller, Werner Brähne, Bundesarchiv-Kornelimünster, Deutsche Dienststelle, Rolf Engel, Heinrich Hertel, Harry O. Ruppe, Eduard Igensbergs, the Institut für Zeitgeschichte, Archiv, Interessengemeinschaft der Ehemaligen Peenemünder, the Militärgeschichtliches Forschungsamt in Freiburg, Albert Püllenberg, Irene Sänger-Bredt, *Der Spiegel*, Colonel Max Wachtel, Carl Wagner, Heinz Grösser, and Horst W. Schneider.

A particularly fascinating part of the V-weapon history was played by the Polish and Netherlands underground movements. Colonel Kazimierz Iranek-Osmecki, Michal Wojewódzki, and Tadeus E. Schnitzer helped us immeasurably from the Polish side, and J. M. J. Kooy from the Dutch side. We thank also the Studium Polski Podziemnej – Study Trust, London.

We are grateful for the aid in Great Britain of Professor R. V. Jones, head of scientific intelligence for the Air Ministry during World War II; the Rt. Hon. Duncan Sandys, who coordinated all British investigations of German missile activities; Colonel Terence R. B. Sandys, who was involved in surveying V-weapon installations along the Channel coast as well as in Poland; Colonel W. S. J. Carter, who was closely involved with Operation Backfire; Sir William R. Cook; Constance Babington Smith; Colonel Kenneth Post; Captain G. O. C. Davies; Jack Hanson; the late

Colonel A. D. Merriman; M. G. J. Gollin; Colonel N. L. Falcon; General Sir Frederick Pile; Alfred Price; David Irving; Joan Banger, and Anthony Cave-Brown. Interviews with S. G. Culliford, who flew into occupied Poland during the war, and Douglas N. Kendall, who coordinated V-weapon interpretation at Medmenham from mid-1943, were conducted, respectively, in Wellington, New Zealand and Toronto, Canada. We are grateful to both, and also to the support of Anees Jung, C. Holley Taylor-Martlew, and Frederick I. Ordway IV.

Among the British organizations that lent us help are the Imperial War Museum and E. Hine, the Ministry of Defence, the Air Historical Branch, the Royal Aircraft Establishment, the Public Records Office, the Central Army Historical Library, the Naval Historical Section, the Ministry of Defence War Office Library and Air Department Library, Ministry of Aviation Supply Central Library, Central Library—Local Studies Library, Norfolk County Council, Norwich; Central Library, Suffolk County Council, Ipswich; Cambridge University Library, and the Science Museum.

The United States came into the V-weapon picture rather late in the war, but soon assumed a dominant role in the exploitation of German missile talent and components alike. We are grateful to Dr. Richard W. Porter, who played a key role in Paperclip events; Colonel Gervais W. Trichel, who, as chief of the Rocket Branch within US Army Ordnance, urged then Colonel Toftoy to locate and ship to the United States a hundred V-2 missiles; Charles W. Stewart, then a lieutenant, who was the first responsible American to interview von Braun, Dornberger and others after their surrender to US forces; and Colonel James P. Hamill, who helped the postwar V-2 program get off the ground at Fort Bliss and White Sands. We also acknowledge the cooperation of Walt W. Rostow, Colonel James Miller, G. Edward Pendray, Charles Hester, James J. Fagan, Tom Moore, Jose L. Gonzales, James Bramlet, Ramón Samaniego, M. S. Hochmuth, Warner S. Ray, Colonel Brian O. Montgomery, and Robert B. Staver.

We are most appreciative of the cooperation of Helmut Gröttrup, the man who rightly or wrongly has been called the "von Braun of the USSR." He worked untiringly to synthesize at this rather late date the complete story of the Germans who cast their lot with the Russians following World War II.

Other organizations and individuals to whom we are indebted include: Alabama Space and Rocket Center, Huntsville, and Edward O. Buckbee; Ambassade de France, l'Attaché Scientifique Jean-Pierre Pujes; CEDOCAR, Paris; Central Intelligence Agency, Washington; Department of the Air Force, Archives Branch, Historical Research Division, Maxwell AFB, Montgomery, Alabama; Air Force Logistics Command,

Office of History, Wright-Patterson AFB, Dayton, Ohio; Air Force Foreign Technology Division Library, Wright-Patterson AFB; Air Force Office of Information, Public Information Division, Magazine and Book Branch, Washington, D.C.; Department of the Army, US Army Intelligence Command, Historical Office, Fort Meade, Maryland; US Army Missile Command, Historical Branch, Missile Intelligence Agency Library, and Redstone Scientific Information Center, Redstone Arsenal, Alabama; Army Matériel Command, Historical Office, Washington, D.C.; Army Matériel Command, White Sands Missile Range, Public Information Office, White Sands Missile Range, New Mexico; Army Ordnance Center, Ordnance Museum, Aberdeen Proving Ground, Maryland; Department of the Navy, Office of the Chief of Naval Operations, Operational Archives Branch, Naval History Division, Washington, D.C.; General Services Administration, National Archives and Records Service, Modern Military Branch, Military Archives Division, Washington, D.C.; Library of Congress, Washington, D.C.; National Aeronautics and Space Administration—Marshall Space Flight Center, Historical Office, Huntsville, Alabama; National Air and Space Museum, Frederick C. Durant and Frank H. Winter; University of Alabama in Huntsville—The Willy Ley Special Collection, Huntsville, Alabama; Michel Bar-Zohar; Richard Bauer, Harold J. von Bechk, Georg Borttscheller, John A.Reinemund, Gabriel Brillante, Eric Burgess,Major General Alexander M. Cameron, Colonel William T. Coleman, Sir Alwyn Crow, Edward M. Cynarski, H. Czarnocka, Dan Driscoll, Eugene Emme, Richard Fein, Charles H. Gibbs-Smith, Irmgard Gröttrup, Foster Haley, Jack Hanson, Francis J. Heppner, Clarence H. Hickman, Lieutenant Colonel Gerald M. Holland, Lonnie B. Hughes, Dieter Huzel, Major Albert F. Jones, Jr., Brigadier General Julius Klein, Willy and Olga Ley, Captain Grayson N. Merrill, Daniel E. O'Brien, William H. Pickering, Colin Ronan, Frederick P. Schneikert, Howard S. Seifert, Christina Skolasinska, E. E. Stott, Professor Mieczyslaw Subotowicz, Colonel Audrey E. Thomas, Colonel Charles N. Toftoy, Lieutenant Commander W. J. Tuck, Colonel Harold R. and Virginia Strom Turner.

Frederick I. Ordway III
Huntsville, Alabama and Washington, D.C.

Mitchell R. Sharpe
Huntsville, Alabama

CHAPTER 1

"Take Us to Ike!"

"Komm vorwärts mit die Hände hoch!"

So ordered Private First Class Frederick P. Schneikert as he clicked the safety lever of his M-1 rifle to "fire" and climbed up from the culvert in which he had been hiding. He pointed the M-1 at a man in civilian clothes bicycling down the road toward him. It was the morning of May 2, 1945. Schneikert had reason to be cautious. Two days earlier, Hitler had committed suicide in his bunker deep beneath the Reich Chancellery in Berlin; though the Third Reich was crumbling rapidly no one knew what to expect next. Among American soldiers rumors were spreading fast: The Germans were going to withdraw into the practically impenetrable Alps of southern Germany and Austria, a region they called the National Redoubt; there, they were to continue fighting for years. Another rumor was that guerrillas, called Werewolves, were already at work behind the American lines. These were said to be fanatic Nazis, mainly SS soldiers, in civilian clothes who specialized in slitting the throats of American soldiers incautious enough to go to sleep at night.

Thus, Schneikert and several of his buddies in Antitank Company, 324th Infantry Regiment, of the 44th Infantry Division, guarding this particular approach to Schattwald, a small village on the Austrian border of Germany, were more than a little edgy at the approach of a well-dressed civilian on a bicycle.

On hearing the command to put his hands up and come forward, the blond young German on the bicycle immediately braked to a stop and did as ordered. He was surprised at being addressed in his native tongue.

True, the German he heard could not be called fluent or colloquial; but he was nevertheless startled to meet an enemy soldier who spoke his language at all. The German phrase had popped almost automatically into Schneikert's head; Fred had forgotten a lot more of the German he had earlier learned from his grandmother, who lived with his family in Sheboygan, Wisconsin, but who had been born and reared in Bamberg, a city some 275 kilometers to the north of his present position.

With his hands above his head, the German approached Schneikert and began speaking rapidly in English and German as well. He said his name was Magnus von Braun and that his brother Wernher had invented the V-2. Wernher and other members of the team that had developed the large rocket were close by and wanted to surrender. It was, he said, seriously important that he and his brother be taken "to see Ike as soon as possible."

Schneikert kept his rifle on the German. The story was an awfully strange one. He knew about the V-2 rocket, of course. However, there was something fishy about the man in front of him. He wore a very clean blue-gray, ankle-length leather coat and underneath it a fresh shirt and tie. He was too well dressed, too well fed—not at all like the thousands of bedraggled and emaciated German civilians Schneikert had seen along the roads as his unit had pushed south through the Fern Pass into the Austrian Tyrol.

"Hey, I've got a nut here! What should I do with him?" he called to his buddies still in the culvert.

Receiving no enlightenment from them, Schneikert spent some thirty minutes talking to von Braun, shifting back and forth between English and German. As each minute passed, he became all the more incredulous. Magnus von Braun kept repeating that his brother Wernher and a group of engineers, scientists, and army officers who had developed the V-2 were at Haus Ingeburg, an inn near Oberjoch, only four kilometers back up the road.

Schneikert and his fellow soldiers had no idea of accompanying the German to Oberjoch. Only three days earlier, their company had lost two men when one of their squads was lured into an ambush by a similar surrender offer. Instead, Schneikert took von Braun first to the command post of his antitank company, and then to the command post of the 44th Division's Counter-Intelligence Corps at the village of Reutte, some 26 kilometers to the east of Schattwald.

There, he turned the young German over to First Lieutenant Charles L. Stewart, one of twelve CIC agents assigned to the division. This was a hard time for Stewart. He was shaken physically and mentally by an event that had happened only two days before.

Stewart's mission was to determine if any and what type of resistance could be expected from the rapidly deteriorating Nazi party apparatus. Logically enough, he had headed for the office of the local *Gauleiter*. He was surprised to find it intact. It was the first party headquarters he had ever found in such a condition and this should have made him suspicious. Every file was in place, the telephone was still connected, the portrait of Hitler in his most bellicose pose was still on the wall, and the cheap and gaudy furniture had not been upturned or looted. Stewart felt almost like Goldilocks entering the house of the three bears.

Upstairs he found a number of radio sets for maintaining contact with both party headquarters in Berlin and secret agents in the field. Downstairs, in a fortified cellar, he found an arms cache that could have outfitted a battalion. There were rifles, machine guns, antitank rocket launchers, and ammunition for different weapons. Again, he should have been forewarned.

But Stewart was distracted by the appearance of looters, mainly displaced persons, who had been recently liberated by American troops. They milled about the building. As fast as he could herd a few back through the front door, waving his .45 automatic pistol in their faces, more came through the back door, dumping files onto the floor while looking for liquor or other booty.

Realizing that he could not cope with the situation, Stewart decided to go back to his command post and collect several men to guard his valuable find. He left the building and then, as he later recalled: "As I was going up the street to the CP, I was flattened against the wall by a horrible explosion and turned around to see that the building was exploding. Stuff was coming out of the windows, and the ammunition in the cellar was going off. . . . Some of the people in the building were blown through the windows and survived—others were blown against the interior walls. . . . A time bomb planted in the chimney had gone off."

After the holocaust, Stewart went back to question the housekeeper of the party headquarters again. She had been reticent to speak before the explosion; now, however, she was eager to talk. The *Gauleiter* had a cabin in the mountains nearby, she said. It was less than an hour's walk. Stewart soon located it and found the door unlocked. He entered cautiously, with his pistol drawn, into a scene that would always remain indelibly ingrained in his memory: "In the tradition of the Götterdämmerung, he had killed his four children and his wife; and then set their bodies afire with a mattress, and then shot himself."

Stewart, recovering from these experiences, looked up as Schneikert came into his room with his prisoner.

"This is Magnus von Braun," said Schneikert. "He claims to be the representative of over a hundred top German scientists located at a mountain inn. They want to surrender to us."

He then handed Stewart a brief résumé that Magnus had written while he was waiting to see the counterintelligence officer:

> On May 10, 1919, I was born in Greifswald [Pomerania], the son of a government administrator, Magnus Freiherr von Braun, and his wife, Emmy, née von Quistorp.
>
> I spent my youth in Berlin, where I attended the Prep School and the French School until Easter 1934. Finally, I went to the Hermann Lietz School in Spiekeroog, North Sea, a boarding school from which I graduated Easter 1937.
>
> After having spent six months in the *Arbeitsdienst* [Civil Conservation Corps], I began studies at the Technical Institute at Munich in the fall of 1937, majoring in organic chemistry. Then, when I received my master's degree, I became the assistant in organic chemistry to Professor Hans Fischer (Nobel Prize winner in chemistry) for 1/4 year, at which time I was drafted into the Air Force (October 1940). After the completion of flight training and a short stay as a flight instructor, I came to the Heimat Artillery Park II [Peenemünde], Karlshagen, in July 1943 [at the request of his brother]. In Karlshagen, I worked with Mr. Gerhard Heller, the director of fuel chemistry, who put me in charge of hypergolic fuels for the newly developing *Wasserfall* project. In this capacity, I worked in conjunction with the I. G. Farben Industry. In October 1943, my brother, Professor Wernher von Braun, requested me to work as his personal assistant.

While the statement gave Stewart the broad outline of von Braun's career, it did little to help in further interrogation. Stewart was neither an engineer nor a scientist, and although he spoke German, he could not ask technical questions related to the V-2 project, about which he knew next to nothing. Furthermore, he had no specific instructions on what to do with such nonmilitary personnel if he did apprehend them. However, Magnus von Braun's story fascinated him.

The young German told Stewart that his brother's group was in immediate danger from SS Obergruppenführer Hans Kammler's troops at nearby Innsbruck. These soldiers were under orders to shoot all the civilian rocket personnel before letting them be taken by the Allies, or, alternatively, to use them as hostages for postwar negotiations. Again, Magnus von Braun pointed up the need to move quickly, and again he asked to see Ike as soon as possible.

At first, Stewart received a mistaken impression of von Braun because of the German's imperfect English. He thought that Magnus had come in with the express purpose of surrendering his brother Wernher in return for some as yet unproposed favor. Stewart became angry and was ready to dismiss von Braun as a man who would betray his own brother; however, between Magnus' English and Schneikert's eighth-grade German, they managed to clear up the misunderstanding, and Stewart pondered what to do.

The story sounded plausible to him, but so had other ruses he had heard lately. There was one thing he could do without exposing American soldiers to an ambush.

"Look, von Braun, I'll escort you back to your bicycle and give you passes for a safe return back to our area," Stewart finally said, "Bring your key people here. I'll talk to them and we'll determine the next course of action then."

At Haus Ingeburg, Wernher von Braun was becoming worried. His anxiety over his younger brother's mission was not lessened by the pain in his left arm, badly broken in an auto accident two months before. In mid-March, during a trip from Leuchtenberg to Berlin, his driver had gone to sleep at the wheel. Their car plunged off the road, down an embankment, and onto a railroad track. The first attempt to set the arm had been a failure, and a month later the arm was rebroken and reset at a hospital in Sonthofen, some 14 kilometers from Haus Ingeburg. When Major General Walter Dornberger, former military commander of the rocket research center at Peenemünde, heard on April 25 that von Braun was in Sonthofen and that the village had undergone three days of bombing and strafing, he immediately sent an ambulance there and had von Braun removed to Haus Ingeburg at Oberjoch. While Dornberger feared for his chief civilian engineer's safety, he was also alarmed because Sonthofen lay directly in the path of the approaching French army.

Von Braun had been at Haus Ingeburg for about a month with some hundred others, mainly officers on Dornberger's staff. The inn had originally been a ski lodge, but during the war it became first a rest center for battle-weary pilots of the Luftwaffe and later a military hospital. Von Braun a few weeks previously had removed some four hundred of his engineers and scientists from the area around Nordhausen, in Thuringia, 400 kilometers to the north, and relocated them in Oberammergau, home of the world-famous passion plays, on the orders of Obergruppenführer Hans Kammler, the SS officer then in charge of V-2 production and field operations. By concentrating the nucleus of the V-2 team in one place, Kammler accomplished two ends: The group was available to continue work, and it was conveniently available for instant liquidation to prevent its falling into Allied hands.

During the first week of May, Dornberger had been in conference with his senior managers. It was apparent that the war was at an end, and plans had to be made for themselves and their colleagues in Oberammergau, which fell to the Americans on April 29. The managers agreed that some way must be found to contact the Americans and negotiate with them as a team. They felt the group had so much knowledge and skill in an advanced technology that it should not be dissipated in the chaos that would follow the war's end. Certainly, Dornberger could foresee no rocket projects of the scope of V-2 in Germany's future.

The group at Haus Ingeburg had gathered there when Dornberger removed his headquarters from Bad Sachsa, in Thuringia, to be closer to von Braun's personnel in Bavaria. Life at the inn was unbelievably calm and carefree, despite Germany's imminent collapse. Snow still lay in patches on the slopes around the inn, but spring was coming to the Tyrol. There was absolutely no work to do. Some team members were herded up into the nearby hills to go through defensive military exercises. Unaware that the preparations were aimed more at fighting off a possible attack by diehard SS troopers under Kammler's command than the soldiers of the Allies, Dieter Huzel objected loudly to such a waste of time. Huzel, an engineer, was at one time von Braun's assistant at Peenemünde. He performed a major service to the team and its later usefulness to the American army by hiding its priceless engineering and scientific papers on rocket research.

Despite Haus Ingeburg's relative remoteness, the inn seemed to its guests to be in danger of capture by the French. Rumor had it that the troops driving into the region were the dreaded Algerian Spahis and that the capture of the inn would be accompanied by looting and slaughter. Rather than face such an end, Huzel conferred with his friend Bernhard Tessmann, who had been an associate of von Braun since before the war. The two men made tentative plans to steal a truck and make their way to the nearest American lines.

When Herr Rother, owner of the inn, heard of the possible arrival of the Spahis, he became almost hysterically concerned about a cache of valuable wines and liquors in his cellars. Many of the bottles dated to prewar times. A bucket—or more properly a bottle—brigade was formed to bring the potables up into the lounge and sunny patio. A monumental binge took place. Many bottles were done away with by the most obvious means of consumption. Yet so much remained that two trucks were loaded and driven to a cliff, where their cargoes were dumped and cascaded with a heartrending crash into the valley of Oberjoch. (A few individuals, opposed to such means of denying the enemy the delicious spoils of war, held back some bottles for future needs.)

Clearly the only one not enjoying life at Haus Ingeburg was Wernher von Braun. His arm was still giving him pain, and he was concerned about the status of his colleagues around Oberammergau. Also, he was worried about members of his family whom he had not seen recently. As he later recalled, "Most of all, I worried about my parents, who were deep behind the Russian lines. Sigismund, my older brother, had been with the German Embassy in Vatican City when the American army took Rome, and was safe out of the war."

On the afternoon of May 1, several groups of army officers and civilians sat on the stone veranda of the inn playing chess. Others sat sipping excellent French wine from the last remaining bottles of the innkeeper's hoard. A few listened tranquilly to the radio, over which the German State Radio Service was broadcasting Bruckner's Seventh Symphony in E Major. Its funereal adagio provided an appropriate threnody for a sudden announcement preceded by a roll of drums: "Our Führer, Adolf Hitler, fighting to the last breath against Bolshevism, fell for Germany this afternoon in his operations headquarters in the Reich Chancellery."

The announcement was wrong about how and when Hitler had died; actually he had committed suicide the day before, April 30, but Dornberger and von Braun had no way of knowing this. They waited, thinking the next news would be of Germany's surrender; instead, the announcer stated that Admiral Karl Dönitz had been appointed by Hitler to continue the war.

The gravity of the immediate situation at Haus Ingeburg moved the two men to take drastic and—since Germany was still at war—treasonable action.

They decided to send Magnus von Braun down the winding road from Oberjoch and east through the hamlets of Schattwald, Zoblan, Tannheim, and Haldensee until he met the advancing Americans. Magnus was chosen because he spoke English better than anyone else in the group. So early the next morning, May 2, Magnus hopped on his bicycle and headed down the road for his eventual rendezvous with Private First Class Schneikert.

About two o'clock in the afternoon, Magnus returned.

"He's back," someone shouted from the veranda. Wernher von Braun and Dornberger hurried out to meet Magnus as he ran up the steps.

Wernher led his brother into the inn and sat him on a large sofa. The others crowded around.

"I think it went well," Magnus said. "I have safe conduct passes, and they want us for further interrogation."

After Magnus' recapitulation of what he had done and his reception by the Americans at Reutte, some of the group went about getting together their belongings. For better or worse, they would not be returning to Haus Ingeburg.

Dornberger and von Braun decided to go to the Americans right away, and selected the group to accompany them in the initial convoy. Bernhard Tessman and Huzel were chosen because they knew where the Peenemünde documents were cached. Magnus was essential because he had made the initial contact. Herbert Axster, Dornberger's executive officer, was an obvious choice. Hans Lindenberg went along because he was the chief propulsion engineer for the team. The choices were not difficult because there were relatively few civilians at Haus Ingeburg. Most of the hundred or so at the inn were military members of Dornberger's organization, but Dornberger did not think it would be wise to show up at the American outpost with a group of German army officers —particularly after Magnus' talk with Stewart about rocket "scientists and engineers."

Once the three field-gray BMW cars, which had been proudly polished by their drivers despite the heavy clouds, were loaded, Dornberger, Tessmann, and Magnus entered the lead vehicle. Wernher von Braun and Huzel climbed into the second one, and Axster and Lindenberg followed in the third. At 4:10 PM they pulled away from Haus Ingeburg; a light sleet began to fall, making the winding road through Adolf Hitler Pass slick and dangerous.

As the lead vehicle pulled into the roadblock at Schattwald, an American soldier waved them to a halt with his rifle, and Magnus got out of the car to show his safe-conduct pass. He was motioned into a nearby house that had at one time been an Austrian customs station, then returned to Dornberger and Tessmann. An American soldier stuck his head through a window of the lead car. He asked a question, in English, which none of them caught. Dornberger turned testily to Magnus and said, *"Der Idiot soll Deutsch sprechen, wenn er was von mir wissen will"* ("That idiot should speak German if he wants to know something from me!"). Magnus and Tessmann were greatly relieved to see that the soldier obviously did not understand German.

It was dark when the group finally reached Reutte, a town that was darker than normal, being without electricity. The team delegation immediately was shown to Lieutenant Stewart's office in a typical Austrian–Bavarian mansion. The American was seated at a table in a stygian gloom, his face only dimly illuminated by a candle. Axster and Dornberger were just barely able to make out, to their surprise, the familiar faces of other German officers of their group. Apparently they had been rounded up independently of the group at Haus Ingeburg.

Stewart's German-speaking intelligence soldiers verified the names of those present but made few other queries. The group was then led by soldiers with flashlights to the rooms assigned them in Stewart's recently acquired headquarters. Almost as soon as they entered their rooms, electric power returned to Reutte. Their new quarters were nowhere near as luxurious as those of Herr Rother at Haus Ingeburg. The bedclothes were rumpled and the rooms were not so clean as those at the inn; but they looked inviting enough after a day of considerable strain. If the rooms were a disappointment, the food was something else. After freshening up a bit, the Germans were taken to a dining room and given what to them was a feast. Fresh eggs, butter, white bread, and real coffee—foods they had only dreamed of for three years or so. (Von Braun was later to recall, somewhat bemusedly, "They didn't kick me in the teeth or anything. They just fried me some eggs.") Following the voluptuous dinner, the group returned to its beds and slept soundly—almost too soundly.

Stewart had remained at his table working on his report of the incident at Reutte for submittal to his commander at Seventh Army Headquarters. What could he really say about what had happened? A group of German officers and civilians, claiming to be the developers of the V-2, had voluntarily placed themselves in his custody. He was not prepared to do more than question them in nontechnical terms and provide them quarters. Being a military man himself, he was more interested in Dornberger and Axster. It seemed suspicious, to say the least, that the little major general, who claimed to be boss, had played a background role, if not actually sought refuge behind, the jovial, loquacious blond civilian with his arm in a cast.

Wernher von Braun was something of an enigma to Stewart. The American found it hard to believe that a man only thirty-three years old could be the technical director of an organization that had produced the V-2 rocket. Von Braun seemed to be "open, forthright, and obviously brilliant," but Stewart listened to him with a curious mixture of fascination and horror. Besides developing the V-2, von Braun said he had on the drawing boards a missile that could reach New York. It also became obvious to Stewart that the German's real interest in rockets lay in the field of space travel.

As Stewart pondered his report, he heard a bump on the stairway outside his door. He reacted automatically, leaping to his feet. By the time he reached the doorway, he heard a second sound, an unmistakable one. It was the click of the safety catch being let off on a weapon. He silently stepped through the door in time to see a man climbing the stairs to the rooms above where the Germans were now asleep. As quietly as possible, he crept up the stairs after him. Seeing the other man's German machine pistol, Stewart reached for his own .45 automatic pistol. *Damn!* He had left it in his room below.

"Where are you going?" he called out.

"To kill those German swine!" an obviously drunken voice replied. "Where are they?"

Stewart knew at once from the heavy accent that the man was one of the Polish refugees who served his unit as cooks and kitchen personnel.

"Give me the gun," he shouted and at the same time grabbed the weapon from the hands of the self-appointed executioner. He then escorted the Pole back to his quarters and posted an American soldier at the head of the stairs. A few days later the Pole was sacked, not for this incident, but for stealing army property.

Early the next day, May 3, the Germans were aroused at reveille by the noise of a rear echelon American army unit getting ready for a new day. Everything was done in a leisurely, protracted and, it seemed to the Germans, distinctly unmilitary fashion. Still, it made sleep impossible. After washing up, the Germans, unaware of their close escape from death the night before, were escorted to the mess hall. Again, they were given fresh eggs, coffee, toasted bread, and a boxed cereal. They assumed the cereal was to be eaten dry and were surprised to see their captors carefully slitting open the wax-paper-lined box and adding sugar and milk.

After breakfast, the Germans met Bill O'Hollaren, a public relations sergeant from the 44th Infantry Division. He had come to interview them.

O'Hollaren, a former newspaperman, soon had his lead for a story in the division's *Beachhead News*, which appeared on May 11:

> A smiling, nonchalant young German scientist sat down in a 44th Infantry CP [command post] and said that if he had been given two more years, the V-2 bomb he invented could have won the war for Germany.
>
> Professor Wernher von Braun, 33, said that he formed the basic idea for his deadly weapon while a student in Berlin. When he received his aeronautics degree in 1932, the Nazis took him under their wing. He and some of his assistants were sent to an experimental laboratory at Peenemünde on the Baltic Sea. Here the weapon was completed.

He also reported that von Braun said that if Germany had been able to produce two hundred V-2s a day, it was conceivable that the course of the war would have been changed.

Accompanying O'Hollaren were a couple of photographers. They spent most of their time "shooting" von Braun, who recognized the value of publicity. He posed obligingly, with his left arm, encased in the cast, raised awkwardly from the leather overcoat buttoned over his shoulders. He answered all questions put to him, talking so freely that one of the American soldiers observed cynically, "If we haven't captured the greatest scientist of the Third Reich, we've certainly taken its greatest liar."

As the bulbs were flashing, Magnus turned to Tessmann and said wryly, "We're celebrating now, but I'll bet they will throw telephone books at us if we ever reach New York." (Wernher also made a small bet with Private First Class Schneikert that he would be in the United States before the American—and von Braun won.)

After the press conference, the Germans returned to their rooms to write rather lengthy autobiographical sketches. Stewart, who hardly knew what else to do with his catch, sought advice from his superiors at Seventh Army Headquarters. The reply was predictable, pragmatic, and wholly irrelevant as far as Stewart was concerned. He was directed by his superiors to find out if the Germans concerned were Nazis. Forgetting military protocol for the moment, he yelled into the phone, "Screen them for being Nazis! What the hell for? Look, if they are Hitler's brothers, it's beside the point. Their knowledge is valuable for both military and possibly national reasons."

Thus, the first American in authority to contact the German rocket team—a group that could have changed the course of war against the Allies—found himself ensnarled in the mindless bureaucratic machinery of higher headquarters, but Stewart had more immediate concerns than the lives of the German scientists he had just saved. Soon, he was told to send them back to Germany, to Peiting, some 35 kilometers to the north. He bundled his charges off in an army truck.

Just before noon, Axster, Tessmann, Magnus von Braun, Lindenberg, and Huzel arrived in Peiting, where they joined some forty familiar faces, among whom were Albert Schuler, Erich Kaschig, Hans Hüter, and Heinz Schanrowski, all engineers from Peenemünde. Huzel remembered his surprise upon entering their new quarters:

> It was the town kindergarten, and the furniture and sanitary installations were scaled accordingly. Even with no usable furniture, no beds, not even a rug, we spent a joyful and interesting evening of storytelling by candlelight. I still had a bottle of sweet vermouth from the cache of Herr Rother in my suitcase, which was just enough for a sip for everybody.

The group stayed in Peiting for five days. They were free to move about the village, small as it was; but there was little to do except swap stories of the recent past and wonder about their future in the hands of the Americans. On May 8, at two in the afternoon, the Germans were loaded into US Army trucks.

"Where are we going?" they asked.

"Garmisch-Partenkirchen" was the reply.

Thus Dornberger and von Braun, two of the men most wanted by US Army intelligence, were now in American hands—and by an incredible stroke of luck they were taken into custody in Bavaria, at a time when the Americans had given up hope of finding them because Peenemünde on the Baltic, where they had expected to catch them, had fallen to the Russians on March 9. And the men of Peenemünde had set the switches themselves for this outcome many weeks before when they decided to evacuate their immense research and development center.

CHAPTER 2

Countdown at Raketenflugplatz, Berlin

The technological roots of the V-2 go back to the amateur rocketry and the resurgent interest in armament of the late Weimar Republic, which preceded Adolf Hitler's Third Reich. Its theoretical antecedents can be traced right to the closing years of the previous century in Russia, and to Konstantin E. Tsiolkovsky, a village schoolmaster in Kaluga. However, the more specific foundations for the V-2 lay in the work of Hermann Oberth, a German-speaking Transylvanian schoolmaster, whose book *Die Rakete zu den Planetenräumen* (*The Rocket into Planetary Space*) was published in 1923.

While the theoretical basis for large, liquid-propellant rockets was being laid out by Oberth in Germany, a number of gifted amateurs in that country had banded together and begun empirical experimentation with such rockets. The nucleus of the largest and best known of the groups was formed on July 5, 1927, in a back room of the Golden Scepter, a restaurant in Breslau. The organization gave itself the imposing name *Verein für Raumschiffahrt* (*Society for Space Travel*) and hoped to do two things: popularize the idea of flight to the moon and planets, and perform serious experiments in rocket propulsion. Oberth was its president.

By September 1929, the society had 870 members. One of them was nineteen-year-old Wernher von Braun, who had just graduated from high school. He came from an affluent family; his father was minister of education and agriculture for the Republic.

Despite its high seriousness, the *VfR* was not above using publicity to attain its own ends where liquid-propellant rockets were concerned. In

1929, the Ufa film organization and famed director Fritz Lang decided to make a movie as a result of the increasing interest in rocketry. Lang's wife, the actress Thea von Harbou, wrote the script, based on material in books by Oberth and Willy Ley, founding members of the *VfR* (Ley was later to become a well-known historian of rocketry), for a science-fiction film called *Frau im Mond* (*Girl on the Moon*). At first Lang approached car manufacturer Fritz von Opel, who liked to be called "the Henry Ford of Germany," as a collaborator, but the latter turned him down, saying he didn't have the time. Lang then approached Oberth, who readily agreed. In addition to serving as technical consultant on the film, Oberth was to build a rocket and have it ready to launch coincident with the premiere of the film. The project was doomed from the start for the simple reason that Oberth, visionary and theoretician, had no engineering aptitude or experience. One of his colleagues later said of him, "If Oberth wants to drill a hole, first he invents the drill press."

Trying to overcome these shortcomings, Oberth placed an advertisement in the newspapers for an assistant. He hired Rudolf Nebel, a former pilot in World War I, who had a degree in engineering but who had never practiced professionally. A second assistant joined him: Alexander B. Shershevsky, a Russian student of aviation who had overstayed his visa and maintained himself in Germany by writing articles for aviation magazines. This team then was to design and build a liquid-propellant rocket. For various reasons, they never got started until some twelve weeks before the premiere was scheduled. That was not much time in which to come up with a rocket that would reach an altitude of seventy kilometers, as the Ufa organization's public relations men had assured Germany it would.

Oberth and company came up with several ideas for rockets that were less complicated than the liquid-propellant variety; but in the end, all came to nought. The date for the film premiere came and went without the heralded rocket launching, and Oberth quietly left Berlin in November for his native Rumania, where he remained until the following spring. The *VfR* was left with a bad dream.

Following the debacle, Nebel became a member of the *VfR*, which more or less marked time while its president was sulking at home. At about the same time, Johannes Winkler, a founder of the *VfR*, returned to the organization. He had left earlier on a six-month secret assignment for the Junkers Aircraft Factory in Dessau. There, he worked on the design for an aircraft that could fly in the stratosphere and also experimented with small rockets, later called *jatos* (for *jet-assisted takeoff*), and used for boosting heavily laden airplanes from conventional runways.

During Oberth's absence, Nebel began to design what he called a "minimum rocket"—or Mirak—to test liquid propulsion systems. Mirak externally resembled the rockets of Sir William Congreve, which helped burn Copenhagen in 1807. The rocket body was attached to a long stabilizer stick. However, there the resemblance ended. The body accommodated a combustion chamber and a small liquid oxygen tank. The stick contained gasoline under pressure from a carbon dioxide cartridge. However, developmental work on the rocket was well over a year away.

The pace of activity for the *VfR* picked up in the late spring and early summer as the members prepared for a very important project. The Chemisch-Technische Reichsanstalt (a scientific establishment similar to the US National Bureau of Standards) had agreed to test-fire the group's *Kegelduse*, a rocket built to Oberth's design during the *Frau im Mond* fiasco, and to certify its performance. The *VfR* felt such a certificate, assuming a successful firing, would open the doors to government, and perhaps some industrial, financial assistance. Surprisingly enough, the motor worked, and the group received a certificate that stated the *Kegelduse* "had performed without mishap on July 23, 1930, for 90 seconds, consuming six kilograms of liquid oxygen and one kilogram of petrol, and delivering a constant thrust of about seven kilograms."

The technical recognition, however, did not lead to funds for further research, as the *VfR* had hoped. Indeed, it merely compounded a setback that had occurred a couple of months earlier. Max Valier, *VfR* member, was killed while working on his own liquid-propellant rocket engine. It exploded May 17, 1930, killing him almost instantly. Since Valier was a well-known writer for the German press, his death received wide notice. Indeed, it instigated a public outcry against such hazardous experimentation; a bill banning this activity was introduced into the Reichstag, but not passed. When the hue and cry had abated, Nebel, with Klaus Riedel, a young engineer, and Kurt Heinisch, a skilled automobile mechanic, retired to a farm near the village of Bernstadt, in Saxony. There they continued checking the Mirak through the autumn of 1930, but with only indifferent success. The testing came to an end September 11, when the rocket blew up, frightening a group of curious Saxon burghers; Riedel and Nebel returned to Berlin.

By this time, the *VfR* had reached the point where it needed its own facilities. The society could not conduct serious research from an open field in Bernstadt while having rockets manufactured hither and yon over Germany. Nebel took it upon himself to find a base of operations for the group. He did so in a remote and abandoned World War I ammunition storage facility of the German army in the northern Berlin suburb of Reinickendorf. It was owned by the city of Berlin, which was happy to rent it for the sum of ten reichsmarks per annum. On September 27, 1930, Nebel formally opened the Raketenflugplatz, Berlin.

Within its four square kilometers were several useful structures. One was a magazine surrounded by a thick, earthen barricade, which made a natural static-test facility and machine shop. The other was a smaller building that served as an administrative office. It also provided bachelor quarters for Nebel and Riedel during the winter of 1930–31, where one of their first and most ambitious tasks was to draw up a ten-year plan. Because the winter was relatively mild, the members of the *VfR* could work outside, clearing land and installing the test stand and its ancillary equipment. The range was ready for operation in March 1931.

By early summer, static tests with a new type of rocket, called the Repulsor, led by Riedel and the others to discuss the possibility of launching one of them.

While the group at Raketenflugplatz, Berlin was ruminating on such a launching, their independent *VfR* member Winkler was preparing to do likewise with a rocket of his own on the drill field at Gross Kühnau near Dessau, in Anhalt. He launched the first liquid-propellant rocket in Europe. It used liquid oxygen and liquid methane, attaining the grand height of 3 meters on February 21, 1931. A more successful launch was made on March 14. These events produced mixed emotions among his colleagues in the *VfR*, who had assumed the honor would eventually be theirs in Berlin.

As it was, the first launching from Raketenflugplatz, Berlin was unintentional and unplanned.

On May 11, 1931, Klaus Riedel made a phone call to Willy Ley. He said, "You know the secret baby we discussed; well, I took it out yesterday to make a test run. I didn't expect anything; you know I used those heavy valves and the fat struts along the fuel lines. And the damn beast flew! Went up like an elevator, very slowly, to 18 meters. Then it fell down and broke a leg."

The Repulsor was easily repaired, and its first planned launching took place on May 14. The launcher consisted of two sections of stove pipe stuck into the ground to accommodate the two propellant-tank legs of the rocket.

The first year at Raketenflugplatz, Berlin was a busy one for the *VfR*. There were 87 launchings, as well as 270 static firings of various rocket motors. On the first anniversary, in September 1931, the Ufa Film Co. devoted its weekly newsreel to a group later called Those Fools at Tegel (as Rudolf Nebel entitled his memoirs in 1972). During the filming, the group proudly launched a Repulsor, only to have it impact on the roof of a building across from the police station, setting fire to it. Retaliation was swift. All rocket flights were banned forthwith. On October 17, however, the police reconsidered and permitted firings to resume under tightened safety practices.

Despite occasional setbacks, the proving ground was eminently successful for advancing the state of the art of liquid-propellant rocket motor development. However, the terrible economic depression also was making itself felt at Reinickendorf. More and more members could no longer afford the dues of eight marks. Frequently heard was the ominous reason that they were needed for "party dues" or, more importantly, for food.

Both Oberth and Nebel had talked and written about the possibility of rockets as weapons. If some of his ideas for rockets as weapons sounded technically fantastic to his fellow members of the *VfR*, Nebel had a ready reply: "If we say it can work, we'll get army money to try it."

The *VfR* was not the only amateur rocket group active in Germany during the period. Others included the *Studiengesellschaft für Raketen* (Rocket Study Society), founded in 1928, in Frankfurt am Main, and the *Gesellschaft für Weltraumforschung* (Society for Space Research), established by Hans K. Kaiser in 1937 in Breslau. Perhaps better known was the *Gesellschaft für Raketenforschung* (Society for Rocket Research) formed on September 20, 1931, in Hanover by Albert Püllenberg and his friend Albert Löw. This group was notable in that several of its members eventually found their way to Peenemünde.

While the *GfR* was never as large or as affluent as the *VfR*, its members engaged in serious theoretical and practical rocketry. Püllenberg had made an extensive search of the literature in the mid-1920s and had come to the conclusion that liquid-propellant rockets were the only ones upon which to expend time and energy, money being in as short supply in the *GfR* as in the *VfR*. Even before the official establishment of the organization, Püllenberg and the others had built several *Gardienenstangenraketen* (curtain-rod rockets) that they tested on the Vahrenwalder Heide, some fifteen kilometers from Hanover. The reason for the name of the rocket was simple: It was made, in part, of old curtain rods, the tubular sections of which served admirably as propellant tanks.

As did the *VfR*, the *GfR* raised funds by charging admission to see static firings and launch attempts. The group at Hanover was every bit as ambitious as the one at Berlin. Among its projects was a three-meter-long rocket designated Diesel-FT Rak. III. As the name implies, it utilized diesel oil rather than gasoline. Püllenberg was nothing if not economical. The former fuel was a great deal cheaper than the latter. The first attempt to fire the rocket on March 30, 1934, ended in an explosion.

In 1935, the *GfR* was visited by representatives of the German army ordnance department, who were curious to see what progress was being made at Raketenflugplatz, Hanover. They were impressed, but Püllenberg absolutely insisted that his organization have nothing to do with the military. His only interest was in space travel. Consequently, it was not long after the visit that Püllenberg was summoned to the head-quarters of the local Gestapo for a warning that rocketry was *verboten*, at least by amateurs. Unintimidated, he continued to test and launch liquid-propellant rockets until 1937, both in Bremen and Hanover. The *GfR* had disintegrated for all practical purposes by then, and in 1939, Püllenberg himself accepted a position in Peenemünde.

The army's interest in rockets dated back to late 1929, and may have been inspired partly by the publicity for the proposed rocket for the Ufa film premiere. Restricted by the Treaty of Versailles to no more than 100,000 men and only to specified defensive weapons, the army was searching for new means of waging war. During the early years of the short-lived Weimar Republic, science and applied technology flourished, but so did pacificism. The intellectuals in the universities and governmental research institutes were simply not interested in rocketry. Meanwhile, German industry was so closely interlinked with counter-parts in other countries through international cartels that industrial and government secrets were impossible. So, for the first time since 1864, the German army itself began investigating the possibilities of the war rocket, turning first to the solid-propellant rocket. The investigation began with a review of earlier literature, especially the work done in that field by Theodor Unge, a Swedish artillery officer who in 1908 had sold some of his patents to the German armaments firm of Alfred Krupp.

The responsibility for the rocket research devolved upon Walter Dorn-berger, an artillery captain who graduated in 1930 from the Technischen Hochschule, Berlin with his MS degree in mechanical engineering. He was assigned to the Ballistics Section of Colonel Karl Becker in the Army Ordnance Corps. Becker was more than a professional soldier. In 1925, he had co-authored with Dr. Carl Crantz the second volume of Crantz's monumental *Lehrbuch der Ballistik* (*Handbook on Ballistics*), still a standard work in the field. It is significant also that in Part 2 of Volume 2 there is a section on the ballistics of rockets. Becker may have been thinking about them as weapons at that time.

A special static-test stand for such solid-propellant rockets as were available commercially was constructed by Dornberger at the army proving ground at Kummersdorf, a suburb some 25 kilometers south of Berlin. The army was looking for a cheap, lightweight rocket that could be fired in numbers against an area target at a range of 6–7 kilometers. It was also interested in the newly emerging liquid-propellant rocket as a poten-tial weapon.

Becker's instructions to Dornberger were simple and direct. They set the goal but left the details and methodology to the captain. He told Dornberger that he was "to develop in military facilities a liquid-fuel rocket, the range of which should surpass that of any existing gun and the production of which would be carried out by industry. Secrecy of the development is paramount."

In this field of technology, Dornberger could not call upon commercial suppliers because there were none. Universities and governmental research organizations evinced no interest in the theoretical problems of such devices, let alone the military ones. Initially, then, Dornberger had to turn to the legion of independent rocketeers about whom he read in the popular press. A few study contracts were let, but the results were unscientific and unreliable.

At this time, Becker received an unsolicited proposal from Nebel entitled "Confidential Memo on Long-Range Rocket Artillery." After studying it, Becker laid it aside with no further action.

By mid-1932, the army had decided, all things considered, that the only answer lay in expanding its research facility at Kummersdorf and staffing it with civil servants. This approach had the advantage of providing the requisite secrecy that was impossible when dealing with inventors enamored of the newsreel, radio, and popular press.

The army's purpose for research at the time was unequivocal. There was not the slightest interest in space travel. Dornberger made it clear:

> The value of the sixth decimal place in the calculation of a trajectory to Venus interested us as little as the problem of heating and air regeneration in the pressurized cabin of a Mars ship. We wanted to advance the practice of rocket building with scientific thoroughness. We wanted thrust-time curves of the performance of rocket motors. ... We intended to establish the fundamentals, create the necessary tools, and study basic conditions.

In the spring of 1932, Dornberger, his commander, Captain Ritter von Horstig, General d'Aubigny von Engelbrunner and Becker, dressed in civilian clothes, paid a seemingly casual visit to Raketenflugplatz, Berlin. While they were impressed with the advanced state of experimental work, they were equally appalled by the lack of documented results: there was not a single thrust-time curve (a graph of rocket performance) on a single motor! Nevertheless, the *VfR* was given a contract for 1,000 reichsmarks to build a "one stick" Repulsor for demonstration launching at the army's proving ground at Kummersdorf. It was fired in July of that year. Von Braun described the event:

> Early one beautiful July morning in 1932 we loaded our two available cars and set out for Kummersdorf. ... As the clock struck five, our leading car with a launching rack containing the silver-painted Mirak II atop and followed by its companion vehicle, bearing liquid oxygen, gasoline, and tools, encountered Captain Dornberger at the rendezvous in the forests south of Berlin. Dornberger guided

> us to an isolated spot on the artillery range where were set up a formidable array of phototheodolites, ballistic cameras, and chronographs—instruments of whose very existence we had theretofore been unaware.
>
> The rocket was erected and fueled by two o'clock in the afternoon. At the signal, Mirak II soared upward for a distance of some 200 feet. Here, however, its trajectory became almost horizontal so that the rocket crashed before the parachute could open.

Dornberger later remarked: "The failure of this demonstration brought home to us in the Army Weapons Department how many scientific and technical questions needed answering before we could hope to construct a rocket that could fly efficiently." More impressive than the rocket to Dornberger was one of the rocketeers:

> I had been struck in my casual visit to Reinickendorf by the energy and shrewdness with which this tall, fair young student with the broad massive chin went to work, and by his astonishing theoretical knowledge. ... When General Becker later decided to approve our army establishment for liquid-propellant rockets, I put Wernher von Braun first on my list of proposed technical assistants.

On October 1, 1932, Dornberger hired young von Braun, who by then had his degree in engineering from the Technische Hochschule, Berlin. While von Braun was pursuing his duties at Kummersdorf, Dornberger arranged for him to attend Friedrich-Wilhelm University for continuation of his studies; he received a PhD in physics on April 16, 1934.

Following the employment of von Braun, Dornberger began to recruit other members of the Raketenflugplatz, Berlin staff and members of the *VfR* in general. Men like Heinrich Grünow, a genius mechanic; Walter Riedel, formerly of the Heylandt Co.; and Arthur Rudolph, who had built a workable rocket for the army in 1931, soon joined the organization at Kummersdorf.

Raketenflugplatz, Berlin continued to operate under the direction of Nebel and Klaus Riedel, who fired a motor designed to produce between 250 and 750 kilograms of thrust in March 1933, but with the departure of von Braun to the army and a general decline in membership in the *VfR*, its days were clearly numbered. Winkler (again within the *VfR* fold) and his assistant, Rolf Engel, left to form a new but short-lived organization called *Raketenforschungsinstitut–Dessau* (Rocket Research Institute–Dessau) and some of the *VfR* machinists and skilled laborers found jobs in industry. In the spring of 1933, a group of nattily attired young men in the powder-blue uniforms of the *Deutsche Luftwacht* appeared at the gate and explained that the rocket range was now their drill ground. For a while, they interrupted activities with their drills. In the end, however, Raketenflugplatz, Berlin was done in by leaking water faucets. Nebel was presented with a water bill for some 1,600 reichsmarks, and could not pay it. In January 1934, the rocket range reverted to the use for which it had been established in World War I—a storage area for ammunition. And about this time—the winter of 1933–34—the *VfR* itself gradually collapsed, due to deteriorating economic and political conditions. The records and miscellaneous equipment of the *VfR* were stored in a warehouse of the Siemens Co. by Klaus Riedel, who went to work with that firm for a brief period before joining von Braun at Kummersdorf. They disappeared during World War II.

In summary, then, by the middle of the 1920s, the time had arrived technologically for serious research and development of the rocket motor. German industry could provide liquid oxygen; a variety of high-energy fuels; strong, lightweight, and easily worked metals; and at least some electronic equipment—all essential elements of a rudimentary liquid-propellant rocket and crude guided missile.

Had it not been for economic depression, Hitler, and World War II, would the *VfR* have grown into an organization capable of great accomplishments in the field of rocket engineering? Willy Ley, in 1943, after nine years of retrospection in the United States, answered the question thus: "The more time I have had to think about it the more have I arrived at the conclusion that the *VfR* progressed as far as any society can progress. . . . Experimentation had reached a state where continuation would have been too expensive for any organization, except a millionaires' club."

CHAPTER 3

From Berlin to the Baltic

The night of December 21, 1932, in Berlin was miserably cold but clear. Shivering in it, Captain Walter Dornberger sought shelter behind a rather inadequate fir tree in his army proving ground at Kummersdorf West, and peered intently at a floodlit structure a few meters away. It was the recently completed static-firing stand for the first liquid-propellant rocket motor developed by his team. He was very proud of it, although to the unprofessional eye the stand was rather prosaic.

Three concrete walls 6 meters long and 4 meters high were topped by a wooden and tar-paper roof. The front was enclosed by folding metal doors. Behind the back wall, unseen by Dornberger, were Walter Riedel and Heinrich Grünow. The former had his eyes on two gauges in front of him and his hands were grasping two large wheels that controlled the main propellant valves. Once the pressure reached a certain mark, Riedel would crack the valves open, permitting liquid oxygen and 75 percent pure ethyl alcohol to flow into the combustion chamber of the rocket motor. Grünow was also watching a gauge. He was controlling the valve that pressurized the propellant storage tanks behind the test stand.

The rocket motor itself was not very impressive either. It was no behemoth belching flame and smoke like the ones that sent the first satellites and men into space. Suspended in the center of the stand was a small, pear-shaped aluminium motor only 50.8 centimeters long, shining in floodlights like a star on opening night of a hopeful hit play, which in a way it was.

Standing in the open door of the stand was Wernher von Braun, gingerly holding a wooden pole some four meters long with a can of flaming gasoline on its far end.

Riedel announced the pressure was correct and von Braun began cautiously moving the blazing can beneath the nozzle of the motor. There was a blinding flash of light and an incredibly loud explosion. Shards of metal sang through the air like bomb fragments, clipping branches off Dornberger's fir and burying themselves in the pines and firs about him. The floodlights went out immediately, and the area was filled with the acrid odor of burning rubber, mixed with smoldering wood, hot metal, and unburnt alcohol.

That there was no smash opening that night disappointed Dornberger and his small team; there would be more disappointments in the years ahead before the team produced a technological hit.

Kummersdorf had begun as a shoestring operation only slightly better off financially than Raketenflugplatz, Berlin. The general penury of the times was reflected in Dornberger's efforts to obtain even the most trivial supplies for running his modest proving ground. As he later remembered:

> ... The bureau of the budget kept a keen and jaundiced eye on us. We were not permitted to order either machine tools or office equipment. ... We learned in a hard school to get everything we wanted. We acquired things "as per sample." For instance, even the keenest budget bureau official could not suspect that "Appliance for milling wooden dowels up to 10 millimeters in diameter, as per sample" meant a pencil sharpener, or the "Instrument for recording test data with a rotating roller as per sample" meant a typewriter. ... And if there was nothing else to do, we entrenched ourselves behind the magic word "secret." There, the budget bureau was powerless.

Slowly, facilities at Kummersdorf grew in extension and sophistication. The officers' mess, in addition to performing its primary function, also served as a home for two of the range's bachelors: von Braun and Arthur Rudolph, who joined the organization in 1933. In some ways, it was an ideal household. Rudolph later reminisced: "We didn't like to get up early; we liked to work late at night instead ... at midnight von Braun had his best ideas. He would expound them on a sketch pad and his ideas led to one thing: space travel. It was at that time that he developed his flight plan to Mars. We didn't want to build weapons; we wanted to go into space. Building weapons was a stepping-stone. What else was there to do but join the War Department? Elsewhere there was no money."

While plans were made for a rocket that would fly rather than merely remain tugging at the restraints of a static-test stand, von Braun continued quietly to recruit the nucleus of a team of engineers and scientists that would remain with him from the development of the A4 (later called the V-2) through the development, in the postwar US, of military rockets such as Redstone and Jupiter as well as NASA's Saturn 5, which sent the first men to the Moon.

In 1935, he approached Bernhard Tessmann, who had a relatively good job as design engineer with Orenstein & Koppel, a firm in Berlin that had contracts with the Army Weapons Department. With considerable charm and logic, von Braun offered him considerably less pay, harder work, and longer hours at Kummersdorf.

Despite the grim economic climate, Tessmann accepted. He found the work stimulating, exciting, and satisfying. This psychic income more than made up for the difference in reichsmarks. It was a time when von Braun, Rudolph, Walter Riedel all pitched in at the drawing board and lent a hand with the wrench and hammer when Grünow needed it.

The fireworks that took place on the night of December 21 were simply one of the hazards to be faced in the growing technology of rocketry. For example, Dornberger was severely injured when a solid-propellant rocket ignited as he was attempting to disassemble it. Unwisely, and apparently ignoring his own safety regulations, he had used a steel hammer and chisel instead of copper tools. A spark set off the rocket propellant and filled his face with thousands of tiny black-powder particles. Removing them took a year of visits to a military hospital, where an orderly did the job by rubbing Dornberger's face with butter and then setting to work with a pair of tweezers.

Von Braun's first effort for Dornberger was a liquid-propellant rocket designated A1 (Assembly 1). It weighed 150 kilograms and had a 300-kilogram-thrust motor. It was 1.4 meters long and 30.4 centimeters in diameter. The propellants were liquid oxygen and ethyl alcohol. To stabilize it in flight, the rocket carried a spinning payload in the nose. The concept was sophisticated, but A1 was destined never to fly. Further study revealed that location of the heavy payload in the nose meant the rocket would *not* be stable in flight.

In 1934, the A1 was redesigned and designated A2. Its dimensions and performance were the same, but the gyroscopic payload was shifted to the middle of the rocket. In December 1934, von Braun took two of these rockets to Borkum Island in the Baltic Sea. They were nicknamed Max and Moritz, the two troublemaking lads in a popular comic strip (known in the United States as *The Katzenjammer Kids*, where they were named Hans and Fritz). Considering the unpredictable behavior of rockets in those days, the names were appropriate. Yet, flight-testing made the names respectable. Each flew perfectly a few days before Christmas, reaching an altitude of about 2,000 meters and proving out the motor design. Additionally, these tests showed that gyroscopic control of large rockets was possible.

The success of Max and Moritz gave a new impetus to the embryonic rocket program by convincing the higher-ups in the Army Ordnance Department that the rocket was practical and desirable as a weapon.

Even as Max and Moritz were being planned, larger and more powerful rocket motors also were in the works at Kummersdorf. These had thrusts of 1,000 and 1,500 kilograms. A test stand for them had been constructed and a third was under way that would accommodate a complete missile to be known as A3.

The A3 took shape on the drawing boards as "a purely experimental apparatus to test liquid-fueled rocket propulsion for missilelike bodies and for trials of the guidance system." Even then it was apparent to Dornberger and von Braun that the range limitations of Kummersdorf were beginning to cramp their operations. Not only was the range too short to fire future liquid-propellant rockets, but it also could not be further expanded. By 1935, the work force had grown to some eighty people. The team was outgrowing its cradle.

In March, 1936, Dornberger persuaded Major General Werner von Fritsch, commander in chief of the Reichswehr, to visit Kummersdorf. There, he was given a briefing by von Braun and Dornberger. As a noisy coda to their carefully orchestrated presentation, von Braun and Dornberger took the general out to their test stands where they fired in succession the 300-, 1,000-, and 1,500-kilogram motors. When the smoke and thunder had cleared, von Fritsch said the one thing Dornberger wanted to hear: "How much do you want?"

Other events took place that dictated a need for larger facilities. In January 1935, the Army Weapons Department had begun a cooperative program of rocket research with the soon-to-be-established *Luftwaffe*, Germany's nascent air force. One of the officers who early foresaw the possibilities of rocket-propelled aircraft was Major Wolfram von Richthofen, Dornberger's counterpart in the new service as well as a past commander of the German Condor Legion in the Spanish Civil War and cousin of the World War I ace pilot. Von Richthofen was imaginative, forceful, and dynamic. What he saw at Kummersdorf merely reinforced his belief that the day of rocket-propelled aircraft was not far away.

The army became interested in this project, even though airplanes were not its usual field of interest, and decided to try a static firing of one of its

300-kilogram-thrust motors attached to the underside of a Junkers "Junior" airplane. The test pilot, if he could be so called, was von Braun. Later, von Braun also rode a simulated rocket plane consisting of a 1,000-kilogram-thrust motor mounted on a cabin attached to the end of a 20-meter arm on a centrifuge. This dynamic simulator boosted him to an angular velocity that produced five Gs of acceleration—approximately that experienced by the Apollo astronauts on their voyage to the Moon thirty-three years later.

Dornberger began consideration of a center that would permit development and testing of the very large rockets of the future as they moved from the drawing board to the battlefield. He established several criteria for a new proving ground. It should be located on a coast, where large rockets could be launched away from the base itself. The firing should be parallel to a long stretch of the coast so that the entire trajectory could be observed by camera and by radio. The land itself should be flat and large enough to accommodate a long runway for the airfield to be used in jato and rocket-powered aircraft testing. Finally, the area should be in as remote a place as possible for reasons of safety, security, and secrecy.

The location finally selected was suggested by von Braun after he mentioned the need for such a site during a visit with his mother, who had been born and reared on a country estate near Anklam, a small town on the river Peene some 140 kilometers north of Berlin. She recalled that her father had often hunted ducks on the northern spit of the nearby island of Usedom in the Baltic Sea. Dornberger approved the island after he had visited it a few days following von Braun's suggestion.

In order to expedite the development of rocket motors for aircraft, von Richthofen generously offered to put up 5,000,000 reichsmarks for the construction of facilities for the army. What followed next is best described by Dornberger:

> His offer constituted an unprecedented breach of military etiquette as between branches of the Wehrmacht. ... Colonel von Horstig solemnly led me into the office of General Becker, who had become Chief of Ordnance. The general was wrathfully indignant at the impertinence of the junior service.
>
> "Just like that upstart Luftwaffe!" he growled. "No sooner do we come up with a promising development than they try to pinch it! But they'll find that they're the junior partners in the rocket business!"
>
> "Do you mean," asked Colonel von Horstig in astonishment, "that you propose to spend more than five million on rocketry?"
>
> "Exactly that," retorted Becker. "I intend to appropriate six million on top of von Richthofen's five!"

On April 2, 1936, Becker, Dornberger, von Braun, and von Richthofen met with Major General Albert Kesselring, commander of the Luftwaffe, now just a year old, to discuss the joint effort. Kesselring approved it for his service and paid 750,000 reichsmarks to the town of Wolgast for the property. Almost immediately a party of military engineers and construction workers moved to the site. The buildings were designed along the lines of the modern Luftwaffe installations rather than the nineteenth-century gothic architecture still favored by the army.

With the build-up of facilities, there also occurred an increase in personnel. Von Braun continued to seek out old colleagues from the *VfR* and to hire bright young university graduates, as well as some of their professors.

However, one of the old hands of the former group at Raketenflugplatz, Berlin presented a problem.

What to do with Nebel? Instinctively, von Braun knew that Nebel would not and could not fit into the pattern at Peenemünde. Klaus Riedel agreed with him. The days of the amateurs at Raketenflugplatz, Berlin were over. And Nebel was a great promoter and salesman, but not really a disciplined engineer. However, both men felt the army owed him something, for it had been Nebel's energy and talent in collecting tools, materials, and money that had sustained the *VfR* and its rocket range and had got von Braun and Klaus Riedel themselves started in their future careers. Von Braun hit upon a scheme to provide some money for Nebel. He convinced the army to purchase the license to a rocket motor patent that Nebel and Riedel had been granted in 1931. The design was of no real use, having been tried and abandoned at Kummersdorf. But its purchase provided a face-saving gesture under the circumstances.

While Nebel was not subsequently involved with research on the A4 at Peenemünde or its production in Nordhausen, he did make a contribution to its technology in an indirect way. As a member of the consulting engineering firm of Bauer and Nebel, his company obtained a contract to assist in the development of the huge underground A4 assembly and launching complex that was built at Wizernes, France, in 1942. Nebel's company designed the intricate erector that was to remove the assembled and checked-out missile from its underground plant, transport it to the surface, and set it up on its launcher. The hustling pioneer of rocketry in the early days of the *VfR* survived World War II. On July 16, 1969, he sat next to another venerable pioneer of the same days. Nebel and Hermann Oberth occupied special places reserved for very important people at the Kennedy Space Center, Cape Canaveral, Florida, and watched the launching of the first men to the Moon.

The removal of personnel and facilities from Kummersdorf to Peenemünde did not take place overnight. Since the construction of the rocket-engine test stands at Peenemünde required considerable time, many of the propulsion engineers and test personnel remained in Kummersdorf until 1940.

In March 1939, the rocket men at Kummersdorf had a visit from Adolf Hitler—his first and only trip to Dornberger's rocket establishment in the nation's capital. (He never visited Peenemünde.) It was a chilly, wet day with an overcast sky, and Hitler's mood seemed somehow attuned to it as he moved unimpressed beneath the dripping firs and pines.

Hitler was given a special tour. First, there was a stop at the test stand where the 300-kilogram motor shot out a light blue jet with an ear-splitting roar. Even though his ears were packed with cotton, the Führer grimaced slightly; he did not say a word when the test was over. Much the same thing happened at the test stand where the 1,000-kilogram motor was fired. The third test stand had a cutaway model of the A3 suspended in it. Von Braun gave his patter, pointing to various components and subassemblies as he mentioned them. After he had finished, Hitler carefully inspected the model, but still made no comment. The grand finale was a Wagnerian firing of an A5 suspended in the large test stand. Everyone was sure this would provoke a response from Hitler, but again, not a word came from him.

Later, as Hitler consumed his vegetarian dinner and mineral water in the officers' mess, Dornberger kept up a steady sales talk. Hitler, with calculated casualness, asked how long it would take to have the A4 move from the drawing board to the firing troops and if it were not possible to use steel instead of aluminum for the fuel tanks. Dornberger said that such a substitution was possible at least for the alcohol tank but that it would delay the development of the rocket considerably. Then Hitler simply said enigmatically, "*Es war doch gewaltig!*" ("Well, it was grand!").

After he left, Dornberger could not fathom Hitler's unresponsiveness—if, indeed, it was that—to what he had seen at Kummersdorf. Certainly he was the first such visitor to be unimpressed. By contrast, when Göring was given the same treatment at Peenemünde five years later, he could scarcely contain his excitement and enthusiasm. He impulsively grabbed Dornberger and hugged him, saying "This is colossal! We must fire one at the first postwar Nuremberg Party rally!" He then went on to predict a great future for rocket power, naming everything but space travel.

Even as the A3 was in the planning stage, the team began thinking of a much larger rocket in the summer of 1936. It would be the A4 (later the infamous V-2). Dornberger, too, felt that it was time to get down to a weapon rather than to continue building research rockets. Somewhat arbitrarily, he decided the A4 would have a range double that of Germany's famed Paris Gun of World War I, which could fire a 10.5-kilogram shell to a range of 130 kilometers. The new weapon would carry a warhead with 100 times as much explosive. Preliminary calculations indicated that the motor for such a rocket should have a thrust of 25,000 to 30,000 kilograms. Realizing that, in a war, the weapon would have to be transported by road and rail, Dornberger dictated that the length and diameter of the new rocket be determined by the tunnels and track curvature of the German federal railway system.

Dornberger also added some accuracy requirements:

> ... the dispersion—that is, the distribution of 50 percent of the impact points around the target point—should be 2 or 3 mils both longitudinally and laterally. This means that for every 1,000 meters of range a deviation of only 2 or 3 meters either too far or too short was acceptable, and the same for lateral deviation. This was stricter than is customary for artillery, where 50 percent dispersion of 4 to 5 percent is considered acceptable.

The preliminary design for the A4 began at once under the direction of Walter Riedel.

It soon became apparent that the A4 was an extremely ambitious project for the current state of rocket technology. Dornberger could see that the problems facing his team might require years to solve. However, he told Dr. Walter Thiel, in charge of engine design, who had remained at Kummersdorf, to continue development of the injector system for a 25,000-kilogram-thrust motor. In the meantime, other components that would eventually be utilized in the A4 would be tested in yet another research rocket, known as the A5.

The A5 was a very sophisticated device, technologically speaking. It certainly was ahead of anything America's Dr. Robert H. Goddard was flying at the time in New Mexico. The German rocket had practically all of the mechanical features that would later be incorporated into the A4. The A5 was essentially a test rocket for proving concepts as well as prototype equipment and instrumentation.

However, the development of the A5 and its subsequent technological success notwithstanding, the rocket team suffered a setback as a result of the sudden victories of the blitzkrieg that had begun in Poland on September 1, 1939. Hitler decided that the A4 would not be needed. Upon the counsel of General Georg Thomas, chief of the War Economy and Armaments Office of the *Oberkommando der Wehrmacht* (High Command of the Armed Forces), appropriations of men, matériel, and money for the center were reduced on November 13. This cutback eventually ballooned into at least a two-year delay in getting the huge rocket into production.

For the time being, however, activity on the remote island in the Baltic Sea continued to expand.

CHAPTER 4

The Flowering of Peenemünde

In 1935, the small fishing village of Peenemünde on the northern end of Usedom Island was not even listed in the German federal railway guide. Indeed, the area north and east of it and some 12 kilometers south to the resort of Zinnowitz, which was listed, was almost a pristine wilderness and a sportsman's dream. Huge Pomeranian deer ran freely through the dense oak, beech, and pine and grazed on the heather and bilberry bushes. Multicolored rabbits, descendants of domestic escapees from farms to the south and original denizens, bounded through the bracken, frightening pheasants, ducks, coots, geese, and grebes that proliferated in the marshes. The meandering river Peene provided good fishing, and the blue Baltic Sea was ideal for sailing enthusiasts.

The sylvan environment was not to last after the decision was made to build a joint army–air force rocket research establishment on some 50 square kilometers of the island. However, during construction, every effort was made to cut as few trees as possible. The reason was not so much ecological or esthetic as practical: The ancient forest provided a natural camouflage.

The venture did not proceed smoothly despite the initial enthusiasm, cooperation, and funding of the Luftwaffe and the army. Costs soared beyond estimates, additional facilities kept being added to original plans, a dike had to be built, the harbor at Peenemünde needed dredging, and more railroad tracks were required. In early 1939, the Luftwaffe withdrew its support, and the project seemed in danger of foundering. With the commencement of war in the autumn of the year, the task of completing Peenemünde was given to Albert Speer, the *Generalbauinspektor* (inspector general of construction) who had won Hitler's favor by building the Third Reich's new Chancellery in only one year; he had escorted the Führer through it just eleven months earlier.

Speer sought not only to complete the project but also to lay magnificent plans for its future. These plans were of the same magnitude as his Arch of Triumph and Great Hall that were to be built on Adolf Hitler Platz in Berlin after victory. Peenemünde was to become to rocketry, upper atmospheric research, and space exploration what Friedrichshafen had become a decade earlier, a well-organized scientific and industrial community devoted to research and development in aerostatics and the manufacture of dirigibles. Speer's planners envisioned a city of some thirty thousand people.

The existing center would be expanded to include even larger rocket engine test stands, launching pads, several supersonic wind tunnels, liquid propellant plants, and over twenty specialized laboratories. A scale model was built for Hitler's approval, which he gave. In many ways, the concept foreshadowed the urban-industrial complex that was to be built only a few years later at Oak Ridge, Tennessee, where the first atomic bomb was developed. It also looked forward to Akademgorodok, the scientific and technological community that appeared in the USSR some eighteen years later.

However, in Germany as elsewhere in the world, "the best laid schemes o' mice and men gang aft agley." Such a future for Peenemünde was not destined.

What did become a reality at *Peenemünde Ost*, the army's facility, was in itself an accomplishment of significance. At the cost of 300 million reichsmarks (over US $70 million, 1 DM = US $0.238), *Heersversuchsstelle Peenemünde* (Army Research Center Peenemünde) became fully staffed in August 1939, when all remaining personnel were relocated from Kummersdorf. By 1942, *HVP* employed 1,960 scientists and technicians and 3,852 other workers. Additional thousands were under contract at universities, research centers, and plants elsewhere. The annual payroll for those at *HVP* in 1942 reached almost 13 million reichsmarks ($3.1 million).

HVP occupied by far the greater portion of the area (*Peenemünde West*, the Luftwaffe center in the northwest, covered some 10 square kilometers). It was located generally along the eastern coast of the island from its northernmost tip to Zinnowitz, 15 kilometers south. Close to the sea, on the northern strand, were a series of test stands. The largest of these was Test Stand 7, from which the A4 was actually launched. With the same foresight shown by Speer, the army designers of this huge test stand built it for motors with thrusts as great as 100 tons, even though the

A4 had only 25 tons. Continuing to the south were smaller structures for statically testing the engines, gas generators, propellant pumps, and jatos for aircraft.

The engineering and research area was somewhat more inland and south of the test stands. It provided the shops and laboratories needed to support *HVP*. Other components included the guidance, control, and telemetering laboratory; materials testing laboratory; military and civilian headquarters; officers' quarters, mess halls, and various shops needed for plant maintenance. Here also were located the liquid oxygen plant and wind tunnel.

About 2 kilometers farther south was the pilot production plant, where some 250 A4s would ultimately be manufactured for research and developmental flights. To the west of the huge assembly building and its associated workshops were three large stands upon which were to be tested the A4s to be built elsewhere in Germany.

Many of the civilian scientists and engineers and their families lived in a special housing compound known as the *Siedlung* (settlement) on the beach at the village of Karlshagen, where there also was a barracks square for soldiers of the center and a camp for foreign workers. A camp at Trassenheide held prisoners of war that were utilized at both *HVP* and *Peenemünde West.*

Other features of the center included radar and optical tracking stations spaced out along the Pomeranian coast east toward the Polish border.

This, then, was the base for the team that had grown from a few dedicated individuals with inadequate resources at Kummersdorf several years earlier. *HVP* was one of five military proving grounds that were a part of the *Waffenamt Prüfwesen* (weapons proof and development office) of the *Heeres Waffenamt* (army weapons office), the former commanded by Lieutenant General Richard John, and the latter initially by General Becker and later by General Emil Leeb. The activities of *HVP* were directed by *Wa Prüf 10 & 11*, one of twelve suboffices of *Waffenamt Prüfwesen* charged with developing weapons of all types needed by the army. Indeed, the office was concerned with developing everything needed by the army except food, clothing, and housing.

Wa Prüf 10 & 11 was concerned with research into both liquid-propellant and solid-propellant rockets, as well as all controlled guided missiles except those of the *Luftwaffe*. The office was headed by Major General Dornberger. The proving ground at Peenemünde was commanded by Colonel Leo Zanssen and had as its civilian technical director Wernher von Braun and his deputy Dr. Walter Thiel. Reporting to von Braun were the heads of the nine major departments of *HVP*: Technical Design Office, Walter Riedel; Aeroballistics and Mathematics Laboratory, Dr. Hermann Steuding; Wind Tunnel, Dr. Rudolf Hermann; Materials Laboratory, Dr. Mäder; Guidance, Control, and Telemetry Laboratory, Dr. Ernst Steinhoff; Development and Fabrication Laboratory, Arthur Rudolph; Test Laboratory, Klaus Riedel; Future Projects Office, Ludwig Roth; and Purchasing Office, a Mr. Genthe.

As things became organized at Peenemünde, it was natural that conflicts between its personnel in various offices would arise. Auditors, accountants, and clerks, who generally feel outside the mainstream in any large research organization, began to flex their muscles at the expense of the scientists and engineers. They found endless interpretations for regulations. But Dornberger brought this practice to a quick halt, once he saw what was happening. He set forth the policy for *HVP* in clear terms: "The engineers in Peenemünde have top priority. All their wishes are to be fulfilled to the extent you can assist. Don't look for regulations that prevent things getting done. Look for all means possible for getting the job done. Regulations are made by men; they can be changed by men." Crestfallen, the bookkeepers withdrew to ponder their codes in this new light.

While construction was still underway at Peenemünde, launching operations began from Greifswalder, a small island some 8 kilometers off the northern coast of Usedom. The launching pad was a simple concrete foundation, and the blockhouse looked like a fort on the early American western frontier. It was made of logs partially covered with earth. From within, rocket engineers peered through chinks between timbers at the rocket not far away.

There were difficulties other than discomfort associated with the offshore launch site. For one thing, it was the natural abode of literally hundreds of thousands of tiny field mice, who quickly became addicted to the insulation of the telephone wire provided them by the rocketeers from across the bay. Launches were often delayed as a result of their appetites. On the other hand, there were compensations. During such long "holds" in launchings, impromptu pheasant shoots by members of the team provided recreation as well as excellent dinners in the officers' club at Peenemünde.

On December 4, 1937, the first rocket launched from Oie, as the island was familiarly known, was the third generation A3. (Between 1938 and 1941 at least seventy large rockets were launched from Oie.) Technologically speaking, the A3 was a quantum leap from the A2. The motor produced 1,500 kilograms of thrust for 45 seconds. However, its secondary purpose was to demonstrate trajectory control of rockets. It had a very sophisticated—for the time—guidance and control system. The guidance section featured three gyroscopes and two integrating accelerometers. These instruments, working in unison, sensed deviations from the established path and fed corrective signals to actuators that turned molybdenum vanes within the exhaust gases of the motor. Ultimately, the system failed simply because it was too advanced a project for the engineering of the day. The instruments were not equal to their assigned tasks.

The A5 followed the A3 into the skies above Oie. In October 1938, the first of these forerunners of the A4 reached a respectable altitude of 8 kilometers and was recovered by parachute, a recovery system that proved so reliable that many of the A5s were refurbished and launched several times over during the following three years. The rocket was soon capable of regularly reaching an altitude of 13 kilometers.

Logistical problems began to beset the growing research center.

A shortage of ethyl alcohol soon appeared in a nation where schnapps had yet to be displaced by Coca-Cola as a leading national beverage. In view of the criticality of this propellant, any loss or wastage was of concern to both Dornberger and the local Nazi Party organization. Explanations to Dornberger that the loss was due to evaporation and spillage fell on deaf ears. He wanted immediate and effective measures to prevent pilferage and to provide punishment that would deter pilferers. (One former engineer at Peenemünde several decades later pointed out that a completely fueled V-2 represented the equivalent alcoholically of approximately 66,130 moderately dry martinis.) Dr. Martin Schilling in the testing laboratory called a staff meeting to discuss the matter. Several suggestions were made, and one was accepted.

An unhealthy pink dye was added to the propellant. Within one week, the thirsty and enterprising imbibers around Test Stand 7 found that alcohol so treated could be filtered through raw potatoes to remove the offensive dye and be converted into "instant schnapps." Karl Heimburg, one of Schilling's engineers, then proposed an alternative. A powerful purgative was placed in the propellant. The result should have been predictable. Absenteeism and frequent rushes to the bathroom threatened to bring testing to a halt altogether.

The situation became more critical later on because there were increasing demands for alcohol in other quarters of the war effort than Peenemünde, and the means of production were beginning to feel the combined effects of Allied air raids and poor potato crops. Methyl alcohol was then cut into the ethyl alcohol, with the result that one employee went blind and another shortly afterwards died.

The state alcohol control and distribution board as well as the local Nazi Party headquarters at nearby Swinemünde began to make threatening noises. Dornberger had always been most anxious not to offend the party and to keep it at arm's length from the operations at Peenemünde.

Believing that desperate times call for desperate measures, the impulsive but crafty Heimburg went to the party headquarters with a suggestion that turned even the most dedicated SS men pale: "Give me the body of the fellow who died. I'm going to nail it to the fence of the main gate for three days. This will really set an example." His carefully contrived ruse worked; the party backed away and allowed the rocket center to deal with the problem on its own, hoping, of course, that the task was not to be assigned to the barbarian Heimburg.

When it was expeditious, the team also called upon the expertise that existed among the universities of the country. Typical of the way German universities contributed to the mission of Peenemünde was a three-day conference, September 28–30, 1939. With the spur of declared war, the scientists and engineers at the large center realized that there were theoretical problems that could better be solved by their former professors while they devoted their talents to matters closer at hand, such as perfecting and testing engines for large rockets and deciding on the optimum design for their structures and the metals they were to be made of. Known jokingly as *der Tag der Weisheit* (the day of wisdom), it began with the arrival of a stellar collection of scientists from the excellent system of *Technische Hochschulen* (institutes of technology) located regionally in major cities throughout the Third Reich. Thirty-six such men had been invited to a round of technical briefings and problem presentations.

Among those attending were Professor Theodor Buchhold, who held the chair of electrical engineering at Darmstadt and would join von Braun in the United States after the war; Professor Alwin Walther, mathematician, Darmstadt; Professor Carl Wagner, physical chemist, who also would follow von Braun to America; Professor Ernst Hüter, another electrical engineer, Darmstadt; Professor Walter Wohlman, electronics engineer, Dresden; Dr. A. Beck, mechanical engineer, Berlin; and Dr. Max Schuler, gyroscope theoretician, Göttingen.

Von Braun, Thiel, Riedel, and other members of the team presented the general types of engineering and scientific problems facing the organization. They stressed that only two years were available for solutions. Then von Braun pointed out that he was looking only for solutions. He was not interested in long, scholarly articles prepared for the prestigious scientific journals.

From the "day of wisdom" developed a new technique and scientific symbiosis at Peenemünde that proved viable and mutually rewarding. Peenemünde had the funds, which the universities needed; but the universities had the knowledge and talent that Peenemünde required. Von Braun described the consequences:

> Cooperation with these professors was extremely agreeable as well as constructive; likewise extremely democratic. There was much discussion and many symposia and visits. Contracts for scientific work were eventually drawn in very broad terms in order that the institutions be allowed a wide latitude of approach. Their members were also thoroughly familiarized with all practical aspects. This stimulated many creative contributions. When some prototype device was working properly, direct contact was established between originator and producer.

By the beginning of 1940, some 550 million reichsmarks ($131 million) had been expended on Peenemünde in its first two years of operation. The budget for 1940–41 was set at 50,400,000 reichsmarks ($12,000,000) and then halved. In March, the steel allotment was cut; by July, the installation was not even on the list of military facilities with any priority for national resources. Unexpectedly, in August, high priority was restored; and just as unexpectedly it was removed in October. Such an oscillation in funding played havoc with contract negotiations in an organization that greatly depended on the support of industry.

Hitler, at best a vacillating supporter of Peenemünde until the first successful launch of the A4, soon spoke to his intimates of firing an initial volley of five thousand of the still untested A4s against England. Only in March 1941, following the loss of the Battle of Britain by the Luftwaffe, was the highest priority possible given to *HVP*.

However, by September an acute labor shortage at the huge installation threatened the A4 schedule. If the first rocket was to be launched by the following February and the missile placed into production by April, as then planned, more technical manpower was required immediately. In December 1941, Dornberger found a solution with the support and connivance of Field Marshal Walther von Brauchitsch, newly appointed commander in chief of the army, who less than two years earlier had accompanied Hitler on his visit to Kummersdorf and had been highly impressed by what he saw and heard there.

Von Brauchitsch authorized formation of a battalion of troops to be called *Versuchskommando Nord* (Test Command North) at Peenemünde. Technically a combat unit assigned temporarily to the homefront, it was one of the strangest military units ever raised by any nation. It consisted of some thousand men divided into six companies, each commanded nominally by a lieutenant but in reality by a grizzled sergeant major. Major Heigel, the battalion commander, was hard pressed to direct the activities of his enlisted men, all of whom were scientists or engineers. Many of them were from combat units either on leave or recently released from hospitals. The unit was formed in late 1942 and rounded out over the following year or so.

Assignment to and duty in the *VKN* was a puzzling experience for many of the men who filled its ranks over the following two years.

Dr./Lance Corporal Ernst Stuhlinger, once an assistant to Professor Johannes Geiger, the eminent atomic physicist at the *Technische Hochschule* at Berlin, was an infantryman among those who had unsuccessfully attempted to relieve the surrounded Sixth Army in Stalingrad. He was ordered to report to Peenemünde, a place he had never heard of. Slinging his rifle on his back, he had walked a good distance back from the eastern front for duty as a "scientist" in some "supersecret project." Lieutenant Krafft Ehricke, former tank platoon leader wounded at Dunkirk and veteran of the bitter fighting in the drive toward Moscow, felt conspicuous in his black Panzer uniform as he stood in the mixed bag of officers and enlisted men dressed in army field gray waiting to be "interviewed for a job" rather than assigned to one.

Others, like Lieutenant Hartmut Küchen, although never in battle, later had equally bizarre "war stories" to tell. He had been a civilian engineer with the Siemens Co. at work on a construction project during the very early days at Peenemünde. In 1941, he was called to active duty as an army reserve officer. Then he was posted back to his same job at Peenemünde, at considerably less pay—a novel method of cost reduction for such an expensive program!

To some soldiers the placement of *VKN* personnel took on the aspect of a slave market. So it seemed to Lance Corporal Willi Mrazek, recently out of a hospital and recovering from a saber wound received when his unit on the Eastern Front was attacked by Soviet cavalry. With more deference than he had hitherto received as a lance corporal, he was shown into the cafeteria at Peenemünde. Seated at a table were four men whom he had never seen but who would become his peers over the next few decades. The group included von Braun, Eberhard Rees, head of prototype manufacturing, and Arthur Rudolph, von Braun's old colleague from Kummersdorf. With them was Dr. Ernst Steinhoff, head of the Guidance, Control, and Telemetry Laboratory.

Mrazek had not the faintest idea of what he was supposed to do. Since the men before him were not military officers, he decided not to salute. He compromised by standing at a somewhat relaxed state of attention.

> "What did you do before you went into the army?"
> "I designed storage tanks for caustics at the Solvaywerke."
> "Good, send him to Patt over in TB/B; he needs someone who knows something about fuel tanks."

Similarly unsettling was the experience of Corporal Walter Wiesemann, of the Luftwaffe, who found himself standing rigidly at attention in front of a Captain König, commander of the *Flakversuchstelle*, a Luftwaffe counterpart to the *VKN*, which he later transferred to the VKH. Seated with his feet on the captain's desk was Dr. Walter Thiel, scanning Wiesemann's military record. Chatting with Thiel was *VKN* Private Hans Lindenmayr, chief of the center's valve laboratory, with his tunic open at the neck and wearing white socks and civilian shoes. König explained to Corporal Wiesemann that he would be working for Private Lindenmayr. The latter glanced at his new employee's sleeve and said disenchantedly, "Oh my God! Another superior officer." (Wiesemann later shortened his name to Wiesman.)

As if this military informality were not strange enough, some new members of the *VKN* and *FVS* soon found themselves working cheek by jowl in a design office staffed entirely by 150 Russian officer prisoners of war. These prisoners were kept in an especially constructed camp at nearby Trassenheide, and they also manufactured graphite jet vanes for the A4 control system!

Another valuable source of personnel for Peenemünde was the *Kriegshilfe* (war service) girl. During the war, women as well as men were mobilized for civilian work. Girls with specialized skills were sent where they were needed most. Thus, many found their way to Peenemünde, willingly or unwillingly. There, they performed not only routine administrative and secretarial tasks but also technical jobs such as drafting, illustrating and mathematical calculations. These latter women, working with mechanical calculators and slide rules, were known as the *Mess-frauen* (measurement girls). They worked together in a large room under the direction of a benevolent but demanding Dr. Paul Schroder. In addition to deriving data from missile test flights, the *Mess-frauen* also reduced the enormous amounts of data flowing continuously from the several static-test stands. A special group of fifty girls was called the *Tapeten-frauen* (wallpaper girls) because of the great number of rolls of paper they produced in calculating the trajectory of missile flights.

As the war progressed and went against Germany, still another source of labor appeared at Peenemünde. It was in the form of pressed foreign laborers and prisoners of war. Among both these groups was a good number of skilled machinists and other technicians. They worked under their own *Kapos* or chiefs, who were very strict on them and demanded performance of a quality acceptable to their German masters. The unskilled among this group were employed in menial work such as maintenance of roads, grounds, and buildings. While such use of prisoners of war was strictly forbidden by the Geneva Convention, the Germans flouted the convention generally. At the Nuremberg trials after the end of the war, Albert Speer admitted that, by 1944, 40 percent of all prisoners of war in Germany were being employed in the production of weapons and munitions. (Indeed, POWs were even used to carry ammunition on the front lines and to man antiaircraft guns!)

The SS maintained a tight control over the POWs, and attempts at sabotage were few. Most of the men were happy to have fairly good meals and to be working rather than marking time behind barbed wire. Among the more enterprising of them, skills such as barbering brought in pocket money from German engineers too busy to take time off the job for a visit to a barber in one of the nearby towns.

In addition to personnel problems, there were technical problems at Peenemünde. An onerous one that cropped up early in the development of the A5 was associated with the steering mechanism. The molybdenum jet vanes of the motor had been replaced by carbon ones at a great saving in both money and a highly critical metal in short supply. With the new material came the problem of excessive erosion and the tearing loose of the vane from its support arm. Like many of the technical riddles facing the engineers at Peenemünde, it was solvable only by empirical means—cut and try, cut and try again. With time and an almost infinite number of experiments, the optimum shape for the vane and the most efficient way of securing it emerged.

As advanced as was the technology at Peenemünde, there was an area curiously lacking that seriously hampered the development of the A4 and added greatly to its cost. Today, the use of radio telemetry is a standard tool of the aerospace engineer. By means of it, he can tell whether a valve opened at the prescribed time, the flow rate and pressure of a fluid through a line, the temperature of a rocket motor wall, the position of a gimballed motor, and the voltages and currents in critical circuits of a guidance system.

At Peenemünde, in the early 1940s, no such sophistication existed. The number of measurements that could be telemetered during early A4 developmental flights was only four. By comparison, during the first flight of the Saturn 5 from the Kennedy Space Center at Cape Canaveral in 1967, a total of 3,552 measurements by telemetry were continuously monitored by the engineers who had designed the huge rocket.

The lack of telemetry imposed upon the Germans the necessity for firing an enormous number of experimental rockets to obtain relatively little overall data on total performance. For example, perfecting the propellant shutoff valves for the A4 required some 20 A4s to be launched. The problem of why the A4 was bursting in the air after reentry into the atmosphere was never really solved largely because the engineers had to depend upon visual observations and limited telemetry.

Flight-testing was particularly hampered by the lack of reliable telemetry when there was a failure, such as when rocket No. 17,003 (the number does not indicate that it was the seventeen thousand and third round produced) was launched in January 1944. This missile was the third to be produced at the underground plant—Mittelwerk—in Thuringia and shipped to Peenemünde for flight-testing from Test Stand 7. It detonated just three seconds after ignition, so technically it was not a flight-test failure since the missile was not then airborne. At the time, test engineer Willie Muenz remarked to Hartmut Küchen, chief engineer, "It sure would be nice to know what happened to 17,003, but I'm afraid the war will be over before we have a decent telemetering system."

"We just blew a million marks," Küchen replied, "in order to guess what could have been reported accurately by an instrument probably worth the price of a small motorcycle."

A major problem on the A4 proved to lie in a guidance system that would meet the specifications for accuracy that Dornberger had established. Two systems were investigated for the missile. One used a simple autopilot to keep the rocket flying true along its course, with superimposed radio control, in which the expensive precision equipment remained on the ground; i.e., the large and cumbersome radar antenna. While this guide-beam system provided extreme accuracy, it was vulnerable to enemy jamming, and it meant complicating the already complex ground support equipment needed merely to launch the rocket. The alternative was an inertial system, in which everything needed to obtain the required impact accuracy would be within the missile itself. It utilized gyroscopes, accelerometers, and an analogue computer to furnish data to position the jet vanes during powered flight for maintaining trajectory control.

In managing Peenemünde, Dornberger led an extremely taxing life, physically as well as mentally. His week began at 6:00 AM Monday morning when he left Berlin for Peenemünde, where he spent most of the week living in the guest house. Fridays he spent at Kummersdorf, where solid-propellant rockets were being developed. On Saturday, he returned for a day to his Berlin office as chief, *Wa Pruf 10 & 11*, from which he directed the operations and funding problems at both Peenemünde and Kummersdorf. Sunday, he spent at his flat in Berlin with his youthful and charming wife, Bunny.

The concept of the team approach to missile development was constantly stressed by both Dornberger and von Braun. When an employee was in doubt about his contribution to the overall program, educating him was a simple matter. Thus, when Dr. Helmut Hölzer, a specialist in calculating trajectories for the A4, complained to von Braun that, because he spent all day in a cubbyhole, he did not "have a good feel" for his work and its relationship to the more practical aspects of the A4, von Braun asked Hölzer to accompany him down to one of the static-test stands.

With Hölzer peeking through a hole in a wooden fence, the A4 engine ignited with a roar. The pressure tore loose the fence and flattened him on the ground.

"I picked myself up, bleeding at the nose," Hölzer recalled. "It was an indelible introduction to the practical aspects of the program."

On March 23, 1942, Hitler once again shifted priorities to the detriment of the A4, putting all Germany's resources into the invasion of the USSR, while allowing the large rocket project to languish.

Despite this, on June 13, the A4 was reády for its first launching. While the crew was busy preparing the missile at Test Stand 7, excited members of the team, most of whom were illegally in the area or there only under the flimsiest of reasons, gathered. They did so at distances from the "bird" in direct proportion to their faith in the rocket's designers and builders or in inverse proportion to their ignorance of what could very well happen upon ignition. The more optimistic and faithful (as well as some of the more ignorant) stood chatting shop in front of the bunker located near Test Stand 1, only 250 meters from Test Stand 7. There was a fairly dense cloud cover, but the A4 lifted off normally enough—or so it seemed. There was a slight rolling about its long axis as the rocket disappeared through the clouds; then, there was a muffled rumble. Thunder, perhaps? No, sickeningly, back through the clouds plunged the cart-wheeling rocket. The A4 crashed into the Baltic little more than a kilometer away, sending up a tremendous fountain of water. Then, just as suddenly, there was nothing to be heard but the gentle lapping of the Baltic upon the shore.

Rocket No. 2 was no more successful than the first. After launching on August 16, it exploded at an altitude of 11.72 kilometers and fell into the sea some 9 kilometers from the launch site. Among the observers of this debacle were four influential men; the armaments chiefs of the three armed forces—Field Marshall Erhard Milch, state secretary of the *Luftwaffe*; Admiral Karl Witzell, of the *Kriegsmarine* (navy); and Colonel General Friedrich Fromm, of the *Heer* (army)—plus Minister of Armaments Albert Speer, who arrived late but saw the test from the air as his plane approached the airfield at Peenemünde West.

The third test, on October 3, was a different story, as Dornberger recalled:

> It was an unforgettable sight. In the full glare of the sunlight the rocket rose higher and higher. The flame darting from the stern was almost as long as the rocket itself. The fiery jet of gas was clear and self-contained. The rocket kept on its course as though running on rails; the first critical moment had passed. Missile A4 had shown itself to be stable about its longitudinal axis. The projectile was not spinning; the black and white surface markings facing us did not change.

The missile followed its programmed trajectory to the target point 192 kilometers down range in the Baltic.

Among those pressing around Dornberger to offer congratulations was Hermann Oberth, who had arrived in Peenemünde in 1941 and would leave in 1943. He shook the general's hand excitedly and said in words highly uncharacteristic for him, "That is something only the Germans could achieve. *I would never have been able to do it*." In a lighter mood, engineer Albert Zeiler erected a monument at the foot of Test Stand 1, from the top of which Dornberger had observed the historic launching. It was a huge boulder with the painted inscription: "A great weight has fallen from my shoulders."

Following this successful launching, von Braun was awarded the *Kriegsverdienstkreuz I Klasse mit Schwertern* (War Service Cross, First Class, with Swords), not the Iron Cross, which was a military decoration.

That evening at the officers' club Dornberger was host to the top-level managers of his rocket team. After recapitulating the success, he said something that was near to the hearts of his listeners:

> The following points may be deemed of decisive significance in the history of technology: We have invaded space with our rocket for the first time—mark this well—have used space as a bridge between two points on earth; we have proved rocket propulsion practicable for space travel. To land, sea, and air may now be added infinite empty space as an area of future transportation, that of space travel.... So long as the war lasts, our most urgent task can only be the rapid perfecting of the rocket as a weapon. The development of possibilities we cannot yet envisage will be a peacetime task. Then the first thing will be to find a safe means of landing after the journey through space.

In early 1943, the war began going worse for Germany on its eastern front. With the surrender of Field Marshal Friedrich von Paulus' encircled Sixth Army in Stalingrad on February 2, 1943, Hitler began reconsidering the A4 as a weapon. But then, a month later, came the most bizarre check yet to development of the A4. Dornberger recalled the incident that touched it off:

> In March 1943, the Führer said in reply to one of Speer's repeated requests for higher priority for the A4 program, "I have dreamed that the rocket will never be operational against England. I can rely on my inspirations. It is therefore pointless to give more support to the project." I personally saw, in Major General Hartmann's office at the Ministry of Munitions, a memorandum of this statement of the Führer's, printed in the large type characteristic of headquarters. Speer ... confirmed it.

Speer, years later in a letter to the author, doubted that Hitler would have stopped the program for such a reason, but Dornberger, at the same time, was adamant about having seen the memorandum. He said it was written on a special machine with very large type, "Führer type." (Hitler's vanity would not permit him to wear glasses while others were around him. The very existence of the typewriter itself was classified as an *offizial geheim Reichsache*, an official state secret.)

Despite his bad dream, Hitler did reconsider the A4, and on July 7, 1943, he granted it, once again, top priority.

Dornberger's peculiar position of being wholly in charge of his project and reporting directly to the chief of the *Heeres Waffenamt* had certain advantages. In particular, he managed to keep the Nazi Party and the SS out of Peenemünde for several years.

The military commander of Peenemünde during most of the period from 1940 to 1943 was Colonel Leo Zanssen, an old friend and colleague of Dornberger's, but unlike him, a devout Roman Catholic and a practicing anti-Nazi. (Dornberger, in the tradition of the professional German officer corps, was apolitical and areligious, at least externally.) Zanssen, also unlike Dornberger, at least in World War II, had been in combat; indeed, he asked Dornberger to release him for duty on the Russian front in 1942, which the latter reluctantly did. Once back at Peenemünde, in 1943, Zanssen's anti-Nazi attitudes and behavior moved beyond mere silent contempt and philosophical revulsion. He began deliberately antagonizing local party officials. For example, when the chief of police of Stettin demanded that Zanssen make available trucks from Peenemünde for moving his furniture and other household effects from one house to another, Zanssen peremptorily refused—one black mark. And when a Nazi official was designated *Burgomeister* of the Peenemünde area, Zanssen redrew the boundaries of Peenemünde to exclude it from his jurisdiction.

"I am responsible here," he said coldly to the aspiring *Burgomeister*, "so you have nothing to say about what goes on here." Another black mark.

Things soon came to a head.

On April 26, 1943, Zanssen was removed from command in a ploy that was at once comic and utilitarian. Dornberger simply reassigned Zanssen to his own job in Berlin, and then himself took Zanssen's post at Peenemünde.

After a few months in his new job, Dornberger received an important phone call from Albert Speer. On July 7, 1943 Dornberger and von Braun traveled to *Wolfsschanze* (Wolf's Lair), Hitler's headquarters in East Prussia, near Rastenburg. With Dr. Ernst Steinhoff at the controls of an He-111 light bomber, the two men took off from the airfield at Peenemünde in a thick fog. With them were the films made of the successful launch of the A4 on the previous October 3, a wooden model of a massive underground launching site that von Braun proposed (over Dornberger's strenuous objections) for the A4, a large sectionalized scheme of the A4, and charts and manuals on the rocket. They were off to brief Hitler on what was already becoming known as a *Wunderwaffe* (wonder weapon).

Arriving with time to spare, they were met with the news that their appointment had been postponed for a few hours. During the wait they went over their briefing, though it seemed hardly necessary, since both Dornberger and von Braun knew it by heart. Later in the evening, the doors swung open and a party herald marched in with a fascist salute and shouted "*Der Führer!*"

It had been a little over four years since Dornberger had seen him:

> I was shocked at the change in Hitler. A voluminous black cape covered his bowed, hunched shoulders and bent back. He wore a field-gray tunic and black trousers. He looked a tired man. Only the eyes retained their life. Staring from a face grown unhealthily pallid from living in huts and shelters, they seemed to be all pupils.

Following closely behind Hitler were Speer; Field Marshal Wilhelm Keitel, chief of the Armed Forces High Command at the Führer's headquarters at Berchtesgaden; General Walter Buhle, chief of the army staff of the Armed Forces High Command at Berchtesgaden; and Colonel General Alfred Jodl, of Hitler's staff, all accompanied by their aides.

Von Braun confidently launched into a briefing that he had given so many times. "The bird will carry a ton of amatol in her nose, but it will hit the ground at a speed of over 1,000 meters per second, and the shattering force of the impact will multiply the destructive effect of the warhead," he said in a matter-of-fact voice.

"I don't accept that thesis," Hitler interrupted. "It seems to me that the sole consequence of that high impact velocity is that you will need an extraordinarily sensitive fuse so that the warhead explodes at the precise instant of impact. Otherwise, the warhead will bury itself in the ground, and the explosive force will merely throw up a lot of dirt.'

(Once back at Peenemünde, von Braun ordered a study of the problem. Later, in recalling the result, he said, "I'll be damned if he wasn't absolutely right. Hitler may have been a bad man, but he surely was not stupid!")

As he was bidding good-by, Hitler shook hands with von Braun and said, "Professor von Braun, I should like to congratulate you upon your remarkable achievements."

Von Braun was startled and puzzled at being addressed as *professor*, but Speer and Dornberger smiled conspiratorially. In an earlier visit to Peenemünde, Speer had told Dornberger that he was suggesting to Hitler that von Braun be given the title, which was an honorific bestowed by the state rather than an academic rank. (Two weeks later, rather informally, von Braun received his certificate of appointment, signed by Hitler: It appeared in his in-tray with the morning's mail.) Having been assured of renewed support for the A4, Professor von Braun and Dornberger took their leave of the Führer with added enthusiasm and personal motivation.

After Dornberger and von Braun departed, Hitler returned to his underground bunker. The eyes, which had so impressed Dornberger, became animated and took on the fire characteristic of a decade earlier: "The A4 is a measure that can decide the war. And what encouragement to the home front when we attack England with it! This is the decisive weapon of the war; and what is more, it can be produced with relatively small resources. Speer, you push the A4 as hard as you can," he told his minister of armaments.

During the development of the A4, the engineers at Peenemünde tended to lose sight of the fact that the rocket was merely one part of a very complex weapon system. Consequently, not until mid-1943 had attention been given to designing the launcher and requisite ground-support equipment. The task was assigned to the one team member who could do the job, the talented and resourceful Klaus Riedel. He undertook it with characteristic energy, taking unheard-of shortcuts in engineering to get the job done. Sometimes Riedel's engineers merely made rough sketches of items needed and then turned them over to specialized German industries for detail design and production. Thus, the Meilerwagen, a combined transporter and erector for the A4, was produced in Munich by the Meiler Co., hence the name, only a month after its design had been approved by Riedel.

Early in 1944, Heinrich Himmler increased his efforts to move into the field of rocketry, an area of weaponry he had long coveted. It was only a matter of time before he made his move to establish an enclave at Peenemünde.

On the morning of February 21 von Braun climbed into the cockpit of his private sky-blue Me-108 airplane for a flight over to Rastenburg. This time, however, it was not the Führer he was to see. He had been summoned there by Himmler, whose field headquarters in a camouflaged train was at nearby Hochwald. It was a command appearance even though it had the external guise of a friendly chat. Himmler had decided to take a knight in his game to win the rocket and missile programs of the Third Reich. He came to the point immediately:

> I hope you realize that your A4 rocket has ceased to be a toy, and that the whole German people eagerly await the mystery weapon . . . As for you, I can imagine you have been immensely handicapped by army red tape. Why not join my staff? Surely you know that no one has such ready access to the Führer, and I promise you vastly more effective support than can those hidebound generals. . . .

But von Braun's loyalties to his team and the army were strong, and he was not about to be tempted. He replied, perhaps somewhat hastily:

> Herr Reichsführer, I couldn't ask for a better chief than General Dornberger. Such delays as we're still experiencing are due to technical troubles and not red tape. You know, the A4 is rather like a little flower. In order to flourish, it needs sunshine, a well-proportioned quantity of fertilizer, and a gentle gardener. What I fear you're planning is a big jet of liquid manure! You know that might kill our little flower!

Himmler smiled at the scatological simile, changed the subject, and after a few amenities—and with characteristic politeness—dismissed von Braun, who flew back to Peenemünde.

But Himmler was not easily rebuffed.

On March 5, von Braun, after an arduous six-day week, attended a party in Zinnowitz, where many colleagues lived. Several of his associates such as Klaus Riedel and Helmut Gröttrup—assistant to Steinhoff—were there, as well as his brother Magnus, a recent arrival at Peenemünde. Von Braun had a couple of drinks, relaxed, and played several classical pieces on the piano. Inevitably, however, the talk turned to space travel, as it always did outside duty hours. While von Braun, Riedel, and Gröttrup chatted about their consuming passion, they were unaware that their conversation was being carefully monitored by a local woman physician and Gestapo agent. In fact, Gestapo agents had begun compiling dossiers on von Braun and his colleagues, who had incautiously spoken too much of space travel since October 17, 1942. They were even suspected, according to the personal diary of General Jodl, of constituting "a refined Communist cell."

At two in the morning on March 15, von Braun was awakened by a loud pounding on the door of his rented room in Koserow, 19 kilometers south of Peenemünde. He opened it and was immediately confronted by three Gestapo agents and was driven to the prison in Stettin. Also brought in were Klaus Riedel; Helmut Gröttrup; Magnus von Braun; and Hannes

Lührsen—an architectural planner—his wife, and his mother (who owned the hotel in which the party had taken place).

Von Braun later described what happened:

> Finally, a court of SS officers charged me with statement to the effect that the A4 was not intended as a weapon of war, that I had space travel in mind when it was developed, and that I regretted its imminent operational use! That sort of attitude was rather common at Peenemünde, so I felt relatively safe, were that the only accusation with which they could confront me. But they went further and maintained that I kept an airplane in readiness to fly me to England with important rocket data! This would be difficult to disprove, for I was in the habit of using a small, government-owned plane which I piloted myself on business trips throughout Germany. How could I *prove* that I had no traitorous intentions?

The situation was far more serious than von Braun and the others assumed.

Dornberger found out about the arrests late that afternoon when General Buhle told him what had happened. Dornberger's reaction was one of absolute incredulity:

> I could not believe my ears. That couldn't possibly be true. Von Braun, my best man, with whom I had worked in closest collaboration for over ten years and whom I believed I knew better than anyone, whose whole soul and energy, whose indefatigable toil by day and by night, were devoted to the A4, arrested for sabotage! It was incredible. And Klaus Riedel, who had worked out the entire ground organization with untiring zeal and absolutely outstanding perception of military needs, who was one of our most devoted followers. And Gröttrup, too, Dr. Steinhoff's deputy. Sheer insanity!

On the following morning, Dornberger reported to the office of Field Marshal Keitel, who informed him that the charges were serious and that the men involved could very well lose their lives. Dornberger remonstrated that he would personally vouch for von Braun, Riedel, and Gröttrup, regardless of the charges against them. He demanded that the men be transferred from the Gestapo to army control because as civilian employees of the army they were subject to military justice, not the civil courts. Keitel, who held his position by steering clear of the SS, was obviously fearful of crossing Himmler. He shrugged his shoulders and said he was powerless to intervene with the Gestapo.

"I can't release them without Himmler's agreement," he carped. "I must avoid the least suspicion of being less zealous than the secret police and Himmler in these things. You know my position here. I am watched. All my actions are noted. People are only waiting for me to make a mistake."

Dornberger then attempted to get an appointment with Himmler himself, only to be turned down peremptorily by the Reichsführer of the SS, who stated that protocol demanded the army general go through SS General Hans Kaltenbrunner, chief of the SS Security Office. Not to be deterred, Dornberger immediately drove to Berlin and then to the bomb-scarred SS headquarters in Prinz Albrechtstrasse. There, however, he had to deal with SS General Heinrich Müller, chief of the Gestapo, who was filling in for the absent Kaltenbrunner.

Müller began quibbling at once. He pointed out that Dornberger's men had not been "arrested"—they had been "taken into protective custody." Dornberger fumed that as far as he was concerned there was no difference between "arrest" and "protective custody." Upon hearing such statements, Müller informed Dornberger that he had quite a thick dossier on him as well as his wayward engineers. Dornberger could tell from Müller's comments concerning statements Dornberger had made earlier about Hitler's dream that no A4 would ever reach England that Müller's agents were on his own personal staff. However, he bluffed Müller, suggesting that if the SS thought the charges would stick, then he should be immediately arrested. It was a dare Müller couldn't accept, with so much ballyhoo having been raised within the Nazi Party about "mystery weapons" that would reverse all military losses.

After several days of delicate and tedious negotiations between the army and the SS, Dornberger secured the release of von Braun "provisionally" for three months. Riedel, Gröttrup, and the others were released a few days later. Technically, however, Gröttrup was still under house arrest at the end of the war.

Von Braun and his fellow "detainees" also received assistance from another and far more influential quarter. Albert Speer later explained:

> When Hitler visited me at my sickbed in Klessheim and treated me with such surprising benevolence, I took this occasion to intercede for the arrested specialists, and had Hitler promise that he would get them released. But a week was to pass before this was done, and as much as six weeks later Hitler was still grumbling about the trouble he had gone to. As he phrased it, von Braun was to be "protected from all prosecution as long as he is indispensable," difficult though the general consequences arising from the situation were.

As 1944 drew to an end, von Braun received another recognition for his work at Peenemünde. In a setting and ceremony that would probably have moved Hitler deeply had he been there, because of his love of things gothic, von Braun was awarded the *Ritterkreuz zum Kriegsverdienstkreuz mit Schwertern* (the Knight's Cross of the War Service Cross with Swords) by Speer. Several days after the launching of the first V-2s against Antwerp, in December 1944, von Braun was summoned to Schloss Burg, a gloomy stone castle, more appropriate as a setting for young Werther than young Wernher, several kilometers outside Remscheid, in the Ruhr Valley overlooking the river Wupper. Dornberger later described the neo-Wagnerian event that took place that night:

> Around the castle in the dark forest were the launching positions of the V-2 troops in operation against Antwerp. The dining room was darkened. Every time a V-2 was launched, a curtain toward the launching site was opened. After each launch, Speer decorated one of the recipients. It was a scene, the blackness of the night, the room suddenly lit with the flickering light of the rocket's exhaust (and slightly shaken by the reverberations of its engine.)

Just before von Braun's brush with the Gestapo, Speer had sought to bring order to the proliferation of rockets and guided missiles in the army and air force because of the demands being made upon the German economy by such expensive weapons. He formed the *Kommission für Fernschiessen* (Long-range Bombardment Commission) to give ministerial direction to the development of all long-range guided weapons. Its chairman was Professor Waldemar Petersen, vice-president for engineering of the *Allgemeine Elektrizitätsgesellschaft* (the German branch of General Electric). Members included representatives from Speer's ministry, the Luftwaffe, heavy industry, the army ordnance office, and several companies having major or supporting contracts for the development of such weapons.

On Speer's order, Petersen made an orientation visit to Peenemünde. Upon arrival, he told Dornberger he was looking to see if the huge base could benefit from the experience of the German electrical industry. Dornberger was dubious of the value of such assistance; however, the man was from Speer's ministry. Accordingly, he laid out the by now well-worn red carpet for another in the seemingly endless concatenation of very important people from Berlin.

After two days of poking about and being briefed by various members of the team, Petersen returned to Dornberger for his exit interview. He made a surprising statement: "I should now like to thank you for the freedom of inspection accorded me. I shall expressly state in my report that if the long-range rocket ever becomes a reality, it can be made only here and under your management."

On May 26, 1944 Petersen returned with the full commission. With him in addition were Speer; Grand Admiral Karl Dönitz; Colonel General Friedrich Fromm, chief of army armaments; Field Marshal Erhard Milch, armaments chief of the air force; and Karl Otto Saur, Speer's chief technical adviser. They were to observe a competitive demonstration between the A4 and the Fi103, the Luftwaffe flying bomb that would ultimately become known as V-1. The designation Fi103 was for its designer, Gerhard Fieseler; later it was known as *Flakzielgerät 76* or *FZG 76*—antiaircraft target device 76, a deception-inspired code name.

At noon an A4 was launched from Test Stand 7 and the results could scarcely have been better for Dornberger. It flew to a range of 280 kilometers, missing the target by only 4.8 kilometers. His pride and hopes were dampened considerably when the second rocket was fired several hours later. It left the launcher and arched out over the Baltic—only to plunge into it in full view of the committee.

Yet Dornberger had little to worry about. The two Fi103s launched at Peenemünde West both plunged immediately into the sea.

Milch, whose weapon was in competition with the army's A4, slapped Dornberger on the back and gave him a hearty, if not heartfelt: "Congratulations! Two to zero in your favor."

After conferring, the commission came up with a decision that pleased both Dornberger and Milch, but solved no problems for Speer. Both weapons should continue development and be placed into production as complementary means of delivering high-explosive warheads on the enemy without risk of human flight crews.

After August 1943, when the Royal Air Force bombed Peenemünde (described in Chapter 7), efficiency dropped, not because of the bomb damage, which was surprisingly light, but because of the resulting decision to disperse Peenemünde's facilities.

The valve laboratory, for example, was set up in Anklam, some 30 kilometers to the south. There, it functioned in a building on the air base that supplied Luftwaffe units in Norway. However, the field was attacked a year later by Allied planes; and Peenemünde's valve laboratory again took to the road, settling near Friedland, 38 kilometers farther south. Likewise, the Materials Testing Laboratory moved into an unused warehouse of the *Weiblicher Arbeitsdienst* (Women's Work Service) at Sadelkow, also some 70 kilometers to the south of Peenemünde.

By the end of September 1943, the ground support equipment people who were overseeing development of the *Meilerwagen* transporter, the launcher, and other complex and expensive matériel needed to prepare and launch the A4, were relocated in a curious place—two unused railway tunnels at Marienthal, near Bonn, 575 kilometers across the Rhine to the southwest.

In November, Peenemünde's supersonic wind tunnel was dismantled and shipped to a safe haven in the Bavarian Alps near the village of Kochel, where vital electric power was available. Its relocation had long been discussed, but the bombing accelerated the move. By January 1944, the wind tunnel was "back on stream" with a new name: *Wasserbau Versuchsanstal G.m.b.H.* (Hydraulic Test Institute). The new name concealed its true purpose, and the facility had been made an independent research organization to support the requirements of any governmental agency needing it.

Because of the decentralization of facilities, managerial problems were multiplied inordinately. At Peenemünde, close personal supervision of operations was a hallmark of the team and had been accomplished with little trouble. A department head could, and usually did, visit all of his shops daily, often just by bicycle. After the relocation, however, Dr. Steinhoff, for example, found it necessary to travel by uncertain railway changes to Berlin, Darmstadt, Eschwege, and Kassel to check on his outlying operations in guidance and control systems. Gone, too, were the days of informal communications between section chiefs and department heads. Before the bombing, technical decisions and their ramifications throughout an operation had as often as not been verbal rather than written, with engineer talking shop to engineer over *ersatz Kaffe* in the plant cafeteria or over wine in the officers' club or screaming to each other over the roar of a rocket engine under static firing.

As activity on the A4 waned when it went into production at Mittelwerk, the huge underground plant near Nordhausen, in Thuringia, at the end of 1943, the team at Peenemünde applied its talents to other weapons. With the increasing pressure brought by Allied air activity and the apparent impotency of the Luftwaffe to combat it, research turned to antiaircraft rockets. Impetus had originally been given to the development of such weapons by a letter sent on September 18, 1942, from General Walter von Axthelm, inspector of antiaircraft artillery, to the heads of all armed forces research and development centers.

In response to this request for assistance, von Braun prepared a study entitled "A Guided A-A Rocket," on November 2, 1942. In it, he described three such rockets, two of which were solid propelled and one of which was liquid propelled. Collectively the project was to be called *Wasserfall* (Waterfall). He estimated (grossly underestimated as it turned out) that only two hundred engineers would be needed for the job of developing the C2 rocket, as the liquid-propellant version was designated; but he stated emphatically that a top priority for the project would have to be issued by Speer's Ministry of Armaments if it were to succeed.

As early as 1941, the Future Projects Office of Ludwig Roth, a kind of miniature "think tank," had been considering a missile such as the *Wasserfall*; but it had not received further encouragement because of the need for putting all resources into the A4. Thus, at the end of 1942, when Peenemünde received the official production order for the new missile, over a year had been wasted. Preliminary design on the new missile did not begin until August 1943.

Technical design problems unfortunately cropped up immediately. It became apparent that the missile as envisioned could not function properly at the speeds and maneuvers required. The propellants would tend to gather around the sides of their tanks during "high G" turns instead of flowing into the motor. To inspire designers, Dornberger offered a prize of 1,000 reichsmarks ($240) for the one who solved the problem. It went to engineer Werner Dahm, who devised an interior funnel-like feed system that ensured the propellants would find their way to the motor under any conditions of flight. Dahm, however, never collected his prize because there was little he could buy with it.

Facilities and personnel were especially troublesome problems; a relatively inexpensive test stand had been designed by November 1943, but it was not until February 17 of the following year that the Luftwaffe sanctioned the construction. By August, the desperately required stand still had not been completed because of a shortage of construction workers. To further compound their troubles, the team itself was also short of engineers. To solve that problem, the Luftwaffe agreed to supply a thousand engineers from its military ranks much in the way the *VKN* had been created earlier. But by September 28, 1943, only 104 men out of 546 requested by name had been made available. The commanders of the others simply refused to let them go.

Despite such drawbacks, engineering design of the *Wasserfall* progressed, and the first complete set of drawings was available by April 20, 1943. Specifications required that the rocket, with its 90-kilogram warhead, reach and destroy a plane flying at 865 kilometers per hour at an altitude of 20 kilometers and at a distance of 50 kilometers. The rocket as designed at Peenemünde could not meet these military requirements under all conceivable circumstances, but it could easily attain the service ceiling of all then existing enemy bombers.

Wasserfall was 7.45 meters long, 0.88 meters in diameter, and had a thrust of 8,000 kilograms. It was to be guided with the help of two radar units. One was to track the target aircraft, and the other was to track the missile. A ground-based computer received instantaneous positional data on both the airplane and the missile and generated commands for the *Wasserfall* so that its flight path and that of the target gradually converged. The missile-tracking radar would also transmit the command to a receiver on board the rocket that would override the basic autopilot control system. This ultimate system was being developed incrementally. By the end of the war, however, the missile was visually tracked and radio-steered to the target by an operator on the ground. At night, of course, as long as the radars were not yet tied into the system, the target would have to be illuminated by searchlights, and the *Wasserfall* had to carry a small colored light. This interim guidance, in turn, necessitated a complex "approach" fusing system, which was never fully developed. It utilized an infrared seeker that homed on the heat radiating from the exhausts of the airplane's engines.

Wasserfall's liquid-propellant rocket engine also posed problems. The propellants were to be nitric acid and a petroleum by-product called *Visol*. The two were hypergolic, meaning they ignited spontaneously upon coming in contact with each other. To simplify and lighten the propulsion system, it was decided not to use turbopumps as on the A4. Instead, compressed nitrogen gas forced the propellants from their tanks directly into the combustion chamber of the engine.

Notwithstanding the increased industrial chaos raised by the Allied attacks on Germany, well over 80 percent of the production of *Wasserfall* was contracted out to industry. Only the tail section was manufactured in Peenemünde's own shops.

The *Wasserfall* propulsion system was static-tested in the river Peene on a floating test stand called *Schwimmweste* (lifebelt), a concept that had arisen as a result of the damage to test stands by the air raid in August 1943. The first fully successful guided flight of *Wasserfall* was not made until February 5, 1944.

In the midst of planning for *Wasserfall* and cleaning up remaining problems of the A4, a curious event took place at Peenemünde that would have brought utter chaos to a less dedicated and motivated team.

On August 1, 1944, high-ranking managers and supervisors were summoned to a meeting in the officers' club.

A man who introduced himself as Paul Storch, of the Siemens Co., announced that he was now their top manager. He added that *Heimat Artillerie Park 11* (Home Artillery Park No. 11)—a code name given to Peenemünde after officials felt that *HVP* had been compromised shortly after the bombing of August 17–18, 1943—no longer existed. The new organization was *Elektromechanische Werke, Karlshagen*, a private enterprise administered by Siemens Co. on behalf of its sole public stockholder, the Third Reich!

As incredible as this news was, the new employees of *EMW* learned that the soldiers of the *VKN* would remain on the job and in their same status. General Dornberger, on the other hand, had become redundant. He moved back to Berlin and devoted himself full time to his former duties as chief, *Wa Prüf 10 & 11*.

According to the new organizational chart, manufacturing and testing were no longer under von Braun, being now separate departments under Eberhard Rees and Martin Schilling. However, things changed little. Von Braun remained the *de facto* head of the team; indeed, he was also technical deputy to Storch, who was *EMW*'s president and tried with moderate success to introduce the techniques of big business into Peenemünde. Storch, to give him his due, was a gentleman, and that helped immensely. He realized that the new situation into which he found himself projected involuntarily was just as delicate for him as it was for the Peenemünde team, who felt lost without their original mentor, Dornberger.

Four months later, in December 1944, Speer took a final step that was a year late in coming. Dornberger was given complete responsibility for a unit that would break up the Allied air superiority. It was named the *Flak E Flugabwehrkanonenentwicklung*, and von Braun was appointed its technical director. One of Dornberger's first tasks was to consolidate his thinly spread and rapidly dwindling resources. Of the missile and rocket systems programmed, planned, or under development at the time, he suggested that all except *Wasserfall*, *Enzian*, *Taifun*, *X4*, and *Schmetterling* be canceled. Even so his action came too late. By the end of 1944, it was clear to all that *Wasserfall*, most promising of the lot, could not possibly be operational until 1946. Still, development continued until February 1, 1945.

In a final desperate attempt to deal with the problem of antiaircraft protection for the Third Reich, Dornberger focused his efforts on the simplest rocket ever designed at Peenemünde, *Taifun* (Typhoon). Only 375 millimeters long and 20 millimeters in diameter, the liquid-propellant rocket weighed but 9 kilograms. Its simple, pressure-fed motor burned hypergolic propellants that boosted the little rocket to a velocity four times the speed of sound. The warhead contained 500 grams of high explosive with a simple contact fuse. Forty-five *Taifun*s were to be launched at a time from cheaply made racks at bombers overhead. Despite its simplicity, the weapon could not have been operational until August 1945. Even so, ten thousand were manufactured at Peenemünde; and in December 1944, plans were made to produce two million rockets per month, beginning in mid-1945.

Thus, a combination of poor decision making at highest levels and questionable deployment of limited human resources at Peenemünde militated against a solution to the problem of Allied bombers. In the final analysis, however, time was simply running out. *Taifun* and *Wasserfall* came too late to save Germany from the Allied air forces.

While the major thrust of research at Peenemünde was in developing the A4 and *Wasserfall*, Ludwig Roth's small group of "far out" designers was considering a series of rockets that looked ahead to such modern space vehicles as the Saturn 5 and the space shuttle. Essentially, Roth's team considered the A4 as a baseline or fundamental vehicle, laying it, or its technology, into a fantastic series of rocket designs. Generally speaking, the team worked between 1940 and 1945 on a series of advanced rockets, despite its official abolition in 1943 by Dornberger following the Allied raid in August on Peenemünde.

Von Braun knew why Dornberger had no other alternative but to direct all Peenemünde engineers to stop dreaming about space and to buckle down to the serious business of supplying weapons for a war effort obviously in trouble. But some of the dreamers kept dreaming and the studies continued *sub rosa*.

One project certainly not stopped was proposed in the fall of 1943. It was suggested that a submarine could tow as many as five A4s in watertight containers to positions off the eastern seaboard of the United States. There, they would be floated into an upright position and become launchers for missiles aimed at New York and other metropolitan areas.

Given the code name *Prüfstand XII* (Test Stand 12), the project was top secret. Only a limited number of engineers was assigned to it, including the talented Klaus Riedel, Bernhard Tessmann, Hans Hüter, and Georg von Tiesenhausen. The preliminary design was done in conjunction with the Vulkan-Werft, a shipyard in Stettin. Each container was about 36 meters long and 5.5 meters in diameter. It displaced 500 metric tons and contained enough propellants for the A4 to last for an ocean voyage of four weeks, including losses because of evaporation. In one proposal, the ballast tanks of the container were to be filled with diesel oil used by the submarine's engines. The novel exhaust system for the launcher consisted of ducts that turned the flaming gases 180 degrees and shot them through the top of the container itself, a feature that would be incorporated into the underground launchers of intercontinental ballistics missiles of the United States and the USSR three decades later.

By early 1945, the Vulkan-Werft had completed all drawings, and plans were made to begin construction in March. But time was against this project as it was against *Wasserfall* and *Taifun*. With the evacuation of Peenemünde in March and the capture of Stettin in mid-April, *Prüfstand XII* became a part of the history of what might have been, given more time by better fortunes for Germany on the battlefield.

Future planning by the team also produced a series of rockets that foreshadowed the development of space vehicles that would occur in the succeeding four decades. They were designated alphanumerically in the system established at Kummersdorf more than a decade earlier.

A4b (*b* for bastard) was a plan for increasing the range of the A4 from 320 kilometers to 750 kilometers. It was to be done with minimum cost. As early as 1940, wind tunnel tests had shown that the range of the A4 could be increased significantly by adding wings to it. However, it was not until October 1944 that serious design and development began on the project, when coastal launching sites for the A4 in Holland were lost. The missile that emerged had the same dimensions as the A4 but two swept-back wings were added.

Two A4b's were launched in January 1944 and controlled in flight with the standard A4 autopilot system. Both missiles reached three times the speed of sound and demonstrated, for the first time, the feasibility of winged supersonic flight by rockets.

The A6 never reached flight-testing, being a theoretical study only. It was a modification of the A4 which would utilize storable propellants—nitric acid and kerosene. This type of propulsion system would greatly simplify the handling, and lower the logistic costs, for such a rocket.

The A7 was a forerunner of the A4b and A9, but was only half as long and half as great in diameter. The thrust was 1.8 tons instead of 25 tons, as with the A4. In fact, it was an A5 of the 1938 vintage with wings. It was considered to be a test bed for studying control systems for winged rockets. A nonpropulsive model was used to study the aerodynamics of the vehicle, being dropped from an aircraft and photographically tracked from the ground. Drop tests were made in 1943, but no flight tests were ever undertaken.

Of the same A5 dimensions generally but having a thrust of 3.4 tons was the wingless A8. It was essentially an A4 of reduced payload and range. Like A6, it would have utilized storable propellants—nitric acid and diesel oil—and also like A6, it never left the drawing board. Had it reached flight-testing, its designers predicted a range of at least 135 kilometers could have been attained.

The A9 must be discussed within the context of the A10. The two together represented the first tangible engineering planning that anyone had done for a multistage rocket. The concept of staging, or discarding unwanted weight while increasing velocity of an ultimate stage, was not new. Hermann Oberth had developed the mathematical proof in 1923.

In 1946, von Braun described the A9/A10 missile for American Army Ordnance officers, who queried him about it because they suspected that such a weapon might be under development in the USSR by Helmut Gröttrup and a group of German engineers (see Chapter 17):

> The A10 was under design in the years 1943 and 1944. This work was discontinued in 1944 in order to stress the development of a simplified A9 rocket, known as the A4b. The anticipated range of the original A9 boosted by an A10 was 5,200 kilometers. The simplified A9 (A4b), in which many improvements provided for in the original A9 were omitted, was supposed to reach a range of 3,000 kilometers when boosted by the A10.... The development of the A10 was

> discontinued early in 1944 because it was foreseen that this work would take two more years. Likewise the development of many improvements, especially in the propulsion system provided for the A9, was discontinued in the summer of 1944. At that time it was decided that the A9 should be prepared for quantity production with the original A4 propulsion unit and as few deviations from the standard A4 parts and subassemblies as possible. In order to retain the A-1 priority issued for the A4 project, this simplified A9 was then rechristened A4b.

The designers of the A9 also drew plans for placing a pilot in a pressurized cabin instead of the warhead. This model of the A9 possessed a tricycle landing gear. Von Braun said of this project, "We computed that the A9 was capable of carrying a pilot a distance of seven hundred kilometers in seventeen minutes. It might have taken off vertically, like an A4, and then landed glider-fashion on a medium-size strip!"

The A11 was described by von Braun this way:

> The A11 was never developed beyond the preliminary design study stage. It consisted of a booster rocket for the A10 and 9 combination with a thrust of 1,600 metric tons. Thorough calculations showed that this three-stage rocket could reach the so-called orbit velocity of 7,800 meters per second at which the third stage, i.e., the A9, could perform a continuous powerless flight of infinite duration around the earth.... Thus the combination A9, A10, and A11 would have unlimited range. It can hardly be considered a weapon.

Here, von Braun, usually a gifted prophet in the field of astronautics, erred. In 1966, the Soviet Union launched its first fractional orbital bomb system, a weapon that could deliver an atomic warhead by exactly the method described for the A9/A10/A11.

During the same period, Roth's team also considered an A12. It would, in von Braun's words, "... have a thrust of not less than 12,800 tons. It could bring the A10 to satellite velocity, but this time not with a lone pilot, but rather with a payload of some 30 tons! A number of such ships, maintaining a regular shuttle service to the orbit, would permit the building of a space station there."

Thus, as early as 1944, the Peenemünde team was considering the ultimate in the design of large space boosters—the space shuttle. Perhaps even more incredibly, they also were considering forms of rocket propulsion that have yet to appear. For example, on October 15, 1942, Peenemünde let a contract with the *Forschungsanstalt der Deutschen Reichspost* (Research Institute of the German Post Office) "to investigate the possible exploitation of nuclear decay and chain reactions for rocket propulsion." Two and a half years previously, on April 24, 1939, physicist Paul Harteck and his assistant Wilhelm Groth, of the *Technische Hochschule* at Hamburg, had written to the German War Office concerning possibilities of nuclear energy for military explosive purposes. The letter passed through General Becker's office apparently without comment. In the following summer, however, funds were allocated for atomic research by the army; and a small laboratory, under Dr. Kurt Diebner, was set up at Kummersdorf where Dornberger and von Braun had eight years earlier developed their first army rockets. Apparently, little of interest to the team at Peenemünde came of Diebner's work.

However, in November 1942, Dr. Walter Thiel passed on to Krafft Ehricke, at Peenemünde, some interesting reports that he had received concerning research in nuclear fission being pursued by Professor Werner Heisenberg and his associate, Professor H. Pose, at Leipzig. It was a steam turbine utilizing the heat of disintegrating atoms of uranium in heavy water. Thiel asked Ehricke to look into the experiment as a possible means of rocket propulsion, since Ehricke had been a student of Heisenberg's.

Ehricke looked over the reports and made his own evaluation to Thiel. Later, he recalled to the authors:

> My rocket-propulsion-oriented recommendations were negative as far as the heavy water moderator or the use of water as a propellant were concerned; but they were positive in recommending the replacement of heavy water by a solid moderator and of water by hydrogen (or at least methane). These were rather far-out recommendations for those days. But Dr. Thiel, who was very forward-looking, concurred.
>
> I saw Heisenberg in late 1944. He was trying [in Berlin-Dahlem] to make a heavy water moderated reactor "go critical." He told me, "Of course one could build an atomic bomb." But this would require huge isotope separation plants that could not be erected under the situation of Allied air superiority. He thought the most immediate military use of a nuclear reactor would be for submarine propulsion.

Dreams of atomic propulsion notwithstanding, the days of *Elektromechanische Werke, Karlshagen* at Peenemünde were numbered. Early in 1945, development on *Wasserfall* and *Prüfstand XII* was in full swing but the tide of battle in Germany crept relentlessly toward Usedom Island from the east. The communications between Berlin and Peenemünde became erratic and undependable, and travel about the country was chaotic. As the Soviet army approached the heart of Germany, the civilians at Peenemünde began to wonder not only about their jobs but about their lives.

The A4, however, was safely in serial production some 400 kilometers to the southwest. In a huge, underground plant near Nordhausen thousand of prisoners of war, slave laborers, skilled technicians and a nucleus of engineers from Peenemünde, were manufacturing the rocket in the hundreds per month by the end of 1944.

CHAPTER 5

Producing the Missiles

By the time Peenemünde had entered into full-scale operations toward the end of the 1930 decade, Albert Speer had become intimately involved in the dreams and aspirations of Dornberger, von Braun, and their associates. The munitions minister admitted later that their probings into the future "exerted a strange fascination upon me. It was like the planning of a miracle."

The friendship and support of Speer proved invaluable throughout the tortuous development cycle of the A4, beginning as early as the autumn of 1939 when Hitler removed Peenemünde's priority claim on men and materials. "By tacit agreement with the Army Ordnance Office," Speer confessed after the war, "I continued to build the Peenemünde establishment without his approval—a liberty probably no one but myself could have taken."

Less than three months after the first successful A4 test flight on October 3, 1942, Speer persuaded Hitler to sign an order directing that plans be made to mass-produce the missile. At the time the order was signed (on December 22, 1942), the munitions minister estimated that the missile would be sufficiently developed for production to begin in July of the following year.

In early January, a total production goal of six thousand missiles was tentatively established, with assembly to be centered at Peenemünde and at the Zeppelin works at Friedrichshafen on Lake Constance. Speer, through his deputy Karl Otto Saur, assigned the task of organizing A4 assembly and component production operations to Gerhard Degenkolb, the well-known director of the Locomotives Special Committee. Degenkolb, whose offices in Berlin were often referred to as "the locomotive works," quickly established a similar coordinating committee for the missile program, the Special Committee (*Sonderausschuss*) A4, using methods described by a close associate as "cruel and inhumane." Particularly vivid is Dornberger's description of the man: His "completely bald and spherical head, his soft, loose cheeks, bull neck, and fleshy lips revealed a tendency toward good living and sensual pleasures, while the restlessness of his powerful hands and the vigor of his movements were evidence of vitality and mental alertness. He was never still." To Dr. Paul Figge, a key figure in the A4 production effort, Degenkolb suffered from a "tremendous inferiority complex, which he tried to overcome by exerting force on his workers."

Whatever his faults and shortcomings, Degenkolb had earlier achieved a production quota of 2,000 locomotives a month, so if anyone could reach the goal of 6,000 missiles he was thought to be the man. To Detmar Stahlknecht, special commissioner in the Munitions Ministry and a member of Degenkolb's new committee, the 6,000-missile goal was too ambitious, and in February he offered a scaled-down program calling for only 5,150 rockets. Starting with an output of five in April 1943, he looked forward to a maximum monthly production of 600 missiles in September 1944. He agreed that production should be concentrated at Peenemünde and at Friedrichshafen.

Stahlknecht's plans did not suit the ambitions of Degenkolb, who in April—with Munitions Ministry blessings—set his sights on monthly outputs rising from 650 missiles in October 1943 to 900 in November and then a peak of 950 in December. The Henschel Rax-Werke at Wiener-Neustadt was to be brought into the picture, bringing to three the number of sites to be involved in assembling the A4. Whatever the merits of Stahlknecht's more limited aims, Figge felt that he was "too set, too rigid" in his thinking. "He wouldn't take the risk of doing new things," Figge added, and "could never compromise what he had previously learned, and consequently could not adapt to the wartime situation, when improvisation was a way of life."

Although Degenkolb, Stahlknecht, and others were thinking of the A4 in terms of mass production, the missile still remained very much in the research and development stage and did not enjoy the top priority rating that Speer, Dornberger, and von Braun knew was indispensable. Following a meeting with Hitler on January 8, 1943, Speer had to report to his colleagues at Peenemünde that "the Führer cannot give your project higher priority yet. He is still not convinced that your plan will succeed."

Six months later Hitler changed his mind. During the course of the meeting on July 7, 1943, when he saw for the first time the films of the

October 1942 A4 test flight his imagination was kindled: Here was an exciting new weapon that could retaliate against the devastating bombings of German cities. Without the benefit of anything approaching an operations research study on the probable effectiveness of the device, and without a realistic estimate of when it could be ready and in what quantities, the supreme warlord of Germany made one of his many essentially capricious decisions: The A4 would receive its coveted priority.

A stroke of the Führer's pen could not compensate for months of irretrievably lost time. Final A4 production drawings were not ready, nor could they be, for the missile had long struggled along hampered by shortages of manpower and materials. And now, suddenly and unexpectedly, Hitler was beginning to pin exaggerated hopes on a problem-plagued, untried weapon.

Although the A4 was months from the production stage, a large number of drawings had been prepared and special tools designed and built. In anticipation of a British air raid on Peenemünde, duplicate sets and copies were prepared and stored outside the complex. This was indeed a fortunate precaution, for on the night of August 17, 1943, the great Baltic coast experimental station was attacked by hundreds of RAF bombers. Whatever other damage was done, production planning was not seriously offset by the attack. Problems more subtle than an air raid were about to befall the Peenemünde production team.

Although Degenkolb had clearly announced his ambitious production goal for the A4, engineers up at Peenemünde were not yet able to deliver a production-ready design to him. Delays building up in the program were seized upon by SS Reichsführer Heinrich Himmler as an excuse to exert increasing control over the A4 development. Realizing that Degenkolb would inevitably face a severe labor shortage in attempting to manufacture nearly a thousand missiles a month, the wily Reichsführer proposed assigning hordes of concentration camp workers and technicians to the gigantic task.

Neither Speer nor his deputy and head of the Central Office at the Munitions Ministry, Karl Otto Saur, nor even the resourceful Degenkolb could offer viable alternatives. And all three found themselves viewing the A4 more and more as a possible savior of the Third Reich. Saur, whom Dornberger had once called his "greatest adversary" in the Ministry of Munitions, by the spring of 1943 had become an avid supporter of the missile program. Fiercely political in outlook, he was, in the words of Figge, "completely unscrupulous." Arthur Rudolph, who was to become a key member of the A4 production team, later remarked that Saur was "full of pipe dreams; he really thought we could beat the Allies with the A4."

A few months earlier, on May 13, 1943, *Gauleiter* Fritz Sauckel, Reich Plenipotentiary General for the Allocation of Labor, had visited Peenemünde and discussed the serious manpower situation with Stahlknecht, Dornberger, Colonel Gerhard Stegmaier (military commander of the development works), and others. Although William L. Shirer called him a "second-string Nazi . . . a pig-eyed little man, rude and tough," and even Goebbels had once described the *Gauleiter* as "one of the dullest of the dull," Sauckel was shrewd enough to recognize the importance of obtaining labor for anticipated large-scale production of the missile. However, even he could make no immediate promises of help. Sauckel's hesitancy in an area directly within his purview provided Himmler with a ready-made opening.

Himmler's suggestion to use foreign labor contradicted an earlier desire of Hitler, as expressed to Speer in July, that only native Germans be employed in plants where A4s were to be assembled. This desire, which was based on security rather than moral considerations, was never fulfilled. Indeed, a week after the British raid on Peenemünde on the night of August 17–18, Hitler came around to Himmler's viewpoint and agreed that concentration camp workers could assist in A4 production. A name that was to become all too familiar to Speer and to the rocket engineers at Peenemünde, SS Brigadeführer (later Obergruppenführer) Major General Hans Kammler, was ordered to coordinate the labor program outside the channels normally used by Sauckel.

At one time or another, almost everyone involved with the A4 complained of troubles they were having in securing reliable components and in minimizing design changes. Stahlknecht lamented that preproduction design drawings "come in so late and are no sooner delivered than they are fundamentally changed again." Von Braun talked often of frustrations arising from attempting to accommodate the A4 to the continually changing raw materials situation. Not being able to rely on deliveries of a given set of materials "keeps on forcing us to make designs with other materials said to be in better supply." (At one time, virtually the only aluminum available came from Allied bombers that had been shot down.)

Eberhard Rees was dissatisfied with the quality of components received from the contractor and subcontractor network; rejects were common. Propulsion chief Thiel reported that "the components [that industry] does turn out are not in any way up to standard. The motor is too complicated and very far from suitable yet for mass production." In a mid-August meeting with General Dornberger, Thiel despaired:

> For months now we have had one breakdown after another. We expected too much of our A4. In present conditions the job just can't be done. Our machine is a flying, fully automatic laboratory. To put it

> into mass production is sheer madness. We aren't through with development by a long way. . . . If you persist in your point of view [to ready the missile for mass production] I must decline all further work. I see no possibility that we shall achieve our aim before the war is over. The project must be abandoned. I have given the whole matter thorough consideration and ask to be allowed to resign. I intend to join a technical college as lecturer in thermodynamics.

A few days later Thiel was killed in the British air raid.

Dornberger was not as pessimistic, but correctly observed that "the snag up here is that we have too many brains and ideas jostling each other in one spot." He ordered his engineering colleagues to desist in making any further design changes that were not "absolutely essential." Even though he personally disliked Degenkolb and his methods, he promised to

> . . . give my agreement to the placing of full-scale orders on the basis of the present stage of development, even at the risk of modifications later or of unserviceability. We have got to get on with the job.

And they did. Armed with coveted top priority rating (coded "DE"), the blessing of Hitler, and the favor of Speer, A4 component production was initiated in companies all across the spectrum of the German armaments industry. Almost immediately, there was reaction from production directors whose programs were being adversely affected by the rising star of Peenemünde. Plants turning out vital components for aircraft and tanks soon found part of their manpower and materials being diverted.

The Air Ministry, and particularly its armaments chief Field Marshal Milch, became alarmed not only at the effect the A4 was having on night-fighter production—massive British bomber attacks against German cities had been going on since March—but on the much simpler and cheaper Fi103 flying bomb that required only a fraction of the labor and materials consumed by its big brother. Speer himself soon became convinced that, important as the A4 was, it should not be allowed to interfere with the aircraft production program. Orders to this effect were circulated in mid-August.

At the time of the RAF bombing of Peenemünde, the Allies suspected that the Luftschiffsbau Zeppelin at Friedrichshafen (an "experienced constructor of large, light-alloy structures," noted an intelligence report) and the Henschel Rax-Werke at Wiener-Neustadt in Austria were somehow involved in A4 matters. As it turned out, these plants were never destined to assemble A4s due to heavy Allied bombings. (Ironically, the Rax-Werke were damaged by accident, for the attacking US bombers were really after the Messerschmitt fighter aircraft plant nearby.) Meanwhile, A4 production planners (fully aware of the devastation rained on Peenemünde and of the fact that both Friedrichshafen and Wiener-

Neustadt were within bombing range and sooner or later would be hit) resolved that at least some of the assembly should take place underground. The factory where this was to take place would be known as the Mittelwerk, or Central Work.

In accordance with agreements later reached between the Army War Office, the Peenemünde establishment, Degenkolb's Special Committee A4, the SS, and others, Mittelwerk would control all component production and assembly contracts throughout the country. The Friedrichshafen and Wiener-Neustadt plants would become the Southern Work, and a proposed new plant near Riga would be known as the Eastern Work. The Demag-Fahrzeugwerke at Berlin-Falckensee was also to be involved. Subsequently, as Allied air raids intensified, it was decided to concentrate all A4 production at Mittelwerk.

Under Kohnstein Mountain near the town of Niedersachswerfen in the Harz Mountains, Paul Figge found a seemingly ideal site that could be modified to assemble hundreds, perhaps even a thousand, of A4s a month—and Fi103s as well. "I was overjoyed to discover the location," he reminisced, "because we would not have to build an entirely new installation." Degenkolb was quick to agree.

What Figge discovered was a network of tunnels whose construction had begun back in 1917 when the firm Badische Anilin Sodafabrik had purchased the property to exploit ammonia, anhydrite, and gypsum resources. Railway connections with Niedersachswerfen were soon completed, and operations got under way, largely, it seems, with the help of World War I prisoner and female labor. The mines remained active through 1934.

In 1935, the War Production Commission of the Reich Ministry of Commerce was given the job of locating sources of strategic raw materials and of establishing a centralized fuel and chemical depot safe from bomber attack. An economic research organization known as the Wirtschaftliche Forschungsgesellschaft—Wifo for short—was charged with the responsibility of locating the depot. At the suggestion of I.G. Farbenindustrie, Wifo management investigated the Kohnstein tunnels in the summer of 1935. Soon afterwards, they established a field office there and went about making plans for extending the old mine tunnels.

Under the direction of a production engineer named Neu, a three-phase expansion program was established, known as Wifo I, Wifo II, and Wifo III. The first phase was completed in 1937 and the second in 1940; Wifo III never got started. Up until 1943, the tunnels served as storage depots for oil, gasoline, and chemicals. It was then that the government decided to convert the Kohnstein tunnels to a production and assembly facility for the A4. Eventually, the Mittelwerk was expanded into an area-wide Mittelbau project for a variety of war production programs. The history of the entire Mittelbau buildup and operations is thoroughly told in Manfred Bornemann's *Geheimprojekt Mittelbau*.

The tunnels also harbored chemical poisons that had been stockpiled for possible use by the Luftwaffe. These, obviously, had to be removed. "I was called into Göring's retreat in Karinhall," Figge reported, "where he forbade me to use the area for missile production purposes because of the danger in moving the chemicals." Ultimately, Göring was overruled by Hitler.

As early as February 1943, Professor Karl Maria Hettlage of the Munitions Ministry had suggested to Dornberger that Peenemünde should become a private stock company. Though his plan was defeated at the time, it was not unexpected that Hettlage should take an early role in establishing the Mittelwerk corporation somewhat along the lines he had earlier outlined for Peenemünde. In August 1943, a planning meeting was held in Berlin to determine how best to finance the Mittelwerk organization. The upshot of the meeting was that Mittelwerk was to operate on funds supplied from Hettlage's Rüstungskontor G.m.b.H, Amt für Wirtschaft und Finanzen at the Munitionsministerium—a sort of war production fund. The following month, Hettlage prepared a memo on behalf of Speer* advising Degenkolb, Dornberger, Saur, and Kammler as well as representative Bucher of the Allgemeine Elektrizitätsgesellschaft and Director Wehling of Wifo that an organizational, economic, and financial meeting for "Project Mittelwerk" would take place on Tuesday, September 21, 1943. Minister Speer was to be briefed afterwards as to its results.

On September 24, Dr. Ing. Kurt Kettler, general director of the Borsig-Lokomotiv-Werke G.m.b.H. in Berlin, dictated a memo to all participants in the meeting (which had also included Heinz Kunze, a member of the A4 committee; Schmidt-Lossberg of the Rüstungskontor [war production fund]; and Kettler himself). He wrote:

> Based on preliminary discussions, it was decided to form a new corporation bearing the name
> "Mittelwerk G.m.b.H."

The new "Middlework Corporation," whose headquarters was to be in Berlin, would be supervised by a board of directors consisting of three persons, one of whom would be Kettler. A personnel director would be named by General Kammler to function in close concert with the commandant of the local Dora SS concentration camp. In addition, a small

* At the time, Speer was minister of weapons and munitions (Reichsminister für Bewaffnung und Munition). Later in the month he became minister for armaments and war production.

council would be established under the chairmanship of Degenkolb. Its other members: Hettlage, Dornberger, Kammler, and Schmidt-Lossberg. Kettler's memo also noted that Wifo would be charged with providing the missile manufacturing plant, and that the Mittelwerk corporation would rent the facility from them. Government-furnished machinery would be used. "Contracts will be awarded by the army; the prime contractor will be the Mittelwerk," added Kettler.

Soon afterwards, the other two board members were named: SS Sturmbannführer (Major) Otto Förschner and Otto Karl Bersch. Somewhat later, the board was enlarged by four additional individuals: Alwin Sawatzki (representing technical interests—he was, in fact, technical director), Bütting (in charge of personnel), Zanker (finance), and Hubert (general organization). Arthur Rudolph was placed in charge of the Technical Division under Sawatzki and Börner ran the Administrative Division under Bersch.

Conditions appeared so promising that in October Saur, Degenkolb, and Kettler were proposing to assemble 900 A4s a month (along with some component and subassembly fabrication) at Mittelwerk, and another 900 a month at the other three planned sites (despite earlier air raids). Saur, for his part, as far back as July had proposed (during the course of a meeting at the Ministry of Munitions in Berlin) that the already absurdly high figure of 1,800 be rounded off to 2,000.

A dismayed Dornberger objected. "Two factors made this program impossible," he said. For one thing, sufficient ground support equipment for mobile batteries would not be available. Moreover, there was the propellant problem: "Underground oxygen-generating plants could not be conjured up from nowhere." And how much alcohol fuel would the potato harvest yield? he asked the assembled group. As it was, sufficient quantities of alcohol could not be fully guaranteed for even 900 A4s a month.

Despite objections from Dornberger and others, on October 1 Kettler sent a request to the Army High Command, Oberkommando des Heeres, Wa Chef Ing 4 branch at Steinplatz I in Berlin:

> With reference to our previous conversations, we herewith request that a contract be awarded to permit the Mittelwerk in Hammersfeld (code name for Niedersachswerfen] to produce 1,800 A4s monthly.
>
> Planning data and cost estimates amounting to RM [reichsmarks] 11,500,000 are enclosed.
>
> In view of the special priority . . . we request the immediate award of a contract.

General Emil Leeb, head of the Army Weapons Department, did not oblige, and on October 19 he prepared a war order for Degenkolb's original 900 A4s a month. Kriefsauftrag Nr. (war contract number) 0011-5565/43 called for the manufacture of 12,000 A4s at a monthly rate of 900 at a unit price of RM 40,000. The total would therefore be RM 480,000,000 or about $115,000,000 (during the war, the reichsmark was officially quoted at $0.238). Later, on November 20, Kettler and Förschner devised a more realistic price schedule:

Series	1 to 1,000 units	RM 100,000 each
Series	1,001 to 2,000	90,000
Series	2,001 to 3,000	80,000
Series	3,001 to 4,000	70,000
Series	4,001 to 5,000	60,000
Series	5,001 on	50,000

For each 3,000 units, the government would pay 30 percent advance on orders, another 30 percent upon completion of production, and the final 40 percent upon delivery.

The basic requirements for a 900-a-month assembly operation were soon laid out for the plant. Management estimated that 10,800 KVA electric energy would be needed, 47,600,000 BTU of heat, 10,600 cubic feet an hour of water, 21,200 cubic feet an hour of compressed air, and 5,300 cubic feet an hour of propane gas. A total work force of 18,000 persons was considered necessary (including the Wifo workers assigned to expanding the tunnels).

The budget called for a tooling and equipment investment of RM 11,500,000 ($2.75 million) a high percentage of which was subject to tax depreciation. The money was to be made available through the Rüstungskontor (the special account for war industries) and was to be drawn out at an interest rate of 3.25 percent. This budget was soon to increase: A financial report dated November 22, less than two months later, stated that the cost of the production equipment and tooling needed for manufacturing the A4 would be RM 31.4 million ($7.50 million). Even this was not sufficient, as records show that approximately RM 35 million ($8.33 million) were ultimately drawn out of the fund. The cost of outfitting Mittelwerk pales into insignificance when compared with the RM 300 million already invested in the Peenemünde complex (over $70 million).

Though Leeb's war order came inopportunely, the formal contract did not. Kettler was advised by the War Office on November 8 that it was their intention to award a contract for the establishment of A4 production based on (1) the assembly of 900 units a month, and (2) "the manufacture of particularly important parts and subassemblies as shall be later determined." It was announced that the contract number would be 4/XL-0900-3071/43H.

> The final contract [the letter went on to say] can only be awarded upon clarification of the scope of the order and the method of financing various cost groups. It has been agreed that the still-pending discussions on ways of financing will have no influence on the accelerated conduct of this project, since the necessary funds will be provided . . . by the Rüstungskontor partly on a final and partly on an advanced basis.

The reason for the delay in writing the contract soon became evident:

> You are asked to start negotiations with the owner of the facilities, the Wifo organization, to determine the rental due, and to submit the draft of the [rental] contract [with them] for verification and approval by the War Office. . . .

The contract had still not been issued a year later, due to discrepancies in cost data.

Just at the time that the Mittelwerk operation was getting under way, the devastating Battle of Berlin began, which lasted from mid-November 1943 to the middle of March of the following year. On the very night of the first attack, November 18, the Special Committee A4 was meeting at the Borsig locomotive administrative building. Figge never forgot it:

> After the first bombs had fallen, all 50 men [participating in the meeting] moved to the shelter underneath. A bomb fell on the adjacent building, and everything above them was burning. Luckily, it did not fall in the locomotive house—this would have delayed the program over half a year!
>
> The direct result of this close call was the dispersal of various departments and organizations to all parts of Germany. I moved my operations to a school in Ilfeld [code named Napola and located near the Mittelwerk].
>
> The bombings hardly affected progress on the A4 program, because our enthusiasm still remained high to accomplish the goal. So actually, the more difficult the conditions became, the more the enthusiasm grew to finish what we had begun. "Enjoy the war—the peace will be terrible" was the motto.

In early September 1943, Sawatzki, his assistant Arthur Rudolph, and a work force of about ten engineers had moved into the Mittelwerk facility from the prototype plant up at Peenemünde. Taking maximum advantage of the high priority that the A4 was enjoying, an improving supply of labor, and general support from many crucial quarters, Sawatzki got on with his job. To help coordinate activities at contractor plants, he was assisted by a team of inspectors reporting to a Captain Dr. Kühle, whose authority rested in a memo prepared by Degenkolb dated 30th October:

> I have established a special committee under Captain Dr. Kühle. He is to act under the orders of the head of the Special Committee A4, and has authority to direct the contractors. All the privileges of the Special Committee A4 are at his disposal.

Kühle immediately set up his staff (Führungsstab), known as the FS-Zentrale, which became principally concerned with electric parts, components, and subassemblies. In order to eliminate difficulties that could not be solved on the local level, this staff worked through a network of plant representatives.

Kühle became involved in another area, one that made him respected and feared by his associates—including Sawatzki himself. As controller of communications, it was his practice to listen in on telephone calls. "One day," relates Rudolph, "I found on my desk a folder with a record of all my recent phone conversations. When I left my office briefly, someone came in and removed it. I could only speculate that it was left there as a friendly warning from Kühle."

Rudolph described the man as very forceful, a Prussian type, realistic and intelligent. "He assigned a young lieutenant to assist me—and presumably to keep an eye on my work." Kühle took his inspection mandate more seriously than his associates may have desired.

Physically, the huge underground plant seemed ideal for the task established by the Munitions Ministry. But the human conditions in it were not. When Speer visited Niedersachswerfen on December 10, he was shocked at the state of what were euphemistically called East workers. "The conditions of these prisoners," he later recalled, "were in fact barbarous, and a sense of profound involvement and personal guilt seizes me whenever I think of them." He immediately ordered that food, living, and sanitary conditions be improved.

Conditions in the tunnels were described by Hannelore Bannasch, who from November 1939 to the time she reported to Mittelwerk had worked first as secretary to von Braun at Peenemünde and then with A4 contractors at Weilheim and the Rax-Werke in Wiener-Neustadt:

> There was much hard work. We all labored for 12 hours a day, and occasionally for stretches up to 72 hours virtually without stopping. Most of the time we didn't see daylight.
>
> We lived in a hotel named Netzkater near Ilfeld. At first the laborers slept in the tunnels—Germans and foreigners alike. Because of the dampness, many died of pneumonia. Actually, as time went on we got to work quite well with the foreigners—it was a veritable melting pot. But they often fought amongst themselves. Remember, many had become prisoners for criminal and homosexual reasons as well as for their political and religious

> beliefs. We needed the laborers, so we tried not to mistreat them. It [Mittelwerk] was a top secret operation, so once you were in you stayed.

As the nearby Dora concentration camp was built up, conditions did improve to an extent. Control was absolute over the prisoners, and their German co-workers could only communicate with them in the presence of SS guards. Rudolph recalls that many were highly intelligent individuals who had formerly occupied distinguished positions in their own countries. A French university professor, for example, was assigned to check out electrical equipment. As time went on, the number of forced laborers was reduced, since they became more inefficient in their work. The table illustrates the trend.

Labor Involved in Mittelwerk A4 Production Program

Period	*German* Engineering, Technical, Supervisory	Administrative	Skilled Labor	Common Labor	Total	Concentration Camp Labor	Total
July 1944	500	1,000	1,500	400	3,400	5,000	8,400
October 1944	600	800	2,000	600	4,000	3,500	7,500
March 1945	800	600	2,500	1,000	4,900	2,000	6,900

Almost from the beginning, conflicts broke out within the various layers of directors, managers, department heads, representatives of the Special Committee A4, and individuals reporting directly to the Speer ministry in Berlin, One of the most serious problems involved Kettler himself. Although effective, intelligent, and fair in his dealings with his associates, he could never get along with Sawatzki—who delighted in displaying the power and prestige his former work on the Tiger tank program had given him. The Kettler-Sawatzki enmity helped spur the arrival, on April 13, 1944, of a new managing director, Georg Johannes Rickhey.

Dressed in Nazi uniform and just arrived from Berlin, he advised the assembled Mittelwerk management that henceforth he would supervise all operations at the underground plant. Kettler, who was told he would remain as a director, reflected twenty-seven years later: "As I was not politically committed, I was replaced by Rickhey, a political activist. I was completely unaware of this change of position—it was very sudden." Fortunately, the two men got along well.

Rickhey's strong sense of public relations proved effective in coercing Berlin to allocate increasingly scarce supplies to his operations. In June, he decided to reorganize Mittelwerk. Serving directly under him were the directors, Kettler, Förschner, and Bersch. Then came the plant managers, division chiefs, subdivision heads, and various office and section heads. Figge, a member of the Special Committee A4, declined an offer of a directorship for personal reasons: "It was clear at the time," he later admitted, "that we could not win the war." In terms of who did what, Rickhey was responsible primarily for purchasing and personnel; Kettler for plant direction; Bersch for sales, finance, and general administration; Sawatzki for planning; Dr. Arnold for central administration and organization, review and statistics; Dr. Schüning for field offices; and Schwön for security, antiaircraft protection, and firefighting.

Production of components and subassemblies began at Mittelwerk in August 1943 when techniques developed at Peenemünde were ironed out for introduction at the underground plant. Then, in early September preliminary efforts were made to assemble a few missiles and Sawatzki was ordered by Rickhey to prepare a delivery schedule. "Sawatzki called me in," recalls Rudolph, "and said that he wanted to assemble fifty A4s in December. I told him if we made five we would be lucky. He was furious and kicked me out of his office. He even threatened to put me in the concentration camp!"

Bright, unscrupulous, and hard driving, Sawatzki was once called a "motor" by Hannelore Bannasch. "He convinced himself," she relates, "that he didn't need to sleep. He was a hard man—as hard against himself as he was with others." Thus it was characteristic that he went out on a limb and assured Speer's Munitions Ministry that A4s would start flowing off the assembly lines before the end of 1943.

Theoretically, at least, he kept his word, for Rudolph miraculously managed to put together four missiles in late December. He loaded them on camouflaged railway cars on New Year's Eve 1943, a couple of hours before midnight. Rudolph remembers that

> . . . it was snowing and at the time I was relaxing at a get-together with a few close associates, enjoying respite from the horrible pressure of the plant. All of a sudden, I was called away to help resolve some loading difficulties. . . . It was very cold, and I cursed at having to leave the party just to get those missiles out before the end of the year so that their timely delivery could be officially reported.

In actual fact, the missiles that Rudolph did manage to put together were not in operable shape. After they had been duly logged out, they were stored for a few days and then brought back into Mittelwerk and disassembled!

Despite this uncertain beginning, Mittelwerk planners insisted on the feasibility of building up A4 production so as to meet the 900 units per month schedule (percentage at plant):

Complete A4 assembly	100%
Hull fabrication and assembly	100%
Tail section fabrication and assembly	50%
Engine assembly	100%
Servo system fabrication and assembly	100%
Sheet metal forming for hull and tail	100%
Turbine and pump assembly	100%

To help assure that such a schedule could be maintained, Heinz Kunze—a member of Degenkolb's staff and representative of the Speer ministry for A4 affairs—arranged to move a group of special investigators into Mittelwerk. Their job: to attempt to discover why production did not move ahead at the ambitious rate established by Degenkolb. Kunze also took over some of the responsibilities of Sawatzki, who had become very ill as a result of overwork and nervous tension.

When Sawatzki fell ill, Kunze asked Rudolph to meet the established goal of increasing production month-by-month to the target 900 figure. In turn, Rudolph asked that Kunze remove him from directly under Sawatzki's command so that he could establish—and have a good chance of meeting—a more realistic A4 production goal. This in due course was done, and Sawatzki (as soon as he recovered) began to concentrate on the Fi103 flying bomb program with the aid of Börner.

The first twenty operable missiles produced at Mittelwerk were shipped off to Peenemünde in early 1944, where they were checked out on Test Stand 7. Later, as production quality improved, fewer missiles had to be shipped north. In all, about fifty Mittelwerk units were launched from Tests Stands 7 and 10 up to the end of 1944. (Some Mittelwerk missiles were reworked at Peenemünde before being tested, so that problems involving cracked seals, misfitting parts, broken struts, burnt solenoids, wrong connections, leaks, etc. could be worked out.)

The first missile to come off the Mittelwerk assembly line was No. 17,001. The first to undergo evaluation testing at Peenemünde was No. 17,003; it was fired on the night of January 27, 1944. Up to that time, a total of forty-six Peenemünde-made V-series missiles had been fired, many successfully. But the Mittelwerk missile was a complete failure: It rose a few feet, then lost thrust, settled back on the launch pad, toppled, crashed to the ground, and exploded. The sight was not altogether unfamiliar.

Every time a major flaw was discovered by Peenemünde, production at Mittelwerk had to be interrupted. Missiles would be pulled out of the line and tested with the desired modifications incorporated. Then assembly would resume. Changes were handled in a rapid and efficient fashion.

Hans Lindenberg would regularly travel south from Peenemünde with a handful of change orders, which would end up in the hands of floor managers. Modifications would be executed almost literally on the spot, solely because of the fact that so many former Peenemünde craftsmen were involved in the assembly program. The assembly line was kept flexible, with supervisors walking up and down the floor inquiring about problems and making immediate decisions as to the best way to proceed. Later, Lindenberg and a staff of experts transferred to Mittelwerk so as to be able to work closely with Rudolph. From then on, changes were mutually agreed to and were introduced in blocks (lots). The documentation for each change was put in a small box attached to each missile, so that the firing batteries would be kept up to date and hence would be able to increase their efficiency at the front.

When full-scale operations got under way in 1944, Mittelwerk was primarily an assembly plant, depending on up to five thousand contractors and subcontractors for components and parts. As the war situation deteriorated, however, and factories were destroyed by bombs or captured by advancing armies, the role of the underground plant changed. Soon, it commenced nearly complete fabrication from stock materials of many elements (except for turbines, pumps, tanks, and some electrical and mechanical control units).

Originally, Rickhey planned to assure himself of three sources of supply for each component, with the capacity of any two adequate to maintain production. A card-indexing system was maintained of all principal contractors and the many subs. Thus, under code T-FA-a (referring to combustion chamber procurement) the name Linke-Hofmann-Werke AG of Grundstrasse 12 in Breslau appears. The names of ten subcontractors follow, starting with Hallesche Maschinenfabrik und Eisengiesserei and ending with A. Ziemann (in Halle and Ludwigsburg, respectively). The prime contractor for the pumps was Wumag, assisted by six subcontractors. A total of sixteen firms worked on the tail section (Code T-FA-ed), while tubes, valves, and accessories for the propellant system required thirty-seven firms.

Companies outside of Germany also participated, examples being Johansen Maschinenfabrik of Copenhagen and N.V. Holland-Nautic of Haarlem in the Netherlands. It is known that some components came in from neutral countries such as Switzerland. For example, the German Federal Railways would issue contracts for servomotors ostensibly for its own "peaceful" uses; but, in reality, they were destined to end up at Mittelwerk. Some Allied machine tools and equipment were reported to have found their way through devious channels into Spain and Andorra and thence to the A4 production program.

* * * * *

Missile assembly at Mittelwerk was accomplished in a facility possessing approximately 1,200,000 square feet of usable floor area, consisting principally of two long main tunnels (Fahrstollen) some 35 feet wide, 25 feet high, and about 500 feet apart. Extending over a mile beneath the overlying hill, they were connected by 47 cross tunnels or "Hallen," each approximately 30 feet wide and 22 feet high. The soft mountain rock above ranged from 140 to 200 feet thick.

Raw materials and finished components from feeder plants were first delivered through one of the two main tunnels, were processed and subassembled in the cross-tunnel moving lines, and then were assembled as complete A4 units in one of the main tunnels on a moving assembly line. Some of these cross tunnels were used for machining, sheet metal stamping, welding, and subassembly without major modification, but others were deepened for special purposes—such as work on major A4 subassemblies and checkout of completed A4 units.

For a distance of about 800 feet from the beginning of the two longitudinal tunnels, the walls and ceilings were faced in concrete. The remaining area was painted white with highly illuminating reflecting-surface paint that at the same time provided waterproof protection against leakage through the rock.

In some tunnels, second floors were installed. Two-by-eight-inch planking was used along with stringers supported by 12-inch "I" beams, laid transversely to the tunnel. These, in turn, were supported by 12-by-12-inch block columns built up against the side walls, as well as by steel-pipe columns running down the center line of the tunnel. Spare parts awaiting use were generally stored in the second floor section of cross tunnels nearest the position where they were required for installation.

Cross tunnels numbers 20 through 47 and corresponding sections of the main tunnels were devoted to A4 rocket production. Standard gauge railway tracks ran through the two main tunnels so that assembled products could be loaded directly onto railroad cars without being observed by Allied aerial reconnaissance.

Ventilation and lighting in the plant were surprisingly well controlled. Circulation was accomplished by large metal ducts that carried a forced draft of temperature-controlled air to the two main tunnels and the connecting tunnels. General lighting came from rows of lamps emplaced overhead and along the walls, while individual lights were provided for those operations that required them. Power came from stations at Bleicherode and Sondershausen; local transformers were installed in and near the tunnels. Although Mittelwerk was bombproof, had the Allies elected to raid the power system they could have interfered with production. They could also have bombed the little villages like Ilfeld, where the

German managers and workers lived. There is no doubt that the Allies were aware that something important was going on at Niedersachswerfen.

R. V. Jones, coordinator of V-weapon intelligence at the British Air Ministry, first suspected the existence of the Mittelwerk during the summer of 1944, "when [he told the authors] I heard that there was an 'M' works somewhere in the middle of Germany that was said to be producing rockets. Photographic reconnaissance was a matter of routine once a prisoner of war had given us its rough location." One ground intelligence report noted that during April 1944, "Sixty flat cars left the plant; three cars had two rockets each in them. The rockets are manufactured by 2,000 civilians and 10,000 prisoners who live in nearby barracks. About 500 to 1,000 prisoners come in each week; they died fast or were killed by mistreatment."

In August and on into September, ground intelligence reports and prisoner of war debriefings trickled in to the Allies revealing that Mittelwerk was indeed the main assembly point for the A4. Photo reconnaissance flights dispatched to Niedersachswerfen in October for the first time began to turn up pictures showing that the shadows cast by loaded railway wagons were suggestive of rockets enveloped in what seemed to be huge canvas covers. These silhouettes soon turned up frequently on the German rail system, allowing Allied intelligence analysts to pinpoint Mittelwerk as the source of rocket production. Despite this, the complex was never bombed, though the Allies did consider a scheme for dropping large quantities of inflammable liquids on the target. The thought was that liquids would enter the main entrances of the underground plant, as well as its ventilation system, then ignite. As the fire consumed oxygen, everyone inside would be suffocated.

Missile output was never directly disturbed by the strategic bombing offensive against German industry, though before the war was over, A4 output was effectively slowed as a result of attacks on component producers and on the transportation system to and from Neidersachswerfen and nearby Nordhausen. Next to Mittelwerk, great efforts were expended in the construction of numerous tunnels under nearby hills to be used for the production of engines for jet fighters, antiaircraft rockets, and other war material. They were however, only partially completed and operational before the end of the war.

Assuming a total output of 5,000 A4s, it was estimated that Mittelwerk and its suppliers would expend the following monthly tonnages of materials: steel and iron, 24,000; aluminum, 5,200; copper, 477; alloy metals, 244; and silver, 3.2. Steel and iron did not become scarce until near the end of the war, but aluminum was always tight: Supplies were derived from the Hungarian bauxite mines, and virtually all the output was demanded by the Luftwaffe. The situation for the alloy metals was always difficult, a high part of the demand being met by Finland and Turkey and by "what fell from the sky" (shot-down bombers). As the Third Reich shrank in size, vital supplies as well as suppliers were inevitably cut off.

Even when bombing did not contribute directly or indirectly to difficulties with components, problems would still arise. Thus, the production of servomotors, established at Mittelwerk under the direction of Magnus von Braun, had later to be expanded in a salt mine some fifty miles from Niedersachswerfen for the simple reason that they were extremely prone to failure. In fact, only a quarter of the finished products were found to function properly.

Occasionally, the lack of acceptable parts would bring the entire A4 assembly line to a halt. Rudolph explains how

> once I had to send an employee off to Alsace to attempt to fetch components from a plant already under Allied artillery fire. We knew, through radio contact, that the driver actually reached the plant and got the parts. Then communications were lost, and we never heard from him again. We sent people with rucksacks on their backs out on bicycles—once even through Allied lines—to bring back fasteners, rivets, and bolts. Even the parts we did bring in suffered from degraded workmanship and substitute materials. Yet, we were always asked to put out better-performing rockets. It was a nightmare.

Sometimes, huge quantities of major assemblies (such as tail sections) would accumulate, while the assembly line would creep along at snail's pace simply because there were not enough fasteners. Once, Rickhey informed Rudolph that some machinery from bombed-out plants in the Ruhr was being sent by train to Nordhausen. Shortly after its arrival, Nordhausen itself was bombed and the railway complex and rolling stock damaged. The flat cars on which the machinery had arrived were needed urgently to carry munitions to the front. Reports Rudolph:

> So I was told to unload. I rounded up a sufficient number of workers from assembly operations, which had slowed down because of the lack of minor parts. We simply dumped the machinery by the railway tracks, and the empty cars went off to collect the munitions. The machinery dumped by the wayside was an incredible sight.

In an attempt to keep supply communication lines open, Figge had installed ten telephone lines and three telegraph lines in his office within the military establishment at Halberstadt. Soon, they became overloaded. "Saur," relates Figge, "thought they were set up especially for him." After a heavy bombing, he established himself in

Blankenburg using Figge's lines to call in to Hitler details of Mittelwerk production.

> This came to be a sort of "hot line," for Hitler's adjutant was always ordered to keep Saur on the line when he called in. When Saur spoke on the phone, most of the workers wanted to tune in in order to hear the Führer's voice. It got to the point that so many listening devices were plugged into the main line that the connection between Hitler and Saur became disrupted. One day, one of them threw the receiver down declaring that the connection was too bad to permit the conversation to continue.

Rickhey reckoned that, at the Peenemünde pilot plant, between 10,000 and 20,000 man-hours went into the manufacture of each missile. At Mittelwerk, it took some 17,000 man-hours per unit for the first hundred produced, declining to 7,500 per missile up to 1,000 units and then 3,500 man-hours per missile up to about 5,000 units. When that stage was reached, he figured that the unit cost would be about RM 56,000 ($13,330), not far from Kettler's and Förschner's adjustment of Leeb's original RM 40,000 ($9,520). Naturally, these prices reflected the use of much essentially free forced labor.

From January to May 1944, production at Mittelwerk increased steadily, but Hitler ordered it cut back in June until engineers solved the problem of air burst—the premature explosion in the air of the reentering missile and warhead after its arch through space. Rickhey's estimate of monthly production is given in Table 1 of Appendix A. If his figures are accurate—and there is reason to believe they are high*—nearly 7,000 A4s were produced during World War II.

Rickhey's figures seem to represent monthly targets rather than the actual number of missiles produced. In the Appendix various estimates are reviewed, ranging from 5,789 to 6,915 missiles produced during the war. Based on the best available analysis, it seems probable that just short of 6,000 missiles were produced at Mittelwerk by the Third Reich before its collapse in 1945.

At the end of the war, an American production intelligence expert, T. P. Wright, made the following observation following his inspection of the Mittelwerk plant:

> A large proportion of the factory was used for the manufacture of V-2s. Most of the rocket power unit was fabricated elsewhere with final assembly and fabrication of the fuselage and tail units carried out at Mittelwerk. The factory was well equipped with machine tools ... with a very well set up assembly line for the rocket power unit. The current output at the conclusion of the war was variously claimed to be from 10 to 20 units a day ... with capacity output at about 20 per

* See Appendix A.

Flames started by a time bomb in the Nazi Party headquarters at Reutte, Austria, on April 30, 1945.

Reutte, Austria, May 1945. Left to right: Charles L. Stewart, Herbert Axster, Dieter Huzel, Wernher von Braun, Magnus von Braun, Hans Lindenmayr.

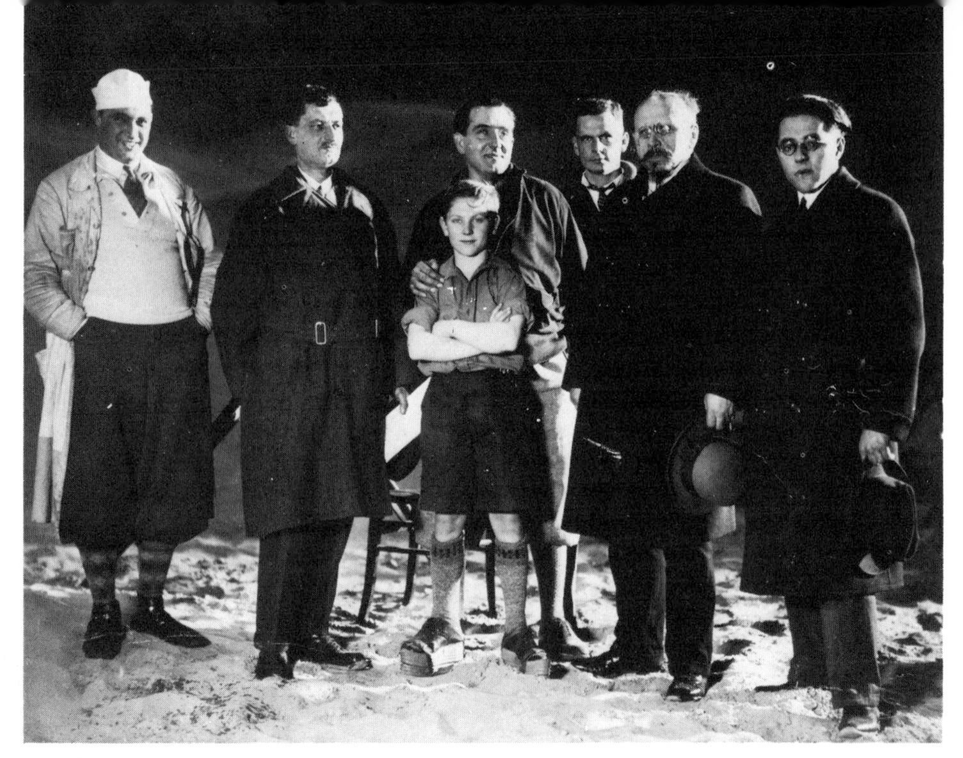

The late 1920s and early 30s witnessed an incredible burst of popular interest in rocketry and space flight in Germany. Here is a set from Frau im Mond, *Fritz Lang's film on a Moon voyage whose scientific consultant was Hermann Oberth. Left to right: Otto Kanturek (who built the scenes), Professor Oberth, Fritz Lang, Gustl Stark (boy), Gestettenbauer (one of the actors), unidentified cameraman, Hermann Ganswindt, Willy Ley.*

August 5, 1930, following a successful test of Professor Oberth's Kegelduse *rocket at the Chemisch-Technische Reichsanstalt, Berlin-Plötzensee. Left to right: Ing. Rudolf Nebel, Dr. Ritter, Mr. Barmuller, Kurt Heinisch, unidentified, Professor Hermann Oberth, unidentified, Klaus Riedel (in white coat, the rocket he is holding is a wooden mock-up of Nebel's Repulsor type rocket), Wernher von Braun, unidentified.*

Albert Püllenberg, another German rocket pioneer, with a model of his proposed postal rocket. Taken at the Bremen post office.

Left, G. Edward Pendray and Klaus Riedel near the rocket testing stand at Raketenflugplatz, Berlin, April 1931.

A vigorous proponent and popularizer of rocketry and space travel, Max Valier was also an avid experimentalist. He is shown with Dr. Paul Heylandt (left) inspecting their rocket car.

Hitler, center, visits Kummersdorf in 1939. To his left are Colonel Walter Dornberger and Dr. Walter Thiel, later killed in a RAF air raid on Peenemünde in 1943.

Wernher von Braun and Oberstabstartzt (staff physician) Bahr sailing in the Baltic Sea off Peenemünde, about 1942.

Potentially of greater threat to the Allies than the V-2 was the Wasserfall, an antiaircraft missile that could easily have countered the RAF and USAAF over Germany. However, it came too late in the war to be a significant factor.

Preparing the A3 rocket for launching on November 29, 1937, from Greifswalden Oie, off coast of Peenemünde in the Baltic Sea. Von Braun is the tall man on the right.

Despite its secrecy, Peenemünde drew a continuous stream of VIPs from Berlin. Dornberger, at left, describes the V-2 launching to a less than enthusiastic Field Marshal Erwin Rommel, center.

Peenemünde West, in 1942. Wernher von Braun to the left of Air Force General Adolf Galland (hands behind back and mustache). First Lieutenant Josef Poehs is third from right.

Rigid housekeeping rules were enforced at Peenemünde. No Smoking (Rauchen Verboten*) was enforced even though the propellant tanks for the A4, in the foreground, were quite empty of their flammable fuel and oxidizer.*

Contributing to the manufacture of A4s at Peenemünde were skilled Russian prisoners of war, always under the watchful eyes of German supervisors. Reports of sabotage from these prisoners were few.

An A4 propulsion unit shown on dolly inside the underground assembly facility at Niedersachswerfen near Nordhausen, Germany.

Stacked A4 combustion chambers in the Mittelwerk plant at Niedersachswerfen, Germany.

The entrance to the Ahr Valley underground Meilerwagen assembly facility, code-named Rebstock, near Bad Neuenahr, 1944.

> day (900 to 1,000 a month). It is the opinion of the writer that such production was reasonable of attainment at this plant.

Rudolph recalls that 700 A4s were completed in December 1944, and close to 800 in February 1945. Then the output dropped sharply because of the scarcity of minor parts that had been supplied from plants in or near the Ruhr Valley, and because acetylene gas for welding was giving out. "We had hundreds of all the major components on hand—engines, tanks, midsections, tails, etc.—but we couldn't make complete missiles! The situation was desperate."

When a missile was completed, it was tested and checked out in a vertical position at the end of the main assembly line. This was done primarily to determine if the servomotors operated properly (or at all); if the electrical resistance was satisfactory; if the missile was properly aligned, and if the electrical and pneumatic valves functioned. This was, in effect, the final systems check before allowing the missile to be shipped to the field. "If a missile failed," Rudolph explains, "the government inspector would not accept it. We would then attempt to make repairs in any of four repair stands. If the difficulty could not be put right there, the faulty A4 would be sent to the repair line which ran parallel to the main assembly line back in the plant."

Estimates vary somewhat as to what happened to the 6,000 A4s that appear to have been assembled at the Mittelwerk plant plus several hundred Peenemünde experimental units. Almost 3,600 rockets were launched against Britain and targets on the Continent, of which some 25 percent failed largely because of airburst (see page 251):

Missiles reaching the United Kingdom	1,115	
Missiles reaching continental targets	1,775	
Subtotal		2,890
Failures (25%)	700	
Number at Mittelwerk when occupied by US Army in May 1945	250	
Number in field storage, May 1945	2,100	
Subtotal		3,050
Expended in trials at Peenemünde, Heidelager, and Heidekraut	300	300
Total		6,240

Until the A4 went into operation against England in September 1944, missiles were regularly shipped from Mittelwerk by rail eastward to the Heidelager test-firing grounds in Blizna, Poland; or from the end of July, from the Heidekraut firing position ten miles east of Tuchel in eastern Germany (Heidelager fell to the Russians early in August 1944, but had been abandoned by the Germans some ten days before). The week-long journey to Blizna was made under the watchful eyes of troops assigned to protect the trains. Amazingly, these trains with their shrouded cargo were rarely molested by Allied aircraft and indeed went largely unnoticed by aerial reconnaissance as they made their way eastward for flight trials. For a long time, the Allies failed to relate rail and truck activity in the Nordhausen area to the A4 assembly program. The underground secret of the A4 was well kept almost to the eve of the V-2 campaign.

Although the Germans never reached the target 900 to 950 missiles per month capacity at Mittelwerk, they approached within two-thirds of it during the September 1944–February 1945 period. If there had not been conflicting priorities coupled to a continuous flow of design modifications, there is little doubt that the plant could have attained the full output. It was, in fact, the literally thousands of design modifications that prevented sustained, large-scale (in excess of 500 units a month) assembly from beginning before August 1944. Thus, during the two preceding months some 200 test missiles had to be sent out to Heidelager in Poland for flight tests so that "living statistics" could be accumulated. Test firings on a scale inconceivable today gradually led to the improvement of the missile. Even so, it never really emerged from the trial stage before the war was terminated.

The worsening war situation and the difficulty of obtaining adequate labor combined to give Himmler and his underling Kammler a widening wedge in Hitler's so-called secret weapon program. In fact, Kammler was becoming one of the fastest ascending stars in the SS power constellation. So much so, that on February 6, 1945, the Obergruppenführer prepared a long, top-secret memorandum in which he outlined means of using guided missiles in the "breaking of the air terror." Using his expanded authority, he proclaimed, "I hereby determine that the following weapons and equipment will be under my jurisdiction." And he went on to list dozens of projects ranging from the V-1 and V-2 (these "vengeance" designations were by now replacing Fi103 and A4) to guided and unguided antiaircraft missiles and their support equipment.

Among the projects that he ordered to be continued at that late date were the *Schmetterling* (8-117), the X-4 (8-344), the modified *Wasserfall*, and the *Natter*. To be canceled immediately were the *Enzian*, the *Rheintochter*, the 8-298 missile, and the *Hochdruckpumpe*—the so-called V-3, a high-pressure pump gun that was to have been installed at Mimoyecques in France. Kammler then advised that "all personnel released through the change [e.g., the canceled projects and programs] will commence working on the remaining programs." Training activities, he added, would continue to be carried out under the direction of Dornberger, while the Special Committee A4, of Speer's Armaments Ministry, would remain in control of production. The haughty Obergruppenführer did not bother to add that he, in turn, controlled this committee. The SS, in the person of Kammler, was in clear control of the vaunted A4/V-2, and all other missiles in development or production.

Surprisingly little has been written about this extraordinarily ruthless and ambitious man and the incredible power he enjoyed during the final months of Hitler's Third Reich. In an interview, Paul Figge described his association with the SS general:

> I came to know Kammler at the first meeting in the locomotive house in Berlin. Since he handled the personnel department, we came in close contact. Kammler warned me that he considered me a political rebel and politically unreliable. I had to submit a signed paper to him declaring that I was a "foreign" German because I had lived in Belgium for fifteen years. Saur and Degenkolb, however, covered for me so well that these accusations hardly came to affect me.

Meanwhile, Kammler developed an ego that was not only unpleasant, but also burdensome. He took over the position of ensuring the safety of Mittelwerk, and with this, the top leadership ring of the Special Committee A4. He then tried to incorporate his SS henchmen into all the positions and jobs using any methods he could.

Before the war, Kammler was a small-town architect who had joined the party quite early. He was a nonachiever in the SA and, consequently, he joined the SS and made a fantastic career out of it, earning quite a reputation. For example, between 1943 and 1945 he was promoted from Oberführer to Obergruppenführer. He was an extreme egomaniac, which accounted for his rapid promotion. But, by the same token, it became a justification for every means required to attain his goal: Near the end of the war he had 250 prisoners killed without reason.

There are many versions of his death, but the most credible is that he gave his adjutant the order not to allow him to be taken alive by the Russians. Toward the end of the war, Hitler had ordered him to stop the advance of the Russian troops in Czechoslovakia. As far as is known he was shot by this adjutant.

Albert Speer's first impressions of Kammler are uncomplicated: "blond, blue-eyed, long-haired, always neatly dressed, and well bred." He also observed that Kammler was "capable of unexpected decisions at any moment," and felt at the time that Himmler had made a "significant choice" in selecting him—the SS chief was always on the lookout for energetic, intelligent, and zealous men even though they might not possess extended party experience and connections. Speer noted other characteristics, including "objective coolness," and saw Kammler not only as a frequent partner but also as a possible rival. "In many ways," recalls Speer, he was "my mirror image. He too came from a solid middle-class family, had gone through the university, had been 'discovered' because of his work in construction, and had gone far and fast in fields for which he had not been trained." Later, the munitions minister grew to despise Kammler, calling him "a cold, ruthless schemer, a fanatic in the pursuit of a goal, and as carefully calculating as he was unscrupulous." To Colonel Herbert Axster, Dornberger's chief of staff, Kammler was absolutely intolerant . . . and ruthless—the most reckless man I'd ever known. I can truly say that we hated him."

As Dornberger got to know Kammler, he—like many others—found that he possessed an overwhelming desire to command. "It was quite out of the question to get him to change his mind," Dornberger noted wryly. "His ambition, lust for power, mistrust, and vengefulness were matched only by his morbid inferiority complex and his mimosalike sensitivity." He tended to put off those of equal or superior intelligence and influence, preferring to surround himself with young followers who were easily "dazzled by his zest and tireless energy" and with weaklings who "applauded his caprices and his brutal jests." To those above him in the Nazi hierarchy, he "was cunningly deferential and amiable" but to those below him he was "arrogant, brutal, overbearing, intolerably haughty. . . . He had no moral inhibitions whatever in getting what he wanted." At the time, Dornberger could watch him as "an interested spectator." He was soon to see him in a different light.

From November 1943 Kammler appeared with ever greater frequency at Peenemünde and at the Heidelager training ground in Blizna, Poland. He pressed hard during the following half year to take over control of the A4 program, and in early July 1944 went so far as to call Dornberger "a public danger." Shortly after the July 20 aborted attempt on Hitler's life his efforts were rewarded. Himmler took over from Colonel General Fromm (then commander in chief of the Replacement [Home] Army and director of armaments), and on August 8 gave Kammler undisputed control over the entire A4 program. This was almost immediately expanded to include all SS armaments and industrial and construction activities. Obviously, to the extent that the influence of Himmler and Kammler increased in these areas, that of Speer decreased. So desperate was the war situation, however, that Speer was not overly perturbed at Kammler's intrusion into the A4 program. In fact, he may have seen in the SS general's overriding ambition the only hope for fielding the missile before Germany's collapse under the Allied juggernaut.

Heinz Höhne, writing in his *The Order of the Death's Head*, saw Kammler's ambition "abnormal even by SS standards." For example, he would

> build gas chambers in Auschwitz or launching ramps for flying bombs with equal attention to detail ... in spite of his exalted position [Kammler] bore no loyalty to the SS; for him, the Order was nothing but a ladder for his own advancement. ... Kammler's meteoric career was an illustration of the central position to which the SS and its leaders had risen among the power groups of the Third Reich.

Besides being a highly complex missile, one that employed many components and parts not common to other contemporary weapons, the A4 also consumed propellants that were quite different from those that powered tanks and airplanes. The missile's fuel, alcohol, did not pose much of a problem. Prepared from potato crops grown in eastern Germany and Poland, it did not become in short supply until the latter part of the war when westward-rolling Soviet armies began to overrun raw material sources. Liquid oxygen, the oxidizer, was another matter. Besides being expensive to produce, it was awkward to transport and to store, particularly in a wartime environment. Of the roughly 9 tons of liquid oxygen needed to sustain a single A4 firing, slightly less than 5 tons was actually pumped into the missile in the field. The rest was lost during transfer, first from storage in tanks at the manufacturing plant, then to 48-ton railway cars, then from these to both 5- and 8-ton tank trucks, and finally to the rocket itself.

At about the same time that Peenemünde's research and development activities were dispersed following the August 1943 bombing raid, Hitler personally decreed that A4 liquid oxygen production be transferred to underground facilities. Not only would this policy make it difficult for the Allies to fit together all the pieces of the A4 puzzle, but it would prevent the missile's deployment from being jeopardized in the event one of the plants should be located and destroyed. That this policy was successful is confirmed by the great difficulties the Allies experienced in neutralizing A4 facilities.

Aside from four plants affected by Hitler's directive—Raderach, Redl-Zipf, Lehesten, and Liège—Peenemünde was to continue producing enough liquid oxygen for its own experimental purposes and two additional plants were to be built capable of producing 6,000 tons of the oxidizer each month. One was started at Mittringen in the Saar and the other at a site near Mittelwerk. Neither was completed.

In the autumn of 1943, the first of the dispersed liquid oxygen manufacturing plants went into operation. Located near Friedrichshafen at the tiny hamlet of Raderach, the plant was officially known as Prüffeld-Anlage Raderach but also carried the code name "Porcelain Factory." The first sketches for the installation were prepared by Peenemünde's Bernhard Tessmann back in early 1941 and construction got under way a year later. In all, some three thousand construction workers and civil engineers were involved, along with between fifty and sixty military and government personnel. General Dornberger's representative at Raderach was a Captain König, while the top Peenemünde civilian representative was Anton Beyer. Manheim Strohmayer was in charge of the installation.

The liquid oxygen production effort went along quite smoothly, but when it was decided to test combustion chambers at the same site, the resulting roar and flames were heard and seen in nearby Switzerland. Security was quickly compromised, and soon Allied intelligence became aware that something was going on at Raderach, something that appeared to involve a large "semi-underground factory" where "strange experiments" were thought to be taking place. One intelligence brief, dated January 1, 1945, made reference to an earlier agent's report to the effect that

> heavy clouds of smoke filled the sky in the day and at night [produced] a red glow. The experiments caused the earth to shake. These experiments are with atoms, and when the experiments proved successful the plant went into operation. Workmen are not allowed to leave the factory.

Even though the Allies could not fully confirm their suspicions as to the potential danger of Raderach, it was bombed by the British shortly after it became operational. Though some damage was done, the liquid oxygen plant continued in operation. Raderach was again attacked on August 3 and August 16, 1944, by the US 15th Air Force as a site suspected of producing hydrogen peroxide—even then thought by some to be an important element of the A4 when in fact it was only the auxiliary propellant used to drive the turbopump system. After the war the Raderach installation was thoroughly studied by a French scientific mission.

The second plant, code-named Schlier, was located at the Redl-Zipf brewery between Vöcklabruck and Vöcklamarkt in Austria. The oxygen was produced in a deep tunnel, above which an acceptance test facility was constructed for combustion chambers manufactured by Linke Hoffmann in Breslau. Rocket tests notwithstanding, the well-known product of the brewery, Zipfer beer, continued to be made nearby. For the test engineers, a special variety, "peace beer," was clandestinely brewed; it was described as "a bit more dynamic than beer normally available during the war."

The test stands were installed directly on top of the liquid oxygen production plant built some 100 feet below ground level. Around them stood a 10-foot-thick concrete wall with spaced outlets for the flames emanating from the combustion chambers to pass. Before being static-tested, the chambers were checked out in an assembly hall three floors below. Communication between the production and assembly area below and the test installations above was assured by an elevator shaft.

Redl-Zipf was struck twice by tragedy. In the first accident, some ten people were killed. It seems that unknown to plant engineers, the extremely low temperatures of the liquid oxygen had embrittled the steel of the pipes through which it flowed, finally causing them to crack and then to burst explosively. From then on, only aluminum and copper tubes were used. As one engineer was later to put it, a lesson had been learned in "a virgin field of technology."

In the second accident, the death toll was higher—twenty-seven, including the daughter of Hermann Oberth, who was working at Redl-Zipf as a test evaluator. Karl Heimburg explains what happened:

> As we expanded operations at the plant, an enormous water hammer problem developed which one day actually ripped the pipe lines from their concrete anchors. The vent line from the alcohol tank passed through 10 feet of concrete on top of the test position; it had been placed there as a protection against aerial bombardment. On account of this water hammer phenomenon, the alcohol vent line ripped apart at the seam.
>
> At the same time, in order to measure the amount of alcohol in the tank, it [the tank] had been placed on a scale. As it turned out, the tank often rocked against the wall during loading, making accurate measurements impossible. The result of this was that the workers overtanked on occasion. Because of the rupture in the tube, the overtanked alcohol would run back through the concrete to the shaft and down to the lox [liquid oxygen] plant below.
>
> On my first visit to Redl-Zipf, I was shocked at what I saw: the whole place reeked of alcohol. In fact, it smelled much like the brewery next door! Realizing its explosive potential, we evacuated all the workers and cleaned the whole place out with water sprays. When I left, I warned the plant manager that the place was a death trap. The plant could remain open, in my view, only if the vents were carefully watched to make certain that there were no cracks anywhere.

At about this time, managerial conflicts began to arise, directly contributing to what was about to happen at the secret Schlier plant. Georg Rickhey had by now taken charge of Mittelwerk and wanted to find a job at Redl-Zipf for a former fraternity brother. So he told the plant manager there that he had to leave. But the manager happened to be a friend of the local *Gauleiter*, and immediately sought his protection. Rickhey, however, insisted that the manager—Strohmayer—be dismissed; ultimately his desires prevailed. Heimburg, nervous about this managerial changeover, told his two top master mechanics, Kohn and Pflanze, that if they should get into any trouble with the new man they must leave immediately. No sooner had Rickhey's man taken over than Kohn warned him of the impending danger. Impressed, the new plant manager teletyped Rickhey asking for permission to close the facility until it could be put right. Rickhey, severely pressed to meet production quotas, refused to comply.

Meanwhile, back at Peenemünde, Karl Heimburg received a letter from Kohn expressing his deep fear that trouble was about to erupt at Redl-Zipf. The very morning the letter arrived, the fatal explosion took place. Even though an escape hatch, an *S*-shaped tunnel, had been installed, all persons working in the affected area were killed. Heimburg's premonition proved terrifyingly accurate.

Immediately upon receiving word of the explosion, Heimburg, accompanied by Wernher von Braun, left Peenemünde for Schlier to determine exactly what had happened and to lay out plans for reconstruction. But first, the dead workers had to be buried. Heimburg was sent to fetch Hermann Oberth and drive him to Austria. Heimburg recalls that when they passed through Munich the city was in utter darkness, "so dark that we had no idea how to find the way south. So Oberth got out of the car and said, 'We'll navigate by the stars.' By 7 AM the next morning we were in Schlier."

Despite the accident, within three or four months Schlier was again in business, now under the direction of Peenemünde's Hermann Weidner. As a result of the loss of Test Stand No. 8 at the Baltic test site during July and August 1944 air raids, his entire test stand crew was released to Redl-Zipf to accomplish the reconstruction.

While the plant was being repaired, plans were laid to bore a tunnel into a mountain south of Vöcklabruck near Ebensee on the south shore of Traun-See. At this close-by site, test stands were to have been constructed to augment testing undertaken at Schlier. It was even suggested that combustion chambers be manufactured at Ebensee. The Allies soon got hold of the plans and filed the following intelligence brief:

> ... the [underground] factory is [to be] built into a hill with the usual tunnel entrances spaced 85 to 100 feet apart. Power lines run into the site and a rather large number of hutments are available for 2,400 employees.

The Traun-See facility was never completed. While Redl-Zipf/Schlier was out of commission, a third facility was established at Lehesten south of Saalfeld in southern Thuringia. To those who located the site—Gerhard Degenkolb and Peenemünde's Martin Schilling and Bernhard

Tessmann—the slate quarry was ideal for straight-down-the-wall combustion chamber testing without the need for flame deflectors. And the huge underground tunnels and caves were just as ideal for a secure liquid oxygen production plant to be built. In fact, everything about Lehesten seemed ideal, including its proximity to Mittelwerk.

At its peak, Lehesten remained "on duty" for twenty-four hours a day, week after week, as it produced its 8 metric tons of liquid oxygen an hour. Normally, it would operate for twenty-seven days a month without stopping, and then would be shut down for three or four days so that equipment could be unfrozen. Toward the war's end, Lehesten provided acceptance testing for combustion chambers earmarked for direct transfer to V-2 troops in the field.

At about this time, Obergruppenführer Kammler moved into the picture and assigned unqualified party administrators to handle Lehesten's management. One of the results of this decision was the use of totally inexperienced pressed laborers to open up new tunnels in the slate quarry. Knowing nothing of the nature of slate, a major collapse soon occurred, burying hundreds of workers in the rubble. When Martin Schilling objected to what was happening, an SS officer growled, "Shut up. Or do you want to join them?"

As the war drew to a close and Peenemünde's position became untenable, the question arose, "Should Lehesten's two static firing stands and other facilities be expanded?" Berlin approved the development of plans for expansion, but would not authorize any construction funds for the time being. Soon it became too late, as American troops were moving in from the west. Just before the installation was captured, Heimburg was quickly sent to Berlin to confer with Rüstungskontor chief Hettlage to seek permission to assign his men to some technical program activity and thereby keep them from forced service in the last-ditch *Volkssturm* fighting units. Hettlage, deeply pessimistic over the way the war was going, simply said, "Heimburg, plan all you want. Do what you can. Don't worry, your men won't be drafted."

The final new plant, built near Liège in Belgium, was primarily used to supply liquid oxygen to operational units at the firing front in Holland. Its 900 tons-a-month capacity compared with 2,100 tons for Redl-Zipf/ Schlier, 5,000 for Lehesten, and 900 for Raderach.

There were, of course, surface plants that manufactured the oxidizer for industrial and military users not involved in the A4 program; in Germany alone, fifteen such plants produced over 6,400 tons a month, while in France thirty-eight produced 1,350 tons. About 420 tons came out of six facilities in Belgium and the Netherlands.

Even if the capacity of these surface facilities had been added to that of the special new plants and of Peenemünde, Germany would never have been able to sustain a firing rate of 1,800–2,000 missiles a month that at one time had been envisioned.

The supply of alcohol was another factor that would have prevented such an ambitious launch program from becoming feasible, even before the fall of Poland and eastern Germany. In 1941, Germany was able to produce an estimated 18,800 tons a month, of which the chemical industry consumed 10,000 tons. About 14,000 tons of alcohol a month would have been needed to sustain an 1,800-missiles-a-month launch rate.

Sodium permanganate and hydrogen peroxide (H_2O_2) were the other two propellants employed in the A4 missile; both were needed in relatively small quantities for the operation of the turbopump system.

Another facility associated with the A4 but not involving propellants was located in the Rebstock railway tunnel near the Ahr Valley town of Marienthal. Limited to wiring, the manufacture of cable harnesses, and the installation of electrical and mechanical equipment in various types of military vehicles and its checkout, the facility grew out of the general order to disperse operations and personnel from Peenemünde following the great mid-August 1943 RAF bombing raid.

The principal vehicle handled by Rebstock was the Meilerwagen, used to haul and erect the A4. Werner Gengelbach, in charge of setting up the Ahr Valley operation, recalls:

> Orders were received from Berlin to seek a new location underground west of the river Rhine, as close as possible to the French border. Alsace-Lorraine had proven unsuitable for accommodating equipment and men [so we began looking] for unused railroad tunnels. We finally found a good place near Marienthal. With two adjacent tunnels and good access roads and personnel accommodations, and with a wine-making school nearby for administrative purposes, the place was ideal.
>
> One of the tunnels had been used by a Spaniard to grow mushrooms. The warm and moist atmosphere was ideal for him. The government paid him RM 200,000 [$47,600] as compensation for his business; so, with the beginning of our operations, the valley was supplied with the best mushrooms. The Ahr Valley is beautiful and the weather conditions are excellent for making good wine. We knew this; and, by the end of September, we had begun to move, code-naming our operation *Lager Rebstock* (Camp Grape Vine).
>
> The main tunnel was to be used for shops, testing, checkout and inspection operations, while the second tunnel would serve as a supply storage area. The area between them could house the personnel and serve as temporary offices.

Personnel at Rebstock increased from an initial 30 to about 180 by November and 300 by the end of December 1943. At the end of 1944, production output reached about 50 percent of the monthly goal and by

March of 1945 was humming along at 100 percent capacity with 600 persons on the employment roster. This, in spite of growing problems associated with the acquisition of supplies. Gengelbach remembers only too well the difficulties brought on by daily and nightly air raids and the effects they had on the railways:

> We sent out eight to ten men to assure that our supplies were properly loaded on railway cars. They then would accompany the shipment through, piloting, so to speak to the destination. These people were quite skilled in finding detours to bring supplies to our locality.

In March, once all construction was completed and the planned production capability had been reached, the entire operation was transferred to the industrial concern Gollnow und Sohn. Production leveled at supplying the equipment necessary to sustain an A4 launch rate of three missiles per hour.

Morale was quite good in the pleasant Ahr Valley surroundings. "We drank the area wine every evening," recalled Gengelbach. "It was a good way to relax and to build up new energy for the next day—which was certain to bring more pressure and more hard work." No one particularly liked work in the tunnels themselves. Dieter Grau, who had been transferred to Rebstock from Mittelwerk and was thus used to underground conditions, complained of poor air circulation and high humidity. He knew there was not much chance of being bombed, however. "It would probably have been somewhat futile to do that," he observed, "though of course the Allies could have gone after staff housing in the area." Grau lived in a small town forty minutes walking distance away, which he considered "fairly convenient in this respect."

It is, perhaps, odd that the Allies did not make a concerted effort to hit housing in the vicinity of known or suspected V-weapon production and assembly sites. But scattered civilian housing of uncertain importance did not invite the attention of mission planners who were far more concerned with a seemingly endless list of defined military and industrial targets.

Considering the rapidly deteriorating situation with which Germany was faced during the final year or so of the war, it is astonishing that the nation was able to produce a new, largely untried weapon in respectable quantities and then introduce it into combat. The feat was particularly impressive when one realizes that the V-2 formed part of a complex weapon system that included not only the Meilerwagen but an array of related ground support equipment. Then, too, one has always to consider the complex logistics chain that such a system implies, a chain that stretched from the Mittelwerk and its supplier plants all the way to the firing positions.

What lessons could be learned from the V-2 production effort? Certainly, the experience demonstrated that a highly disciplined, technology-shrewd nation could accomplish near miracles under the combined incentives of sheer will and utter desperation. Beyond that, Mittelwerk demonstrated that a new and major weapon of war could be assembled and fielded if society was willing to pay what many considered excessive financial and above all human costs. But perhaps most important of all from the long-term perspective is this: except under compelling circumstances, the missile, like any other product of high technology, should not be committed to the mass production line until it has passed through a rigorous research, development, and flight readiness cycle. The war ended before the cycle was completed.

CHAPTER 6

Intelligence and Reconnaissance

During the early years of rocket experimentation in Germany, Willy Ley faithfully kept his foreign colleagues informed of progress made by the rocket pioneers. In his awkward but quite intelligible English, he reported to American and British rocket enthusiasts of one firing after another. On May 25, 1931, for example, he wrote how a rocket "knocked" against a 20-meter-high tree "and was absolutely destructed." Undeterred by such mishaps, he assured American Interplanetary (later Rocket) Society president G. Edward Pendray in New York that "now we'll construct a new and a little greater Repulsor and, we hope, it will reach an altitude of 1 km. Then the Mirak will be shotten."

The rapid deterioration of conditions in Germany caused Ley to leave his country in early 1935. He went on to become the world's foremost popular writer and lecturer on space travel, making his base in New York. (Ironically, he died there only a few weeks before Apollo 11 reached the Moon in July 1969.)

With Ley's departure from Germany, the flow of rocket society news virtually ceased—and hardly anyone noticed. This is not surprising, for outside of small amateur societies operating on shoestring budgets and occasional isolated dreamers and experimenters typified by America's Robert H. Goddard, interest in rocketry was virtually nil in the United States and the United Kingdom. The broken link failed even to cause concern among intelligence experts in those countries. After all, what possible meaning could be attached to the disappearance from public view of a handful of struggling rocketeers, including a young college student named Wernher von Braun?

To be sure, the American and British rocket/space flight societies expressed editorial disappointment at the drying up of news. Thus, in September 1934, the US journal *Astronautics* lamented that:

> Following the Hitler revolution in Germany, and partly due to internal dissension, the famous Verein für Raumschiffahrt; at one time the largest society of astronauts and rocket enthusiasts in the world, was dissolved.

And in December of 1937 the *Journal of the British Interplanetary Society* observed, "Astronautical activities seem to have ceased around the world lately, nothing having been done in the U.S.A., Germany or Austria." A couple of years later, in November 1939, *Astronautics* speculated as to what might be going on in Germany:

> In a recent speech [Danzig, September 19], Adolf Hitler, Chancellor of Germany, remarked that if the war continued four or five years, Germany would have access to a weapon now under development "that will not be available to other nations." Rocket experimenters naturally wonder whether this isn't an allusion to military rockets. Prior to Hitler's rise to power in Germany, that country was among the world's leaders in rocket development, and the German rocket society, the Verein für Raumschiffahrt, at one time had more than 1,000 members. About 1934, the Verein was broken up and some of its leaders left Germany. Since then, news of rocket experiments in Germany have been vague and contradictory. Many American experimenters believed that the German work had stopped. Possibly it merely became a part of the country's general preparedness for war.

During Hitler's rise to power in the mid and late 1930s and well into World War II, the secrets of Kummersdorf and later Peenemünde were well kept. Hitler's boast of new weapon developments was taken by most authorities to mean attack bombers perhaps coupled with long-range artillery. Even Ley, who had been with the German rocket group from the beginning and who knew von Braun from the time he first joined the society, speculated only in very general terms about what went on within the army immediately prior to and then in the years following his departure from the Third Reich.

Thus, in the first edition of his book *Rockets* published in New York in 1944, Ley wrote of a burglary attempt at Raketenflugplatz in Berlin that he felt was faked. Why, he asked himself? Probably as "an excuse to fingerprint everybody present." He also reported that in July 1932—several years before he left Germany—a VfR Mirak II rocket was flight-tested at Kummersdorf. Then, several months after the test, he picked up the rumor that the army had "discarded liquid fuels at once because of their nonstorability." This was simply not true. Shortly after, von Braun vaguely mentioned that the rocket had attained some 30,000 meters, but he did not specify whether he meant vertical altitude, length of arc traversed, or horizontal range. The story also reached Ley that the army had decided to cut off all research on rockets and, effective November 1932, had given von Braun a "dull routine job" which "he did not like!"

It is clear that Ley was not privy to what was going on within German army ordnance circles during his final years in Germany. Indeed, he did not learn the full story until he could talk to his former colleagues after World War II. However, word did get to him back in mid-1936—he was by then in the United States—that "the press had been forbidden [in Germany] to even mention the word 'rocket,' no matter in what connection." Ley also spoke of a telephone conversation allegedly overheard at the War Ministry in Berlin during which an official boasted, "Now I've got all the rocket people safely on ice around here and can watch what they're doing."

After the war, Ley lost no time in reconstructing his early associations with von Braun and the events that led to von Braun's accepting employment with the German army. Ley explained what had happened in a letter to the British Interplanetary Society (published in its June 1947 *Journal*):

> Although I handled the membership at that time [1929–1930], I cannot recall whether von Braun was on the list or not; either case is possible. But I do remember that he visited me in my home one day, presumably that was during the summer of 1929. Mostly because I have had many young visitors with the same request since then I can recall what he came to see me about. He told me that he had graduated from high school some time before that and was going to study physics and/or engineering. He was greatly impressed with the possibilities of liquid fuel rockets and had decided to devote his life to that problem. . . .
>
> During the Spring following the Winter 1929–30, Prof. Oberth returned to Berlin and the experiments at the *Chemisch-Technische Reichsanstalt* near Berlin were made. During these experiments Oberth had three helpers, one who occupied himself mainly in driving around and getting tools and materials, Nebel, and two who settled down at the work bench: Riedel and von Braun. I know the date of the final experiment because of its official certificate; it was 23 July 1930, and I have a photograph taken a few days earlier which showed Oberth, Nebel, Riedel and von Braun. Wernher von Braun must have appeared on the scene between March and May of that year. Just how this happened I do not know, but I suppose that he

> may have visited Oberth the way he had visited me a year earlier. I also remember for some reason that he was about 18 years then.
>
> From then on I saw von Braun on and off at the rate of about once a month until he was swallowed up by German military secrecy. Physically he happened to be a perfect example of the type labelled "Aryan Nordic" by the Nazis during the years to come. He had bright blue eyes and light blond hair and one of my female relatives compared him to the famous photograph of Lord Alfred Douglas of Oscar Wilde fame. His manners were as perfect as rigid upbringing could make them ...
>
> Did we discuss politics? Hardly, our minds were always far out in space. But I remember a few chance remarks which might be condensed into saying that in von Braun's opinion (as of that time) the German Republic was no good and the Nazis ridiculous.

However he felt about Nazi philosophy, von Braun was a realist. Amateur rocketry was ending in Germany, and prospects for the peaceful development of the rocket were nil. Only through association with the military could he carry on his work—and realize his dreams.

As Hitlerian Germany prepared for war, scant attention was at first paid to the fact that the army had taken over what the *VfR* had begun. America, feeling remote from the growing power of Germany and plagued by persistent Depression-related problems, exhibited total indifference to secret-weapon developments. By the late 1930s in Britain, however, a few slight misgivings were beginning to be expressed.

Early in 1939, Sir Henry Tizard (the head of the Committee for the Scientific Survey of Air Defence of Great Britain) was asked to recommend a scientist to be attached to the Air Ministry's Intelligence Branch. The name Dr. R. V. Jones was put forward. After the war, Sir Henry would say that "I never had any reason to regret that recommendation."

Jones, who had come into government intelligence service from the Clarendon Laboratory, took up his new duties on September 11, 1939, shortly after the outbreak of war. His task: to look into rumors of the development of new German weapons.

In his first intelligence report, made following a survey of Special Intelligence Service files, Jones recommended that the following weapons be taken seriously: gliding bombs, pilotless aircraft, long-range guns, and last but not least, rockets. His search through the files was a sobering experience, teaching him "how primitive was our Intelligence service compared with what, from a schoolboy onwards, I had imagined it to be." He recollected:

> At times of alarm, such as followed the outbreak of war and Hitler's speech [in Danzig], casual sources crop up in large numbers.

> These are mainly people who, under the stress of the situation, think that they have information of value to the country. Much of the information is useless, but in the days following Hitler's speech, one casual source came up whose information was of remarkable interest. It happened in this way. Our naval attaché in Oslo received [on November 4, 1939] an anonymous letter telling him that, if we would like a report on German technical developments, all we need do was to alter the preamble of our German news broadcast on a certain evening, so as to say: *"Hullo, hier ist London,"* instead of whatever we usually said. The writer would then know that we wanted the information and would send it to us. We duly altered the preamble, and the information arrived. It told us that the Germans had two kinds of radar equipment, that large rockets were being developed, that there was an important experimental establishment at Peenemünde and that rocket-driven glider bombs were being tried there. There was also other information—so much of it in fact that many people argued that it must have been a plant by the Germans, because no one man could possibly have known of all the developments that the report described. But as the War progressed and one development after another actually appeared, it was obvious that the report was largely correct; and in the few dull moments of the War I used to look up the Oslo report to see what should be coming along next.

The British naval attaché at Oslo, Captain (later Rear Admiral) Hector Boyes, confided to Jones after the war that he had not realized the Oslo Report's value. He said that this was principally because the Admiralty had assumed it to be a "plant" by the German intelligence service.

Jones told the authors that because of this belief, no one besides himself took any notice of the report. He, however, felt it was much too valuable to have been a hoax, in part because of the proximity fuse firing tube that it contained. "I had the job of opening the parcel," he said, "and I can remember speculating on whether it was going to be a bomb and would burst in my face!" In his book *Most Secret War*, Jones devotes an entire chapter to the Oslo Report. He observed that some of the developments described in the report were to be carried out at Peenemünde—"The first mention we had ever heard of this establishment."

About three inches thick and covered in brown paper, the Oslo Report contained a note inside reading "From a German well-wisher." But who might that well-wisher have been? To Sir Robert Cockburn, head of British countermeasure efforts within the War Radar Department of the Air Ministry, the diversity of the report's components indicated that it must have been assembled by someone high up in the Abwehr (German military intelligence). Some speculate that it was the Abwehr's top man himself, Admiral Wilhelm Canaris, who was not only disenchanted with

Hitler but had made concentrated efforts to keep Germany from invading Poland and thus sparking World War II. Canaris traveled extensively outside Germany during the war (to France, Italy, Switzerland, North Africa, Portugal, Spain, and Turkey), and it is known that selected Abwehr personnel were in rather regular contact with the Allies in a number of neutral capitals. Finally, the Admiral joined the anti-Hitler conspiracy toward the end of the war.

As one British expert pointed out to the authors, "It was during these 1942–45 transactions [e.g., the meetings with Abwehr personnel in neutral countries] that we got real information on Peenemünde, though we already had much of it through our wireless taps. But because Peenemünde was such a secret base, we couldn't bomb it right away on just this intelligence information. Otherwise, it would have betrayed the fact that we were in touch with their agents and were reading their ciphers—and they would change it all." (In other words, the Germans would have taken precautionary measures, which of course the British wanted at all cost to avoid.) Despite this, R. V. Jones told the authors that "whether Canaris had anything to do with [the Oslo Report] is a matter of speculation."

Whoever, inside or outside the Abwehr, made the Oslo Report available, in Cockburn's view the motivation was simple: By revealing the direction of German weapons research at any early stage, the British should be able to blunt the eventual results of the research by developing countermeasures. But the "well-wisher's" effort was in vain for, as we have seen, virtually everyone exposed to the report felt it to be a planned deception.

This is partly due to the lamentable state of scientific intelligence in the United Kingdom back in 1939. In fact, the only qualified man in position was Jones. Yet the scientist lamented to the authors that "I failed to get any help at all, not even a secretary." Jones was deeply concerned that a failure of scientific intelligence might well lead to national disaster. He was particularly conscious of the necessity to act upon intelligence, to be able to recommend that countermeasures be undertaken. Such countermeasures could be active (for example, the destruction of a suspected weapons installation) or they could be preparatory (alerting troops, for example, to the probable deployment by the enemy of new weapons or the unorthodox deployment of conventional ones). But this was not to be; in the early months of the war Jones could do little more than review and assess whatever intelligence material might come his way.

What really confirmed the authenticity of the Oslo Report was an incident concerning German radar. "We got some intelligence reports from Berlin in 1941," a source indicated,

> from a Chinese scientist who happened to be visiting the Berlin zoo. There, he saw a large paraboloid reflector on top of a tower. He passed this [information] on to the American Embassy, who then passed it to us [the British] as we had a reciprocal arrangement. . . . The paraboloid reflector confirmed exactly this unknown correspondence description of the principal German radar system. . . . Until then [we had not accepted it]. Here we were, getting scientific research worth billions of pounds for free. No one would give that away except for some very good political motive.

("This unknown correspondence description" referred to the description of the system contained in the Oslo Report.)

The years 1940, 1941, and 1942 passed with only bare notations in intelligence files of anything connected with rockets. Although Peenemünde had been mentioned in the Oslo Report as a key site, British military intelligence was unable to secure confirmatory evidence. As to why the British did not conduct extensive photoreconnaissance of the research and development center during those three years, Jones replied that such a reconnaissance would not have been practical early in the war. For one thing, hardly any photorecon aircraft existed and most of those that were available did not have the required range to reach Peenemünde. Then, too, Jones was not in a sufficiently authoritative position to order a recon mission even if adequate planes had been available. Finally, with the discovery that the Germans possessed radar, British intelligence placed virtually all its limited reconnaissance capabilities on the growing German radar network.

As the war progressed, reports on Peenemünde began to appear, and by the spring of 1943 aircraft with sufficient range to reach the Baltic site were in regular operational service. During the entire period, Jones was well aware that in the world of scientific intelligence, "there is an enormous premium placed upon going off at half cock," that preference seemed to be given to "a busy show of promptness" rather than to "quiet, mature action, however timely." He tried never to succumb to the temptation of releasing reports on a given aspect of enemy weaponry until he felt at least reasonably certain of the validity of his conclusions.

By the spring of 1943, it was becoming reasonably certain to the British that the Germans were working on rocket weapons of one kind or another. On April 15, for example, the Chiefs of Staff were able to advise Churchill that no less than five reports had been received covering some aspects of a suspected long-range rocket program. Even if details in any one or even in all of them turned out to be inaccurate, it seemed probable that these reports were based on a foundation of fact.

> The Chiefs of Staff [they wrote] are of the opinion that no time should be lost in establishing the facts, and, if the evidence proves reliable, in devising countermeasures.... In addition, the Chiefs of Staff propose to warn the Minister of Home Security of the possibility of such an attack, and of what is proposed. It is not considered desirable to inform the public at this stage, when the evidence is so intangible.

As the existence of German long-range rocket weapon developments became increasingly certain, many high-level figures came into the picture. Among them were General Sir Hastings Ismay, chief of staff at the Ministry of Defence; Lieutenant General A. E. Nye, vice chief of the Imperial General Staff; Dr. Alwyn D. Crow, controller of Projectile Development, Ministry of Supply; and Professor Charles D. Ellis, scientific adviser to the Army Council. It was Sir Hastings who first advised Churchill of the Chiefs of Staff's concern over German work in the field of rocketry.

That the Germans might be working on rocket weapons did not seem illogical to the Chiefs of Staff, though inexplicably reference was never made to the *VfR* experiments conducted in the late 1920s and early 1930s, nor to the fact that open rocket testing had suddenly and mysteriously ceased after Hitler's takeover. But as the conflict wore on, the strength of Göring's Luftwaffe declined, making it appear ever more likely that substitute offensive weapons would become attractive to German military planners. The point was not overlooked that Germany controlled France and the Low Countries from which London, Bristol, and the south coast ports could be bombarded from fairly short range. A new rocket weapon would be one way of satisfying Hitler's desire for revenge against the British for their growing bomber command attacks on German cities. And, even in the spring of 1943, more than a year before the Normandy invasion, it was not lost on the Germans that a surprise weapon could be useful to thwart any plans the Allies might be hatching to invade the Continent.

If the threat of German secret weapons was indeed real—and there was growing evidence that it was—the Chiefs of Staff decided that a single, high-ranking coordinator should be selected to manage the investigation and to recommend such countermeasures as might be considered necessary once the extent of the danger was finally established. The person proposed was Duncan Sandys, who not only possessed excellent credentials in the rocket field (he had headed an experimental rocket battery in Aberporth, Wales), but had risen from financial secretary of the War Office to the post of joint parliamentary secretary, Ministry of Supply. In itself, his supply position had given him the responsibility, and the overview it made possible, for weapons research and development; but

Sandys enjoyed another unique advantage: he was Winston Churchill's son-in-law and close adviser.

Although a seemingly excellent choice for the crucial new position, Sandys was soon to feel the antipathy of Professor Frederick Alexander Lindemann, Viscount Cherwell, who occupied the post of paymaster-general and scientific adviser to the prime minister. During the course of the long-range rocket investigations, Lord Cherwell refused to accept what gradually came to be irrefutable evidence that a large rocket missile did, in fact, exist, and that it posed a distinct danger to the Allied cause in general and to the southeast portion of the United Kingdom in particular.

Actually, the rocket debate got started when Lindemann (made Viscount Cherwell in 1941) disagreed with Nobel laureate Sir George Thomson, scientific adviser to the British cabinet. Sir George quite correctly argued that if the Germans had set their minds to developing the rocket, there was no technical or scientific reason why they could not achieve success. Cherwell, however, held that the large amount of fuel necessary to propel a rocket-powered missile from the Continent to southeastern England would make the whole scheme unfeasible.

In his biography of Lindemann, *The Prof in Two Worlds*, the Earl of Birkenhead observed that

> Prof [Professor Lindemann/Lord Cherwell] was criticized by Dr. Crow ... and by other scientists for the obstinacy with which he refused to believe in a German rocket attack, and it cannot be denied that he seriously underestimated the German preparations for the V-2 rocket, and advised the Prime Minister that its construction would be so difficult that it was almost inconceivable that the Germans would divert precious man-power and material from the manufacture of bombers which were far more effective than any rocket....

Birkenhead partially defended Lord Cherwell, however, by advising that "It should be remembered in his favour that in his minutes to the Prime Minister he never stated that it was impossible that the Germans were making rockets, or that they could not be made," only that it was not likely that "they would be so foolish." As far as his relationship with Duncan Sandys was concerned, Birkenhead minced no words:

> Given the two protagonists ... it was inevitable that the controversy would become partisan. In the Prof's circle what "they" were doing with "their" rocket referred not to the Germans but to Sandys and his Ministry. At a meeting convened by Prof of such important men as G. P. Thomson he was led into overstating his case by his rivalry with Sandys for the Prime Minister's ear.

It got so bad by 1944 that "nothing could alter the Prof's *idée fixe*, except the arrival of the missile itself."

The Cherwell–Sandys rivalry was of more than human interest; it was important to the entire V-weapons investigation in that it affected the speed with which protective—and counter—measures could be taken. Despite the disdain of Cherwell, evidence accumulated that was to vindicate Sandys'—and Jones'—position. British intelligence took maximum advantage of prisoner-of-war interrogations, and did not hesitate to eavesdrop on conversations as prisoners talked among themselves. Direct and indirect reports from agents on the Continent were combed to determine if a coherent pattern of information could be detected. Journals were studied, radio broadcasts monitored, and coded cyphers intercepted. Moreover, so-called secondary intelligence was developed to determine what light it could shed on the long-range rocket mystery. Such intelligence was concerned not with an on-site observation of a missile taking off, but rather on the effects its development had on, for example, German manpower movements, on the rail system, or on industry. A second-order indicator was the knowledge that, in addition to military personnel, a large number of civilian scientists, engineers, and technicians were employed at the Peenemünde site—whose name, incidentally, was appearing ever more frequently in intelligence briefs.

As the 1943 spring wore on, it became increasingly apparent that something important was going on at the Baltic Sea base and that this "something" was in all probability related to rocketry. Accordingly, orders were passed through the Air Ministry for intensified aerial photographic reconnaissance in an effort to seek confirmation of ground intelligence and to provide such additional information as might be necessary to plan, and later to execute, countermeasures.

Such reconnaissance was a vital adjunct to conventional ground intelligence in unraveling details of the German missile program. As R. V. Jones emphasized to the authors, "Even if [aerial reconnaissance] had only been confirmatory, there was always the point that a photograph was easily the most convincing way of presenting information to senior officers who had to take the operational decisions."

To permit aerial photography to be exploited to the fullest, in mid-1940 the British established the Central Interpretation Unit at the Royal Air Force station at Medmenham. Overflight missions were placed under the operational control of the RAF, while personnel from all services were employed in interpretation and evaluation activities. Following its entry into the war in December 1941, the United States provided support to the CIU. Eventually, five British, five American, and four Canadian squadrons provided aerial reconnaissance for the facility, backed up by some 3,000 ground personnel.

Medmenham's largest section was devoted to investigating the potential threat posed by German long-range secret weapons. In fact, according to Douglas Kendall—who in 1943 coordinated the missile investigations—80 percent of all Allied intelligence information (concerning the threat) came from this source. He told the authors that because the eighty officers involved represented many nationalities, it was decided to direct the effort by a board consisting of a British army colonel, a naval officer, an American colonel, and Kendall himself. "Normally," he explained, "the section commander would report to me directly." (He was chairman and carried the title chief technical officer.) "Eventually, however, the V-2 [at the time designated A4 by the Germans] came to hold such priority that the joint chiefs of staff and the Air Ministry came to take personal charge of the division".

The Peenemünde establishment was first photographed on May 15, 1942, when Flight Lieutenant D. W. Stevenson was on his way to cover Swinemünde following a photographic mission over Kiel. When he reached the island of Usedom, on which Peenemünde is located, an airfield and nearby construction aroused his curiosity. So he turned on his cameras.

Upon return to England, Stevenson's photography was analyzed by the Central Interpretation Unit. Constance Babington Smith, head of the Aircraft Section, recalled the event:

> ... Second Phase interpreters puzzled over some strange, massive ringlike things in the woods near the [Peenemünde] airfield, and they worked out the pinpoint and noted down "heavy constructional work" and then turned their attention to destroyers off Swinemünde. The sortie then went on to the Third Phase sections as usual, for interpretation on different specialised subjects. I remember flipping through the stack of photographs and deciding the scale was too small to make it worth while looking at the aircraft. Then something unusual caught my eye, and I stopped to take a good look at some extraordinary circular embankments. I glanced quickly at the plot to see where it was, and noticed the name Peenemünde. Then I looked at the prints again. "No," I thought to myself, "those don't belong to me. I wonder what on earth they are. Somebody must know all about them I suppose." And I then dismissed the whole thing from my mind.

When the Stevenson sortie photography had finished its round, no one at Medmenham claimed responsibility for the Peenemünde rings so the photos were filed away in the print library. And there they would remain for the rest of the year. (Kendall told the authors that the Peenemünde rings, which he described as "oval hollows with shallow banks around them," were tentatively interpreted as explosive testing zones.)

The various phases of intelligence referred to by Babington Smith need some explanation. First Phase was the immediate reporting of intelligence, such as movements of aircraft, location of ammunition dumps, severity of a just-completed bombing attack, and the like. Second Phase meant reports due out within twenty-four hours of a mission, providing a coordinated look at the day's operations rather than just the immediate or individual postmission report. Finally came Third Phase reporting, which was aimed at specialists interested in certain aspects of photoreconnaissance, such as ports, airfields, industrial plants, and secret-weapon installations.

By the end of 1942, the latter was assuming greatly increased importance. And the name Peenemünde was cropping up with ever greater frequency in relation to the rumored German long-range rocket program. Kendall told the authors that serious investigations of the rumors got under way at CIU when a

> small group came down from the MI [Military Intelligence] Department of the War Office to state that the Germans were developing long-range rockets to fire on Britain. Little information was available, but we were advised that they were so heavy that they had to be carried by rail, and it could be expected that some rails would be erected to guide the weapon [during the initial launch phase]. Subsequently, some rumors reached us that the Peenemünde area on the Baltic coast had been closed to the public, and that some sort of military activity of a secret nature was taking place there.

This led Medmenham to strengthen its team dedicated to the long-range missile program. In part, this was accomplished by assigning a group of army interpreters under Major Norman Falcon. Their job, according to Falcon, was to investigate reports that the Germans were planning "some form of long-range projectiles capable of firing on this country from the French coast."

As the army group reviewed Stevenson's observations of heavy construction work at Peenemünde together with later photo-cover carrying such CIU notations as "huge elliptical embankments" and three circular earth banks that were "not unlike empty reservoirs," they failed to find anything that brought to mind rocket-powered missile projectors. In fact, as Babington Smith wryly observed, "Nothing seemed to tally with anything."

Soon after Duncan Sandys' appointment, Group Captain Peter Stewart, Medmenham's station commander, was instructed by the Air

Ministry to give highest priority to the secret weapons program investigation. Stewart made Wing Commander Hugh Hamshaw Thomas chief of the investigation, and ordered Falcon and his army interpreters to look for possible launch sites on the French coast, and Flight Lieutenant André Kenny and his RAF associates to observe activities at Peenemünde itself. At the same time—April 1943—the RAF station at Benson and the satellite American base at Mount Farm were put on the alert to undertake aerial reconnaissance of the French coastal area from the Belgian border in the north to Cherbourg in the southwest. "No one really knew what they were looking for," wrote Babington Smith after the war in her *Evidence in Camera*, though the Air Ministry had suggested long-range guns; pilotless, rocket-powered aircraft; and even "some sort of tube located in a disused mine out of which a rocket could be squirted" (in the words of the Air Ministry).

In a meeting with Duncan Sandys and his advisers in London near the end of April, Kenny pointed out the existence of a large power station at Kolpin with its network of distribution lines—the kind of installation an important development and/or productions works would require. Attention then focused on the earthworks at Peenemünde, which Kenny believed might be rocket test stands. Before the meeting was over, Sandys had come to the same conclusion. On May 9, during the course of a visit at Medmenham, he asked Babington Smith if she had ever come across any unidentified aircraft at Peenemünde. Although her response was negative, she henceforth kept an extremely careful watch on the airfield.

Late spring weather over the Baltic seacoast can be very pleasant, and so it was in 1943. British intelligence took maximum advantage of it, flying many photoreconnaissance missions during the months of May and June. In May, Kenny had spotted a road vehicle that seemed to be carrying a cylindrical object estimated to be 38 feet high and 8 feet wide, almost certainly an A4 rocket. Moreover, a captured German general had talked about rockets being fired from the nearby Greifswalder Island toward the Baltic island of Bornholm, providing further confirmation of what Peenemünde was doing.

Sortie N/853 taken on June 12, 1943, seemed to resolve matters as far as the A4 was concerned. Medmenham had immediate access to the photographic cover, but somehow failed to capitalize on it. Six days later, on the 18th, R. V. Jones received the cover and, as he perused the photographs, spotted on a railway car something that seemed to be a "whitish cylinder approximately 35 feet long and at least 5 feet in diameter. It had a 'bluntish' nose at one end and fins at the other. A rocket!

"I experienced," he later would reminisce in his *Most Secret War*, "the kind of pulse of elation that you get when after hours of casting you realize that a salmon has taken your line—especially when someone else has had an exhaustive first chance at the pool." He immediately reported his find to Cherwell, who advised him to inform Sandys without delay. This he did the following morning. Four days later, two rockets showed up clearly on another sortie, flown by Flight Sergeant E. P. H. Peek.

Following the discovery, Babington Smith was instructed to review thoroughly all Peenemünde photo cover taken to date, and to be on the lookout for anything "queer" or different. Soon after embarking on her task, she spotted four tailless aircraft that "looked queer enough to satisfy anybody." Dubbed "Peenemünde 30" by CIU, they were in fact experimental Messerschmitt Me163 rocket-powered fighters. Armed with this discovery, she once again went over earlier, more blurred cover and, sure enough, more of the little planes turned up in hazy outline. As she later reminisced, "Each new find was likely to throw new light on earlier photographs which had meant nothing when they were first examined.... 'First photographed' and 'first seen' often did not refer to the same cover at all."

The appearance of the Me163s further complicated an already puzzling situation. Some CIU interpreters thought the whole matter of German secret weapons was a hoax, and a few went so far as to suggest that Peenemünde was a decoy and the rockets merely dummies. Others suspected that the little planes were unmanned missiles of some sort powered either by rocket or by jet engines. Still others accepted both missiles and manned aircraft at face value.

As far as Peenemünde's being a hoax was concerned, R. V. Jones was very skeptical. For one thing, he pointed to the many signs of activity there revealed by aerial reconnaissance. Also, agents insinuated into the army of foreign workers at the installation further substantiated its importance. And, of course, there was Peek's photo of two large rockets, which to Jones "almost clinched the evidence."

> But since some of our own experts had hitherto thought that such a large rocket was impractical [Jones told the authors], they argued that it was a hoax to distract our attention from more important developments. Now if it were a hoax and it succeeded, we should probably be led to bomb Peenemünde; the Germans would presumably only tempt us to do this if Peenemünde were not a genuine, serious experimental station. I finally managed to show that this was extremely unlikely from an apparently insignificant piece of evidence gathered in quite another field. This was a circular to various German Air Force experimental stations, signed by a petty clerk in the German Air Ministry, giving revised instructions for applying for petrol coupons. Now all the experimental stations were on the list of addres-

> sees, apparently in order of importance, and Peenemünde was shown on the list above some other stations of whose importance we were certain. The clerk, who could hardly have known that his little circular would come into our hands, was in fact an unconscious witness to the importance of Peenemünde.

However important Peenemünde would ultimately prove to be as a secret weapons research and development center, it seemed clear to the Allies that if such weapons were in the offing they would have to be launched from somewhere. And "somewhere" was presumably the coastal region of northern France and Belgium, and possibly even Holland. This supposition led to intense photographic reconnaissance of the cross-Channel region beginning in April 1943 and carried on throughout the summer and autumn of 1943.

CIU interpreters did not wait long for results. As early as May, photoreconnaissance showed that a large clearing was being made in the Bois d'Eperlecques a mile west of the northern French coastal town of Watten. That in itself did not cause undue concern, but when an undercover agent reported in early July that something strange was going on there, new photographic cover was ordered. What came to light was a massive concrete structure serviced by rail lines. It did not take CIU long to surmise that such a structure might be a supply, staging, and/or firing bunker for large rockets. Subsequently, during the course of the spring and summer of 1943 further "heavy site" activity became evident at such places as Wizernes, Lottingham, and Siracourt between Calais and the river Somme; at Mimoyecques near Cap Gris Nez; and at Sottevast and Martinvast not far from Cherbourg. In some cases railway spurs had been or were being installed at the sites, while in others it was evident that excavations and tunneling were occupying the German construction crews.

The discovery of these sites added further to Medmenham's perplexity. For one thing, opinion there was by no means unanimous as to whether there was a single secret weapon associated with the sites or if there were two. Also, agreement could not be reached as to the nature of the weapon nor how close to deployment it might be. Complicating matters further was the lack of information that could lead to reliable estimates as to the range, accuracy, and warhead power.

Duncan Sandys and his small investigative staff under Colonel Kenneth Post were by now thoroughly convinced of the existence of a rocket, code-named Bodyline, which was believed to be in an intermediate development stage. As expected, Lord Cherwell refused to believe in any kind of large rocket threat, speculating that a winged, pilotless aircraft was much more likely to be the vaunted secret weapon. He talked of an "air mine with wings," which he suggested might be close to operational readiness. Thus, his recommendation that the strange sites being discovered in northern France be dealt with rapidly and effectively came as no surprise.

To Sandys, it was no longer a matter of deciding whether the German secret weapon threat was a rocket *or* a pilotless aircraft. In a report dated May 17, 1943, he formally proposed that *both* weapons were in development, though he was reluctant to estimate which was the more advanced or when either would be placed into operational service. Further information was clearly needed from aerial cover, from agents on the Continent, from prisoner-of-war interrogations, and from other sources.

The secret-weapon investigations were complicated by a series of often conflicting agents' accounts of large objects they believed to be either rockets or the cylindrical shapes from which they were fired coupled with POW stories of powerful new propellants. Cherwell, of course, disputed the large rocket claims altogether while Alwyn Crow and his deputy William Cook tended to underestimate progress that the enemy might have made in liquid fuel and oxidizer technology. This was due largely to the fact that both Crow and Cook were heavily engaged in solid rocket propellant (cordite) developments and had had little or no experience with liquids. Yet, at that very time a Shell International Petroleum Company group under the direction of Isaac Lubbock was conducting successful experiments with gasoline and liquid-oxygen rocket motors at its Langhurst experimental station. Although this work, which was funded by the Ministry of Supply, had been witnessed by Crow, Cook, and others on May 7, 1943, it seems not to have colored their opinions as to the efficiency of liquid propellants and hence the probability that missiles powered by them would have been developed by the Germans. Prewar German—and American—liquid-propellant rocket research and testing were ignored completely.

In part because of their belief that the Germans were using solid propellants and in part because of exaggerated or simply incorrect agents' reports and POW interrogations, Crow and Cook conjectured that if a large rocket existed it would likely weigh up to 100 tons and possess a warhead of at least 8 tons. If such an enormous rocket were being developed, the havoc it might rain on England would be grave indeed—so grave that the Ministry of Home Security felt obliged to draw up plans for the possible evacuation of large numbers of Londoners. Indeed, one estimate warned that four thousand casualties might result from a single rocket impact on the British capital. Jones was much more conservative in his estimates of the German rocket, preferring a figure of between 20 and 40 tons.

Liquid- or solid-propelled, long-range rocket or pilotless aircraft (or both), immediate or not-so-immediate threat, the British and their American allies decided that there were enough indicators to justify taking offensive action against both the Peenemünde establishment and Watten, the most advanced of the heavy sites near the French coast. On June 11, Sandys advised the Air Staff that

> the latest reconnaissance photographs provide evidence that the Germans are pressing on as quickly as possible with the development of the long-range rocket . . . and that frequent firings are taking place. There are also signs that the light anti-aircraft defences at Peenemünde are being further strengthened.
>
> In these circumstances, it is desirable that the projected bombing attack upon this establishment should be proceeded with as soon as possible.

The final decision came about largely as a result of a major War Cabinet Defence Committee (Operations) meeting in late June 1943, at which Cherwell, Sandys, Jones, and others presented their positions. Peenemünde would be attacked first, followed by Watten—it was desired to hold off on the latter so that further ground and air intelligence might establish just what its purpose might be.

During the month of July and into August, reports continued to pour in from agents on the Continent; one suggested that the warhead of the long-range rocket might be atomic in nature and another that liquid-air bombs and phosphorus would soon be dropped on London. Buildup at Watten continued to the point that it was even suggested to send in a force of paratroopers to capture, examine, and then destroy it. Irving quotes Cherwell's reaction to that particular scheme: "No doubt before sacrificing one hundred and fifty highly trained men, the Chiefs of Staff will assure themselves that the evidence connecting these particular sites with the putative L.R.R. [long-range rocket] is consonant." He ended his memorandum, directed to General Ismay, with the admission that he was "biased by the fact that I do not believe in the rocket's existence."

Meanwhile, Peenemünde was subjected to infrequent but searching air reconnaissance—only a few sorties were authorized as the 1943 summer wore on so as not unduly to alert the Germans of British interest in the installation. It soon became apparent that antiaircraft defenses were being increased and that smoke generators were being installed, evidence of preparations against possible air attack. In the light of the crushing British raid on Hamburg on July 24, 1943, the Germans were fully aware of what might be in store for their prize Baltic research and development center.

And so it occurred. On the night of August 17–18, 1943, under a full moon, nearly six hundred British bombers unleashed by Air Chief Marshal Sir Arthur Harris' Bomber Command hit Peenemünde. Then, ten days later, the US Eighth Air Force raided Watten just when freshly poured concrete was beginning to harden. (The details of these and other Allied raids on secret-weapon-related sites are given in Chapter 7.)

Shortly after the August bombing raids, intelligence activities were shifted somewhat at the Central Interpretation Unit. Hamshaw Thomas turned over the unit's entire secret weapons investigations to Douglas Kendall, who was by then a wing commander. At the same time, recently promoted Lieutenant Colonel Norman L. Falcon and his army interpreters were directed to expand their efforts to identify secret-weapon-related construction activities along the northern French coast. During the late summer and autumn they observed continuous construction work at the heavy sites, but even so it was far from clear what the Germans were up to there. In the light of this, a whole new cover of coastal France was ordered in late October. Perhaps it could make some sort of sense out of what was going on.

Medmenham was in for some surprises. Out of the mountain of photographs secured from a hundred or so photoreconnaissance sorties, CIU's Captain Robert Rowell and Captain Neil Simon were startled to find something quite different from what they had seen before. Following up ground intelligence leads, a number of areas were examined on November 3 where the Germans were clearly engaged at some sort of construction. But what did it all mean? The two interpreters could only guess. They noticed that construction was standardized at the sites and that of the nine buildings being set up at each, three definitely resembled, of all things, *skis*. Rowell and Simon also noted rows of six pairs of 2-foot-high concrete studs set 20 feet apart. At the location where work was most advanced (a wooded spot known as the Bois Carré near Yvrench) two identical ski-shaped buildings were found, while the third was a little shorter. Careful analysis of these and other buildings brought forth little clarification as to purpose. "These new sites might be for launching projectiles of some sort," wrote Babington Smith, "but they bore no relation to anything else that had been found so far." The mystery along the French coast thus deepened.

Whatever the purpose of the ski sites, they did not appear to have anything to do with large, vertically launched rockets. To make some sort of sense out of a confused situation, Churchill appointed his minister of aircraft production, Sir Stafford Cripps, to conduct an independent assessment of the secret weapons threat. Shortly after, on November 8, Duncan Sandys, Douglas Kendall, Neil Simon, André Kenny, and others presented their findings to the specially assembled group of military and civilian experts that made up the Cripps committee.

"The meeting called by Sir Stafford Cripps was interesting," recalled Kendall, "in that it produced rather startling results." After taking into account the available evidence, it was concluded that a rocket-powered secret weapon did indeed exist. But German propaganda efforts seemed to indicate something more. (For some time, the Goebbels machine had been leading the public to believe that new super weapons were on the way.) Toward the end of the meeting, Sir Stafford asked Kendall if he had any evidence of activities other than the rocket. "It so happened," Kendall later told the authors,

> that the previous day we had had our aircraft out, taking photographs of an area known as Pas de Calais.... We had found on these photographs sites which were under construction, and I told Sir Stafford that they might be connected with [secret] weapon activities because they did not correspond to any former installation which we had seen. On the other hand, they did not fall in with the intelligence briefing which had been given to us by the War Ministry. Each of these sites was in the woods, connected only by second-class roads. There were no rails anywhere near them. They did not, in other words, look suitable for heavy rockets. I explained that I had been up all night studying the photography, but that we had come to no conclusions as to what they [the sites] were. Sir Stafford asked, therefore, how long it would take to form some opinions and I asked, in turn, for the next forty-eight-hour period.

Within that short period of time, two Canadian aircraft were dispatched over the northeast French coast on a low-altitude (500 feet) recon mission. An examination of the cover returned raised to twenty-six the number of "ski" sites discovered. Although still evidently under construction, each site was similar, with three ski-shaped structures thought to be for storage, a square hangarlike building oriented in the general direction of London, and what seemed to be a sloping ramp parallel to the hangar and also pointing toward the capital. Medmenham thus decided that the sites were designed to handle ramp-launched pilotless aircraft rather than vertically launched rockets. Furthermore, it was believed that these aircraft had a wing span of nearly 20 feet, a couple of feet less than the estimated width of the hangar doors. Also, it was suspected that the craft were fitted with automatic pilots, which would account for the direction of both the hangar and the ramp. (The auto pilot was presumably "set" in the hangar prior to the craft's being placed on the ramp.)

Medmenham's conclusions regarding winged pilotless aircraft were based on more than aerial photography of the ski sites. During the summer and autumn of 1943, agents' reports coming in from the Continent referred increasingly to them. And, of course, Lord Cherwell had long talked of such weapons and his opinion carried considerable weight. Even more important was Air Chief Marshal Sir Charles Portal's revelation in late August that on the twenty-second of that month a pilotless aircraft had been seen flying north of Peenemünde. Moreover, this same craft had later crashed on Bornholm Island and been photographed by Danish agents. To clinch the association of the Baltic Sea installation with the pilotless aircraft on the one hand and the French coast ramps on the other, a photorecon flight undertaken in late November turned up one of the little craft sitting on its launch ramp not far from the Peenemünde airfield.

By November 1943, then, the Allies were virtually certain that one of the long-range secret weapons was a pilotless aircraft and that preparations were actively under way to deploy it against targets in the United Kingdom. Moreover, few disputed that a large rocket-powered missile was also being readied, one that Duncan Sandys suspected might weigh some 60 tons. He estimated that if the missile's range was 200 miles, the warhead could weigh in the neighborhood of 12 tons. For a shorter 130-mile range, a figure of up to 20 tons for the warhead was suggested.

On November 18, Sandys' responsibilities officially passed to the Air Staff with Air Marshal Norman H. Bottomley, its deputy chief, named as coordinator. The time had come, the Chiefs of Staff felt, when strong countermeasures should be taken against the secret weapons threat and coordination through the Air Ministry, would be the most effective way to make this possible. Moreover, it was reasoned that Sandys had successfully completed his original assignment. The existence of the pilotless aircraft and long-range rocket had been established and at least something about their nature and the sites from which they were likely to be launched was known. Nevertheless, Sandys and others involved in the investigation were to be on hand for consultation whenever the Chiefs of Staff might desire.

CHAPTER 7

Attacking the Missile Installations

Three months earlier the first serious blow against Hitler's vaunted wonder weapon program was carried out by the massive Royal Air Force raid on Peenemünde. The event was notable not only for being directed at a top-secret research and development facility, but because it initiated the introduction of a new technique of bombing: control by radio telephone link.

Since spring, the Baltic Sea establishment had created great consternation among British intelligence and military experts. However, all realized that it could not be safely bombed until the short summer nights had begun to lengthen. Detailed planning began on July 7 at the headquarters of the RAF Bomber Command's commander-in-chief Air Chief Marshal Sir Arthur Harris and continued to the middle of August. The results of the great Battle of the Ruhr were assessed and the tactics studied of the three devastating fire raids on Hamburg (July 24–25 and 27–28 followed by one on August 2–3). Planners were elated at the confusion caused by the newly introduced "window," metallic strips of paper dropped from planes to degrade the effectiveness of both the Würzburg ground radar systems used to control German night fighters and antiaircraft artillery. In addition, the same strips often disturbed airborne radar systems carried by many night fighters.

The raids on Hamburg followed up by three on Berlin proved beyond doubt the capability of massed bomber formations to penetrate deep into Germany. These great cities, like many in the Ruhr, were spread across large areas and thus made relatively "easy" targets. Such would not be the case with Peenemünde, which, as targets go, was small. True, the RAF had already hit other small targets, but usually during the day, in raids involving tens rather than hundreds of aircraft. Peenemünde, however, would be a notable exception to this practice.

So distant was Peenemünde from Britain that the bomber force could not be guided to the target by ground-based radar. At the same time, the H_2S airborne radars of the period could not always be relied upon to identify and mark targets. These technological limitations led planners to choose a moonlit night for the attack, though even then it would not be easy for the pilots to discern all the features of the target area.

Harris and his staff decided that a pathfinder force of bombers should be sent ahead to locate and then mark with flares several preselected aiming points. This done, the master bomber of the force would come in low to evaluate the position of the target indicators. Using a newly devised radio-telephone link technique, he would then send back instructions to the main bomber force. Taking advantage of this "on-the-spot" information, the main force would execute the attack.

Such an innovative tactic had several drawbacks. For one thing, the master bomber could be shot down and the force circling above would then not know precisely how to proceed. Moreover, the delay introduced while the master bomber made his investigatory run might provide just the time needed for German night fighters to arrive and break up the bomber formation. And, of course, the radio link might malfunction or be jammed. But these were chances that Harris felt were well worth taking.

The night of August 17–18, 1943, was selected for the raid upon which so much depended. The weather had to be at least relatively clear, for in the event that the Germans generated heavy smoke over Peenemünde, time and distance runs from Rügen Island to the north would have to be made. (If the island were shrouded by mist or clouds, it would obviously be useless as a time and distance marker to the target.) Moreover, only beginning in mid-August were the nights sufficiently long to permit a raid of this penetration and complexity to take place.

As a security measure, the crews were not informed of the true nature of the mission lest information leak out that the British government was unduly concerned over the secret weapons threat. Instead, briefing officers explained that Bomber Command was going after a facility engaged in the development of specialized radar gear destined for German night fighters. This in itself provided a special incentive for the approximately four thousand men who would be involved in the raid.

The attack on Peenemünde, code-named Operation Hydra, was preceded by a small force of eight Mosquito light bombers dispatched on a diversionary raid to Berlin (Operation Whitebait). The Hydra force, meanwhile, consisted of nearly six hundred Lancaster, Halifax, and Stirling bombers.

An hour and a quarter after the Mosquitos had departed, the pathfinder force under Master Bomber and Group Captain John Searby winged its way toward the Baltic coast to mark three aiming points within the main target area. The first of these, aiming point *F*, was Peenemünde's *Siedlung*, or housing estate, where top scientific, engineering, and administrative personnel were believed to be living. R. V. Jones informed the authors that it had been "... our intention to wreck the establishment itself [but that] the aiming point was moved nearer the housing estate at a very late stage in the raid planning, and not with our [e.g., Air Ministry intelligence] approval." Duncan Sandys personally convinced Harris that the housing area should be included as a subtarget for the raiders.

The second aiming point, designated *E* on target maps, was to attract the second wave of bombers. It was centered on Peenemünde's workshop and preproduction area. The third bomber wave would hit aiming point *B*, the so-called *development* works, where various laboratories and offices were located.

Upon approaching Peenemünde, Searby and his pathfinders dropped red marker flares along the north shore of Rügen Island. As a result of the fact that the island did not show up clearly on H_2S radar and small clouds made visibility difficult, some of the markers were dropped inadvertently over the northern edge of Usedom Island. This error later caused nearly a third of the main bomber force to bomb the Trassenheide labor camp instead of the *Siedlung*. What the subsequent history of rocketry and space travel would have been if this error had not been made is a matter of conjecture.

As a direct result of the initial error made with the red markers, white parachute flares and red target indicators were subsequently released over a mile south of aiming point *F*. Recognizing what had occurred, Searby ordered yellow target indicators to be placed nearer the true aiming point. Just before 00.15 hours on August 18, green confirmatory indicators were dropped; and, at 00.17, Searby ordered the attack over radio telephone. More than two-thirds of the bomber force were then able to aim at the correct green indicators rather than at the red flares farther south. Within two minutes, the first wave of bombers (attacking at 8,000 feet) had completed its work.

As phase I ended, the Luftwaffe night defense fighters had still not shown up, having been diverted south to Berlin to defend what Colonel General Hubert Weise's Home Defense forces believed to be the real target. All the while, Peenemünde's flak was relatively light and searchlights few in number.

As the second wave of bombers moved in to hit the workshop area,

Searby made maximum use of his radio telephone link to advise on which target markers to accept and which to ignore. As a result, the second wave's mission was efficiently and accurately carried out. The situation for the third wave was less favorable, for not only were many target indicators incorrectly placed, resulting in overruns, but minutes before the bomb run began the Luftwaffe finally appeared.

During the course of the subsequent air battle, forty-one British bombers were shot down by fighters of Gruppe II and Gruppe III of NJG 1 and Gruppe II of NJG 5.

As the bomber fleet approached Peenemünde, General Dornberger sat in the Hearth Room of the officers' club with Wernher and Magnus von Braun, Ernst Steinhoff, Werner Gengelbach, missile launch expert Kurt Debus, and Hanna Reitsch, the famed woman pilot who had learned gliding with Wernher a decade earlier. There was no need for a fire in the Hearth Room on that warm August night; and the conversation had turned to the subject of flying, a topic of common interest to the civilians but not to Dornberger. (The two von Brauns and Steinhoff were rated pilots.) At about 23.00 hours, Dornberger excused himself and left for the nearby guest house in which he was staying. He was tired after a long day's hunting. The others followed shortly.

Von Braun later recalled to the authors the events that transpired:

> When the party broke up, I escorted Hanna Reitsch to her car that was to take her to her quarters at Peenemünde West. [She was to make a test flight in an Me163 rocket-powered airplane there on the following day.]I then retired to my room in *Haus 30* [the dormitory in which von Braun and many of the center's bachelors lived] and went to bed. I had just fallen asleep when the sirens sounded. Upon hearing aircraft noises, I hastily dressed, went outside, and walked over to the air raid alert and communications center (500 meters from *Haus 30*) to get a status report from the officer in charge. Waves of bombers coming from England were reported approaching across Denmark and Schleswig-Holstein. As this was a frequently traveled "minimum flak exposure route" to Berlin, the consensus in the communications center was that this was just another of those rather frequent raids on Berlin.
>
> Soon we heard on the radio that German night-fighter wings had been dispatched to the Berlin area to intercept the raiders, which seemed to confirm this assessment. Confident that this was not "our night," I strolled back to Haus 30 with the idea to retire again.
>
> As I walked I noticed that the artificial fog system enshrouding the Peenemünde facilities had been activated. Through the thin fog shone the pale reddish disc of a full moon. Suddenly I saw a flare

> lighting up through the fog, and within a minute the sky was covered with what we called "Christmas trees."

Similarly, Dornberger had returned to his quarters and turned in for the night. He was jolted from his bed by the thumping of guns. At first he assumed that it was the test-firing of a tank-mounted antiaircraft rocket, the trial of which he had authorized earlier. Fully awake, he realized the noise was coming from the antiaircraft guns at Peenemünde West, Lake Kölpin, and farther to the south at Karlshagen. Almost immediately, he heard the staccato of the 20-millimeter guns on the tops of the taller buildings around him.

Von Braun, gazing into the swirling mists manufactured by the smoke generators stationed along the periphery of the huge installation, heard the roar of the pathfinder force that was directing the target-marking operation.

While Dornberger was struggling into his tunic and breeches and hunting frantically for his boots, von Braun decided to head for cover. "When the first bombs fell, I entered the concrete blockhouse near Haus 30, where some fifty or a hundred people had already come for shelter," he later recalled.

Dornberger, remembering that his orderly had taken the boots away to polish them, put on a pair of bedroom slippers and began walking gingerly through the broken glass and debris towards the air raid shelter. Once inside, a near-miss ripped off the thick steel door, but no one was injured. The report from the chief air raid warden was ominous:

> Soon after midnight, the first wave flew south over Peenemünde without dropping any bombs, headed for Berlin. The AA [antiaircraft] transmitter gave that as the probable target. About twelve fifteen, wave after wave arrived over Peenemünde, coming from Rügen. The first bombs fell into the [river] Peene in front of the harbor. No damage reported to the test stands so far. As far as I've been able to see, the Measurement House is on fire, the assembly workshop blazing, and the component and repair shops starting to burn. I haven't managed to get farther south yet. The lines to the preproduction works and the settlement went dead after the first report. . . . The power station on the Peene is supposed to have been bombed, too. . . . The scrap dump behind the tool and fixture shop is burning.

By then it was 00.35 hours, and Dornberger despatched von Braun with all the men in the shelter to try and confine to the top floor the fire in Building 4.

Later, one of von Braun's secretaries, Hannelore Bannasch, recorded in her diary what then occurred:

> Haus 4 [where von Braun had his office] burns brightly. Haus 5 [a dormitory for men] is in flames. Everywhere I look it's fire. Fire, everywhere fire—horrible beauty! We run past a building completely burnt. Here and there it still crackles. Smoldering beams fall, a gable suddenly crashes down. Where is my building? I see nothing but flames. Even the shrubbery is on fire. We are in the midst of it. My hair singes.
>
> I am frightened back, I cannot go through a pool of blood in front of me. In it lies a leg torn from a body still in a military trouser leg and boot.
>
> There is my professor [i.e. von Braun]! We must save the secret documents!
>
> But the building is burning. The roof is already down. The gable is ready to fall at any moment. The second story is gone. Can we risk the stairs? The professor takes my hand, quickly and cautiously. There is a roaring, crashing, and crackling.
>
> Groping along the wall we reach the second floor: one, two, three, four, five—yes, this was our door. We must go in there! There are no more doors here now. Fire is to the left and in front of us. We press ourselves tightly against the still standing wall. The other half of the room has collapsed.
>
> We are now in the room. Out with the safe and out with the other furnishings! I run up and down the stairs with the secret papers as long as I am able. The professor is still upstairs with a few men. They throw things that are still usable out of the window. Down below, I gather things together, put them into the upturned safe.
>
> The heat is dreadful. A soldier comes and stands by the safe with his rifle loaded. Slowly, dawn breaks. I return to the air raid shelter. The secret papers are secure.

Dornberger left the shelter and rushed to the Measurement House, the center of Peenemünde's vital guidance, control, and telemetry work and clearly one of the most important structures for the continuation of his rocket development program. As he did so, the bombers passed over repeatedly. With two men assisting him, Dornberger managed to save the building by using fire extinguishers located on each floor.

After stopping to put out an incipient fire in the diesel locomotive shop, he returned to the guest house to collect his belongings. On the way, he met Frau Zanssen, wife of Colonel Leo Zanssen, the former military commander at Peenemünde, and her five-year-old son Gerhard, who greeted Dornberger with laughter and, "I say, Uncle Sepi, what a lovely fire!"

Another employee who actively assisted von Braun and Dornberger during the raid was Hermann Oberth, who had arrived at Peenemünde in 1941. He was awarded the *Kriegsverdienstkreuz I Klasse mit Schwertern* (War Merit Cross 1st Class, with Swords) several weeks after the raid.

Dornberger later said of Oberth's medal, "A decoration for Professor Oberth on account of his . . . outstanding, courageous behavior was due. During the attack, he organized groups to salvage important equipment and files out of burning offices, buildings, and plants. Oberth was always one of the first, one of the most active during and after the raid." Shortly afterward, Oberth requested that he be transferred. Although he had spent two years at Peenemünde, he had played no major role in the development of rocketry there. Technology had simply moved far beyond his early theoretical studies.

Shortly before the alert sounded, Friedrich Dohm and a number of air raid wardens had just finished a meeting to discuss a recent order to tighten up precautions should a raid take place. (Dohm, an electrical engineer who helped develop radars used in the A4 program, rejoined von Braun in Texas after the war. Subsequently, he worked with the German-led team in Huntsville, Alabama, where, in the 1960s, at NASA's Marshall Space Flight Center, he became involved with the design of the astronomical telescope mount element of the Skylab space station.) Incredibly, detailed planning for the dispersion of Peenemünde personnel had begun but a few months before the attack. It included a survey of nearby villages to determine how many persons could be accommodated in them.

Following the survey, villagers were asked voluntarily to share their homes with personnel that might have to be displaced from the Peenemünde center. Not unexpectedly, many objections arose. In fact, only a quarter of those approached indicated their willingness to house center personnel. The result was that on August 15, only two days before the British air raid, the remaining three-quarters received military orders to open their homes in case of emergency. It was not surprising, then, that most center families were still in Peenemünde's housing estate on the night of the raid.

When Dohm returned to the settlement after the air raid warden meeting, he briefly assumed sector warden duties and then went home to prepare for bed. As did Dornberger, Dohm felt that the sirens were merely signaling the passage of British bombers en route to Berlin or Stettin. However, just in case, he checked the readiness of the emergency kit that everyone in the settlement kept close by.

Minutes later, he heard the sound of a single plane flying overhead, the first of the pathfinders dropping flares. Realizing that Peenemünde was under attack, he hustled his wife and baby into his assigned bunker (one for every two families). Scarcely had they taken their places when the shelter's door was blown open, tearing the buttons off Dohm's jacket and dumping the baby onto the floor. Incredibly, no one was injured.

With the first explosions, Dr. Rudolf Hermann, director of the supersonic wind tunnel, hurried his wife and children into their shelter. He then returned to his home, which had been hit by incendiaries and was beginning to burn. Rapidly he made a mental list of priorities for removing things from the house. First would be food, then clothing, followed by furniture, and finally, alas, his books. As he busied himself in saving what he could, his neighbor and assistant at the wind tunnel, Dr. Hermann Kurzweg, pedaled up on his bicycle. He briefly told Hermann as much as he knew and then said, "They destroyed a lot, but they couldn't destroy our minds."

As Hermann's wife sat in the shelter, a man's figure loomed for a moment in the darkness, silhouetted in the door frame by the burning house behind him. In the gloom of the shelter, Frau Hermann could see clearly the burning particles of phosphorus sizzling and popping in the torn flesh of his hands dangling helplessly in front of him. Then he disappeared, leaving behind a faint aroma that, in a nauseating way, reminded Frau Hermann of her kitchen now in flames a few meters away.

Some of the bombs that leveled the housing settlement fell near the barracks of the *Versuchskommando Nord* at Karlshagen, although only one hit.

Literally blasted from their beds, Lance Corporals Ernst Stuhlinger and Hans Maus ran quickly to the *Splittergraben* or slit-trenches behind the barracks. For Stuhlinger, it was nothing new. As an infantryman on the eastern front, he had accepted air raids as a more or less usual way of military life. However, it had been over a year since he had experienced one; and it took him some time to get readjusted to what had once been a familiar occurrence. Not far from him, Maus preferred to lie on his back in the trench, watching the Christmas trees cascade to the ground. (Maus joined the von Braun team in Fort Bliss, Texas, in 1945 and later transferred to Huntsville, Alabama, where he served as director of the Fabrication and Engineering Laboratory. He later became chief of the Executive Staff of NASA-Marshall, retiring in 1971.)

Crouched in a trench close by was fellow soldier Lance Corporal Hans Hosenthien, who was almost buried under debris when a bomb fell only five meters away. It took him several minutes to dig himself out with his hands, after which he helped free others who had been similarly buried. At Peenemünde he worked with Helmut Gröttrup in designing antennas for the A4 ground radars. Two decades later, he was chief of the Flight Dynamics Branch, Guidance and Control Laboratory at NASA-Marshall.

As soon as the last bombs had fallen, some of the *VKN* soldiers left their trenches and went to the civilian housing area about a kilometer away to assist in rescuing personnel and removing bodies trapped in bunkers or the wreckage of houses. Others hastened to Peenemünde to assist in firefighting and damage control.

Fräulein Ingeborg Kame had worked late into the night of August 17—she was secretary to Walter Riedel, head of the Technical Design Office—and was concerned about the five-kilometer bicycle ride she was about to take to the old Hotel Karlshagen, now the dormitory of the so-called *Kriegshilfe* girls. She was particularly afraid of passing the foreign laborers' camp at Trassenheide. Her fiancé, Konrad Dannenberg, Riedel's Propulsion Section chief, accompanied her as far as the center's gate and then returned to *Haus 30*. Ingeborg took a deep breath and sped off on her bike, passing Camp Trassenheide without glancing left or right.

Having arrived at the Hotel Karlshagen, she was about to turn in when the sirens began to scream. As she wondered whether it was another sham raid or the real thing, a bomb suddenly detonated in the garden below, collapsing part of a wall of the hotel. Her inner debate was over. With half a dozen girls, she rushed into the dining room as bombs crashed and incendiaries popped, sputtered, and harshly lit up the garden like klieg lights on a motion picture set. The girls huddled together in confusion and terror until a bomb blew in the wall between the kitchen and the dining room and incendiaries burst in the kitchen, showering incandescent particles of burning phosphorus into the room. Ingeborg groped her way for the door that led into the garden and then blacked out completely.

When she regained consciousness a few minutes later, she found herself half-buried in rubble that had fallen from the ceiling. With the help of several other girls, she freed herself, and together they ran for the beach, only 75 meters away. Her hair was smoldering from several tiny phosphorus particles caught in it, and some of the other girls were screaming in pain as similar particles burned into their flesh. Not knowing that water does not extinguish burning phosphorus, they all reacted instinctively by running toward the Baltic. Once they reached the strand, they found the bodies of a dozen or so of their friends floating in the water under a pier. They had made it into the water only to be killed by the strafing planes that had already dropped their bombs.

Ingeborg made her way to Zinnowitz by early morning, where she sought out the Helmut Zoikes, who would otherwise have been celebrating their wedding anniversary. She was frantic with worry about Dannenberg until she met Dr. Helmut Schmidt, his roommate at *Haus 30*. Putting her fears quickly aside, Schmidt said that her fiancé had slept through the attack in the air raid shelter adjacent to his dormitory. On the following day, Dannenberg spent a grim and grisly time searching for Ingeborg among the corpses of the *Kriegshilfe* girls before learning that she was still alive. Despite her wounds, Ingeborg reported for work on August 20 and helped organize the administrative support for the still burning Peenemünde center. For her dedication, she was awarded the *Kriegsverdienstkreuz II Klasse mit Schwertern* (War Merit Cross, second Class, with Swords). Another of von Braun's secretaries, Fräulein Dorette Kersten, also received the medal "for setting an example for others" during the same period.

Not all members of the Peenemünde installation were caught in the raid. Thus Sergeant Heinrich Schulze, an A4 troop training instructor, had advised his wife that he would be taking the last train from the center to his home in Koserow. A last-minute decision, it was responsible for his missing the raid. Another who escaped was Dieter Huzel, who had been on a scrounging expedition. With electrical components such as motors, switches, transformers, and power distribution panels in critically short supply, he had taken a team of men to the island of Rügen. There, on an idyllic strand between the villages of Sassnitz and Blintz, was the resort center of Bad Mukran, which had been established by Cologne's *Gauleiter* Dr. Robert Ley for the *Kraft durch Freude*—"strength through joy"—movement. Designed to accommodate twenty thousand people at a time, the building was six stories high and almost 2½ kilometers long. Though it had been completed just as the war broke out, its joyful workers had never used it for the purpose Ley intended—cheap vacations for German Labor Front members. But to Huzel it was a treasure house for things he needed for Peenemünde across Greifswalder Bay to the southeast.

Shortly after midnight, Huzel was awakened by Mukran's caretaker, who advised him that British bombers were overhead, apparently headed for Berlin or Stettin. It occurred to Huzel that they might also be headed for Peenemünde, but he went back to sleep. It was the next afternoon before he learned of the bombing and the following day before he could get back to Peenemünde. Once there, he found the heaviest damage had occurred at the housing settlement near Karlshagen and at the foreign labor camp at Trassenheide. The latter place was being cleaned up as he arrived by train. He later described what he saw:

> The grimmest task of all was that of removing the victims from the camp. The bombs had struck while most of the people were asleep. Many apparently had tried to escape [the bombing], but were caught while running. The gates of the chain-link fence, although open, were too far away for most, and many who tried to climb over the fence were killed as they climbed. Here and there the bombs had opened gaps in the fence, destroying at the same time those hapless souls climbing to safety.

> It was too early to establish the exact number lost, but estimates were that half of the one thousand or so inhabitants of the camp had been killed. Expediency did not permit elaborate methods of retrieving the bodies. Groups with stretchers shuttled between the wreckage and waiting trucks. On these, between latticed sidings and under tarpaulin roofs, bodies were stacked like sacks of flour. Someone's grim sense of orderliness had determined that they all be placed with feet towards the inside, and heads poking out the rear. This presented a sickening sight of disarranged hair and dangling arms.

As on previous occasions when air raid alerts had sounded, some of the inhabitants of Camp Trassenheide made straight for the tank cars loaded with alcohol at the railway siding. Dr. Martin Schilling, chief of the Testing Laboratory for Thiel's Propulsion Development Office, later marveled at the physiologic and metabolic differences between Russians and Germans. "They seemed to thrive," he recalled,

> on the 40 percent methyl/60 percent ethyl alcohol mixture [found in the tank cars]. . . . They used bucket brigades to recover the fuel. The Russians were really amazing; they could survive any encounter with the stuff. Mr. [Hermann] Weidner, the head of one of the test stands, one day had a Russian cleanup crew arrive to help with a test. About 11 o'clock he called me to report that the entire crew was "stiff as a board—completely drunk and useless." They were stacked in two layers in lorries and taken back to their camp. None of them died, and there were no permanent side effects. *There is a difference between Russians and Germans*.

(Schilling went to the United States in 1945 and remained with the Peenemünde team for five years in Texas and eight in Huntsville, during which time he became chief of Redstone Arsenal's Research and Development Division. Later he joined the Raytheon Corporation in Lexington, Massachusetts, becoming vice-president for R&D in 1964.)

On Saturday, 21 August, Peenemünde buried its dead.

The small cemetery that had for years fulfilled the needs of the equally small fishing village of Karlshagen was burdened several fold as the 735 victims were interred. Civilian dead from the housing settlement were provided single graves, while VKN soldiers and foreign laborers were placed in mass graves. The scene was almost Wagnerian, funeral pyres flickered atop stanchions placed before the nearest grave but casting little light in the semi-gloom among the towering pines, some of which were shorn of branches to serve as poles for pendant national standards. From a swastika-draped speaker's platform the *Gauleiter* of Stettin Schwede-Koburg gave a Nazi oration and stood stiffly at attention with his arm raised in the "Heil Hitler" salute as the Kurkapelle-Promenade Band from Zinnowitz played *"Das Horst Wessel Lied."* Then Dornberger gave a touchingly personal and humane farewell to the dead, and the band played appropriately somber selections from Beethoven. The religious protocol was performed by a Catholic priest and a Protestant minister.

Despite the number of deaths, the only heavy damage was to the housing estate and the development works. In the former, destruction was total, while in the latter area fifty of the existing eighty buildings were destroyed or severely damaged. The foreign labor camp at Trassenheide had eighteen of its thirty huts destroyed. However, in target area *B*, the second wave of bombers had wrought little damage to the two large assembly buildings. The supersonic wind tunnel was untouched, and damage to the all-important Measurement Building was light. The liquid oxygen plant was also unscathed.

The raid produced several decisions that would effect the operations at Peenemünde. Huzel summarized them:

> First, there would be no rebuilding that could be detected from the air. The housing project was completely evacuated, and the families living there were distributed over the northern portion of the island. Since most of the villages were resort areas, these refugees were easily absorbed. Certain administrative offices moved into the abandoned but still intact project houses, and care was taken not to disturb the "aerial photogenicity" of the entire area.
>
> Next, any plans to mass produce, or even pilot produce, at Peenemünde were abandoned. All machinery being readied for that purpose was transferred to a subterranean plant at Niedersachswerfen, near Nordhausen, in central Germany. . . . Numerous offices and laboratories were evacuated to hotels in the vacation sites south of the plant. . . . Immediately, construction was started on several sturdy air raid shelters and, needless to say, air raid warnings were now taken seriously. . . . For some time prior to the raid, evacuation of Dr. Hermann's supersonic wind tunnel to another location had been considered. Thus, although it had been unaffected by the raid, this important facility was now moved to the area of Kochel, Bavaria, in a beautiful mountain setting. . . . The *VKN* had evacuated to secondary hotels in Zinnowitz, and . . . permission to live in individual rented rooms wherever available was reluctantly granted to engineers and scientists being retained in uniform.

The extent to which the 1,593 tons of explosives and 281 tons of firebombs delayed the A4 program is debatable. On September 16, 1943, less than a month after the raid, a German damage assessment found its way into Allied hands. Carrying a "most secret" classification, it was routed to the manager of the Wilhelm-Schmiddig works at Cologne-Niehl.

> According to available eye-witness accounts, the W plant (E-Stelle) was not hit and the power station and oxygen plant were not damaged. Important testing equipment escaped damage at the E plant, but the administration building, drafting offices, and the main assembly workshops for A4 units were completely destroyed, as were barracks and auxiliary buildings. Between 600 and 800 persons were killed, including Dr. Thiel, in charge of propulsion development at the E plant.
>
> Experimental work in the W plant was fully resumed about two weeks after the attack and the Schmiddig firm's research interrupted for about three weeks.

After the war, Air Chief Marshal Sir Arthur Harris assessed the importance of the raid in the following terms:

> ... it was noticeable that after Bomber Command's attacks on Peenemünde the enemy became much less definite in the threats he uttered about the secret weapons he was preparing for England: in particular he ceased to mention any specific dates when the V-weapons would be expected. There was never, of course, any question of putting a complete stop to the use of V-weapons by bombing Peenemünde; we knew very well by then that if the enemy chose to give first priority to the production of anything, a single attack on any plant could only cause a delay of a month or two at the most. But in the war of V-weapons time was everything and every delay we could cause the enemy, however brief, was thoroughly worth while.

Perhaps the greatest failure of the raid was that it did not achieve one of its major objectives: killing or incapacitating as many of the scientific and technical personnel as possible. Only two important people were killed: Walter Thiel and Erich Walther, chief of maintenance for the shops within Eberhard Rees's Production Department. Since the A4's propulsion system was already in production, Thiel's death did not affect the development of the weapon. However, his talents were lost to such later weapons as the *Wasserfall* surface-to-air missile.

Subsequent to the RAF raid, the great Peenemünde establishment was left unmolested for a year. Allied intelligence discerned that testing and production had been transferred away from the site and that many of the development activities associated with it had dispersed. For the time being, at least, Allied bombers would seek more productive targets.

Meanwhile, attention turned to other industrial and experimental targets of importance to Hitler's wonder weapon program, targets that were hit by the RAF's Bomber Command as well as by the US Eighth and Fifteenth Air Forces. Major attacks are summarized in the table on page 125. Thus, in bombing Wiener-Neustadt and Friedrichshafen, the Allies unwittingly disrupted plans to assemble A4s at Rax-Werke and at the Zeppelin plant.

As for Peenemünde, it was revisited three times in the summer of 1944 by the US Eighth Air Force, not because of its continuing importance in the weapon-development cycle but rather because it was then suspected to be a hydrogen peroxide manufacturing facility. Indeed, as late as July 17, 1944, an intelligence report took note of the supposed fact that hydrogen peroxide production in Germany and in occupied countries was considerably in excess of prewar levels. This led to a search for sites where it might be produced:

> Plants with the expected characteristics (and for which there is no explanation) have been found by air reconnaissance at Peenemünde [and at] Raderach, which are known centres of "CROSSBOW" and/or "BIG BEN" research and development.

The two Fifteenth Air Force raids on Raderach were for the same purpose. Oddly enough, Britain's foremost authority on liquid-propellant rocketry, Isaac Lubbock, had first pointed the finger at hydrogen peroxide as the suspected propellant for the enemy rocket simply because he did not believe the Germans could cope with pumping large quantities of liquid oxygen into missiles under field conditions.

All the while, intelligence tended to explain away known large-scale production of liquid oxygen on the Continent by saying it was being earmarked for Hitler's planned scorched-earth policy. Fortunately for all concerned, this policy was never carried out.

After the war, Munitions Minister Albert Speer, Field Marshal Erhard Milch, Luftwaffe fighter forces commander Major General Adolf Galland, and Mittelwerk Director Georg Rickhey were asked what effect the three Peenemünde raids on top of the earlier RAF attack had had on the V-weapon program. The consensus was perhaps two months as far as the A4 was concerned, though other projects such as surface-to-air missiles suffered more, due to the loss of personnel and damage or destruction of drawings and laboratory installations. British and American raids on Friedrichshafen and Wiener-Neustadt were bothersome, they admitted, but other facilities and sources of supply were substituted so that no net loss of production was experienced. In the light of its own investigation, the United States Strategic Bombing Survey concluded:

> Bombing was not a direct factor in interfering with the production of V-2 weapons . . . Allied intelligence failed to discover the vulnerability of the Jenbach plant, which produced the pumps. If the electric plants which served the Mittelwerk had been destroyed production would have been interrupted for a while. The workers'

Research, Development and Industrial Targets Related to German Missile Program

Location	Objective*	Date of air raid	Air Force Involved
Friedrichshafen—Zeppelin and Maybach plants	A4 assembly	March 18, 1944	US 8th
		April 27/28, 1944	RAF/BC
Fellersleben—Volkswagenwerke	Fi103 assembly	April 8, 1944	US 8th
		June 20, 1944	US 8th
		June 29, 1944	US 8th
		August 5, 1944	US 8th
Rüsselsheim—Adam Opel	unspecified	July 27, 1944	US 8th
		August 12, 1944	RAF/BC
Wiener-Neustadt—Rax-Werke	A4 activities	November 1943	US 15th (first attack)
Weimar/Buchenwald	gyroscopes	August 24, 1944	US 8th
Peenemünde	operations and personnel	August 17/18, 1943	RAF/BC
	hydrogen peroxide facilities	July 18, 1944	US 8th
		August 4, 1944	US 8th
		August 25, 1944	US 8th
Zinnowitz	experimental establishment	July 18, 1944	US 8th
Hollriegelskreuth	hydrogen peroxide	July 19, 1944	US 8th
Raderach	hydrogen peroxide	August 3, 1944	US 15th
		August 16, 1944	US 15th

* Objectives not always accurately assessed. For example, Friedrichshafen was also raided on June 22, 1943, at a time when it was unknown that the city was connected with the A4 program. However, at the time the Germans planned to assemble 300 A4s a month there, but Bomber Command damage in that raid caused plans to be canceled. The 1944 raids thus did not attain a realistic objective. Similarly, the hydrogen peroxide objectives at Peenemünde and Raderach were unrealistic.

> barracks were vulnerable, but none of these installations was touched by bombs.

The survey then went on to observe that the

> ... indirect effect of strategic bombing in the production of V-weapons can be estimated as being about 20 percent caused by sporadic shortages of sheet metal, obtained from mills in the Ruhr, and by difficulty in shipping components to the assembly plants. ... The final collapse of production was caused by failure of the transport system early in April 1945. By that time neither the weapons nor the fuel could be shipped to the launching sites.

Once the relationship between the ramps at Peenemünde and the ski sites on the French coast had been established (Chapter 6), the Allies decided to commence bombing attacks on them as well as to continue to hit the large installations. From December 1943 onward, the offensive campaign against both A4 and flying bomb installations, supply dumps, and factories came under the umbrella designation Operation Crossbow. Crossbow-designated attacks against ski sites began on the fifth of the month, when the RAF's Second Tactical Air Force and the American Ninth Air Force went into action. The US Eighth Air Force followed up with heavy bombers later in the month.

The intensity of Crossbow operations was colored to a great extent by Allied estimates of the numbers of flying bombs the Germans could fire from a given site each day. The director of Air Ministry Intelligence, for example, talked of 2,000 tons of explosives falling on London every twenty-four hours, assuming 100 launch sites were operational. This figure was generally accepted as probable by the Joint Intelligence Committee. Lord Cherwell, on the other hand, disagreed with the postulated launch rates (some fifty Fi103s per site), partly because French agents had reported that a given site could store only twenty missiles (in actual fact, twenty-one), and partly because Allied countermeasures were not being considered.

As the Crossbow bombing campaign got under way, photographic interpreters under Kendall at Medmenham developed a method of recommending which sites should be hit and when. The idea was to let the Germans almost, but not quite, complete a site, and then damage or destroy it, rather than expend bombs, planes, and crews on sites only, say, 50 percent ready for operations.

As the Allies continued pounding the large installations and the ski sites during January, February, March, and April 1944, those responsible for assessing the degree of damage being inflicted and the German response to it began to notice that repairs at the ski sites were made only to the square buildings and to the firing points—but not to the ski-shaped structures. Then, near the end of April, Captain Robert Rowell made an important new discovery: Close to the village of Belhamelin on the Cotentin peninsula a concrete platform with studs was discovered between two farmhouses; and, not far away, the familiar square building, well camouflaged, appeared. Thus, the first of the so-called "modified" sites was revealed. (Later, it was learned that prefabricated catapults were brought to such simplified sites only about a week before launching operations were to begin.) To a total of ninety-six ski sites plus the heavy installations were now being added new sites, whose number could not yet be estimated and which, in any event, did not make very promising targets.

Although by the end of January 1944 the Germans had decided definitely against using their vulnerable ski sites, they covertly did some repair work, camouflaging carefully what they had done and cleverly not filling in bomb craters. The purpose was, of course, to make the Allies believe that the Germans were desperately trying to repair their sites while taking elaborate precautions that minimal and deceitful "reconstruction" would not be noticed—knowing, however, it would be. The ruse worked, and bombs kept falling on the ski sites that otherwise would have rained down on the fatherland.

For several years the British had been deeply concerned with flying bomb and rocket intelligence and the effects the secret weapons might have on England. By early 1944, however, the Americans—mindful of the approach of the Overlord invasion of Europe—had become increasingly active in assessing the same threat in terms of their own interests. Most important was to determine what effect on troop and ship buildup in southeast England a concentrated flying bomb and rocket attack might have, and how gravely overall planning might be disrupted by an onslaught on the British capital.

Concern for London and the Overlord invasion led Allied Supreme Commander General Dwight D. Eisenhower to give Crossbow target bombing overriding priority at one period, eclipsing even the strategic bombing offensive against the German aircraft industry and the major German cities. From December 5, 1943 through to June 12, 1944, just over a day before the flying bomb finally got into operation, the Eighth Air Force and the Tactical Air Forces were chiefly involved. The principal aircraft flown were B-17s, B-24s, B-26s, A-20s, Mitchells, Bostons, and Mosquitos.

During May and early June these planes were called upon to attempt to neutralize the sixty-eight modified sites that had turned up—a very difficult assignment. Medmenham had learned, from studies of site preparations at Zinnowitz near Peenemünde, that the Germans were putting

6-meter-long rail sections on the concrete bases, yielding thereby completed launching ramps. They would then erect the prefabricated square building. It was estimated that only a few days would be needed to activate a site; in fact, by the end of the flying bomb campaign, the Germans could set up a modified site in only eighteen hours.

To alert British defense and home security forces that a modified site was about to be activated, the code signal Diver was introduced; it was first used late at night on June 11, five days after D-day at Normandy and barely more than a day before the first V-1—the propaganda name for the flying bomb—impacted on British soil.

Looking back at the whole Crossbow bombing campaign against both the V-1 and the A4 (which became the familiar V-2), four distinct phases can be recognized:

1. August 17–18, 1943	The initial RAF attack on Peenemünde.
2. August 27, 1943 through June 12, 1944	Began with the Eighth Air Force attack on Watten and other large installations, and then ski sites.
3. June 13 through September 3, 1944	Period of V-1 offensive against Britain from sites in France. Main bombing attacks against modified sites, supply dumps, and large sites.
4. September 1944 through April 3, 1945	Period of combined V-1 and V-2 attacks on Britain and on targets on the Continent already occupied by the Allies.

From August to October 1943 RAF Bomber Command and the US Eighth Air Force were exclusively involved in what became known as Crossbow targets. From November of that year to May of the next year, most of the work was handled by the Eighth and the US/RAF Tactical Air Forces. Then in the summer of 1944 Bomber Command and the Eighth together were principally engaged, with some support from the US Fifteenth Air Force. Finally, from September 1944 to the end of the campaign, RAF Fighter Command and the Second Tactical Air Force bore the brunt of the attack.

The vital role of photoreconnaissance and interpretation in this campaign was underscored in the official US Strategic Bombing Survey "Report on the 'Crossbow' Campaign—The Air Offensive Against V-Weapons":

> Approximately 4,000 reconnaissance sorties flown by British and American aircraft contributed directly to "Crossbow" intelligence. Brilliant interpretation of the photographs confirmed the reports of experimental work on the Baltic, pinpointed and identified launching and supply sites in France, definitely established the purpose of the "ski-sites," provided the material for the targeting of more than 300 objectives, analyzed the results of attacks and, in general, provided the framework for the planning and execution of offensive counter-measures. In all, 4,070 interpretation reports were issued on these subjects.

Other than the 7 heavy, 96 ski, and numerous (61 by D-day, 66 when V-1 attack began) modified sites, 7 specialized V-1 supply sites were damaged (3 severely), and storage dumps in caves and tunnels at Nucourt, Saint Leu d'Esserent, Thiverny, Méry-sur-Oise, and Rilly-la-Montagne and in several wooded areas were raided. (A later attack on Nucourt on July 6, 1944 would prove to be particularly effective, as the intensity of firing against England fell shortly afterwards.) Beyond this, bombers hit many targets in Germany of both industrial and experimental importance.

As successful as the British and US attacks had been, and as brilliantly as the CIU in Medmenham and Air Ministry Intelligence had performed, by May 1944 the latter had been lulled into a false sense of security: With the heavy sites apparently out of commission, the ski sites either destroyed or heavily damaged, and the supply sites harassed, it did not seem that the danger could be great. Anyway, there was little Allied planes could do against the elusive modified sites, and, of course, D-day was fast approaching when even Crossbow would have to take backstage.

CHAPTER 8

Events in Poland and Sweden

The chronicle of German missile development and the Allied response was by no means confined to Germany, the United Kingdom, and the launching sites in western France. It developed that both Poland and Sweden were to play important roles in the unfolding of events.

Four days after the mid-August 1943 RAF raid on Peenemünde, the final decision was made to move most flying bomb and rocket flight-testing activities from the Baltic to a remote site in southern Poland. The location selected was the Truppenübungsplatz Heidelager, an SS training camp located within a roughly 12-mile square bounded by the four villages of Brzeźnica, Rzochów, Kolbuszowa, and Sędziszów. At the center of the site was tiny Blizna and close by was Kochanowia. Rail service was provided by the Mielec–Debica–Tarnow rail line, connected directly to the main Krakow–Lwów route.

When news reached the SS that missile forces from Peenemünde would be moving into the Blizna area, preparation immediately got under way to accommodate them. New living and working quarters were constructed, vehicle shelters set up, and the rail spur into Heidelager improved. The camp was surrounded by a double barbed-wire fence and heavily patrolled by SS security. The launch area was prepared in a large open space with forests all around. Most of the buildings were erected along the fringes of the forest. On the west side of the test area was Blizna itself, almost completely demolished by site preparation crews.

The army ordered Colonel Gerhard Stegmaier, former military commander of Peenemünde's development (preproduction) works and now in charge of A4 troop training at Köslin, to transfer his Experimental Battery 444 to Heidelager during the month of September. Peenemünde engineer Albert Zeiler recalled that, as the troops moved in, six mobile firing stations were quickly readied. Troops would set up an A4 launch platform, he said, and then bring in the Meilerwagen and propellant tank cars. "We lived more or less as a soldier would anywhere. We got up early and were kept busy all day." At first, Zeiler lived in a barracks; but, upon becoming a warrant officer, he moved into a nearby parked train with its kitchen and other coveted facilities. Beer was available, he related, but "we mostly drank cheap vodka" obtained from the Poles.

Even though launch site preparations were still under way, Stegmaier and his principal aide, Major Weber, were able to organize the initial launch attempt early in November. On the morning of November 5 the temperature was well below freezing, complicating missile setup operations. Moreover, the plate of the launch platform's gas jet exhaust deflector was not properly adjusted, with the result, according to General Dornberger, that the

> ... gas jet thawed out the ground and burrowed down into the sand. One leg of the firing table sank down slowly into the soil during the preliminary stage [buildup of thrust]. The rocket rose at a slant, went out of control, and crashed into the forest 2 miles away.

As luck would have it, General Erich Heinemann (the commander designated to operate the weapon under combat conditions) was on hand and witnessed the debacle. He falsely concluded that the missile could be launched successfully only if special concrete pads were constructed, with the result that manpower and materials were soon to be squandered along the Atlantic coast of France in building such emplacements. In actual fact, A4s were subsequently fired from unprepared paths in the Polish forests and from ground strengthened only by logs.

Some A4s suffered from launch phase vibration, with the result that relay contacts would break, the motors would shut down, and the missiles—which had risen only a relatively few yards—would settle back on their firing table and then topple over. Others would rise a mile or more into the air and then explode, while still others would blow up a few thousand feet above the target area (airburst). According to Dornberger, during the initial months of testing at Blizna, only between 10 and 20 percent of the missiles were able successfully to impact on or near their targets. (Flight times to these targets, incidentally, were slightly over five minutes.)

Despite these and other problems, remarkably few casualties were suffered. On one occasion, a rocket exploded just after takeoff, killing twenty-three members of the experimental launch battery. Generally, however, the crews escaped serious injury or death. Zeiler recalls his first launch experience:

> I had just got my new uniform when I returned to Blizna from [a trip to] Peenemünde. We were ready to fire . . . with a light warhead. We had some trouble . . . and all day long worked [to correct it]. Shortly before dark, we finally got ignition in the main stage followed by an almost instantaneous cutoff. The vehicle promptly tipped over, the lox [liquid oxygen] and fuel tanks burst open, and of course there was a big explosion. I told everyone to stay under cover because I wasn't sure whether the warhead had gone off or not. I went out to see what had happened and I saw that the fire was slowly moving towards it. I quickly stepped back when the warhead went off. I wound up in the mud—you can imagine in Poland after the rain what it was like. That was my first experience with my new uniform.

Since 1941, the Polish underground had maintained sporadic watch over the Blizna area. Early in the war, Russian prisoners of war had been interned there, joined, for a short period, by French, Belgian, and Dutch Jews earmarked for extermination.

With the arrival of autumn in 1943, suspicions were aroused that Truppenübungsplatz Heidelager was being, or at least was about to be, used for something more than SS troop training. Immediately, members of the Armia Krajowa (AK), the Polish Home Army, stepped up their surveillance. They observed that large numbers of concentration camp workers had moved in and were building new roads and installations in the area and putting in a rail spur.

AK's Intelligence Bureau chief Colonel Kazimierz Iranek-Osmecki organized a field staff under Józef Rządzki (code-named Boryna) to maintain a round-the-clock vigil. All suspicious activities were reported through underground channels to Warsaw for processing, interpretation, and then transmittal to London.

Rządzki assigned the Heidelager watch to two surveillance teams, one known as Kefir, which was to operate out of Kolbuszowa, and the other as Deser, with headquarters in the town of Debica. Within each team a number of field units was established, each with its own code name. The most productive of these turned out to be Slawa, commanded by Mieczyslaw Stachowski, alias Sep and Maciej. He was an engineer employed by a local German office responsible for forest administration in the Blizna area.

As important as it was for the Polish underground to know what the Germans were doing at Heidelager, unless information could be transmitted to the British even the most effective and heroic efforts would have been in vain. Fortunately, means of transmitting intelligence from Poland were devised, the most common of course, being radio; it was used repeatedly to signal technical and operational information from Warsaw to London and to receive instructions and requests the other way around.

Often, however, British intelligence required more details than could physically or securely be committed to wireless. This usually meant that messengers with their microfilmed reports would be dispatched across occupied Europe to a neutral country and thence on to the United Kingdom. Either of two routes would be selected for this purpose. A courier traveling along the first route, which was the longest but least dangerous, would go from Warsaw through western Poland, Germany, France, Spain, and Portugal and finally to England. It might take up to two months to traverse. The shorter and more dangerous route through Denmark to Sweden might take no longer than two weeks to traverse.

Even the courier technique had limitations. As more and more information was supplied by the Poles on the characteristics and performance of the long-range missiles being tested at Blizna, the more important it became for British intelligence to obtain actual parts and fragments.

Since the end of World War II, debate has gone on as to the extent and importance of the Polish underground's secret-weapons intelligence gathering. Poles often complain that the British failed to give credit where credit was due, and—at least to an extent—their complaints are understandable. In the opinion of R. V. Jones, the man responsible for assessing much of the information and matériel coming from Poland, the debate arose because

> . . . it happened that their information was distinctly confused in the early stages, and the reports that came in from them were very often disjointed. Sometimes it seemed as though there was a "scrambling office" somewhere on the lines of communication so that one rarely received an entire report, but usually got the first half of one report along with the second half of another and often on a different subject, and it was not easy to unscramble the result.

Communications later improved. Tadeusz Bor-Komorowski, commander of the AK from 1943, states that "our intelligence reports were regularly dispatched by radio to London and in the years 1942–44 numbered 300 per month. . . . Apart from radio transmission, the essential facts of our intelligence material were microfilmed and sent every month to London by courier."

That the British were pleased with the information they received is attested by a Ministry of Aircraft Production commendation: "Reports from Polish sources are considered as the most valuable of all we received. They contain entirely trustworthy information. . . ." Josef Garliński, author of *Poland, SOE, and the Allies*, observed that British and American Intelligence officers in London "clamored" for details on the Blizna missiles. "Dispatch-riders," he writes, would stand by "their motorcycles in front of Polish headquarters [in London] and [then dash off] with each new message as they received it; sometimes the liaison officers would grab half a telegram, without waiting for the rest to be deciphered."

What kind of information were the Poles securing for their British allies in London? As the 1943 autumn wore on, Slawa agents reported a continued buildup in and around Blizna, including new road construction and the repair of a branch railway from Kochanowka. Some six hundred additional SS troops were seen to enter Heidelager, and the polygon-shaped camp was given perimeter protection by barbed-wire emplacements, machine guns, and antiaircraft artillery. In addition, Slawa reported new machine shops and a variety of newly constructed barracks. The tightening of security and entrance-exit controls further demonstrated that something important was soon to happen at the Blizna site. (The extent of German concern for security is illustrated by a Slawa report that "... prefabricated houses and barns made out of thin wood [were brought in] and placed within the camp for camouflage purposes." The SS went so far, it added, as "to put artificial flowers around the camp so that it would appear to a casual visitor like a normal country village.") The AK speculated that the arrival of army troops, distinguished by the red insignia of artillery, together with a small number of civilians might well represent the key to what was going on inside Heidelager.

Extremely tight security made it impossible for Polish agents to get close to the railway lines, much less gain access to the camp or to the nearby forest. The SS not only did their job well in protecting Heidelager (which for a time was called Artilleriezielfeld-Blizna) but—with the help of defectors—the Gestapo penetrated AK headquarters and made many arrests. In the face of a potentially disastrous situation, Colonel Iranek-Osmecki ("Makary," "Heller," and "Antoni"), assisted by Antoni Kocjan ("Korona"), reorganized the various units, transferred many members, and redoubled efforts to secure a steady flow of technical and operational information. One of the key members they enlisted was Captain Jerzy Chmielewski ("Rafal").

To be sure, the AK was able to gather some information. Thus, an agent succeeded in buying for 2,000 zlotys a map of the Heidelager compound prepared by a German soldier. Then the month of November

(1943) arrived and with it the beginning of missile firings. The Armia Krajowa, realizing the extraordinary importance of what was happening, immediately began to record the spectacular events. They talked of "tremendous rumbles and explosions," of "frost on torpedo-shaped bodies," of "vapors" escaping into the air, and of "strange objects." Trained Polish engineers suspected that liquid air or liquid oxygen was being employed as a propellant for "aerial torpedoes," and in London, intercepted Ultra (deciphered radio) signals suggested that rockets were being tested in southern Poland.

AK agents soon noticed that aircraft were regularly sent out to spot the impact points of the missiles being fired by the Germans. Though the Poles were not yet able to describe accurately the missiles, they did find out that the pilots would radio instructions to ground recovery crews who would move in quickly and retrieve whatever remains could be found. The AK also learned that a missile had exploded inside the launch compound, killing twenty-three German soldiers and causing an undetermined amount of damage. This piece of information along with other facts were obtained by the first Pole to penetrate Heidelager.

Boleslaw Mikoś, an architect-technician from Kraków, gained entrance to the polygon compound as an employee of the Hermann Osso firm of Vienna. Associated with the construction of a tower-shaped building and of a 150-foot-long assembly hall served directly by rail, he had the opportunity to gather some other useful intelligence. Perhaps the most important item he learned was the simple structure of the launching pads. Seeing them carefully camouflaged with nets and branches of trees, he marked their locations in relation to the tower and assembly building with which they appeared to be connected.

In early January 1944, AK intelligence chief Iranek-Osmecki prepared an assessment from his Warsaw base of activities carried out at and around Blizna during the preceding two months. Fully aware that advanced weapons were being tested, he was nonetheless perplexed by a number of seeming contradictions. For example, some witnesses talked of "cigar-shaped" missiles, while others insisted they had seen small, pilotless airplanes. Piecing together these and many other descriptions, he and AK's aircraft and missile intelligence expert Antoni Kocjan, soon perceived that not one but two missiles were being flight-tested from Blizna. London was so advised.

By the end of January 1944, Iranek-Osmecki noticed that the launch rate of the rocket-powered missile had accelerated. A rough flight record maintained by Colonel Michal Wiśniewski showed that whereas a single such missile had flown in November and another in December, a total of six were fired during January (one of which blew up within the Blizna compound). Typical entries are listed:

Sequence number	Date of launch	Approximate time of launch	Direction of flight and general estimate of the range
1	Nov. 25, '43	0200	toward Kolbuszowa; short
2	Dec. 5, '43	0200	same
6	Jan. 6, '44	0200	probably toward Częstochowa; long
8	Jan. 29, '44	2230	exploded inside Blizna; very short!
24	Mar. 31, '44	2350	toward Niwisk; short

February followed with five firings and March with eleven. During this period, some efforts were made to determine precisely where the missiles were impacting. The AK did not have much luck with A4s, but on one occasion the neighbor of the owner of a house in Rejowiec (some 250 kilometers northeast of Blizna) told agent "Grzegosz" how a motorized German patrol had arrived shortly after a flying bomb had badly damaged the house, apologized for what had happened, paid compensation, and departed—but not before enjoining the owner not to mention to anyone what had happened. The more polite the Germans, the more suspicious the Poles became—and the more determined to recover pieces of the missiles that were dotting the Polish countryside with craters.

In early April 1944, about forty German soldiers moved into the village of Sarnaki south of the river Bug and occupied a schoolhouse. They brought along a few artillery pieces, two radio transmitting and receiving units, and some trucks. Advised of this unusual activity, AK headquarters in Warsaw ordered Colonel K. Tumidajski ("Marcin"), the Lublin area commander, to keep an exceptionally close watch on the little settlement. Local partisans did not have long to wait for events to unfold.

The action began in mid-month, when a large explosion occurred at Mezęnin to the northwest of Sarnaki. Prepared for any eventuality, the AK reacted smartly and were able to recover pieces of the exploded missile (including a hydrogen peroxide container) before German recovery forces could reach the scene. This was but the prelude to a rain of some sixty rockets that fell through the end of May. Opportunities for parts collection were excellent. Partisans formed bicycle patrols, which soon recovered such components as electric motors, coils, containers, gyros, and radio parts. Pieces too large to move were promptly photographed.

AK soldiers had to reach the scene of a given impact with all possible haste if they were to beat the Germans at the recovery game. This was by no means a simple matter, as the Germans were always forewarned of an impending impact and received rather precise prediction notices by radio. From observation positions just to the northeast of Sarnaki, troops would move to the impact point to gather up fragments—most of which resulted from airburst—before the AK could beat them to it. Occasionally, according to Josef Garlinski, partisans and German recovery forces met in armed clashes; however, the Poles avoided confrontations whenever possible.

Another complicating factor was the natural barrier that existed between farmers and the AK: Fear of betrayal by false representation was a real and permanent problem, one that seriously impeded the transfer of information and hardware. Nevertheless, the AK were persistent, with the result that farmers cooperated more and more readily as time went on. Typically, a farmer would go out into his field following an impact, cover up as many fragments and parts as he could, and perhaps carry some into his house or barn. Potato sacks and haystacks made good hiding places for the remains of A4s.

Quite naturally, the Germans tried to discourage this practice, and even distributed leaflets explaining to the populace that from time to time they might come across "airplane containers" used for fuel. Sketches of the containers with their dimensions were prepared, accompanied by the notice:

> Attention! This is not a bomb. Immediately notify the nearest police station or the airport. Warning! This is a fuel.

Once parts had been recovered and hidden, the AK made arrangements to transfer them from the Sarnaki, Rejowiec, Sawin, Konskie and Częstochowa recovery areas to Warsaw where they could be examined by experts. This was accomplished in ingenious and daring ways.

A Polish doctor named Zygmunt Niepokój ("Norwid") took advantage of his assignment to care for German military personnel. Possessing a motor car and a pass that permitted him to travel to Warsaw to fetch medical supplies, he was able to carry along recently recovered missile parts and fragments. One day while en route to Warsaw, he was stopped at the outskirts of the city by SS guards—a high-level German official had just been killed and unusual security precautions had been ordered. As the guards began to search the car, the doctor protested vigorously. "Since when does the SS enter the car of the doctor who serves the Germans so well?" The intimidated guards withdrew and simply checked his papers before allowing him to enter the city.

On other occasions, co-operating Poles would stow bottles of spirits, loaves of bread, pieces of sausage, and other wartime delicacies in the pockets of the inside doors of their vehicles. Should the Germans decide to search, they would soon turn up the "contraband," confiscate it, and then allow the driver to continue on his way. Whenever possible, Poles would pick up German hitchhikers, who were considered excellent "cover."

Once the parts reached the center of Warsaw, they had to be hidden until engineers and technicians could commence their investigations. A favorite hiding place for gyros and radio parts brought in by Norwid was the apartment of a German army officer; this unique site was made possible because of the cooperation of the landlady. If the officer had come under suspicion and his apartment were checked, she confided after the war, "he wouldn't have lasted an hour."

Professor Dr. Josef Zawadzki of the Warsaw Polytechnic Institute coordinated the movement within the city of all missile components delivered by the AK, and, with Kocjan and Professor Janusz Grosskowski (who became chairman of the Polish Academy of Sciences after the war), organized the technical investigations. The fact that missile parts could be examined and drawings of them made within the Warsaw Polytechnic while it was controlled by the Germans is as incredible as the ability of the AK to move them around undetected. It was Professor Marceli Struszyński, incidentally, who analyzed the A4's auxiliary propellant and established it to be concentrated hydrogen peroxide. (London was informed of this discovery on June 12, 1944 in message No. 366/1176.) At the same time, Grosskowski tested the A4's radio system, learning that the missile received signals on a frequency of about 20.5 megacycles and transmitted at approximately 40 megacycles. The British were particularly interested in the radio equipment used by the Germans and constantly pressed the Poles for details.

Supplementing the acquisition of parts was the continuous probing for information on the nature of the A4 and flying bomb *before* they were launched. Again, the AK took advantage of every possible opportunity. On one occasion, a German soldier obligingly sketched an outline of the A4 for a resistance member (as they both drank heady Polish beer in a local tavern), stating that it, and a pilotless airplane as well, were developed at a secret station near Stettin (e.g., Peenemünde). Another source pinpointed Camp Dora as the production site (referring, of course, to Mittelwerk at Niedersachswerfen in Thuringia).

At one point it was proposed that Polish Kedyw (Commando) units attack Blizna to determine once and for all what was going on. The strength of German security forces was such, however, as to make this impractical. Attacks on rail transport were also contemplated, but again the AK's strength was deemed inadequate. One or two night forays against motor transport convoys were repulsed by the Germans, so for the time being at least the underground had to be content with passive means of acquiring information on the secret weapons.

Typical of information so acquired and passed on to London was the repeated occurrence of the airburst phenomenon, so bothersome to the Germans. "The pieces spread out over a 3-kilometer [nearly 2-mile] area when they fell," one report stated. Another piece of information had to do with the sizes of A4 impact craters, which were reported to be "between 11 and 14 meters deep [about 12 to 15 yards] and 27 to 29 meters wide [29½ to 31½ yards]." The size of an individual crater was found to depend on the nature of the flight. Often, experimental A4s would carry little or no explosive and hence would produce smaller craters than operational missiles. AK agents observed that Blizna-launched A4s would cluster in an impact area approximately 5 to 6 kilometers (2 to 2⅓ miles) square—in other words the presumed target area.

In mid-May, Iranek-Osmecki learned that partisans had found an almost complete A4 rocket thrust chamber in a soft river bank near Meźenin. (Depending on the trajectory followed, whether or not airburst occurred, missile attitude upon impact, and the physical nature of the point of impact, an A4 could suffer a greater or lesser degree of damage.) Captain Józef Legut was dispatched immediately to the site to take photographs and to supervise the pushing of the chamber into the river Bug before the German recovery troops arrived. This chamber was never again found, but at least it had been seen and photographed. Shortly after, on May 20, an entire A4 impacted in a riverside marsh with relatively little damage—a gift from heaven to members of the local AK. Soldiers of the 22nd Infantry Regiment promptly rolled the missile into the river and arranged for a farmer to drive his cattle into the water to stir up the mud. When German recovery forces arrived on the scene, all they saw was a herd of peaceful watering cows.

As soon as the Germans gave up trying to find the missile, the Poles went into action. Two recovery groups were established. One was assigned the job of making a complete investigation of the surrounding area, of assessing the underground forces available, and of developing the organization, courier services, passwords, etc. needed to handle the extremely difficult security problem inherent in manipulating an entire rocket. The second group, meanwhile, was to start locating specialized personnel and to collect the tools needed to dismember the A4 once it had been brought back out of the river. Among the group's specific tasks were to obtain a tractor, find steel hauling cable somewhere in Warsaw, and locate hard-to-find torches. The efforts were successful: During the ensuing weeks, personnel and their equipment were slowly moved into the Sarnaki area, using every subterfuge in the resistance book.

Fortunately for the AK, the river bank adjacent to the point where the missile lay hidden was steep, affording relatively good security while the tractor was in operation. Once the A4 had been hauled out of the river, it was quickly hidden in the nearby forest. Working night after night with limited tools, the AK was able to break it up and then to start shipping off key components and parts, first to an old barn in Holoweczyce-Kolonia and thence on to Warsaw. London, of course, was promptly notified as to what was going on. This phase of operations was headed up by Jerzy Chmielewski, assisted by Antoni Kocjan. Soon thereafter, on June 1, 1944, Kocjan was arrested by the Gestapo because of his association with clandestine printing activities and grenade manufacture. Although put through severe torture, he revealed nothing about his missile intelligence activities; nevertheless, he was executed on August 13. Kocjan's work was continued by his deputy, Stefan Waciorski ("Funio"), who, before the war, had been active in glider construction. He was later killed during the Warsaw uprising.

The news of the capture of a virtually intact A4 reached Colonel Marian Utnik of the Sixth Bureau, Polish Army Headquarters, in London late in May. Initially, he doubted the accuracy of the astounding report he had received, but he passed it on to the British straight away. Reconfirmation was immediately requested by Major General Sir Colin Gubbins, head of the Special Operations Executive (SOE) and a member of the Crossbow Committee. An intact A4 in the hands of the AK could not be ignored. As the news of the momentous event circulated through Crossbow and Big Ben circles, requests for additional details poured into Utnik's office. Soon he was processing up to three hundred messages a day to and from Warsaw—in itself a remarkable achievement. Moreover, the Research Commission of AK's Intelligence Bureau soon prepared a four-thousand-word report on the A4, including, according to Garliński, "nearly a score of diagrams, 80 photographs, a sketch of the Blizna camp, a table showing the number of rockets fired and where they had landed, and a list of factories producing the various parts."

Within days, Tadeusz Bor Komorowski, head of the entire Polish underground movement, and his Allied colleagues developed plans for Operation Wildhorn III (known to the Poles as *Most III*), whereby components and parts from the A4, several Polish engineers and intelligence experts familiar with the missile, and the report would be flown out of Poland and brought to the Royal Aircraft Establishment in England for study. However, the plan did not win the immediate approval of Supreme Headquarters because all AK forces were in the process of being organized to destroy bridges and otherwise disrupt German communications in support of Operation Overlord. The aim: to hinder the movement of German troops and equipment that might be ordered westward from the Russian front to help contain the Allied landing on the French coast. Moreover, as fate would have it, just a week after Overlord began, a Peenemünde-launched test A4 crashed in southern Sweden—in Britain's backyard compared to faraway occupied Poland. This event, which is covered later in this chapter, contributed to the lowering of priority assigned to the retrieval of the Polish A4, to the keen disappointment of the resourceful and daring AK.

It was not until late June 1944, when the Allies felt in command of the ground situation in western France and an assessment (albeit preliminary) had been made of the harvest of Swedish A4 parts, that the SOE in London requested that the Poles proceed with Wildhorn III. Since previous Wildhorns (on April 15 and May 29–30, 1944) had worked, Jósef Garliński, author of the procedures, felt he could repeat the performance.

A number of requirements were laid out by Garliński. For one thing, he demanded at least a kilometer (0.6 mile) square field (later code-named "Motyl" or "Butterfly"), the ground of which must be firm enough to sustain the weight of a bimotor airplane. The pilot and copilot must be thoroughly trained in secure night-landing operations, and the navigator would have to have an intimate knowledge of the route across occupied Europe as well as the local countryside. Last, but far from least, AK ground forces would have to be organized to support the landing, loading, and takeoff operations and provide perimeter security to the landing field.

At one time, it was planned to fly the mission directly from England in a Hudson aircraft but, with the Allied advance into Italy, the flight of a lighter Dakota [C-47/DC-3] from Brindisi became feasible. Accordingly, SOE headquarters advised Iranek-Osmecki to

> ... prepare yourself for Bridge [the overland flight element of Wildhorn III]. Define the landing place. Plane involved will be Douglas "Dakota." Matter is urgent and very important. The landing must take place at all costs, to return all the radio equipment, parts of the rocket, and the rocket propellants.

"Rafal" (Chmielewski) was ordered to take charge of assuring that the appropriate parts were made ready for the trip and to gather information on matters pertaining to construction, propellants, radio system, guidance, and missile operations. Overall area command was exercised by Stanislaw Wolkowiński ("Lubicz").

The field selected for the landing site was located at Zaborow near Tarnow (it had been used once in the *Most II* operation). Chmielewski assigned Marian Dembiński the job of seeing that all A4 parts to be sent to London were delivered near the field at the proper time. Dembiński was very well trained for his duties, having already brought key pieces of equipment from the field into Warsaw under the noses of the Germans. "This became my way of life," he said, "for I had been assigned this type of work often." Dembiński (and all other AK members involved) was also given a backup in case of compromise or capture during the course of the operation.

By the time the AK was ready to receive the Dakota, the weather had deteriorated so that the flight had to be postponed and then postponed again. This posed severe security problems to Colonel S. Musialek-Lowicki ("Miroslaw"), in charge of preparing the landing site. AK members could not be allowed to concentrate in the area, so every time the operation was postponed some were forced to relocate—a movement in itself fraught with danger. To lessen risks of exposure, teenage boys and girls were used as couriers between waiting groups of partisans.

While the waiting time lengthened, the Germans suddenly started using Motyl for pilot-training exercises! One day, a Hungarian plane even crashed there. To all concerned, the uncertain holding period was as nerve-wracking as it was dangerous.

Among those standing by in the Tarnow area were the five passengers who were to fly to London. The most important to the A4 intelligence mission was, of course, Chmielewski ("Rafal"), whose "unfailing memory"—in the words of Garliński—supplemented the vital report he was to carry to London. Others were Tomasz Arciszewski (Council of National Unity's candidate for Polish president) and Jósef Retinger ("Brzoza"), a political emissary sent at Churchill's request to determine what the chances were for Communist and non-Communist underground movements to work together after the war. Their waiting continued as the middle of July passed into history, the rains persisted, and the Dakota remained in Italy, where the weather was also poor. Depression affected everyone, as AK leaders worried lest Wildhorn III be exposed. After all, some three hundred Poles were now directly involved in the operation.

As if Iranek-Osmecki, Musialek-Lowicki, Wolkowiński, Chmielewski, Dembiński, and other leaders did not have enough on their minds, the Luftwaffe began moving personnel into private houses in the area until a total of some eighty men had been spotted. The number of training flights at Motyl also increased, principally because it was fairly long and flat, and ran parallel to a small, easily identified river—which was why it was selected by the AK in the first place.

Early in the morning of July 25, 1944, the AK received the long-awaited word that weather conditions would permit the Dakota to make the Brindisi (Italy)-to-Motyl flight, and that it would arrive that very midnight. A secret radio message from Italy announced that reconfirmation would be broadcast, in code of course, from the BBC in London later that morning. The whole complex interplay of hundreds of individuals began. Wildhorn/*Most III* was on.

The British pilot in Italy, RAF Flight Lieutenant Stanley George Culliford, and his crew had become as impatient as the Poles themselves. If a dangerous mission was going to take place, the sooner the better. The

Dakota they were to fly had been stripped down at Bari of all nonessential gear, and a long-range gasoline tank had been added. Culliford told the authors that he knew nothing more than the fact that "a Polish trip was coming up."

"It was the ordinary kind of 'need-to-know' business; they weren't going to tell us anything more than was absolutely essential," he explained. "Anyway, we had no burning desire to know [more than minimal details] for if we did fall into the wrong hands and didn't know we couldn't embarrass the people on the ground."

The crew, which included second pilot Kazimierz Szrajer from a Polish Liberator force, an English navigator, and a Welsh wireless operator, had been meticulously trained for the mission that would take the unarmed transport over the enemy-patrolled Adriatic, across Yugoslavia, up along the Hungarian–Rumanian border, through eastern Czechoslovakia into southern Poland, and if they were lucky, back along the same route.

Culliford, postwar assistant principal at the Victoria University of Wellington in New Zealand, and his crew flew down from Bari to Brindisi on the morning of July 25 and received a thorough briefing in the afternoon—"course, weather, height, where the antiaircraft is, night fighters, and that sort of thing." Culliford continued:

> They were not able to give us a great deal of information about the nature of the landing ground but we were told we would be escorted by two Liberators.... We knew we would be picking up bodies and ... something else—some type of secret weapon.... We knew [that the mission] was important, of course, but we knew nothing about details.

Even though the mission began at 20.00 hours, it was still light. One of the two B-24 Liberators that was to provide escort until darkness was unable to get off the ground, and the other could not climb as fast as the Dakota and was soon ten to fifteen miles behind. Culliford was not perturbed: "We didn't reckon there was too much chance ... if anything suspicious appeared, we would have turned around and headed back to the Liberator as fast as we could go."

Alone after crossing Yugoslavia (the Liberator had proceeded on its own mission following the brief escort assignment), Culliford and his crew

> went smack up along the side of Rumania and then [eastern] Hungary [to avoid the central Hungarian plains where night fighters abounded], nipped across Czechoslovakia and the Carpathians and came over ... the Czech-Polish border ... It was a dark night ... and the first check we got was one of the [Tatra] peaks that still had some snow on it. We then nipped down ... and Willy said, "OK, that's about it," and we flashed them [a Morse code letter *K*; the partisans on the ground promptly answered with an *M*] ... and there smack,

> bang, dab, directly beneath us there they were. This was a superb piece of navigation for I guess we had traveled some 600-odd miles following a roundabout course. Perhaps more.

Under the best of circumstances, the partisans would have been apprehensive over the arrival of an Allied aircraft deep within German-occupied territory. But that particular night called for exceptional precautions and nerve. Incredibly, on the very day of the Wildhorn III mission some hundred members of a German antiaircraft battery had wandered into the Motyl area! And late that afternoon a couple of Fieseler Storch light aircraft landed on the field itself, and their crews chatted with local soldiers. Fortunately, both planes took off just before dark. But, of course, the soldiers were still there.

When Culliford returned to Poland on the twenty-eighth anniversary of the Wildhorn III landing, former partisans told him that the soldiers, thinking nothing amiss, had retired for the night. Immediately thereafter, a group of thirty or so partisans surrounded their quarters, sten guns and pistols ready. Culliford learned that all had remained peaceful until about 23.00 hours,

> when I overshot my first landing [attempt] and had to go around again. My landing lights lit up the houses as bright as day, shone in the windows, and woke the Germans; and, to their horror, lit up the partisans as well. This was only momentary but was followed by the noise of two 1200 hp engines at full power as we struggled to clear the trees. The Germans were aroused and started to move. When they stuck their heads outside they heard the distinct and unmistakable sound of sten guns being cocked in the darkness outside, so they returned again and took no further interest in the proceedings. Jolly good show, that pilot was a bit lucky. Hell, it was me!

But we are getting a bit ahead of the story. When Culliford was approaching Motyl, he saw the partisan lights come on all around the field's perimeter. "I sort of swung out," he recalled, "and came around for a quick approach just to get the thing [airplane] down so as not to hang around in the air any more than I needed." But, he added,

> I found I had come in too high, but just marginally so. . . . I [was] more than halfway down the runway when I [found] that I [couldn't] get her in . . . I had my landing lights on, and I had to whack her open and holding the aircraft down, I tried to pick up sufficient flying speed so that it was maneuverable . . . I was that close to a stall. So while I was doing this . . . trees suddenly loomed out at me. I pulled up, and Szrajer the Pole screamed, "Look at your speed!" We were going so slow that it didn't even mark. But we didn't stall, and we made a hell of a bloody din and round we went and the second time the landing was fine.

On the ground the four crewmen got off, some twenty cases of equipment for the AK were unloaded, and then loading operations commenced. Landing procedures developed by Captain Wiodzimierz Gedymin worked, and the first half of Wildhorn/*Most III* was over. The landing, from the point of view of those on the ground, is described by Jósef Garliński:

> [It] was naturally intended to take place in secret and with as little noise as possible. But all the neighbouring villages knew what was afoot, as they had helped to man the reception committee and patrols, while the great machine's approach could be heard over the slumbering field for miles around.... As soon as it landed it was surrounded by a noisy throng of soldiers in the most varied attire, some of them barefooted, who shouted halloos and greetings and dragged off the heavy suitcases to the carts in readiness near by. The cover patrols were of course keeping watch, but the scene would no doubt have been very different only a year or two earlier, when the German troops were not harried by the pressure of the approaching front [i.e. the Russian advance from the east].

Within fifteen minutes the A4 parts were aboard (first priority), followed by Chmielewski, Retinger, Arciszewski, Second Lieutenant Tadeusz Chciuk (a political courier), and Czeslaw Micinski (in command of the party). Culliford revved up his engines and attempted to taxi into position for the takeoff. He explained what happened during his debriefing after the mission:

> I experienced some difficulty in unlocking the parking brake; then, when I had done so, opened the throttle for takeoff to the northwest. The machine remained stationary, though at approximately 50 inches of boost [amount of pressure in the manifold] the tail left the ground. I sent the second pilot to see if we were bogged, and he returned to say that he didn't think so, so I got out myself to have a look. The wheels had sunk slightly into the ground which was softish underfoot, and the marks where we had taxied were clearly visible, but in view of the boost used, the ground in front of the wheels did not appear sufficient to stop it moving. I concluded that, although the brakes were off in the cockpit, the mechanism might have broken somewhere and therefore the brakes were still on. My second pilot came up to tell me that the Germans were only a mile away, and that unless we could take off at once we would be forced to abandon the aircraft and go underground with these people. With the aid of a knife supplied by a Polish gentleman on the ground, we cut the connections supplying hydraulic fluid to the brake drum. In spite of all boost used the machine still refused to budge.
>
> I stopped the engines and reluctantly prepared to destroy the machine. We managed to persuade the people on the ground to delay

> a little, and on investigation it was found that the wheels were deeper in the earth, although they showed no signs of having revolved. Flight Lieutenant Szrajer managed to produce a spade and each wheel was dug out. The passengers were reloaded with their equipment, the engines started, and we tried again. At 50 inch [boost] the machine slewed slightly to starboard and stopped. We again stopped the engines, and once again prepared to demolish the machine—the wireless operator tore up all his documents and placed them in a position where they would surely burn with the aircraft, we unloaded our kit and passengers and again looked at the undercarriage. The port wheel had turned a quarter of a revolution.
>
> Knowing that the personnel and equipment were urgently needed elsewhere, we persuaded the people on the ground to dig for us for another thirty minutes. This time the machine came free, and we taxied rapidly in a brakeless circle, and found that the people holding the torches for the flare path had all gone home [unknown to Culliford, the partisans had returned to defensive posts nearby to protect the airstrip against the possible arrival of German troops known to be billeted nearby]; we came round again with the port landing-light on and headed roughly NW towards a green light on the corner of the airfield. After swinging violently to port towards a stone wall, I closed my starboard throttle, came round in another circle, and set off again in a NW direction. This time we ploughed along over the soft ground and waffled into the air at 65 mph, just over the ditch at the far end of the field.

Once airborne, Culliford and his crew discovered that they could not raise the undercarriage, all hydraulic fluid having been lost. So they poured water from their emergency rations into the hydraulic reservoir until they were able to pump up the undercarriage by hand.

Since they were running late—Culliford recorded a one hour and five minute stay-time at Motyl—they set course direct for Lagosta Island from the Carpathians, a path that took them through an area near the Danube known to be infested with night fighters. Despite the danger, they had to clear Yugoslavia by daylight. Though several Ju88s were spotted, the Dakota was unmolested and was able to overfly Lagosta on schedule. As they approached Brindisi, a strong crosswind was reported blowing across the single runway. "But," recalled Culliford,

> there was a runway under construction that was well positioned into the wind. Because of our Special Operations status, permission was given to land on it. I think I did the most perfect precautionary landing I have ever done in my life. Fortunately, the Dakota was so designed that if you lost hydraulic pressure, the undercarriage would fall down. And with water in the system, we were able to pump the flaps down. We did a beauty! So that was that. There we were.

The hazardous mission officially ended when they touched down at 05.50 the morning of July 26. Chmielewski, accompanied by Micinski, left as soon as possible for London and on July 28 delivered the A4 parts and accompanying report to the Polish General Staff. Following preliminary study and translation, everything was turned over to British intelligence.

Until they actually had their hands on A4 parts and on complete reports based on AK observations, and mindful of the fact that much information reaching them from Poland was incomplete and misleading, the British understandably looked for confirmation of happenings in and around Heidelager. Throughout 1943 and well into 1944 there was nothing to be done: UK-based Mosquitos could not fly the thousand-mile distance to Blizna and return. By the spring of 1944, however, the war situation in Italy was such that they could operate from a captured airfield at San Severo north of Foggia, only six hundred miles from Blizna.

The first Mosquito recon flight was made on April 15 and the second on May 5. Air Ministry intelligence chief R. V. Jones explained that

> When the photographs became available, we found, as we expected, all the symptoms of flying bomb activity, but we did not know that the site was identical with that used for the rocket trials. All the available photographs had, of course, been through Medmenham, but they had not spotted anything other than V-1 activity, and I had therefore asked for further photographic cover of the whole area in the hope of locating the rocket launch sites. It happened that there was delay in carrying out this wider reconnaissance; and I therefore started to think, almost desperately, about the possibility that the rocket might be on the photographs that we already had. It was a natural tendency for the Germans in unfriendly territory to concentrate their activities within defensible perimeters (we had already noticed this tendency with radar equipment); and so I argued that there was a sporting chance that, although the [V-1] was being developed by the Luftwaffe and the V-2 by the German army, the threat from the Polish Resistance might have drawn them together into one perimeter.
>
> It happened that on the night of, I think, July 16 or 17, I decided to sleep in my office because my wife had taken our children to Cornwall to get some respite from the V-1 bombardment [which had begun in June]. After dinner, I went over the argument that the rocket might be on the Blizna photographs; and I therefore scanned them as intensively as I could. After some time, I saw the outline [on the May 5 cover] of what I believed to be a rocket; and I checked the dimensions. Within the limits of photographic definition, which was relatively poor, they agreed with those of the missile that I had [earlier] found at Peenemünde. . . . When I realized that I had probably found

> the rocket I wrote the information down, giving the sortie reference [of the May 5 flight] and the position of the rocket as *x* and *y* coordinates in millimeters from the edges of the photograph. I put this in an envelope, to draw other people's attention to it if, as was quite possible during the V-1 bombardment, I myself was killed before I could tell anybody else about it. I also telephoned my deputy, Dr. F. C. Frank, at his home in Golders Green, giving him the same information, so that there were dual channels in case either the envelope or Dr. Frank did not survive the night.

War is full of ironies, and the Blizna episode turned out to be one of them.

By mid-July 1944, the British possessed indisputable photorecon evidence of the A4's presence in southern Poland. And less than a fortnight later the pieces of an actual missile were en route to London. The impact of these and related events should have been spectacular.

But it was not to be.

On June 13, 1944, a month before Jones had positively identified the Blizna A4 and six weeks before Chmielewski delivered the captured missile parts to London, A4 flight No. V89 took off from Peenemünde with components of a *Wasserfall* surface-to-air test missile radio guidance system aboard. The purpose of the flight was to check out the response of ground joy-stick control over the missile and to determine what influence engine exhaust might have on radio signals passing to and from the ground transmitter.

To his consternation, the ground controller lost sight of V89 as it passed through a cloud and was unable to reacquire it. Out of control, the missile—or rather parts of it; the familiar air-burst problem was still unresolved—ended up scattered in and around the village of Knivingaryd, not far from Kalmar, in southern Sweden. Ernst Steinhoff explains how the mishap came about:

> In order to test [missile V89] we worked out a program taking into consideration such factors as Earth rotation. I told the man in control to memorize it, meaning that he operate the joy stick so many seconds to the right, so many seconds to the left. We had it all figured out that it [the missile] would fall into the line of fire [along the line of a preplanned trajectory]. But this man became so excited that he applied the corrections the wrong way. So I told him, "You know what you just did? You did all the commands with the wrong sign!" I thought at first that he had done it intentionally, but when I realized how excited he was [I knew] he couldn't have.
>
> We had set our target so close to the Swedish coast because we could determine from landmarks where [the missile] was. We said, "It's gone too far!" On that day, we did not actually change the

> azimuth, we only rotated it by 90 degrees so that it would go in that direction [in other words, toward Sweden] rather than along [Germany's] Baltic Sea shore; because if it had strayed the chances that it would fall on land were much greater than doing it the other way [i.e., the way it was done].... So I told my people to make an immediate evaluation of all the data.

Uncertainties in the data analysis left some doubt as to where the missile had come down. While Steinhoff was flying over the island of Oie shortly afterwards, he received a call from his commanding general: "Steinhoff, did you fire into Sweden?" Steinhoff's lame answer was, "I must have," even though preliminary analyses had shown that the missile might have terminated its flight just off the Swedish shore.

Although the V89 wreckage might have come down in the sea, fortunately for the Allies it did not. The main crater dug into the Swedish soil was some 5 feet deep and 15 feet across. One man, thrown from his horse by the explosion, looked up and saw aluminum-colored pieces raining down on the countryside, presumably fragments from the propellant tanks. Some scattered parts were snatched up by souvenir hunters while others were lost in nearby lakes and ponds. In all, however, about 2 tons of missile wreckage were gathered up by Swedish authorities and transported to Stockholm.

According to a contemporary description, among the parts recovered were the propulsion unit body; main burner unit; fuel supply system—including one almost intact pump; directional control—the jet vanes; elliptical propellant tank; main structure and fins; and radio equipment. One British report on the Swedish incident would later comment that "to judge from the complexity of the radio equipment no effort has been spared to ensure the greatest possible accuracy [of the missile]."

Within a week enough information had become available to British Air Intelligence for a preliminary description of the A4 to be prepared:

> Diameter is probably about 6 ft., length unknown. Rocket propulsion definitely indicated.... Large Venturi-shaped object is certainly main body of propulsion unit, and aluminium object with 18 cups appears to be main rocket burner unit.... Turbine driven compressor, believed to be for fuel pump, outside diameter 19″, estimated horsepower 300, turbine of exhaust type, two stage and apparently driven by gases generated from separate tank fuel or fuels.... Directional control may be adjustable vanes projecting into jet stream.... Actual weight of wreckage so far recovered exceeds 2 tonnes... Radio control 1 F unit similar to that in HS 293 and FX unit [German air-to-surface guided weapons].

Once the Swedes had completed their studies of the A4 wreckage, everything was placed into storage. But only temporarily, for British Air

Intelligence's A.I. 2(g), the technical branch, arranged for two of its officers—Squadron Leaders Burden and Wilkinson—to go to Sweden in late June to examine carefully the remains of V89 and to make arrangements for their shipment to the Royal Aircraft Establishment at Farnborough. Negotiations progressed rapidly, and by mid-July the first lot of parts reached the United Kingdom. The rest followed within a fortnight. Colonel Bernt Balchen, well-known Norwegian-born aviator and Arctic explorer and later a citizen of the United States, would tell von Braun after the war how "in the spring of 1944, while I was operating out of Scotland into Sweden, I had my first experience with rocketry when my organization retrieved a V-2 test-fired from Germany. We flew that V-2 out of Sweden to England.... [It] gave us information on the tremendous power of this weapon borne out by my personal observations some months later when I chanced to be in London when the first one landed there from the other side of the Channel."

In the light of preliminary reports prepared by Swedish military and scientific experts and by the visiting RAF intelligence officers, on July 14, 1944 (a few days before the first shipment of wreckage reached Britain), the Big Ben Sub-Committee felt it prudent to assume a total missile weight not greatly in excess of 30 tons, down considerably from earlier estimates but still high. Based on observations of "eighty-foot craters at Peenemünde and in Poland the warhead is probably between three and seven tons," the Sub-Committee report concluded, though the members admitted that "we cannot yet decide whether a single or multistage is used; the present slender evidence is conflicting, but on the whole points to a single stage."

The report went on to note: "Officers in Sweden report that control is effected through internal movable surfaces deflecting the efflux gases in the jet, while the Poles [observations made by resistance members at Blizna and at the uprange impact sites] state that there are external rudders. Perhaps both sources are correct."

External command was assumed to be by radio link. The British knew that controllable jet steering "would make slow launching feasible," following vertical takeoff. The "launching apparatus" was felt to be a "closed, smooth tub mortar, weighing 3–4 tons, capable of use as container to protect rocket (empty or filled) during transit." The rocket motor was thought to burn for about 40 seconds, during which time it would accelerate the missile to over 4,000 feet per second. The peak of the suspected 150-mile-long trajectory was believed to be between 30 and 40 miles.

Once the parts reached Farnborough from Sweden, the assembly job began. A member of the RAE team happened to be an American,

Thomas F. Dixon, recently transferred to the United Kingdom from Washington. A production expeditor on the US Navy's 5-inch Holy Moses solid-propellant rocket program, his technical experience was put to good use as Farnborough experts sought to make sense out of the dozens of pieces collected across the Swedish countryside.

> They were strewn all over the floor—pumps, thrust chambers, nozzles, valves, controls, sections of the missile ... I was amazed at the complexity of such a missile.
>
> Day and night we pieced together the parts. Some of the systems were so different from our experience or so complex in design that at times we felt like we were putting together a huge three-dimensional jigsaw puzzle, with only faint clues and hunches as to which pieces fitted where.
>
> One important fact came to us quite early—and through our noses. When we gathered together the fragments of the tanks, I was assailed by an old odor with which I was very familiar—the odor of alcohol. We immediately agreed that alcohol must be the fuel.
>
> By spreading out the assembled fragments on our wires and supports, we finally began to see the length of the missile and the configuration. The rocket scaled out about 46 feet long and appeared to weigh something less than 14 tons when fueled. Measuring the size of the warhead compartment, we calculated for a standard explosive and estimated that the warhead weight could go up to two tons.... Through the size of the pumps and pipelines, we calculated the engine thrust and the chamber pressure. The design of the valves and pipes told us that liquid oxygen must be the oxidizer used with the alcohol propellant.
>
> With this in mind, we laid out the missile. We sized the thrust and the specific impulse from the data acquired in the early days of [U.S. rocket pioneer Robert H.] Goddard's work and that of the Jet Propulsion Lab and Cal Tech. After we had pieced the tanks together as best we could, we estimated their size. This fuel capacity we then converted into an estimate of the duration of the rocket, using the necessary standard equations we thought proper for the design. Figuring the weight of the rocket from our previous work on the various components, we calculated the V-2's range with a warhead. This was most important to us, for this would tell us at least the circular areas in which the launching sites would be located. We calculated the range of the V-2 to be 175 to 200 miles.

The recovery of the Swedish wreckage helped bring the A4 down towards its actual size and weight. In the first column of the accompanying table are estimates of missile characteristics advanced in late November 1943 in a highly classified report compiled by engineers of the Armaments Design Department at Fort Halstead. In the second column values compiled in early July 1944 are summarized; these were prepared after the V89 missile crash but before the parts had been assembled at Farnborough. Finally, the third column shows actual values, which are quite close to the figures given by Dixon. Even in early July 1944, most British intelligence experts tended greatly to overestimate the A4 weapon's weight and warhead. Incidentally, Bodyline engineers back in late 1943 thought the A4's thrust was something like 150 tons and propellant consumption about 0.9 tons a second. Actual values were less than 30 tons and less than 0.15 tons a second, respectively.

Estimates of Bodyline/Big Ben Characteristics

Characteristics	*Nov. 1943*	*July 1944*	*Actual*
Length, feet	40.5	35 to 45	46.9
Diameter, feet	6.5[a]	5 to 6[b]	5.4
All-up weight, long tons	48	31 to 37	12.6
Empty weight, long tons	7.8	5 to 7	3.9
Propellant weight, long tons	31	20 to 22	8.8
Warhead weight, long tons	9.2	6 to 8	1

[a] Diameter over fin estimated to be some 10 feet.
[b] Believed to be tapered toward the tail.

The Swedish parts, followed up by the arrival in Britain of the components brought over from Blizna, permitted the Big Ben Sub-Committee to report that

> the Intelligence case for the Rocket has . . . so grown in strength that there is now little doubt that the Germans have developed a technically impressive missile which they call the A4, and whose performance is good enough at least for a desultory bombardment of London.

In the light of the Swedish bonanza, the value of the parts so gallantly retrieved from Poland by the AK and Culliford and his crew was considerably reduced. R. V. Jones commented to the authors that by the time Chmielewski delivered the A4 parts from Poland, "we already had some of [those] from the rocket that fell in Sweden, and so the effect [of the former] was not very great. Had he landed here [in the UK] a month before," Jones continued, "the impact of his information and the components that he brought with him would have been spectacular. What it comes to is that the Poles deserve great credit for what they did, and it was just a matter of luck that we had most of the information somewhat earlier by other means."

Back in Blizna, meanwhile, time was running out for the Germans. Realizing that the Soviet armies were inexorably pressing westward, plans were made late in May for evacuation, and in June—well before Wildhorn III—some Heidelager elements had moved out. Rocket firings continued, however, into early July when the full exodus got under way. During the final days of July and in early August the AK briefly occupied the camp, only to be driven out by German forces retreating through the area from the eastern front. British reconnaissance planes observed that all signs of activity had ceased at Heidelager by July 27, and on August 6, 1944, the Russians announced its capture.

The deteriorating war situation on the eastern front notwithstanding, the Germans were determined to continue to improve their missiles. They moved their *Vergeltungs-Express* (the ten-car train that had provided sleeping, eating, and working quarters for ranking launch and supervisory personnel at Heidelager) and transport vehicles of many types westward to a new position known as Heidekraut. Located in a heavily wooded area about 10 miles to the east of Tuchel in Germany's eastern Tuchel Heath forests (Tucheler Heide), it was a primitive site with virtually no fixed installation. Flight observations were usually made from the train, which was parked at nearby Lindenbusch. Firings from Heidekraut were made toward either of two targets some 120 miles to the south, Ziel 1 or Ziel 2, so that, as Dornberger remarked, if one were to come "under question and investigation we could continue operations by firing on the other." "Question and investigation" meant, of course, incursion or compromise by enemy regular or probing underground forces.

Testing at Heidekraut continued during the autumn and early winter of 1944 until, in mid-January of the following year, that site too had to be abandoned. The missile training group moved eastward to Wolgast and then, toward February 15, transferred to Rethun on the river Weser. No test firings were made from either of these locations, in part because of the general war situation and in part because the five months of Heidekraut operations had resulted in significant improvements in A4 accuracy, in the sensitivity of the warhead's fuse, and in overcoming the irksome air-burst problem.

When the Russians occupied Heidelager in Poland in August 1944, no reports of major demolition of buildings and test structures were forthcoming. It was evident, however, that all rockets and flying bombs along with their mobile support equipment had been removed. Or so it seemed to the Polish AK forces. And so reported the Russian occupiers to their English and American colleagues.

But the Western Allies were not so sure. There still might be something left of interest. Before the Soviet occupation of Blizna/Heidelager, the Allies had been able to examine the remains of the Swedish rocket as well as the parts and documentation brought out of Poland by Culliford and Chmielewski. As a backup to these extremely important sources of scientific and technical information, the Allies resolved to dispatch a technical team to study both the launch and impact areas in hopes of turning up additional information and parts before the Germans could unleash the expected A4 offensive against the United Kingdom.

What became an Anglo-American-Soviet mission to the former flight-test ground was instigated by R. V. Jones and initially placed under the command of E. G. Ackerman. Not impressed with the way the mission was shaping up, Ackerman asked to be released. The mission that finally evolved was headed by Colonel Terence R. B. Sanders (of the Ministry of Supply), with US Lieutenant Colonel John A. O'Mara as his deputy. Sanders' British team members included Lieutenant Colonel A. D. Merriman and Technical Captains G. J. Gollin and Standish Masterman. Flight Lieutenants C. H. Burder and G. Wilkinson, technical representatives; Captain L. H. Massey, the Polish interpreter; and Flight Lieutenant T. J. Durkin and Flight Officer D. Floyd. Russian interpreters, also went along. On the American side, O'Mara was assisted by a radio specialist, Captain E. M. Usher; and a technical officer and interpreter, Lieutenant Colonel Steven J. Zand. Four officers under Lieutenant Colonel K. Sinozersky made up the Russian contingent; two were interpreters—English and German—one a foreign relations specialist, and one (Lieutenant Colonel J. B. Shor) a technical man.

On July 29, eight days before the fall of Blizna, the mission left England. A number of delays ensued, in part caused by the fact that, although Blizna fell on August 6, heavy fighting continued in the immediate area until Debica was captured on the twenty-third. Even after that, up to early September the fighting lines were but five or so miles away.

G. J. Collin, a British fuel expert, related the story of the trip to Blizna to the authors. Leaving London, the Sanders group flew to Casablanca, then to Teheran and on to Moscow. After some ten days in the Soviet capital, the great moment came, and

> we assembled at about 6 AM at Moscow Air Station. At this moment we were introduced to our Russian opposites, of whom only two could speak English and were included as interpreters. We took with us an Air Force type called Floyd who could speak fluent Russian. We left Moscow and flew to Poltava, where we were assured we would pick up our American colleagues and be given breakfast.
>
> Although we had a snack before leaving Moscow at about 5 AM, by the time we got to Poltava air field we were hungry and for a long time we sat in the shade under the wings of our aircraft dreaming of breakfast. About an hour later a fleet of jeeps took us into the camp where a general entertained us to what was called breakfast, but was really a luncheon lasting from about 11.30 AM to 2.30 PM, complete with caviar, vodka and champagne, loyal toasts and speeches. Our

host, the general, proposed the king's health accompanied with dazzling smiles as all his teeth were stopped with or constructed of platinum [actually, probably stainless steel], a common Russian feature. After lunch, having eaten too much and some of the expedition having clearly drunk too much, we took off with a fighter escort on either wing for the front. Here we landed about forty miles behind the front, where the Russians were bogged down before Warsaw. Our nearest major spot in the fighting line was at Debica. Having changed into jeeps, we were driven to our village where a row of villas had been connected by a specially constructed pipeline to a water supply. In the villas we had formidable Russian lady soldiers to act as batmen to us and to wait on us at table. When they performed the latter duties, their medals clinked strenuously and we were assured that these medals were given to these ladies on account of the number of Germans they had killed in open combat. The ladies were what you might call "First Aid Orderlies," who took their places in the front line armed with rifles. In the evening we were entertained with very good concerts, films, and occasionally dancing to piano accordions played by a blind soldier. Our waitresses and batmen acted as dance partners. I seem to remember I pleaded a sprained ankle.

We got down to work and, surrounded by a bodyguard of boys aged seventeen or eighteen, each armed with a tommy gun, surveyed Blizna Experimental Station, its empty workshops, its long runs of small-gauge railway, and its firing sites. Senior members of the party (in spite of Russian obstruction) toured the countryside, getting eye witness reports from peasants.

This existence was rather fatiguing because we left early in the morning and got back to the mess late at night. Our Russian colleagues were indefatigable. The countryside was littered with a fair number of dead horses, the human corpses having presumably been interred. On our arrival at the dining room it appeared as though the tables were covered with a black tablecloth, but on entry this rose into the air and revealed itself as several thousand flies. We explained to the Russians that food should be covered and this they did with muslin, on which the flies thereafter browsed in peace. The result of these insanitary conditions was that the whole expedition suffered from food poisoning, an inconvenient disease when one considers that our villas were surrounded throughout the night by young men armed with tommy guns and that progress through the gardens [to the field latrine] was frequently interrupted by commands of "*Stoi*" for purposes of identification with flashlamps. However, in spite of the disease none of us died and we completed our survey, recovering from swampy ground important fragments of the quarry. The most important exhibit I had the honour of finding ... was a test sheet giving the actual loading of alcohol and liquid oxygen before firing. The circumstances of the finding of this document [in an old, fouled latrine] caused much merriment....

> One evening at dinner one of the two colonels in charge of the expedition announced, "Tomorrow we move." The next morning, having celebrated our move in champagne as was our custom, we got into the jeeps and drove twenty miles to our airfield. From there we flew at treetop height in a very bumpy flight to near Lublin. This is the only occasion on which I have looked through the opposite window of a Dakota and seen buses and motor cars going past. We were assured that at this low level no German fighter could attack us. On arrival near Lublin we found our airfield occupied by cows and our pilot said that even if he was shot for it, he could not land us there, but took us to an airfield [even] nearer Lublin. This caused consternation to the Russian ground commissars, who locked us in the railway station for the rest of the day and drove us about midnight into Lublin, where we were given a magnificent banquet by the governor. The Russian discomfort at our landing in Lublin was because they had in the previous few days set up a provisional government there, and this government was in opposition to the Polish émigré government in London. The opposition was so intense that actually a civil war was in progress in Poland, the London underground army fighting the Lublin Committee (Russian) underground army, both sides of course forgetting their war against Germany. It was due to this open warfare between two underground armies that we came under fire when collecting souvenirs at Blizna—a very unpleasant experience when one imagined oneself to be twenty miles behind the lines.

The next day, the mission moved into the target areas, including the neighborhood of Sawin and Sarnaki, about 100 and 155 miles, respectively, from Blizna. Colonel Sanders picks up the story at this point:

> In the Sarnaki area area evidence was quickly obtained that this had been the main target area for rockets launched during May and June. The area in which the craters were found consisted of a stretch of ground on either side of the river Bug, extending to about eight miles wide by eight miles deep and including the villages of Sarnaki, Meźenin, and Ogrodniki. The majority of rockets arriving over this area had burst in the air, fairly high up, but the warheads had generally landed intact and had blown up, to form craters of varying size. Some warheads had been dummies, filled with sand. In a few cases the whole rocket had landed intact and had blown up on the ground, forming much larger craters, of the order of about 70 feet diameter by 25 feet deep.
>
> Both at Sawin and Sarnaki the Germans had been very thorough indeed in collecting the fragments. It appeared that they knew exactly when and where to expect the projectiles to fall and SS troops were always on the spot instantly to collect the fragments. Nothing at all was left in the neighbourhood of the craters, and although the inhabitants did collect some pieces, the Germans generally found out and

> forced them to give them up by threatening to burn their villages. Nevertheless, some parts were produced which had been hidden in houses and farmyards over the intervening period, and some parts which had been under water when the river was in flood in May and June were discovered lying on the river bank.

Was the mission worth the trouble? Gollin didn't think so.

> Politically and scientifically [he confided] the expedition cannot have been a success. Nevertheless, we all felt we owed a debt of gratitude to the prime minister and Mr. Duncan Sandys for sending us so far, for enabling us to learn a lot about the Russian army and their habits, and finally for setting us up as connoisseurs of the Russian ballet.

More seriously, Jones had this to say to the authors:

> I can still recall the signals coming back to London from the mission, including accounts of their attacks of dysentery while they were waiting on the Russian border for permission to enter. The Russians, of course, having been alerted (if they did not indeed know already), made quite sure that they had the mission and Blizna itself well covered, and I suspect that the only positive result of the mission was that various loyal Poles who came forward and declared themselves to the mission thereby prejudiced their own chances of survival. I can also remember the triumphant signal from the head of the mission reporting that the Russians had at first been incredulous about the alleged rocket developments at Blizna but that they would no doubt be very impressed when the mission returned to Moscow with the rocket parts that it had discovered. This signal itself seemed to tempt Providence, and the Russians evidently made sure that Providence succumbed. For, when the cherished packages arrived in Britain, it was found that the Russians had abstracted the original contents and had substituted old aeroplane parts.

The leader of the Blizna expedition recorded a number of basic findings and/or confirmation of what was already known or suspected:

(1) The evacuation of Heidelager began in early June, though firings continued until the first week in July.

(2) Blizna was captured by the Russians on August 6.

(3) A4 launching was vertical and mobile. The launch platform was about 18 feet square, and was often made of logs.

(4) Liquid oxygen and ethyl alcohol were the oxidizer and fuel, hydrogen peroxide the auxiliary propellant.

(5) Missile reliability was poor.

(6) About four hundred persons were directly involved with launch operations.

Among the parts found in some ten craters inspected during the course of the mission were the rear and forward parts of a tail fin, most of an alcohol tank, a complete combustion chamber, elements of the radio system, servomechanisms, and many pieces of fuselage. Though crated for shipment to the Royal Aircraft Establishment at Farnborough, the A4 parts were removed by the Russians (probably in Moscow) and, as Jones noted, old airplane pieces substituted. The chagrin of RAE engineers was only somewhat mollified when they read Sanders' description of the launch sites:

> The rockets were fired from a corner of the forest. Two areas were used which have been termed the "North Firing Sites" and the "South Firing Sites" respectively.... There is ample evidence that both areas were used extensively for actual firing of rockets.
>
> The method of launching was made quite clear from an inspection of the firing sites, and from a cross-examination of local inhabitants, including one who had worked at Blizna as a bricklayer.
>
> The actual platform normally used was about 18 feet square, and consisted simply of rough squared baulks of timber, approximately 7 inches deep, laid side-by-side on a square of ground, from which the sod had first been removed, and which had then been roughly levelled up. On this platform was placed a steel base-plate, the exact shape of which is not known. It is known, however, that the base-plate was carried on a special four-wheeled trailer, the wheels of which could be wound up when lowering the base-plate on to the platform.
>
> The rocket was brought to the platform on a special lorry which had hand-operated winches, by means of which the rocket was raised into the vertical position, and lowered on to the base-plate.
>
> A little distance away, say forty yards, from the platform, was a dugout heavily revetted with timber, and roofed over with timber. This was presumably the control post. Accommodation was provided in shallow dugouts for at least two fair-sized lorries and for various signalling posts.
>
> A shallow trench about six inches wide by nine inches deep connected the control post to each of the firing platforms and presumably accommodated the electric wires by means of which the actual launching of the rocket could be performed from the control post. Several platforms were served by one control post, and it is believed that attempts were sometimes made to fire salvoes of three or four rockets at once.
>
> When firings were about to take place, a siren was sounded and everybody not engaged in the actual work was cleared away from the area. Sometimes after the siren sounded, nothing would happen for about three hours. Sometimes when firings were taking

> place, several rockets would be launched at fifteen- to thirty-minute intervals, which indicates that several platforms were being used.

As leader of the expedition, Colonel Sanders was, of course, quite aware of the principal shortcoming of the mission: timing. When he issued his preliminary report from Moscow on September 22, 1944, V-2 rockets had been falling on London for a fortnight and almost everything his Anglo-American team had learned was already known by Allied intelligence.

Once again events in Poland had been overtaken by the tides of war.

CHAPTER 9

A4 to V-2: The Final Months

During the months preceding the end of the Blizna episode, British intelligence was desperately attempting to grapple with two basic questions. When would the rocket and the pilotless aircraft/flying bomb missiles enter into combat service? And which would be deployed first and from where? Directly related to these questions was another: Which of the sites cropping up along the Channel coast were designed for which weapon? Or could some, or even all, accommodate both?

Medmenham's Neil Simon thought that "a 38-foot rocket, like the ones at Peenemünde, could just be maneuvered into a ski building, at least without its fins on."

R. V. Jones disagreed. To the scientific intelligence chief at the Air Ministry, the ski sites were not for A4s but rather for pilotless aircraft. To substantiate his position, Jones hoped that photos might be located showing the pilotless aircraft at its Peenemünde-West development base. Looking over coverage then on file, Constance Babington Smith did, in fact, turn up a midget aircraft.

> The absurd little object was not on the airfield, but sitting in a corner of a small enclosure some way behind a building which I suspected, from its design, was used for testing jet engines. . . . I named it "Peenemünde 20," as its span was about 20 feet, but there was precious little I could say about it.

In the light of this discovery, Medmenham decided that photography of Peenemünde should now be reexamined to ascertain once and for all if the rocket was so large that it required rail logistics; or, should it be relatively small in size, if it might rather be handled by mobile road transport. As Kendall, Rowell, and others at Medmenham pored over photoreconnaissance coverage, they found that what had previously been interpreted as a 40-foot-high column was, in fact, an A4. Moreover, the fan-shaped stretch of foreshore near Test Stand 7 had been covered by what seemed to be asphalt and presumably was being employed for launching exercises. The idea of a huge 40- to 100-ton rocket launched from a monster mortar faded away and a more manageable missile appeared in its stead.

While Peenemünde was being photographed anew, confirmatory information on A4 flight-testing was being received through ground intelligence. Thus, on April 15, 1944, a "first class source" reported (through the Czechoslovakian underground):

> A rocket was fired from the ground. It rose with a loud noise and retained its firing angle without any curvature of the trajectory until it reached a height of 8,700 meters.... Then, after curving over sharply, it continued its flight in a horizontal direction.
>
> The torpedo-shaped body displayed two small "carrying" surfaces. At the bottom of the rocket body a pipe was noticed, out of which (from the moment of firing) a flame approximately 30 meters in length issued with a hissing sound. It was blue-white in colour, and only directly behind the pipe opening was it red. During the ascent of the rocket, successive strong explosions of considerable intensity were heard.

Although Crossbow as a name had existed since December 1943, and myriad Allied bombing raids had taken place under the aegis of Operation Crossbow, it was not until the afternoon of June 19, 1944, that the Crossbow Committee of the War Cabinet met for the first time in formal session. This was followed the next day by a Big Ben (new code name for the A4 rocket, replacing Bodyline) Sub-Committee meeting in room 350 of the War Office. Officially participating were:

Professor C. D. Ellis, Scientific Advisor to the Army Council, Chairman
Professor J. E. Lennard-Jones (Chief Superintendent Armament Research), Ministry of Supply
Professor D. M. Newill, Scientific Advisor, Special Operations Executive
Professor L. Rosenhead, Ministry of Supply
Major General S. F. Rowell, Director of Tactical Investigation, War Office

Mr. F. E. Smith, Chief Engineer Armament Design, Ministry of Supply
Sir Geoffrey Taylor, Trinity College, Cambridge
Sir George Thomson, Scientific Advisor, Air Ministry
Sir Robert Watson-Watt, Deputy Chairman, Radio Board

The minutes of the meeting show that the following were also present:

Mr. I. Lubbock, Asiatic Petroleum Company
Colonel K. G. Post, Ministry of Supply
Lieutenant Colonel R. M. Pryor, representing the Director of Military Intelligence, War Office
Dr. H. P. Robertson, representing the Assistant Director of Intelligence (Science), Air Ministry
Mr. T. R. B. Sanders, Ministry of Supply
Dr. W. H. Wheeler, representing the Comptroller of Projectile Development, Ministry of Supply.

Major G. Essame, War Office, Army Council Secretariat served as the secretary. Two members of the Sub-Committee unable to attend that day were Dr. R. V. Jones, assistant director of Intelligence (Science) of the Air Ministry, and Sir Alwyn Crow, controller of Projectile Development, Ministry of Supply. Because of Lubbock's research into fuels, the Sub-Committee recommended that he become a full member. Also recommended for full membership was Lieutenant Colonel O. M. Solandt, superintendent, Army Operations Research Group, and representative of the scientific adviser to the Army Council.

On the same day that the Big Ben group met, Churchill brought Duncan Sandys once again onto stage center by directing that he organize a study of the probable effects on the United Kingdom of a flying bomb and Big Ben offensive. Moreover, he was to determine what defensive efforts and countermeasures could be taken, and so, after seven months officially unattached from the now less-than-secret weapons investigation, Sandys returned

Crossbow's two subcommittees—Big Ben (A4), under the chairmanship of Professor Ellis, and Diver (Fi103, the flying bomb), under newly arrived Sir Thomas Merton—moved rapidly into action. In concert with the military commanders, offensive operations against Diver sites in France continued unabated on a top-priority basis, and the defenses under General Sir Frederick Pile (antiaircraft) and Air Marshal Sir Roderic Hill (Fighter Command, later known as Air Defence of Great Britain) stiffened.

Meanwhile, Big Ben Sub-Committee members got on with their mission "... to bring scientific experts together, to allocate work between them, and to act as a reporting center through which a continuous flow of agreed technical information could be rendered to the main [Crossbow] Committee." As the Sub-Committee commenced to function, its attention was focused on the methods likely to be used by the Germans to determine the impact points of the rockets they were readying for the bombardment of southeast England; the design and capabilities of Big Ben itself; and the possible chain of command and types of organization at every level in its operational employment. The Sub-Committee suggested that the 11th Survey Regiment of the Royal Artillery might be able to locate Big Ben launch sites by means of sound-ranging and flash-spotting. Using the latter technique, these sites could probably be located within about 4 miles, whereas sound-ranging accuracy for direction was expected to be ± 1 to 2 miles and for range ± 6 to 7 miles.

The second Big Ben Sub-Committee meeting took place in London within a week of the first, this time in room 114 of the Hotel Victoria, Northumberland Avenue. It was Tuesday, July 25, 1944. Preliminary reports on Farnborough's examination of the Swedish rocket remains were in, leading to the strong belief that three types of propellants were involved: liquid oxygen, "an alcohol compound," and hydrogen peroxide.

Intelligence operates in many and curious ways. In assessing the probability that alcohol of some sort was the A4's fuel, prisoner-of-war interrogations proved useful. Thus, one prisoner, a former member of the VKN at Peenemünde, had landed up in jail in Stettin because he had been caught with two pints of A4 alcohol in his possession. From there he went to the front, where he was captured by the British. It seems that he had collected his precious alcohol by drilling a 4-mm hole in an alcohol tank car. A forty-two-year-old anti-Nazi Czech, he was inspired as much by accomplishing an act of sabotage as he was to enjoy what he termed his "A4 cocktail." The intelligence report covering the incident noted:

> The stolen alcohol, which cost P/W his job, was taken from the tank car not only as an act of sabotage but also because it was favoured as the basic ingredient for a drink. When heated slightly, and mixed with anise, it produces a very palatable cocktail. This drink was regarded with especial favour by the Russians.

A consensus of the second Sub-Committee members held that the A4 was lighter in weight than heretofore believed. And so was the warhead, though estimates as to what it might be were uncertain. In fact, two warhead models were postulated, one "heavy" and one "normal":

> The latter might be of the nature of a proof round for experimental purposes, but despite the improbability that so complicated a mechanism as the rocket would be used to fire so small a charge [about a ton] the possibility of the smaller head being used operationally

> should not be overlooked. A reduction in the weight of the charge would increase range materially.

The smaller size increased the Sub-Committee's willingness to accept the mobile operational concept. Some missiles might be ground launched, while others could be fired from railway flat cars and even from barges plying canals and rivers. To strengthen this viewpoint, the Polish AK was reporting that A4s were routinely being launched from unreinforced concrete platforms accessible only by road. Control was still believed to be exercised by radio, the missile presumably receiving on the 20 megacycle band and transmitting on the 40 megacycle band.

On July 17, the chief engineer of armament design at Fort Halstead observed that, because of the Allied bombing of five of the supposed heavy sites in France and the capture of two in the Cotentin peninsula, "it was clear that any immediate attack [by Big Ben] could only be launched from some form of 'semi-mobile' projector." He felt that semi-mobile sites "must be associated with good communications and nearby supply bases [and] that the present state of the French railways would indicate a greater dependence on road transport." Indeed, by the end of the month the big bunkers were abandoned. Watten had received a direct hit from an RAF 12,000-pound Tallboy bomb on July 6, while another destroyed Siracourt. The high-pressure pump gun (later designated V-3) site at Mimoyecques that might have been revamped to handle either the flying bomb or the A4 rocket was also knocked out by Tallboys, while the Wizernes site was so generally pummeled that it was no longer serviceable as an A4 launch bunker.

When the Armament Research Department released its interim report No. 27/44 on July 30, they found strong evidence not only that "large rockets, probably of the A.4 size, are being made by full production methods" but that the high explosive charge should weigh about 4 tons, the propellants 16 tons (9 tons of liquid oxygen and 7 tons of ethyl alcohol), and the structure 4 tons. "It is known," the report continued, "that the A.4 rocket is fitted with elaborate radio control and that it can be guided and stabilized by carbon vanes projecting into the [exhaust] gas stream." Mortar-type launchers were finally and definitely ruled out: "Under the conditions now known to exist there is no reason why the rocket should not start from rest under its own power."

Evolving intelligence on the long-range rocket led the British to assume first that the Germans would seek to develop a means to load the A4 with propellants in a sheltered area—e.g., a tunnel—and then bring it out for launch. But, Armament Design Department engineers suggested, "there now appears to be strong evidence that the rocket is, in fact, filled at the firing point and that it is fired without any form of special projector, other than a simple stand which is designed to take the weight in a vertical position and to deflect the [exhaust gas] blast so as to minimise its effect on surrounding objects and vegetation." This surmise was given added weight by descriptions flowing in from the Polish AK and from prisoners of war who had seen—or at least had heard first-hand stories of—the vaunted rocket. Photographic coverage corroborated this view.

Between June 2 and July 23, a total of twenty-six photographic sorties were conducted over Peenemünde, some of which clearly revealed the "special trolley" transporter "backed up to the vertical object" (Big Ben). As a by-product of this photographic cover, flight failures were revealed, for "... on two occasions large craters appeared ... which [did] not appear to have been connected with any bombing of this area." A typical POW report revealed that "the rockets are loaded onto special trolleys which are towed by a heavy tractor."

By August 8, at the time of the fourth Big Ben Sub-Committee meeting, it had been learned that the chief engineer of Armaments Design, the chief superintendent of Armament Research, the comptroller of Projectile Development, the Asiatic Petroleum Company, the Royal Aircraft Establishment at Farnborough, A.I.2 (g) (the technical branch of Air Intelligence), and the director of Operations (Special Operations) had all agreed that "the combination of ethyl alcohol–liquid oxygen is not incompatible with the combustion chamber designed by A.I.2 (g), nor with the rates of flow calculated by Mr. [Isaac] Lubbock." Moreover, the Armament Research Department had concluded "... that the fuel for driving the main turbine will be hydrogen peroxide with a permanganate initiator." The question of what propellants the A4 employed seemed satisfactorily answered.

Knowledge of these propellants was important for several reasons. Most important, weapon performance in general and range in particular could be determined with greater accuracy. Also, assessments could be made of possible limitations in the ability of the nation to manufacture and deliver the three propellants, and hence on the number of rockets the Germans might deploy. And for bombing mission planners, the sources of supply would become inviting targets for Allied strike forces. The Allies could also consider the degree of mobility German launch teams might enjoy in terms of constraints imposed on them by the nature of the propellants being handled.

Professor C. D. Ellis (chairman of the meeting) and other members present learned from the director of Intelligence (R) that just the previous night captured documents had arrived from Normandy seeming to confirm "the general hypothesis upon which the Sub-Committee had worked." A blueprint of the A4 was found which, though it did not include all dimensions, led to the following rather precise measurements:

rocket length, 45 feet 10 inches; maximum body diameter, 5 feet 7 inches; diameter over fins, 11 feet 8 inches; and length of fins, 12 feet 7 inches.

The flying bomb (upon deployment known as the V-1, or first vengeance weapon) offensive against the United Kingdom began in mid-June 1944. As it grew in intensity during the summer, the belief was strengthened in the minds of many that a rocket attack was imminent. The Swedish incident, coupled with reexamination of Peenemünde photography and that of the Heidelager proving ground, showed that the A4 was by no means a rare bird. Intelligence had also pieced together scattered reports from agents on the Continent and from POWs leading to the conclusion that the A4 was in mass production, and that perhaps as many as a thousand missiles were already in existence. One thing that puzzled intelligence, however, was the lack of more reports of movements of large rockets toward the Channel coast. Some scattered, and largely unconfirmed, accounts on A4 movements did come in. A typical report, received May 30, was to the effect that flatcars were seen traveling through Belgium toward northern France transporting what were described as "six 15-ton 'torpedoes.'" Intelligence had concluded that "the possible arrival of small numbers of rockets in France at the end of May points to the advanced state of the enemy's preparations for rocket attack."

Although known to exist in large quantity, ever more detailed information showed the A4 to be a far less dangerous weapon than previously believed. Estimates of the weight of its warhead dropped from 10 (or more) tons in 1943 to more like 1 ton (the actual value) in August 1944. Heavy bombing attacks early in the month had once and for all put Watten, Wizernes, Mimoyecques, and Siracourt—all suspected to be involved in A4 preparations—out of action. The Allied breakout from Normandy and advance toward the northeast promised soon to overrun all French and even Belgian actual or potential coastal launch sites, further reducing the rocket threat.

As the RAE continued to study the remains of the Swedish rocket and evidence coming in from Blizna and from Normandy, the solid conclusion was reached that "... the 'Swedish' rocket is the type intended for use against this country." Missile weights between 7 and 20 tons were now being postulated, with the Royal Aircraft Establishment at Farnborough inclining toward 14 tons and Swedish experts between 8 and 15 tons.

The RAE revised its earlier 4-ton figure and now put the warhead "at somewhat less than 1 ton," while other experts estimated between 1½ and 2 tons. At the same time, RAE engineers emphasized that while the operational range of the A4 was probably between 150 and 160 miles, "the Swedish incident shows that it is capable under circumstances of attaining more than 200 miles."

Churchill was not present at the War Cabinet Crossbow Committee meeting on August 10 at which R. V. Jones decided officially to announce the new, low estimate of the A4 warhead's weight.

> Mr. Churchill had expected to be in the Chair at this meeting, but had to depart for, I think, Italy and Mr. Herbert Morrison deputised for him. Until halfway through the meeting we were expecting Mr. Churchill to come in and take over, but it gradually appeared that he would not be able to do so. I particularly wanted him to hear of my conclusion that the warhead would only be of one ton weight because I wanted him to see the general incredulity with which this conclusion would be greeted. I therefore made myself as inconspicuous as possible so as not to be called upon to speak until the last possible moment. When it became clear that Mr. Churchill was not going to be present, I finally put my cards on the table, and got the incredulous reception that I expected.

At that time—a month before the A4 bombardment of England was to begin—captured maps and documents from Normandy, POW reports and other intelligence provided the following assessment of the situation:

> From these sources it can be inferred that the rocket is brought, without fuel but with warhead, to the launching site on a long trailer. It is halted on the firing platform, at some distance from the actual firing base, and oxygen is then introduced through a hose from a tank trailer on one side of the middle section of the platform. One of the drawings suggests that the trailer is then backed 6 meters nearer to the firing base, where the other principal fuel (alcohol?) and the main pump fuel (hydrogen peroxide) are pumped into it from tank trailers drawn up on the other side of the road. The entire process is estimated to require a period of the order of 2 hours. The rocket is erected vertically on a steel firing base, which is placed on the heavily reinforced broad end of the firing platform. The evidence here under review throws no further light on the method of hoisting, or on the question of whether the erection is carried out after the completion of fuelling. The firing is done from a special vehicle which is driven into a control hut some distance off the side of the road. This and other evidence point to a vertical, or nearly vertical, takeoff, unassisted by auxiliary launching devices.

The growing prisoner-of-war harvest also brought to light the historical origins of German rocketry: the old Raketenflugplatz amateur tests, Kummersdorf, the development of the A-series, and the experimental difficulties with the A4. One POW reported that half of the ten flights he witnessed were failures, helping reinforce the growing Allied opinion that the A4 was far from a reliable, field-ready weapon.

Never intimate, American participation in the Big Ben intelligence picture grew as the A4 threat became better understood. On the technical level, the US armed forces were kept informed of the general situation. As early as October 25, 1943, Churchill had summarized the general situation to President Franklin D. Roosevelt:

> 1. I ought to let you know that during the last six months evidence has continued to accumulate from many sources that the Germans are preparing an attack on England, particularly London, by means of very long-range rockets which may conceivably weigh sixty tons and carry an explosive charge of ten to twenty tons. For this reason we raided Peenemünde [in August], which was their main experimental station. We also demolished Watten, near St. Omer, which was where a construction work was proceeding the purpose of which we could not define. There are at least seven such points on the Pas de Calais and the Cherbourg peninsula, and there may be a great many others which we have not detected.
>
> 2. Scientific opinion is divided as to the practicability of making rockets of this kind, but I am personally as yet unconvinced that they cannot be made. We are in close touch with your people, who are ahead of us in rocket impulsion, which they have studied to give aeroplanes a send-off, and all possible work is being done. The expert committee which is following this business thinks it possible that a heavy though premature and short-lived attack might be made in the middle of November, and that the main attack would be attempted in the New Year. It naturally pays the Germans to spread talk of new weapons to encourage their troops, their satellites, and neutrals, and it may well be that their bite will be found less bad than their bark.
>
> 3. Hitherto we have watched the unexplained constructions proceeding in the Pas de Calais area without (except Watten) attacking them in the hope of learning more about them. But now we have decided to demolish those we know of, which should be easy, as overwhelming fighter protection can be given to bombers. Your airmen are of course in every way ready to help. This may not however end the menace, as the country is full of woods and quarries, and slanting tunnels can easily be constructed in hillsides.
>
> 4. The case of Watten is interesting. We damaged it so severely that the Germans, after a meeting two days later, decided to abandon it altogether. There were six thousand French workers upon it as forced labour. When they panicked at the attack, a body of uniformed young Frenchmen who are used by the Germans to supervise them fired upon their countrymen with such brutality that a German officer actually shot one of these young swine. A week later,

> the Germans seem to have reversed their previous decision and resumed the work. Three thousand more workmen have been brought back. The rest have gone to some of those other suspected places, thus confirming our views. We have an excellent system of Intelligence in this part of Northern France, and it is from these sources as well as from photographs and examination of prisoners that this story has been built up.
>
> 5. I am sending you by air courier the latest report upon the subject, as I thought you would like to know about it.

A couple of months after the V-1 was placed into combat operation (mid-June 1944), but before the advent of the A4, Brigadier General H. M. McClelland, US Army Air Forces air communications officer, directed a memorandum (dated August 7) to General H. H. Arnold informing him that rumors of the A4 being radio-controlled were confirmed as a result of the Swedish incident, and that the missile was believed to weigh 30 tons and to possess a 15-ton warhead! He also referred to P-38 "Droopsnoot" planes, equipped with search gear, that would soon operate out of Russian bases, attempting to intercept control signals from the Blizna proving ground in southwest Poland "to determine exactly the frequencies and characteristics of the signals." Yet, at the very time Blizna was being overrun by Soviet armies!

Another American, Walt W. Rostow, became closely involved with Big Ben affairs when he was assigned to Air Ministry Intelligence in London. Rostow, who later became special assistant for national security to President Kennedy, concentrated initially on studies of German aircraft production and the locations of plants with the aim of generating target material for precision attacks by the US Eighth and Fifteenth Air Forces. He told the authors how he became involved with Big Ben.

> Sometime in the spring of 1944 I was called in by Sir George Thomson, then senior scientific adviser to the Air Ministry. He told me that a conflict was raging in the British bureaucracy over the probable size of the warhead to be mounted on the V-2. There were those who believed it would be as much as 10 tons; others thought it would be 1 ton. He wished me to form a wholly independent estimate and to convey personally my conclusion and the arguments which led me to it.
>
> I proceeded to talk bilaterally to the key figures in the engagement. Among those whom I quietly interviewed was R. V. Jones, then located, if I remember correctly, in a marvelously disheveled office in Broadway. He argued persuasively, on the basis of intelligence, that the V-2 would have a 1-ton warhead. I also spoke with those holding a contrary view. They were mainly men working on rockets in England. Their arguments were, essentially, prima facie. They felt that a 10-ton warhead was technically feasible and that the German

> rocketeers (whose importance had been so much more fully recognized in their country than their opposite numbers in Britain) would surely be able to produce a rocket with a 10-ton warhead. After examining the notes on my bilateral sessions, I tried to define the critical issues of fact and hypothesis which were in dispute and to form up the agenda for a meeting which would constitute a systematic and orderly confrontation of the two views. That meeting was held sometime in late spring of 1944 in, I believe, a basement conference room in the Air Ministry at Monck Street.
>
> Professor Jones's memory is quite correct: It was a "disputatious" meeting. Jones led the party arguing for a 1-ton warhead. I do not recall the names of those on the other side. Although I was at the time relatively young [27], I had acquired some experience with both academic and government bureaucratic structures and their capacity for bloodless tribal warfare. But I had never been present at, let alone presided over, a meeting with more emotional tension than that centered on the size of the V-2 warhead.
>
> What emerged was a reasonably solid intelligence case for a 1-ton warhead against a deeply emotional conviction among the British rocketeers that if they had been backed by their government, they could have produced a rocket with a 10-ton warhead.
>
> I concluded that the evidence Jones had mustered was essentially correct; [I] called on Sir George Thomson and informed him of my conclusion: The warhead would be about one ton. He looked up at me with a twinkle in his eye and said: "You are a lucky young man. A few days ago a V-2 [then known as A4] misfired from Peenemünde and landed in Sweden. We flew it back in the bomb bay of a Mosquito. We have now measured the venturi [combustion chamber nozzle of the rocket engine]. Obviously, it could not develop more thrust than that required for a one-ton warhead." After some exchange on the curious way that essentially rational problems of intelligence and science could generate emotional attachments of great strength, I departed.

Once the Swedish rocket had been thoroughly analyzed, events moved ahead rapidly. The name and purpose of the A4's Meilerwagen—multiple-erector and axis transport vehicle—became known from documents and drawings captured in Normandy; and the missile's propellant servicing vehicles were soon identified. By the time of the fifth Big Ben Sub-Committee meeting in mid-August, 1944, the experts "inclined to the view" that the Swedish rocket and the missiles that the Germans had prepared for operations in France were the same. It was estimated that a rocket firing battery would require an area of about 400 square miles in which to operate—a 15- to 25-mile front with a similar depth. "These figures would probably require readjustment," the

Sub-Committee concluded, "but it should be remembered that the enemy would be reluctant to offer the concentrated targets that undue contraction of the area would entail."

At the time of the sixth meeting on August 22, attention was being focused on the nature and composition of the A4 ground organization. Allied intelligence assumed that the German LXV Korps would control a number of regiments, divided into batteries and firing troops—probably three firing batteries and a technical battery to a regiment. It was also believed that most rockets would be supplied to the firing troops directly from railheads. The British knew the approximate scale of job to be accomplished—based on Normandy and other information—and hence could judge roughly how the Germans would go about doing it. A hypothetical organization was established to provide "a reasonable framework to which countermeasures [might] be related."

Understandably, rates of fire greatly concerned the Sub-Committee, so every effort was made to assess the factors thought likely to influence them. One would be the number of servicing vehicles that could be made available at a railhead at any one time. Another would be the desirability—if not the necessity—of spacing the launches to frustrate Allied spotters and hence countermeasures, while at the same time enabling German spotters to advise on accuracy by observing impact points. The British assumed that railhead facilities would be adequate for the simultaneous fueling and servicing of only one troop at a time: too many vehicles would clearly invite air attacks.

The subject of the possibility of detecting A4 vapor (condensation) trails also came up when Air Commodore J. A. Easton, director of Intelligence (R), quoted a POW report to the effect that the missile climbed "... rapidly at a steep angle, during which it made a loud roaring noise and left a white condensation trail." The existence of such a trail could be of importance in detecting the early ascent phase of flight.

Directly related to rates of fire was the matter of accuracy. "Although the precise nature of the radio control was still a matter of conjecture," the Sub-Committee recorded, it was generally believed that "the necessary equipment could be housed more economically [in operational A4s] than in the Swedish rocket." This could lead to the employment of larger types of warheads. Over a range of about 150 miles, the probable error of the missile was estimated to be 3 miles for range and 5 miles for line (direction).

Aeronautical specialist W. G. A. Perren felt that gyroscopic controls might be employed in the A4, either in conjunction with radio during certain stages of flight or as a standby in the event that Allied radio countermeasures proved successful. He showed Sub-Committee members a memo he had prepared in which a scheme was advanced demonstrating how (correctly, as it turned out) gyroscopic control alone would hold the missile to the desired trajectory. He felt that ground control might be employed only to terminate thrust at such a point on the trajectory as would lead to the attainment of the planned range. Sir Robert Watson-Watt agreed in principle, but again pointed to the radio equipment salvaged from Sweden. Easton saw no reason from the intelligence point of view to reject the suggestion. Perren explained that

> in the initial phase of the motion one gyro will provide roll stabilisation and also measure the angle of pitch, and the other will apply a predetermined rate of pitch to the rocket, and at the same time, maintain line control. The roll gyro, in addition to determining when the rocket has reached a preset pitch angle, which would correspond to maximum range, i.e. about 45°, would maintain the rocket at this angle until the jet thrust is cut off. Simultaneously with cutting off the jet thrust, the external controllers would be set over to apply spin to the rocket during the final phase of its trajectory.
>
> Ground radio would be used to check the range and apply corrections to both the pitch and line controllers, so as to keep the rocket on the ideal trajectory. Ground control would also be used to cut off the jet thrust at some point on the trajectory after the flight at steady angle has been established.

Three days after the sixth meeting, the controller of Projectile Development, the chief engineer of Armament Design, the Royal Aircraft Establishment, and Isaac Lubbock issued their formal report on the Swedish rocket. They drew attention to the fact that by the end of July the last of the A4 parts had arrived at the RAE from Sweden, and that intelligence derived from the Normandy battleground was helpful to the conclusions being drawn. Details were given on missile structure, the combustion chamber and turbopump system, radio equipment, etc., and a very accurate schematic was prepared. Compared with early July estimates, the A4 was turning out to be a much smaller, lighter, less awesome weapon. Based on the Swedish parts, the missile shaped up approximately as follows:

Length (feet, inches)	45, 10
Body diameter (feet, inches)	5, 6
Weight, (tons)	
overall	13.5
warhead	1 (or slightly less)
propellants	9.6
Sea level thrust (tons)	27

By the time of the seventh meeting on August 29, members gathered in room 156 of the Montagu House Annexe possessed a rather accurate idea of the appearance of the A4 missile, what its general characteristics were, and the threat it appeared to pose. In making their assessment, they had taken advantage of photoreconnaissance of Peenemünde, Blizna, and the French coast, and had reviewed reports on the Swedish incident and material supplied by Polish AK sources, POWs, agents and friendly nationals. Moreover, the Sub-Committee carefully considered information gained as a result of continuous Ultra radio intercepts (see Appendix B). And members were fully aware of the role played by Allied scientists and engineers, who had provided the technical analysis that had become such a necessary ingredient to the Sub-Committee evaluations. Information coming in daily from the Normandy campaign could not be overlooked, nor could German propaganda—which in the summer of 1944 was keeping home expectations alive to the existence of a new wonder weapon yet was not making specific commitments as to when it would be unleashed. The Sub-Committee was particularly cognizant of a memorandum from the director of Operations (Special Operations) at the Air Ministry to the deputy supreme commander of SHAEF to the effect that the enemy had planned his mobile system well and that, "unless intelligence reveals some unexpected weakness, air attack against the rocket organization can only yield limited results."

Although most members of the intelligence community now realized that the A4 was capable of attaining a range of up to 200 miles, right to the end of the first week of September 1944 a few responsible individuals connected with Crossbow felt that the dangers of a rocket attack had all but evaporated. There were several reasons for this attitude. On September 1, the V-1 offensive from French launch sites came to an end, as they were either already overrun or in immediate danger of capture. Also, German documents falling daily into the hands of advancing Allied troops conclusively proved that the main rocket offensive was to have been unleashed from France. With the Allies deployed along the Somme by the end of August and with Belgium soon to be captured, London and other targets in southeast England seemed out of danger.

This short-lived euphoria was rudely broken the evening of September 8, 1944, when the first A4, which the Germans now called the V-2, the second vengeance weapon, exploded on English soil five minutes after takeoff. It was launched neither from France nor from Belgium, but from western Holland. The shock of the impact reverberated around the world. A new era in warfare had begun.

CHAPTER 10

V-Weapon Offensive

By the end of May 1943, the German Long-Range Bombardment Commission found itself unable to determine which, if either, of the two missiles—the Fi103 or A4—was superior, or at least gave evidence of greater future promise. Accordingly, the commission decided to recommend to Hitler that both be continued through the development cycle and prepared for joint military operations. It was also felt that the mass deployment of two different types of unusual new weapons would not only confuse enemy intelligence but would place an added burden on his defensive measures.

Preparations and troop training for the Fi103, or V-1, the simpler of the two missiles, got under way within the structure of Colonel Max Wachtel's 155th Flak Regiment, whose headquarters were located at Zempin, not far from Peenemünde. At the same time, a mammoth effort was begun to develop an arc of V-1 launching sites from Calais near the Belgian border southwestward to Cherbourg on the Normandy peninsula. Initial plans called for sixty-four principal sites and thirty-two backup (or reserve) sites, for a total of ninety-six. Each site was to consist of a launching ramp and assembly and storage structures. An additional eight sites would be built to serve as supply centers, with up to 250 V-1s stored at each.

The Organization Todt* assigned forty thousand French and other

* The construction organization built up by Dr. Fritz Todt (1891–1942), who preceded Albert Speer as Minister of Armaments and Munitions and was responsible for such works as the autobahn system, the Siegfried Line or Westwall, and the submarine bases along the French coast.

Assembly of A4 motorized ground equipment (Meilerwagens) in Rebstock railroad tunnel during the summer of 1944. The tunnel was bombproof.

A4 engine undergoing test firing near Lehesten, some 25 kilometers south of Saalfeld.

Professor Frederick A. Lindemann, Viscount Cherwell. As scientific adviser to Winston Churchill, he was loath to admit the existence of a serious German missile threat.

Duncan Sandys, Churchill's son-in-law, servec commanding officer of Britain's first roc (surface-to-air) regiment, as financial secretary the War Office, and as joint parliamentary secret at the Ministry of Supply before being named ov all coordinator of Allied investigations of the C man secret weapon menace and of appropri countermeasures.

Flight Officer Constance Babington Smith, foreground, who, as member of the Women's Auxiliary Air Force, worked at the Central Interpretation Unit from 1940. She located the Fi103 (V-1) flying bomb on an aerial reconnaissance photograph. Taken in her RAF office at Medmenham.

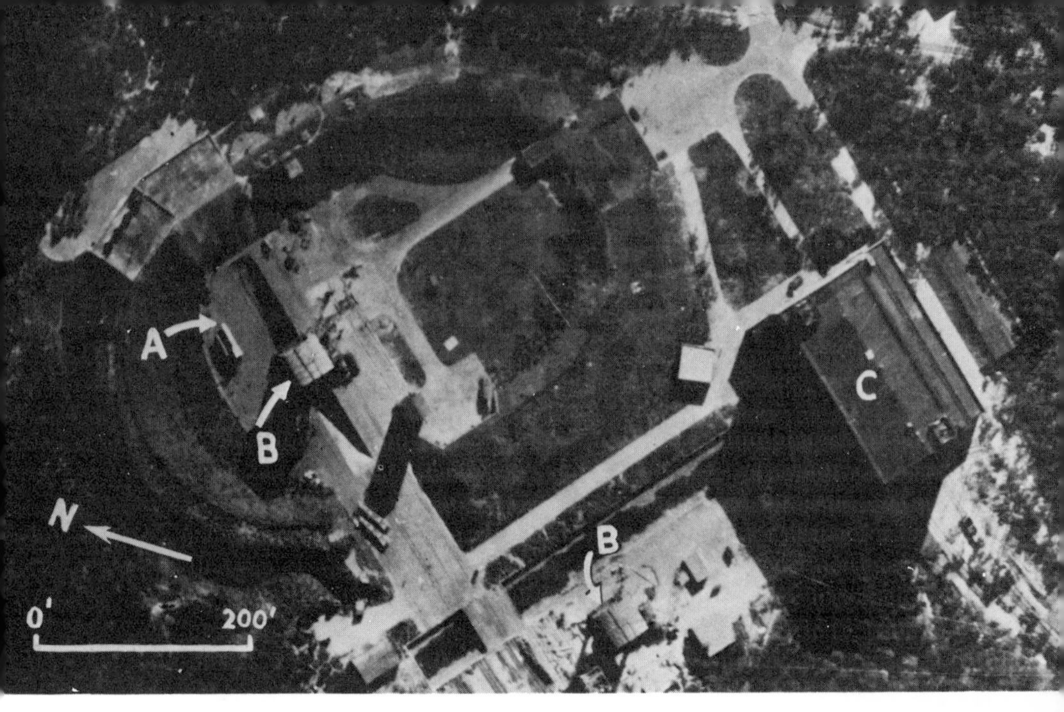

Aerial photograph of Peenemünde Test Stand 7 experimental area, showing an A4 rocket at A, two mobile service towers at B, and the missile assembly hangar at C.

The Fi103 (V-1) seen on its launch ramp at Peenemünde. Photographed by Squadron Leader John Merifield on September 28, 1943. Its interpretation provided the link between Peenemünde and the Channel coast "ski sites" used for launching the flying bomb.

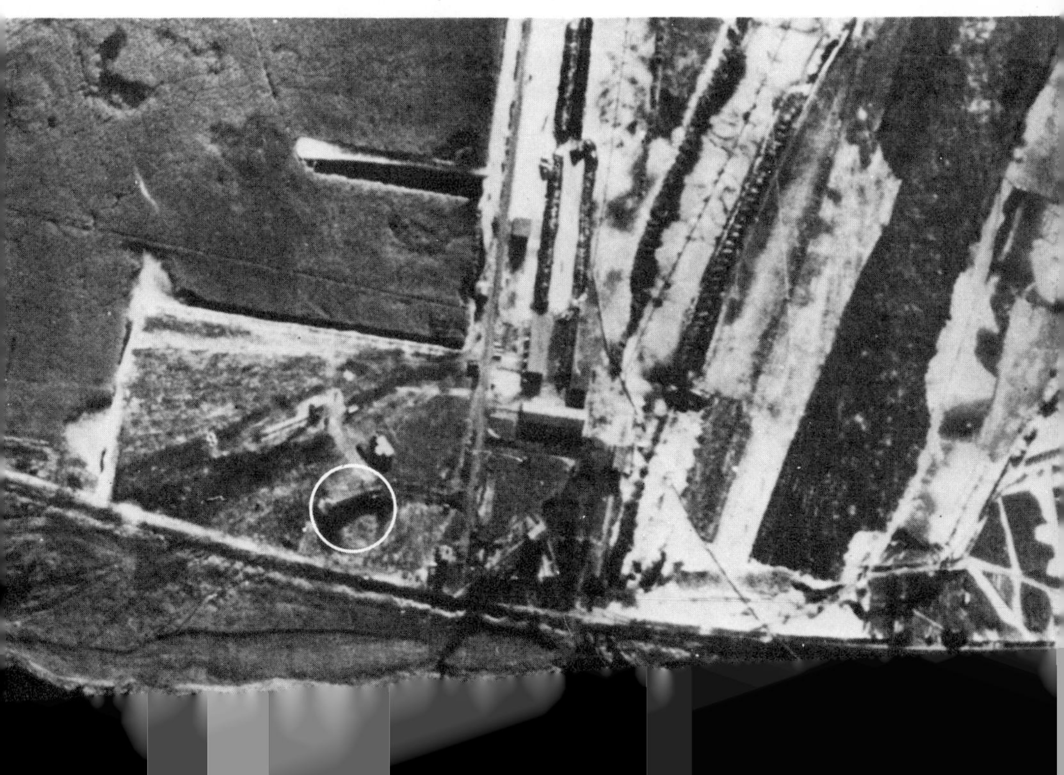

A section of Peenemünde experimental R & D establishment before RAF attack in August, 1943.

A section of Peenemünde, where some forty detached huts were destroyed and about fifty others were gutted by fire or partly demolished by bombs.

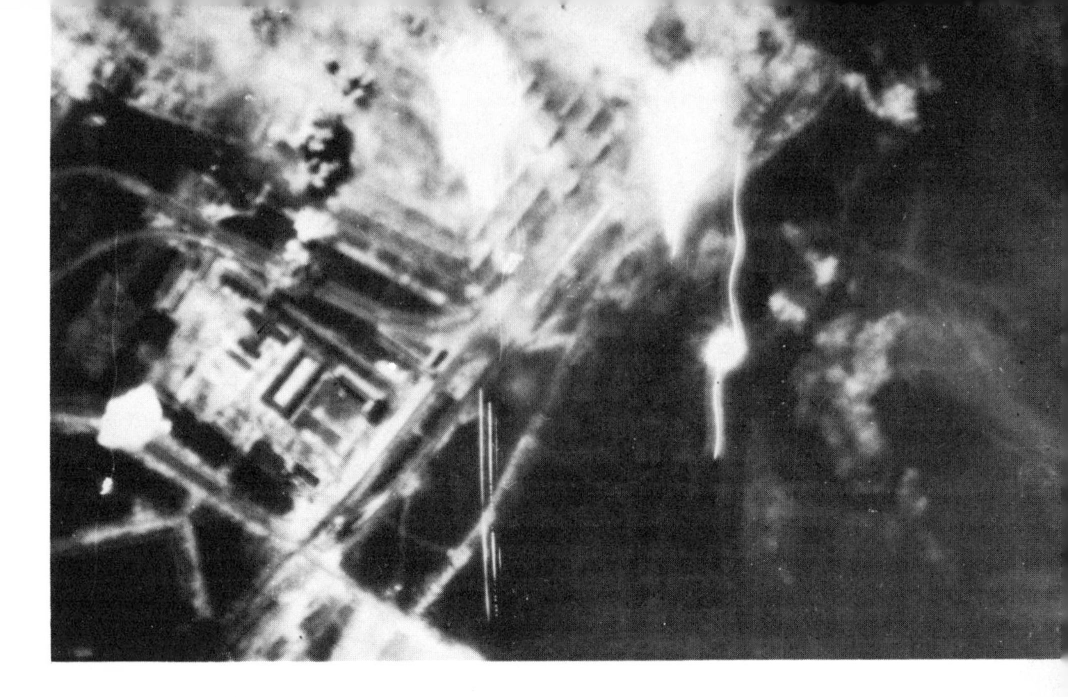

A section of the plant during the attack. Bombs can be seen bursting among buildings.

Peenemünde seen by reconnaissance aircraft in September 1944, showing how the Test Stand 7 experimental area has suffered from aerial bombardment. The annotations A reveal that light flak positions were no longer found on the roof of the damaged hangar.

A V-1 flying bomb launching pla form in France.

This is the reconnaissance phot graph from which photographi interpretation experts forecast th the first flying bombs would fall o Britain within 48 hours. The made this forecast on June 1 1944, basing their deduction on th fact that a launching ramp ha been erected on this site at Vi nacourt, France. On June 1 1944, the V-1 attacks started fro a site at Saleux. The launchin ramp can be seen in the trees, to left.

Lunarlike landscape created by Allied bombing of the heavy installation at Wizernes.

The heavy long-range missile installation at Watten, near Saint-Omer, France, after having been hit by thirty-three RAF attacks. The structure is some 300 feet long, 200 feet wide, 50 feet high, and penetrates some 20 feet underground.

S. G. Culliford, RAF pilot from New Zealand, in the cockpit of his Dakota DC-3 aircraft in Bari, Italy, prior to embarking on his daring mission to Poland in 1944.

RAF reconnaissance photograph of Blizna, Poland, made in June 1944. The train is loaded with camouflaged missiles. The A indicates rocket cradles.

workers to the job, and progress was so rapid during October and November that plans were made to commence the V-1 attack on London on December 15. Overall responsibility for the sites rested with Regional Air Force Headquarters for Belgium/Northern France located in Brussels and with the Regional Air Force Headquarters for Western France in Etampes; both operated through instructions issued at the Air Force Construction Headquarters in Berlin.

At Zempin training was progressing, and starting on October 16, 1943, the 155th Flak Regiment launched its missiles northeastward along the Baltic coast of Germany toward Rixhoft (on the northernmost point of the old Polish Corridor). On October 21, troops from the 155th transferred to the Calais region not only to familiarize themselves with the forward zone but to assist in making launch sites ready for the forthcoming offensive.

On December 5, ten days before the Germans had originally hoped to get the V-1 campaign under way, Crossbow-designation bombings of the sites began. At the same time, technical problems continued to plague the new weapon, making a postponement of the attack on London to mid-January 1944 seem advisable to the Armed Forces Operations Staff. The LXV Korps commander, General Erich Heinemann, realizing that the situation was far more serious, was not so sanguine. He found that launch sites were in many cases located without regard to exposure to enemy attacks, and many were much too complex. Heinemann was certain that simplicity and mobility were the keys to success, not the massive concrete structures favored by the Todt organization (and, initially, von Braun).

Two planning meetings were held in Paris, on December 28 and on January 2, 1944, to attempt to iron out difficulties and differences of opinion and to set forth a realistic set of priorities tied to a workable timetable. All concerned understood that, for the time being, the LXV Korps would deal operationally only with the winged Fi103—soon to become V-1; the A4 (V-2) was still a long way from being ready. Following the meetings, General Heinemann prepared a report for Hitler, read to and transmitted on January 10 through General Keitel (chief of staff of the Armed Forces Staff) at 15th Army headquarters in Tourcoing. It is uncertain whether or not this report ever came to Hitler's attention.

Heinemann continued to press for simple, mobile launch and launch-support facilities, relying on caves and tunnels to the rear for matériel and propellant storage. Nevertheless, work continued on the huge concrete structures, some of whose sides were 16 feet thick. So easily identifiable were they that Allied bombers rather quickly put them out of action. Moreover, as soon as ground or aerial intelligence was able to identify a cave or tunnel, it too was hit and generally knocked out of service. As US Army Lieutenant Colonel M. C. Helfers of the office of the Chief of

Military History would observe after the war, the corps commander "more or less assumed an attitude of not doing what his superiors ordered but doing what he felt his superiors would have ordered had they been familiar with the actual situation" by proceeding as follows:

> (1) Positions and servicing installations hitherto constructed or being constructed were abandoned as impracticable in respect to camouflage and as useless in respect to protection against aerial bombardment.
>
> (2) Work was nevertheless continued as a deception measure on the positions under construction, the French workers being gradually and unobtrusively pulled out. Work on positions in the Cherbourg peninsula was particularly carried on for deception purposes only.
>
> (3) Entirely new positions were constructed along simple lines, ignoring the purely theoretical instructions of the Air Force Construction Department in Berlin.
>
> (4) No servicing installations were constructed. Instead suitable caves were reconnoitered and prepared further to the rear and all servicing work was transferred there.
>
> (5) All new construction was done exclusively by the units assigned to LXV Corps and by aviation construction engineer battalions brought forward.

Stringent security measures were taken throughout the site construction and readiness period. French workers were gradually phased out to be replaced by German personnel. Additional guard and counter-espionage forces were assigned to each site. The 155th Flak Regiment headquarters was shifted for deception purposes to Paris and then moved to a new, code-named site. Furloughs were cut. V-weapon troops could not even receive mail once they had been transferred to firing and logistics locations.

Germans closely involved with the technical and operational aspects of the V-1 were all too aware of its inherent inaccuracies. The missile had been repeatedly tracked during test firings in Germany and Poland, and impact areas pinpointed. Once operational firings against England commenced, however, there was no foolproof method of determining which missiles reached their targets and which did not. Failures, of course, could be due to technical malfunctions or to enemy countermeasures. In both cases, information was essential to assess the effectiveness of the bombardment.

In order to do this, a number of measures were planned. First, forward observation posts were to be set up along the Channel coast to report on the flight behavior of the flying bombs, the extent and effectiveness of enemy counteraction, and weather conditions en route. Insofar as possible, existing army and navy artillery observation posts were to be utilized; otherwise, new ones were put up. To augment ground posts, aircraft warning and air reconnaissance services of the Third Air Fleet and the Ninth Air Force Corps were to be called upon. The most precise impact information was to come from agents located in England and from sound locating and seismographic instruments operated under the supervision of an SS observation battalion deployed on the northern French coast. Finally, it was planned to fit some missiles with miniature radios that would transmit flight information right up to the time of impact. After the war, Colonel Eugen Walter, General Heinemann's chief of staff, stated that only occasionally did aerial observation of impacts prove successful; and, in any event, it ".. gave only a general picture which did not assist in the adjustment of fire nor permit observation of the reaction of the civilian population." Moreover, he said, "all German attempts to observe the effects of the V-weapon campaign against London by means of acoustic devices failed. The distance was too great, the instruments inadequate." Nor did the radio transmitter scheme live up to expectations. "The only reliable observations were made by spies."

Observational difficulties notwithstanding, the first combat missile had still to be launched. And it was the crucial matter of launching that continued to worry Heinemann. No matter how stringent the security precautions, and how simple V-1 launch sites might become, he realized that some would inevitably be detected and destroyed before and during the early phases of operations. This led him to order the construction of many more sites than ordinarily would have been necessary to conduct the offensive against England.

The earliest installations all involved 150-foot-long inclined ramps 16 feet high at their far ends, plus as many as a dozen associated buildings. As noted, these suspiciously large and complex structures drew Allied bombers like honey attracts ants, with obvious results. The simplified sites that Heinemann and his staff soon ordered to be built consisted basically of (1) concrete piles on which the launching ramp A-frames were mounted, (2) small pillboxes or slit trenches; (3) simple two-lane, log-surfaced approach roads, and (4) minimal concrete platforms on which compass settings could be made. Maximum advantage was taken of existing buildings in the vicinity and of the protection afforded by woods, hills, and other natural features. When necessary, camouflaged tents were erected.

As time went on the Germans learned to build their sites more and more rapidly, so that by May 1944 what had once taken from a month and a half to two months to accomplish was being completed in ten days, or even less. By October, a site could be put up in less than a day, the record being 18 hours!

Redundancy and simplicity in site buildup and the dispersal of forward elements certainly helped matters, but Heinemann still had to consider the movement of equipment and supplies from manufacturing and staging areas back in Germany as well as in France. Before opening his attack, he wanted 1,000 missiles on site, and 250 new ones moving daily through the logistic chain. Since most V-1s came in by train, they not only had to be heavily camouflaged but had to move safely through the German and French rail networks with the highest possible priority. To enable missile trains to proceed with minimal delays, a railway repair battalion was established so that repairs to bomb-damaged track and junctions could be made immediately.

As the time for operations approached, the 155th Flak Regiment was brought up to its complement of 6,500 officers and men, organized in four battalions of three firing batteries and a single maintenance and supply battery each. Fifteen men were required to make up a single firing crew. Another 2,500 men operated outside the 155th, bringing to 10,000 the total number of men involved in V-1 operations from the Belgian border down to the river Seine. Firing sites were located a minimum of 10 miles and a maximum of 35 miles inland. The main body of 155th Flak Regiment troops was given intensified training on or near the launch sites, while some crews were returned to Zempin for checkout. At their forward locations, crews simulated all preparatory operations except the actual launch. Supply and transport routines were improved, and a method of reporting crashes was established, enabling troops to locate and advise on missiles that either failed to reach the Channel (aborting between it and the launch site) or strayed off course. Crash report charts were to be prepared so that LXV Korps headquarters would have timely and accurate information on flight failures. Among the many reasons: V-1s that crashed on French soil would have to be destroyed or removed so that resistance groups could not report details of the secret weapon to the Allies.

It appeared that the V-1, its launch crews, and the installations they were manning would be ready for operations by late spring 1944. Accordingly, schedules were revised and security was augmented by special rail and motor transport escort companies trained in preventing sabotage. These companies were stationed near or at assembly plants, supply dumps, and servicing depots. The Germans were taking no chances.

With security in order and other preparations made, Colonel Walter officially informed the Armed Forces Operations Staff of the state of readiness of the LXV Korps. He reported that a full 95 launch sites were set for the offensive, and that between 60 and 70 of them could launch simultaneously. (A maximum of ten missiles were located at any given site.) He reemphasized the necessity of keeping supply lines open, adding that antiaircraft guns were being mounted on trains to provide additional protection.

The arrival of June 1944 found V-1s stockpiled at both their assembly plants and at dumps, from where they were being transferred to servicing stations located in caves and thence to the operations zone. (In some cases, these servicing stations were bypassed, the weapons moving directly to rail unloading sidings at the "front," where they were made ready by maintenance personnel prior to being truck-transported to the launch site.) The servicing stations were staffed with personnel trained to check out the missiles and to make necessary repairs and adjustments to components. Fuel was stockpiled at dumps located within or near general supply depots. Motor transport was being used to move fuel to the front. All appeared to be ready.

On June 6, 1944, which was, coincidentally, D-day in Normandy, the LXV Korps received orders for committing its flying bomb weapons on June 12. In the intervening six days, work on the launch sites had to be completed and the supply chain put into full operational working order despite continuous Allied harassment of rail communication lines as part of their general post-D-day "softening" tactics. To maintain an unbroken supply line, Heinemann had to be content with nighttime rail service—though growing Allied attacks on junctions and shunting stations made even that difficult—plus motor transport secured from any sources his staff was capable of exploiting.

Five days later, on June 11 at a preoperations meeting called by corps headquarters for 155th Flak Regiment officers, everything was felt to be more or less ready for the following day. All four battalions had at least their full authorized issue of missiles (192; two actually had 240 each, 48 in excess of the allowance), and more than adequate gasoline, hydrogen peroxide, and sodium permanganate. Corps headquarters then called the final meeting for the next morning, attended by the regimental commander, supply officer, motor transport officer, and key missile technicians. The chief of staff emphasized that the initial commitment must be successful so that maximum advantage could be taken of surprise. Although some problems had cropped up, it was felt that all firing positions would be set for action by early evening.

The attack was scheduled for 22.40 hours, with simultaneous launchings to be made from all serviceable sites. Just before the attack, the Third Air Fleet was to raid London, then retire so that "all clear" signals would be given and Londoners would emerge from their shelters. Also, German fighters were to be positioned around the British capital so that they could observe the effect of the impacts of the first flying bombs. As part of a deception plan, radio transmitters on the Channel coast were to be lit up to fool the enemy into believing that the missiles were radio controlled. Finally, cross-Channel artillery was to open up on several English coastal towns.

General Heinemann traveled to Saleux near Amiens during the afternoon, where the regimental command post was located. No sooner had he advised his chief of staff that all was proceeding smoothly than a rail target near the post was raided by the Allies, knocking out the telephone network to the firing batteries. Though radio was subsequently relied upon for command communications, the disruption led to a one-hour postponement of operations. It was too late to call off the raid by the Third Air Fleet.

The clock moved slowly forward, and by 23.00 hours all but 9 of Colonel Wachtel's 72 serviceable firing positions had confirmed readiness. The exact time of the first firing did not survive the war, but it appears that it occurred very shortly after midnight; that is, early morning of June 13, 1944. In terms of all the preparations made and the months of training, the opening salvo was ludicrous: Only nine of the missiles managed to get into the air, and none reached England, much less London.

Nor did Wachtel's second round fare too much better: Out of ten missiles airborne at 03.30, only one impacted on London, though three did get as far as southeast England.

In the light of this miserable debut by the vengeance weapon, the rest of the firings planned for the thirteenth were cancelled and an inquiry was made as to what had gone wrong. Certainly the troops were having difficulties with the new weapon, and their morale may not have been the highest—most had not been on leave for a year nor had they been allowed to receive mail. The disruptive effects of Allied Normandy-related air raids must certainly have affected the initial operations. Moreover, it was discovered that vital missile parts had been misshipped or destroyed by air raids en route to the firing positions. Beyond that was the fact that the V-1 just wasn't a foolproof weapon; it was still in the advanced development state and should not have been declared operational in June 1944. As Air Force General Werner Baumbach was to put it later, "The V-1 never got beyond the diaper stage." Yet, diapers or no diapers, the situation had become so desperate in Germany that Hitler was willing to grasp at any means to punish the enemy for all the damage it was wreaking on the fatherland.

Despite extraordinary difficulties and shortcomings, the LXV Korps and the 155th Flak Regiment worked around the clock and, in consultation with the Armed Forces Operations Staff, made ready to renew operations during the nights of June 15 and 16. A total of 244 V-1s got off from fifty-five sites, and the offensive continued without letup. By the end of the month of June more than two thousand missiles had been launched, mainly against London.

Hitler personally decreed that the V-1 be used only against London, passing his instructions through the Armed Forces High Command. In Message No. 772141/44, dated June 25, 1944, the Commander in Chief, West, was advised:

> Commitment of the V-1 on the beachhead [in Normandy] does not promise success. The V-1 is intended primarily as a terror weapon against inhabited localities, whereas the beachhead contains only ground troops against which artillery and aerial bombardment will have a like effect.

Both the Commander in Chief, West (Field Marshal Gerd von Rundstedt) and General Heinemann wanted to hit the Allied troop embarkation ports of Southampton and Portsmouth as well as targets on the Normandy peninsula at and near Cherbourg, but Hitler was adamant. Nevertheless, Heinemann did set off between sixty and eighty missiles against Southampton in the middle of June, with the result that he was reprimanded by higher headquarters (fifty-three arrived on or near target).

Once the V-1 campaign was in full swing, it was never seriously hampered at the launch front by Allied bombing: The Germans could always keep ahead of the game by establishing new sites quicker than old ones were knocked out. The main disturbance was caused by the gradual destruction of the logistics chain, which never permitted the hoped-for per-site launch rate of one V-1 each 26 minutes to be met. The best the Germans could do was a missile every 1½ hours, leading to an average of 102 V-1s every day from June 13 to the beginning of September 1944. The top 24-hour performance saw 295 flying bombs take to the air. Had Wachtel's men been able to employ their ski sites, with the 150-foot-long inclined ramps, they would probably have been able to launch at the rate of twenty missiles per site every 24 hours. Postwar interrogation of personnel responsible for the tactical planning of V-1 operations revealed that a first-rate crew operating from a modified site could get off two missiles an hour—or slightly better as noted above if conditions were ideal. The highest rate actually achieved in one night was eighteen, though the Germans believed that over an extended period no more than fifteen missiles per site could be counted on. This would have yielded 960 missiles a day for a regiment consisting of sixteen troops of four crews each.

Flying bomb incidents in the campaign against England are given in Chapter 11. On the Continent, between February 19 and 26, 1945, 586 incidents (term used to indicate impacts on Allied territory) were recorded out of 704 launches. In order to make their missiles less prone to defensive measures, the Germans began to try out the so-called angular firing tactic. This consisted of programming their missiles to make one or more course changes following launch, confusing enemy tracking stations and thus complicating countermeasures.

Not all V-1s were launched from the ground. For some time before the June campaign started, air crews had been training at Peenemünde with missiles mounted under the wings of He111 medium bombers. After launch, at approximately 1,500 feet, the V-1 would descend slightly and then regain altitude and fly off along its preset course. By adding the host aircraft's range to that of the missile, targets up to 800 miles distant could be hit. By early July 1944, all arrangements had been made to commence the air-launch phase of the V-1 offensive, under the following Air Force Operations Staff orders:

> 111 Bomber Group, Third Air Force, remains assigned to Third Air Force. Headquarters LXV Korps will forward air requirements to Third Air Force, which will ensure smooth cooperation between LXV Korps and III Bomber Group. Commander in Chief, West, has overall command.

The He111 carried its V-1s slung under the right wing between the fuselage and the engine nacelle, lateral stability being maintained by two vertical metal struts. The plane normally flew over the ocean at 325 feet altitude until just before the flying bomb release point. Then it climbed quickly to 1,500 feet. The time of starting the V-1's propulsion unit and the release from the He111 was determined by an instrument containing a tachometer. The reduction to zero of an initial preset number, which took from 15 to 20 minutes from a certain position on the course, indicated the bomb-release point. Both propulsion and the automatic pilot were started about 10 seconds before this point was reached, typically about 120 miles from the target.

Operational orders specified London as the prime target, with Southampton and Bristol as alternates after "sufficient experience [had been] obtained." Plans called for "the destruction of enemy defenses, the bulk of which are now located southeast of London, and the deception of the enemy as to the range and effectiveness of the V-1." The V-1 aerial operation started early in the morning of July 8, planes taking off at first from the Venlo airfield on the German–Dutch border and later (in January 1945) from airfields in the Hamburg-Bremen region. Between September 1 and December 13, 1944, a total of 875 missiles were air-launched against England, mostly from aircraft staging off the Dutch and Belgian coasts. From two to eight planes at a time were generally involved. The overall tactical control of air firings rested with General Heinemann, who also supplied supporting ground personnel and the missiles themselves from his own inventory.

The V-1 campaign was less than two months old when the military situation along the river Seine so deteriorated that the LXV Korps was forced to retreat from its French sites north into Belgium and Holland. Beginning on August 12, Wachtel's 155th Flak Regiment found its left flank beginning to fold as Allied armies began to fan out from Normandy and push northward. While Montgomery was urging his troops onward ("Every officer and man must understand by a stupendous effort now we shall not only hasten the end of the war; we shall also bring quick relief to our families and friends in England by over-running the flying bomb launching sites in the Pas de Calais") and the Americans were advancing briskly, Wachtel's troops sought new positions on Belgium's Channel coast. At the same time, the LXV Korps moved its headquarters to Waterloo, immediately south of Brussels, followed a week later by the chief of staff and the air operations and supply sections that had temporarily remained behind at Maisons-Lafitte near St. Germain-en-Laye. According to Colonel Walter, the withdrawal of the 155th was inefficiently carried out. Much equipment was unnecessarily destroyed, and the morale of the troops was low. As Otto Cerny was later to put it, most of those connected with the program still believed in their Führer but few expected any miracles, least of all those aware of the technical realities of the V-1.

As the left flank of the 155th redeployed in Belgium and Holland, the right flank continued to fire on London, its final salvo going off at 04.00 the morning of September 1. It, too, soon crossed into Belgium, locating for a while at Camp Maria ter Heide near Antwerp. It then retreated to Derwenter in Holland. The entire regiment suffered a serious loss of equipment, much of which had either been demolished or left in cave depots whose entrances had been dynamited by the retreating forces.

The LXV Korps had hardly established itself in Derwenter, when von Rundstedt ordered it to fall back into Germany to take up positions on either side of the river Rhine for action against Belgian and French targets occupied by the Allies. Firing batteries were positioned to the east, south, and west of Cologne. Two headquarter groups operated the divided forces, one at Alfter near Bonn and the other at Wipperfürth. The former was composed largely of army elements, the latter of Luftwaffe personnel, including the air operations section. Firing positions were located on high land east of the Rhine, in the Eifel area, along the river Erft from Möderath west of Cologne to north of the Rhine town of Remagen, and in the area around Mayen west of Koblenz.

Situation maps relating to V-1 operations in early and mid October 1944 show that, while positions east and southwest of Cologne remained virtually unchanged, by mid-month sixteen new firing locations had been added to the north to enable Antwerp—then considered the prime target—to be hit. On October 19, in an effort to deceive the enemy, the LXV Korps changed its designation to XXX Armee Korps z.b.V. This organization had the same mission as its predecessor but a reduced headquarters personnel allotment. During the month, the 155th Flak Regiment's battalions were brought up to strength, and by October 21, firings from Holland and Germany recommenced and continued under the same organizational structure through to January 19, 1945. At that time, the XXX Korps lost control of the V-1, and the 155th Flak Regiment became the 5th Flak Division under the command of Colonel Walter. Then, on January 28, Kammler's Armee Korps z.V. took over the 5th, which passed under the control of Wachtel (who was code-named Colonel Wolf) until the end of the V-weapons campaign.

During the final months of V-1 warfare, Brussels, Antwerp, and Liège were the main targets, with fewer than seventy-five missiles being fired on London from the air. The final V-1 impacted on Antwerp on March 30, 1945, nine and a half months after the flying bomb campaign had begun and a little over a month before Germany surrendered.

Phases of the V-1 Offensive

Phase	*Remarks*
I.	Weak attack during night June 12–13, pause, then heavier attack shortly afterward. Peak reached during first week in July when 902 were fired. Allied attacks on modified sites had little effect, so attention was shifted to the V-1 logistics organization—supply sites, underground depots, railways. By mid-August, firing units began to be pulled out of northeast France; last firing on September 1, 1943.
II.	Air launchings, which had begun sporadically in July, intensified from mid-September until mid-January 1945.
III.	Heavy campaign against Antwerp and other continental targets from October 1944 to March 1945.
IV.	Final long-range air-launched missiles fired from Holland.

If the V-1 flying bomb had problems passing from the research and development stage directly into combat, the A4 rocket had even more. This was hardly due to inadequate planning; rather, there was simply inadequate time for the revolutionary new weapon to become a proven element in army inventory. The missile was not just another tank or field artillery piece or machine gun.

As early as the spring of 1942, well before the first A4 flight tests in June, August, and October, a "Proposal for the Operational Employment of the Long-range Rocket A4" was released. For security reasons, however, the chief of the Armed Forces High Command had all but three copies destroyed, none of which seems to have survived the war. The "proposal" contained descriptions of the rocket-powered missile, of its ground support equipment, and of the recommended personnel organization. It also included some preliminary ideas on launching and firing tactics. At that early date, planners estimated that a force of five thousand missiles would be required to mount an effective campaign against southeast England.

The A4's relatively low standing for several crucial years meant that only limited work could be done toward preparing the missile for field use. The on–again, off–again attitude of Hitler could easily have meant death to the project. Indeed, it was only the tenacity of its developers that assured its survival. Back in September 1939, Field Marshal Walhter von Brauchitsch* had arranged high military priority for army ordnance programs at Peenemünde, but by the spring of 1940 it was removed at Hitler's insistence. Despite Dornberger's repeated appeals to Speer, nothing could be done; indeed, as late as January 1943, the munitions minister told Dornberger that "the Führer cannot give your project higher priority yet; he is still not convinced that your plan will succeed." Hitler's purported dream in March certainly did not help matters, yet only four months later, his "dream" apparently forgotten, the warlord of Germany restored the almost forgotten project to a top-priority status.

How did this come about? Partly as a result of Dornberger and von Braun's being allowed to show him, at his "Wolflair" headquarters (*Führerhauptquartier "Wolfsschanze"*) in Rastenburg, spectacular films of the successful A4 flight of October 3, 1942; partly due to their own powers of persuasion, and partly—in the face of a deteriorating military situation—from the urgent need for new and devastating weapons. The A4 just might fill the bill.

For some time Dornberger had debated the potential military value of his weapon. After severe aircraft losses in the Battle of Britain and the night blitz attacks starting in 1940 and continuing into the spring of 1941,

* A good friend of the A4 program and supporter of Dornberger and von Braun, his order of September 5, 1939, included the opinion "Project Peenemünde . . . , of special defense significance, is to carry on with a high priority and with all [available] means in case of mobilization" (OKH, Feldzeuginspektion, Organisation und Gliederung der Truppe, 27. Aug. 1939 bis 27 Nov. 1939, H37/96). Brauchitsch was sacked as commander in chief by Hitler in December as a result of his failure to capture Moscow. It has often been speculated that he would have been the ideal person to oversee not only the A4 program but the Fi103 as well, including the crucial field-deployment phase. That he was esteemed even in 1945 is shown by a New Year's greeting from Himmler, who wrote, "You will surely be happy to know what a powerful weapon we have in the V-2, the former A4, which was developed at Peenemünde. The Führer often mentions that it was only because of you that the necessary research and development could be undertaken over a period of ten years" (Document No. 1959, Office of the U.S. Chief of Counsel for the Prosecution of Axis Criminality, October 24, 1945).

German air operations over England declined. By 1942, he was arguing that if

> a bomber was shot down after an average of five or six flights over England, if it could carry only a total of six to eight tons of bombs during its active existence, and if the total loss of a bomber, including the cost of training the crew, were estimated at about three times the price of an A4 (38,000 marks), then it was obvious that the A4 came off best.

Dornberger hoped that sooner or later these facts would be recognized at high, decision-making levels. Yet the following year he began to harbor doubts, realizing that by mid-1943 the outcome of World War II could not be decided by launching even a thousand rockets a month, each carrying but a single ton of destructive charge. Too much time had been lost, and Germany was no longer the master of its own destiny.

During the same month of his famous A4 dream and several months before he restored the missile to top priority, Hitler authorized the Organisation Todt to commence construction of the huge storage, missile servicing, and launch complex at Watten on the northern French Channel coast. An explanation for this action has never been given; perhaps it merely demonstrated a subconscious confidence in the Peenemünde program, coupled with a fascination for bunkers of any type.

Whatever the reasons may have been for the Watten decision, a full year earlier—back in April of 1942—a training command had been established at Peenemünde to develop procedures for launching and handling the A4. As tables of organization and equipment were worked out and training and launch methods studied, two basic philosophies evolved. One was predicated on a relatively few, heavily protected, fixed launch sites in which dozens of missiles could be stored and serviced and from which they could be launched. The other favored a highly mobile approach, wherein missiles would be carried into the launch area by rail or truck transport for rapid servicing and firing. Once a number of missiles had been set off toward their targets, the rail or road array would move to another position to recommence preparations for the next salvo. This practice would make it difficult for Allied fighter-bombers to scramble in an attempt to neutralize a firing site, for by the time they had reached it, the firing group would have moved on. Fixed firing locations like Watten, however, would draw not only fighter-bombers but huge, four-engined Lancasters and Flying Fortresses that, as events turned out, were ultimately to dump thousands of tons of bombs on them.

As the Organization Todt focused its attention on Watten, which in modern parlance would be called a "hard" site, plans were drawn up by Dornberger to investigate the feasibility of the alternative launch approach, particularly motorized operations. In the summer of 1943, a Lieutenant Colonel Hohmann and members of the staff of the 760th Artillery Regiment began to put this philosophy into action. Hohmann first made contact with von Rundstedt to advise him of his mission, and then moved into the field to survey potential locations for motorized launch operations. In doing so, he had to keep in mind that each battalion would be made up of three batteries, each of which in turn would consist of three firing platoons operating from as many launch positions.

Although he was primarily concerned with the V-1 weapon, Heinemann's chief of staff, Colonel Eugen Walter, was thoroughly aware of the importance of mobility in missiles of any type and of the need for care in selecting launch sites. He observed that these sites would have to be well camouflaged, "since they constitute the enemy's principal target."

> In the era of aviation and photography [he continued], every weapon whose commitment necessitates complicated constructions and positions, demanding long-time preparations, is inappropriate. Firing installations must therefore be as simple as possible.
>
> The more speedily and more often that changes of positions can be made, the more difficult it will be to identify and eliminate them. Launching installations for long-range missiles must therefore be mobile. The simpler they themselves and their positions are, the easier they can be camouflaged against identification from the ground and the air.

Walter also insisted on taking all conceivable security precautions. "Secrecy," he maintained, "and effective security measures . . . are the two fundamental requisites for successful commitment of guided missiles." Launching sites, supply routes, and supply dumps, being obvious targets for spies, sabotage, and air attacks, must be made as secure as possible, principally by camouflage. He also felt it important to prevent duds from falling into enemy hands in the target areas. "Therefore, care must be taken to provide the missiles with effective automatic self-destruction devices . . ." As for espionage, he recognized that most missiles would be deployed in occupied countries when he ordered that:

> Security measures will comprise the identification and capture of enemy agents, preventing them from gaining access to any agency connected with the construction of long-range missiles, from manufacturing, development, testing, storing, and supply routes to launching sites. It is therefore appropriate to divide the security between one service for the rear and one for the troops.

After the war, Walter took some pride in the fact that LXV Korps headquarters' views on mobility had been thoroughly vindicated. "None of the big structures [the heavy installations like Wizernes], some of which were highly interesting and admirable, was ever put to any practical use ... in spite of concrete ranging in thickness up to five meters."

Meanwhile, at Peenemünde, the 444th Training and Experimental Battery (*Lehr- und Versuchsbatterie 444*) was set up under the command of Colonel Gerhard Stegmaier. Assisted by Lieutenant Colonel Georg Thom, then Dornberger's chief of staff, Stegmaier supervised the development of a missile handling and launch manual and established a training school at Köslin. At that time, it was planned that the field organization be made up of two mobile battalions and a single fixed battalion stationed at the Watten bunker. Dornberger proposed that he run the field operation under the title *Beauftrager z.b.V. Heer* effective September 1, 1943, meaning in translation that he would become army commissioner for the special deployment of the A4. From that time on, he proposed to relinquish his post of chief of the Weapons Research Station.

This was, in fact, done; and, along with his special title, he was named Artillery Commander 191 and placed under the organizational structure of General Fromm's Replacement Training Army.

Under the new setup, Dornberger and his staff had to recruit experienced men from artillery replacement centers and use them to help train and otherwise make ready the missile troops. Colonel Herbert Axster, who became Dornberger's chief of staff later in the war, spoke of the problems that often arose between missile technicians and engineers at Peenemünde and officers being trained to deploy the A4 weapon in the field. He recalls orders he once issued:

> You will work here tomorrow morning, and you will wear the overalls that will be given to you. Before you enter any of the buildings or laboratories you will button them up to your chin. You will also wear a cap so that no one can recognize you as officers. For as soon as the workmen and technicians realize that you are military superiors they will become reluctant to share their knowledge with you.

While training was getting under way, surveys of potential firing locations had to be conducted all along the Channel coast, and inland field trials had to be accelerated to bring the A4 nearer the operational stage. The very fact that the missile was still in development was a major drain on Dornberger's time; and, of course, production goals and methods had to be established, a logistics plan worked out, and a reliable resupply chain developed. By the autumn of 1943, the magnitude of the job was almost overwhelming.

In order to supervise field readiness preparations, Dornberger decided to operate neither from Peenemünde nor from his office in Berlin. Rather, he chose Schwedt on the river Oder because, at the time, "it was the only city free from military dominance." He needed all the freedom he could get as he sought to organize a campaign that soon was to involve expert personnel in training, firing, artillery reconnaissance, signals, flak protection, seismic ranging, supply, transport, construction, and security.

Once he had moved to the new location, Dornberger allocated staff assignments in terms of three major functions: command, supply, and engineering. Thom served as chief of staff, Hohmann as deputy for operations (in northern France), and Stegmaier for training under Kommandostelle S (Headquarters S—for Stegmaier). Other officers concerned themselves with development, storage, propellants, and procurement. An important evaluation office was set up under a Lieutenant Colonel Moser, who served as a focal point through which all information pertinent to field deployment of the A4 passed and from which recommendations for corrective action were made. Moser was also responsible for compiling range tables.

The all-important Long-Range Rocket School at Köslin grew rapidly and was soon augmented by the 271st Replacement Training Battalion at Schneidemühl. After the Peenemünde air raid of August 1943, the 444th Training and Experimental Battery was moved to southern Poland to the SS training camp at Blizna (SS Truppenübungsplatz "Heidelager"). However, all technical units responsible for A4 development test flights—but not training flights—continued to receive their instruction in Peenemünde under the direction of a Lieutenant Colonel Basse.

By January 1944, LXV Korps had been in existence for about a month and Dornberger now found himself promoted to Senior Artillery Commander 191, with headquarters at Maisons-Lafitte near Saint Germain. At this point, Walter recalls, negotiations were opened with the War Office (*Oberkommando des Heeres*) and the Armed Forces Operations Staff requesting Korps authority to "carry out the organization and training of the troop in accordance with its own intentions and principles." Moreover, Korps wanted control of the entire supply system from Mittelwerk to the firing front as well as "a decisive influence on the construction of the structures for use in combat or at least [permission to] determine their position and their employment in combat." These requests were met with resistance. According to Colonel Walter:

> The army was still reluctant to relinquish its weapon to an OKW [Oberkommando West] Korps headquarters ... desiring to retain control of it and use it under command of army officers. Korps headquarters was simply intended to act as a screen. The moving power behind OKW was the Army Ordnance Office, which was the driving force and at the same time the stumbling block in everything that happened. ... For this reason, Korps headquarters found itself compelled to act on its own initiative in taking part in the construction

work and to exert as much influence as possible on the troops [training in Germany].

For a long time Heinemann had been disturbed by Dornberger's dual position as representative of the Army Ordnance Office and of the War Office, and particularly by his complete lack of front-line experience. Heinemann soon arranged for Dornberger's replacement by General Richard Metz, who had excellent combat qualifications as an artillery commander on the Russian front, and whose sole concern would be to field the A4. Dornberger, meanwhile, would continue to exercise his heavy responsibility in seeing the A4 through the development cycle and in supervising training and logistics.

Whereas Heinemann had come to the LXV Korps from his post as Senior Artillery Commander 303 under the Commander in Chief, East, Metz had been Senior Artillery Commander 309. Three years younger than Heinemann—who was born in 1891—he was from Aachen and had seen service during the First World War. Like Dornberger and Heinemann, Metz realized the importance of simplicity in planning and executing field launch preparations and maneuvers. The three generals and their staffs saw little need for hard-surfaced site-access roads, preferring that the launch area be disturbed as little as possible. All recognized the need for utmost secrecy that had been imposed by *Oberkommando West*. Even the commanders of the 7th and 15th armies, in whose territories both flying bomb and rocket sites were being prepared, remained unaware of what was going on.

General Metz had been in charge of the A4 field program as Senior Artillery Commander 191 for about three months. He had become thoroughly familiar with the status and the potential capabilities of the weapon by that time, and had received briefings at Peenemünde by technical personnel. He had watched flight and training tests at the Heidelager field in Blizna and he had visited Stegmaier's training operations at Köslin and at a new site in Grossborn. Toward the end of March, he gave a gloomy status report on the program.

He noted that the 836th Artillery Battalion (Motorized),* training at Grossborn, was in fairly good shape in terms of men and matériel, but that it was by no means ready—though Dornberger promised it would be by mid-April. Its head, Lieutenant Colonel Jannusch, was variously described as "useless, idle, talkative, self-important, and a fool." Nor was Major Rossberg, the commander of the 485th Artillery Battalion, also motorized, much more highly regarded. Major Rossberg had two of his

* Other than its headquarters battery and three firing batteries, the 836th had three supply columns, three security groups with motorized antitank and infantry platoons, and—following its organization in September 1943—three 20mm quadruple antiaircraft guns to help protect the supply columns.

understrength batteries near Naugard while one was flight-training at Blizna. All were promised to be at full strength by May 15. Out of eleven rockets prepared for firing over a five-day period, three burst shortly after launch while one exploded at 3,000 feet at the end of a 90—instead of the planned 100—mile trajectory. General Metz did not visit the final element, the 962nd Artillery Battalion (Motorized), since its troops were to be transferred to bring the other two battalions up to strength. Although plans at the time also called for a fixed, bunker-operation battalion to be organized, plus an SS battery, Metz was not optimistic about the prospects of creating a first-rate, operational force from the "ill-trained and heterogeneously collected" lot he had inspected in the field.

At Köslin, Stegmaier's operations appeared to be progressing more favorably than the field groups, but Metz quickly saw that training doctrine had not been adequately developed and that centralized tactical and organizational control was lacking. He felt that, in time, soldiers could learn to operate the A4 weapon, but only when—and if—it was ready. By the time he got out to Heidelager, he found that only twenty-six missiles had even got into the air (out of fifty-seven attempts), and most of them had air burst. Of the seven that did reach the ground, only four were within the target area. Worse, not a single missile fitted with its warhead had managed to survive through to the ground, and as yet the reasons for airburst had not been determined, much less corrective methods to cure it inaugurated. The enthusiasm for the missile exhibited by Peenemünde developers was not shared by this sober, experienced artillery general on whose shoulders the unwelcome responsibility for fielding an uncertain weapon had been laid.

Then, there was the matter of production. Despite optimism in some quarters, Metz could not fathom how combat-ready missiles could be delivered on schedule and in sufficient quantities to feed the 836th and the 485th, both of which were to grow to two battalions with three batteries per battalion and each with two firing sections. He was so dismayed by all this that he ended his report with a request that he be given a new command. Although Heinemann was understanding, he did not allow his A4 field commander to move to a more conventional operation. Metz tried again, later, and the matter was settled only by the intervention of General Walter Buhle, chief of the Army Staff in the Armed Forces High Command. The most Metz could do was to extract a promise that his forces would be enlarged to four battalions operating within two regimental staffs. Like it or not, Metz was destined to remain with the untried weapon until the SS, in the person of General Hans Kammler, took over.

Between May 18 and 20, 1944, following a series of test shots at

Blizna, plans unfolded for Operation Penguin—the code name for the A4 offensive against Britain that it was hoped would start in August or, at latest, early September. The essential element of the plan was to "shoot and run," fire from one location and then quickly move to another. Missiles were to be launched from firing points along the north French coast, from the Cotentin peninsula northeastward to the Belgian border.

With a new, unfielded weapon Metz realized that his only hope for at least partial success lay in streamlining preparatory and launch procedures at the "front", in building up the strength of the existing battalions, and in attempting either to eliminate or to reduce to a minimum all factors that would divert commanders from their prime responsibilities of placing as many A4s on target as possible.

What concerned Metz most, in assessing the status of his rocket forces (other than the obvious operational reliability of the missile) was the matter of supply, of assuring a strong, uninterrupted logistic chain from assembly plants and propellant manufacturing sites to forward firing positions. Up until the time that the A4 actually went into operation as the V-2, it was rail-transported (under high priority) from the production lines at Mittelwerk in the Harz mountains to supply depots near Germany's western border. All rail movements were coordinated through a Transport Liaison Office at General Dornberger's headquarters in Schwedt. Dornberger's chief supply officer, located near Koblenz at Bad Ems, monitored movement to the border, at which point the LXV Korps took over. From there, A4s generally wound up at the main dump in caves near Méry-sur-Seine, where the 511th Field Workshop Company had installed the necessary equipment to receive and store them.

By June, the 836th Battalion had been transferred to Baumholder close to Koblenz for additional training, and site preparations near the Channel coast were going forward at full speed, Allied air attacks notwithstanding. Moreover, troop buildup was progressing more or less on schedule and word from Blizna was becoming more favorable. Then came the invasion of Normandy on June 6, 1944.

The rapid progress made by Allied armies led to the decision to abandon all A4 launch points to the west and south of the Seine. On June 27, a key planning meeting was held at Bansin not far from Peenemünde, at which time it was further decided to pull back all A4 positions to north of the river Somme. This pullback was accompanied by what US Army military historian Lieutenant Colonel M. C. Helfers describes as "a confusion of orders resulting from the overall military situation and from the ill-defined responsibilities of LXV Korps; Commander in Chief, West; the Special Army Commissioner for V-2 Matters;* and the

* The A4 began to be called the V-2 as early as mid-June, even though it had not yet been fired operationally.

Armed Forces Operations Staff, not to mention Hitler's intervention from time to time." Under the circumstances, it is a wonder that anything got done.

As the military situation deteriorated in France, the LXV Korps received permission (on July 3) from the *Oberkommando West* to give up on rocket positions that were in clear danger of being overrun by the Allies, and also to stop construction on the heavily damaged Watten and Wizernes sites. Hitler, on July 18, reaffirmed the basic decision but said that some work should continue to outfit Watten for a liquid-oxygen production plant and that labor forces should be left in the Wizernes area to deceive the Allies and cause them to continue to waste bombs on it.

Two days later, on July 20, Lieutenant Colonel Count Klaus von Stauffenberg and his accomplices made their unsuccessful attempt to assassinate Hitler at Rastenburg and take over control of Germany. One of the results of this failure was an immediate enhancement of the power of Heinrich Himmler and his SS organization, coupled with a growing distrust of the Wehrmacht on the part of the Führer. The confused situation on the A4 front was a point of obvious army vulnerability, one made to order for an ambitious SS officer like Kammler who was already in control of the V-1 and V-2 production. Drawing power from Himmler's new position as commander in chief of the Replacement Army, Kammler went into action.

First of all, he needed a title. Special Commissioner for V-2 operations would do as well as any. Then, he would have to step into the Dornberger-Heinemann-Metz command structure, which he resolved to do as quickly as possible. The fact that he had no relevant military experience, much less front-line combat duty, did not dampen his determination to move into operational control of the V-2 program.

He carefully laid his plans during August, a month that witnessed the gradual pullback of V-2 units to firing positions in the Ghent-Tournai area in northern Belgium in reaction to Allied crossings of the Seine. Realizing that the bombardment of London would now take place only from firing points in Belgium and Holland, Kammler set up his headquarters in Brussels, just one day after Hitler declared (on August 29) that Operation Penguin should commence as soon as possible. Two days after Kammler had installed himself in the Belgian capital, changes in command were ordered resulting in the removal of the V-2 weapon from the LXV Korps and the SS general's taking charge of "the immediate, improvised commitment" of the missile. Metz would remain in command of the actual firing operations, while Dornberger would continue as Special Army Commissioner for V-2 Affairs in Germany—training, continued weapon improvement, and supply being his major spheres of activity.

Walter recalls that on August 31 he was informed by phone that a conference "dealing with the commencement of action" would take place that night in Brussels, "presided over by SS Gruppenführer Kammler, and with Major General Dornberger present." The Armed Forces Operations Staff thereupon advised him that Korps headquarters was to be "solely responsible for combat action," that Kammler "had nothing whatever to do either with combat action or with the entire V-2 program," and that Dornberger was to continue to be responsible for improving the weapon but that he was to have no say in combat planning.

> Armed with this decision [I] proceeded to the conference.... Besides ... Kammler and Dornberger ... to [my surprise I] met Lieutenant General Metz. Without preamble, Gruppenführer Kammler immediately commenced issuing orders relating to combat action, to the bringing forward of the batteries, moving into position, etc.... [When Metz and Dornberger observed that] the Armed Forces Operations Staff had made the LXV Korps solely responsible, that the action taken by ... Kammler was entirely unfounded and that ... Metz [was] under the sole command of Korps headquarters ... Kammler stated that he would immediately take steps to obtain new directives from Reichsführer-SS Himmler to whom alone he was subordinate. He requested Chief of Staff, Korps headquarters, to report in similar manner to the Armed Forces Operations Staff.

Two days later, the Armed Forces Operations Staff gave in, lamely announcing that "for the time being" Kammler was to conduct combat action, "but that Korps headquarters is to control him tactically and is to remain fully responsible for combat action."

This proviso proved to be meaningless, for as Walter soon witnessed, "the entire V-2 program had, to all intents and purposes, been wrested from the army and transferred exclusively to the SS, who from then on had the decisive voice in all matters pertaining thereto." Korps headquarters promptly lost all contact with Kammler's staff and the V-2 forces.

September 5 was established as the target date for the V-2 offensive to begin. To accomplish this from secure positions, firing units were to be transferred to an area between Antwerp and Malines, where they would assemble as Group North under Lieutenant Colonel Hohmann and Group South under Lieutenant Colonel Wolfgang Weber as follows:

Group North: 1st and 2nd batteries of the 485th
Group South: 2nd and 3rd batteries of the 836th,
plus the 444th

Consequently, Group North, which had come out of the Cleve area in Germany northwest of the Ruhr, moved instead to positions north of The Hague. Meanwhile, Group South, which had first moved from Baumholder to a site to the west of Venlo and then withdrawn to the west of Koblenz in the Euskirchen area, established operations near Laroche so as to fire on French and Belgian targets. Its 444th element, meanwhile, split off to a firing point west of Vielsalm in the Ardennes with instructions to open fire on Paris (which had fallen to the Allies on August 25, 1944). In all, over 6,300 men and nearly 1,600 vehicles were involved.

Despite the rapidly advancing Allied armies coupled with continuous aerial harassment, Kammler's rocket forces stood at the ready on September 5. On that date, the SS officer decided to delay the firing date by one day for the attack on Paris and by two days for the attack on London. The first two launches were made on September 6 by the 444th from a point near Fraiture, 8 miles westward of Vielsalm. Both rockets failed because of premature cutoff of fuel supplies. Transferring to a point 10 miles to the south of Houffalize, the battery made its next—and this time successful—attempt on September 8. At 08.34 the world's first long-range combat rocket was fired against Paris and minutes later reached its target nearly 180 miles away. It produced modest damage when it impacted close to the Porte d'Italie.

Considering the newness of the weapon and the unrelenting enemy ground and aerial offensive against Germany, it is nothing short of miraculous that Kammler got his V-2s into operation against Paris or anywhere else. The ruthlessness of this SS general, coupled with his almost unlimited power and incredible personal energy, undoubtedly helped, though it is, of course, impossible to say what would have happened if the army had been left in charge.

Kammler's drive was almost unbelievable. He spent much of his time traveling up and down the logistics chain, trying to untie local knots and otherwise keeping the flow of supplies open. His own operations staff would work closely with the new Bad Ems supply organization under Lieutenant Colonel Wilhelm Zippelius.

Changes in supply procedures were evident from the time the LXV Korps relinquished control of the weapon. Earlier, Dornberger's supply officer in Bad Ems had monitored rail movements only up to the German border, at which point corps personnel would take over. Once combat operations had started, however, the supply operation was changed somewhat and coordinating procedures improved. First, the Transport Liaison Office in Schwedt moved to Berlin to improve its coordination with the Armed Forces Central Transportation Directorate. At the same time, the responsibilities of Bad Ems increased to the point that Colonel Zippelius was able to control, in consultation with the Director General of Transportation, West, the movement of V-2 trains from the assembly plant in Germany right to firing-unit railheads. By mid-September, Bad Ems was able to maintain an average of one trainload of twenty V-2s per day through the rail network. Because of the exceptional difficulties in handling liquid oxygen, its movement from production plants in Germany to forward storage areas was coordinated by a separate office in Schwedt. Zippelius had to maintain constant contact with the Transport Liaison Office, now located in Berlin. He explained to the authors some of the details:

> Since there was no special transportation equipment for the V-2, I used my connections in the Transport Ministry to contact the heads of the transportation industry. Trains were allocated to us and the cooperation of stationmasters was assured. Since we could not display our wares to the public [if anyone inquired what was on board our guarded trains, we would say circus elephants], we camouflaged the wagons with boxes, wood, and hay. . . . We made several test runs with these trains, traveling all across Germany. Then we loaded them with rockets. They were camouflaged so well that no one ever suspected what they were. Throughout the whole war, we never lost an entire load, though we were attacked by fighters that occasionally did some damage. Every train we sent out reached its proper destination.

In the evening of the day that the first V-2 was fired on Paris, the first and second batteries of the 485th Artillery Battalion opened up against London from positions between The Hague and Wassenaar in Holland. Aiming at a point 1,000 yards east of Waterloo Station, they continued to operate from Holland until their entire battalion was forced to regroup to positions near Burgsteinfurt in Germany as a result of the paratroop landings in the Arnhem area on September 17. To the south, meanwhile, the 836th Battalion was firing at Lille and Mons from positions near Euskirchen in Germany. The 444th Battery, which had engaged Paris on September 8, moved northwestward to the Dutch island of Walcheren where, under Kammler's personal direction, six missiles were launched against London. Following the Arnhem landings, the battery retreated to Zwolle, while Kammler's headquarters at Berg an Del—after barely escaping capture by Allied troops parachuting into the Nijmegen area —shifted to Darfeld, 31 kilometers west and slightly north of Münster.

As with the V-1, phase one of the V-2 campaign got off to a slow start. During the period September 8–17, only forty-five missiles were launched, thirty-one by Group North and fourteen by Group South. These launchings, summarized in the table on page 203, were in a sense desperation shots. The missile and its firing and supply organizations were just not ready. Add to this a confused military situation and continuous aerial harassment and one comes up with an almost chaotic "cocktail of the day." Nevertheless, Kammler readied his troops for what can be termed phase two, a period of firing that began on September 18 and lasted until October 12. To a great extent, this phase was influenced by events arising from the Arnhem landings that commenced on September 17.

These landings were an integral part of Field Marshal Bernard Law Montgomery's plan to thrust northward through Holland into northern Germany, outflanking strong river and artificial defenses along the central and southern fronts. He was also motivated by a desire to relieve the British people of punishment they were receiving at the hands of the V-1 and V-2 weapons. In his *Memoirs* he wrote:

> On the 9th September I received information from London that on the previous day the first V-2 rockets had landed in England; it was suspected that they came from areas near Rotterdam and Amsterdam and I was asked when I could rope off those general areas.

Arnhem, of course, was a disaster for the Allies; the British First Airborne Division—or rather the 2,400 men that remained of it after severe fighting—was forced to withdraw from the town on September 25. Nevertheless, the whole atmosphere was so uncertain in the Arnhem-Nijmegen area that Kammler decided to move the entire Group North out of Holland—and thus out of range of London. However, at all costs he intended to maintain his rocket offensive against closer targets in southeast England. To do this, he redeployed the 444th to Stavorem in Friesland, from where forty-four missiles were fired on Ipswich and Norwich (of which thirty-seven reached the UK). Meanwhile, the 485th's two batteries brought Liège, Louvain, Tournai, and Maastricht under fire from positions near Burgsteinfurt northwest of Munster.

As it turned out, the Allies did not press into Holland for the rest of the autumn and throughout the winter. By September 30, Kammler felt it safe to move the 485th's Second Battery back into the area of The Hague, where once again London was brought under fire. Meanwhile, Group South stayed in Euskirchen until late September and then moved to firing positions north of Montaburg in the Hachenburg area. Phase two firings, summarized in the table on page 204, averaged 6 missiles per day; out of 162 launched, 110 were aimed at targets on the continent and 52 against England. Of the 43 aimed at Norwich, Joan Banger, who made a study of Norwich at war, told the authors that none actually impacted inside the city boundaries. One hit in the direction of Ranworth on September 26, and others the next day in the general direction of Great Yarmouth and Brammerton. One also landed on the Royal Norwich Golf Course at Hellesdon, several miles from city center.

After the war, Field Marshal von Rundstedt admitted, "As far as I was concerned, the war ended in September [1944]." Yet during that very month, he felt obliged to tell his troops "I expect you soldiers to defend

Germany's sacred soil . . . to the very last." To reach that sacred soil, the Allies realized that they must secure Channel ports through which vital military supplies could be funneled. This led Eisenhower to advise Montgomery, "While we are advancing we will be opening the ports of Havre and Antwerp, which are essential to sustain a powerful thrust deep into Germany. No reallocation of our present resources would be adequate to sustain a thrust to Berlin."

Hitler understood the supply problems facing the Allies, leading him to transmit an order to Kammler on October 12 to the effect that except for London, V-2s would henceforth be fired only on the Belgian port of Antwerp. Although Hitler's orders were not carried out to the letter—when the V-2 campaign ended on March 30, 1945, more than two hundred missiles had been launched against such other targets as Paris and Liège—the British capital and the Belgian port did receive the brunt of the attack from October 13 onward. In fact, more missiles rained down on Antwerp before the war was over than on London.

Many of the firings from Holland were observed by the Dutch. Typical of the intelligence summaries of V-2 sightings is that relating the experience of one C. van Fliet, who successfully made his way through the front lines to Allied occupied territory on October 25, 1944:

> He recalled that at about 6:30 PM sometime between 6 and 10 September, the day after a strong storm, he heard "a terrific noise," and looked up to see "a pointed projectile about 10 metres long rising slowly and majestically above the tree tops. A cloud of black smoke billowed slowly, after reaching a height of about 15 metres. The projectile gained speed rapidly with a flame emerging at the rear extending to about half its length." A second rocket went off 15–30 minutes later. Location: Wassenaar.
>
> The subject [C. Van Fliet] later visited the launch sites, finding them to be "in the middle of a roadway passing through a wood." There were circular patches at each from which the tar had been melted or burned. Diam. of patch about 9 m. "In the centre of each burned patch was an unburned part in the form of a regular pattern which suggested that some form of stand had been used."
>
> Trees on side of road badly burned up to height of 1 m and less so to their tops. Evidence of violent low blast. Grass was flattened out on thatched roof of a nearby small house [that] had been lifted and blown off.
>
> Source also described vehicles that had gone to launch site about two hours before firings, and fact that population [was] kept from within half a mile of site. One of the vehicles described as "a long truck having sixteen wheels and with some lifting tackle stowed underneath." Crew consisted of from 16 to 20 soldiers "who at the

> time of launching were completely clad from head to foot with asbestos helmets and overalls." Source reported that inhabitants of Wassenaar left town afterwards for fear of Allied air strikes.

The Germans knew that hundreds of Van Fliets were watching their movements and that reports inevitably were being transmitted to the Allies by clandestine radio. Colonel Zippelius told the authors that tactics developed to avoid aerial attacks were very successful, regardless of the time and conditions of launch. He related procedures undertaken when he served with the V-2 forces located at Wassenaar:

> In the daytime we would measure off certain areas within the city and make our calculations. At night we would block off these areas and move in our convoys of Meilerwagens, missiles, and parts. After we had launched five or six V-2s, we would quickly move our men and equipment away. In the morning, when the search planes flew over in a futile attempt to locate our rockets, all they would see were streetcars and motor cars.

While target priorities were changing, Kammler decided to make some internal changes of his own. For one thing, General Heinemann finally released General Metz, making the following obliquely cautious entry in his records:

> Because of the change in V-2 command organization, which places the weapon directly under the commander in chief of the Replacement Army, [General Metz] missed out in conducting its field commitment.

The exit of Metz forced Kammler quite unwillingly to rely more heavily on Dornberger, and even to accept at his Cleve headquarters members of the latter's own supply staff. Kammler faced problems in other areas. For example, his operations officer, a Major Merthin, committed such absurd errors as having gasoline loaded into V-2 alcohol propellant trucks, causing severe damage to them. According to Zippelius, Kammler immediately sacked the unfortunate major—who also, it seems, was held responsible for losing warheads somewhere between supply centers and the front! Merthin promptly had a nervous breakdown. Nor was that all. Kammler soon had a run-in with Dornberger's new chief of staff, Colonel Herbert Axster, who was acting as a sort of buffer between the development team at Peenemünde and troops training in the field. One day, while Axster was inspecting V-2 firing positions, Kammler drove up in a large car and screamed out, "You have nothing more to do with this program. It's my business now. I don't need you anymore." Axster described Kammler to the authors as "the worst man I ever met in my life. My intention henceforth was always to stay away from him." As a result of these and similar difficulties, Kammler set up an improved staffing system under the divisional designation Division z.v., with Colonel Hohmann taking over as chief of staff.

The phase three period from October 13, 1944 to March 29, 1945 saw more than a thousand V-2s fired on London. Antwerp received nearly fifteen hundred missiles, launched variously from The Hague, Hellendoorn-Zwolle, and Burgsteinfurt. Others were launched against Liège, Paris, and miscellaneous French and Belgian targets, including 11 directed against the Remagen Bridge in Germany (the first and only time V-2s were launched in a tactical mission). Hitler himself ordered the employment of the V-2 in the Remagen mission, apparently without the knowledge of either Field Marshal Albert Kesselring (the C. in C., West) or his chief of staff, General Siegfried Westphal. The firings took place from northwest of Hellendoorn with the following results: One rocket fell 40 miles short some 7 miles east of Cologne; one near the Allied front lines east of the Rhine; one less than a mile from the bridge after a flight of over 130 miles; and eight in the general area of the bridge. If the defective round that fell near Cologne is discounted, the other ten rounds deviated only 1.1 miles in range and 2.5 miles in line according to a SHAEF report of April 9, 1945.

Statistics on these and all other V-2 rounds that fell within Allied-held territory on the Continent were carefully maintained for later analysis. Typical of the dozens of dispatches from the field is one from the 601st AAA Battalion directed to the commanding officer of the 17th AAA Group, US Army:

> V-2 explosion report by all Btrys of this Bn. Time and date of explosion, 27 Nov 1944 0514. Results of investigation by Bn S-2: V-2 exploded at J941798, 100 yds from a farm house, making a hole in a plowed field 54 ft in diameter. Depth could not be determined as hole filled with water to within 4 ft of ground level. Farm house was damaged to the extent of broken windows and about 50% of tile roof was either blown or jarred off. No one was killed or injured as a result of explosion. Small pieces of metal were the only parts that could be found in surrounding area.
> For the Commanding Officer: Pershing J. Nadeau, Capt., CAC S-2.

Phase three movements and activities of Kammler's V-2 forces are summarized in the table on page 205. In line with a reorganization that took place on January 1, 1945, the several V-2 battalions were upgraded to regiments, and new designations were given. Thus, the 836th became the 901st Artillery Regiment z.V. (Motorized); the 485th, the 902nd Artillery Regiment z.V. (Motorized); and the 500th SS Rocket

Battery* became a battalion under the 902nd Regiment. By the end of January, all were placed under Kammler's revised headquarters, which became known as Armee Korps, z.V. Despite all these changes, operations continued more or less as before with the notable exception of supply—always a critical and certainly never a fully solved problem.

Dornberger and his supply staff had observed for some time that V-2 components suffered from a rather short storage life, and that electrical equipment and pipes would often deteriorate well before the V-2 could be moved to the front. A plan known as the "Hot Cakes System," devised late in October, required that missiles henceforth travel by rail from the assembly plant at Nordhausen straight to the firing area, with only a short en route stop at an outfitting station where warheads, fuses, vanes, etc. were loaded aboard for later installation at the front-line technical battery testing and assembly point. From there, the V-2s were transported by truck to the firing position. On-site rejection rates were reduced from 12 to 2 percent as a result of this new system. Meanwhile, some five hundred partially defective rockets located in dumps in Germany were cannibalized for their parts, which were returned to Mittelwerk.

The last V-2 of the war was fired on March 29, 1945. The following day, Himmler ordered all of Kammler's rocket troops to be released from their respective units and to become part of the Provisional Army Blumentritt (named after General Günther Blumentritt, von Rundstedt's chief of staff). Even as the war's end approached, as Dornberger related in his book *V-2*:

> Kammler refused to believe in an imminent collapse. He dashed to and fro between the Dutch and Rhineland fronts and Thuringia and Berlin. He was on the move day and night. Conferences were called for 1 o'clock in the morning somewhere in the Harz Mountains, or we would meet at midnight somewhere on the Autobahn and then, after a brief exchange of views, drive back to work again. We were prey to terrific nervous tension. Irritable and overworked as we were, we didn't mince words. Kammler, if he got impatient and wanted to drive on, would wake the slumbering officers of his suite with a burst from his tommy gun. "No need for *them* to sleep! I can't either!" Fixed working hours and leisure had long been things of the past.... Kammler still believed that he alone, with his Army Corps and the weapons over which he had absolute authority, could prevent the imminent collapse, postpone a decision, and even turn the scales.

But this was not to be. Even the fanatical SS general ultimately had to admit that the V-1 and V-2 could not affect the tides of war. So he

* A new element that was added to the V-2 field forces, it was the only one to employ a radio beam (*Leitstrahl*) for trajectory control.

arranged that Himmler transfer him to a new position, that of General Commissioner for Turbojet Fighters.

With or without Kammler, the Germans could never have fired the large number of missiles planners had dreamed of. Assuming either Watten or Wizernes had been available (but not both), and that the regular mobile units had operated at normal rates, about one hundred missiles might have been launched during a "maximum effort" 24-hour period. Taking half that rate as an average, about 350 missiles a week probably could have been fired. Adding them to the V-1 rate, some 5,390 missiles theoretically could have impacted on England every week. If the V-3 high-pressure pump guns at Mimoyecques (described earlier) had become operational, some 600 rounds an *hour* might have been added to the explosive tonnage.

As it was, during the first week in January 1945, nearly 130 V-2 incidents occurred, 59 in Britain and 80 on the Continent. The best firing rate against an individual target was achieved twice, with 107 missiles involved in each case. The highest rate for a 24-hour period also occurred twice, on December 23 and 26, 1944—both against Antwerp.

Phase 1 V-2 Firings September 8–17, 1944

Firing Organization	Target	No. of V-2s	Remarks
	GROUP NORTH		
485th Battalion, First Battery	London	12	After only ten days of operations from The Hague–Wassenaar area, the Battalion moved to Burgsteinfurt, Germany as a result of Allied paratroop landings in Arnhem, Holland.
485th Battalion, First Battery	Liège	1	
485th Battalion, Second Battery	London	12	
444th Battery, assisted by a battery from the 91st Technical Artillery Battalion (Motorized)	London	6	Under the personal command of SS General Kammler; redeployed later to Zwolle, Holland, after firing from Walcheren September 16–18.
	Subtotal	31	
	GROUP SOUTH		
444th Battery	Paris	3	After firing from 10 miles south of Houffalize in the Ardennes, moved north on September 10.
836th Battalion, Second Battery	Lille	4	Conducted first firings on September 14 from an area near Euskirchen to the west and slightly to the south of Bonn in Germany.
836th Battalion, Second Battery	Mons	3	
836th Battalion, Third Battery	Lille	3	
836th Battalion, Third Battery	Mons	1	
	Subtotal	14	
	TOTAL	45	

Phase II V-2 Firings September 18–30 (with addenda to October 12) 1944

Firing Organization	Target	No. of V-2s	Remarks
	GROUP NORTH		
444th Battery	Norwich Ipswich	43 1	Firings from Stavorem in Friesland, beginning September 25
485th Battalion, First Battery	Tournai Hasselt Maastricht Liège	1 2 8 4	Firings from Burgsteinfurt beginning September 19
485th Battalion, Second Battery	Tournai Hasselt Maastricht Liège	6 4 4 8	Firings from Burgsteinfurt up to September 30. This battery was then transferred to The Hague, from where firings recommenced at midnight on October 3–4. See note below
	Subtotal	81	[through September 30, 1944]
	GROUP SOUTH		
836th Battalion, Second Battery	Lille Arras Cambrai St. Quentin Tourcoing Liège Hasselt Maatstricht	7 5 5 1 5 3 1 2	Firings from Euskirchen area until the last week in September, thence to Hachenburg area north of Montaburg where firings were continued. On September 29 move made to Merzig, north of Saar, from where firings commenced on October 2 (see note below).
836th Battalion, Third Battery	Lille Liège Arras Maastricht Tourcoing Tournai Hasselt	3 5 1 2 1 2 3	As above, except that the Third Battery transferred to Merzig on October 6 (see note below).
	Subtotal	46	[through September 30, 1944]
	TOTAL	127	[through September 30, 1944]
Note: From October 1 to October 12, 35 missiles were fired on London, Paris, Tourcoing, Lille, Hasselt, and Maastricht.		35	[October 1–12]
	GRAND TOTAL	162	[September 18–October 12, 1944]

Phase III V-2 Firings

Firing Organization	Activities during October 13, 1944 through March 29, 1945
	GROUP NORTH
485th Artillery Battalion (Motorized). [From January on, 902nd Artillery Regiment, z.V. (Motorized).]	First and third batteries located near Bursteinfurt. Engaged Antwerp, except for one firing section of the third battery that transferred to The Hague on October 20 to join the second battery already stationed there. In November, all batteries were consolidated in Burgsteinfurt area for concentrated operations against Antwerp. In mid-December, the first was transferred to The Hague, remaining there to the end. It was joined by the third battery the following month. The last missile fired by the 485th was on March 29, 1945,
500th SS Rocket Battery. [From January on, 4th Battalion of the 902nd.]	Commenced activities on October 16 near Burgsteinfurt, aiming its missiles at Antwerp. Later, the battery transferred to Zwolle-Hellendoorn-Enchede area, from where it operated until V-2 firings ended in March. From that area, the 500th, renamed the 4th Battalion of the 902nd, fired 11 missiles toward the Remagen Bridge on March 1, a target more than 130 miles distant.
444th Battery	Remained in the Stavorem area until October 20 to fire on Antwerp, then moved to The Hague for the remainder of the campaign.
	GROUP SOUTH
836th Artillery Battalion (Motorized). [From January on, 901st Artillery Regiment (Motorized).]	In mid-October, bombarded Antwerp from Merzig area, except for first battery that operated from Hermeskiel 20 miles to the northeast. Toward the end of the month the second and third batteries also transferred to area near Hermeskiel. In late November and into December, the 836th redeployed to the Hachenburg/Montaburg area. The loss of the liquid-oxygen-producing plant at Wittringen on December 7 deprived the battalion of most of this essential oxidizer, forcing a reduction in the rate of fire to one missile every several days. The V-2 firings ended on March 16, 45.

CHAPTER 11

Defense Against the V-Weapons

As the V-1 offensive got under way during the night of June 12–13, 1944, the British quickly put into action the defensive measures they had been planning for at least six months. The purpose of these measures was to minimize the effects of flying bombs en route to London and other parts of southeast England by: (1) providing radar and optical warnings, (2) scrambling fighter patrols, (3) opening fire with antiaircraft artillery, and (4) taking maximum advantage of barrage balloons.

On November 17, 1943, Air Marshal Sir Roderic Hill received a directive from Air Chief Marshal Sir Trafford Leigh-Mallory to organize the Air Defence of Great Britain (ADGB) within the framework of the newly created Allied Expeditionary Air Force. Under Leigh-Mallory's command, the AEAF was to exercise overall control of aircraft to be committed to the support of the invasion of France the following June. To Overlord planners, it was evident that the air defense of the base (the United Kingdom) was an integral part of the assault on fortress Europe.

In addition to his direct command of seven fighter groups and the responsibility to develop air interception methods and apparatus for eventual use in ADGB and other theaters, Hill took over operational control of AA (Anti-Aircraft) Command, the Royal Observer Corps, Balloon Command, and other static elements of air defense formerly controlled operationally by his Fighter Command. To carry out his air defense responsibilities, he assigned forty-eight squadrons of aircraft, about half the number involved in defending the country several years earlier. Of these, fifteen were on loan to the Second Tactical Air Force.

The rapid decline in the fighting strength and the will to win of the Luftwaffe, coupled with what Hill calls the "extraordinarily ineffective" navigation, target marking, and bombing of those few German planes that did continue to harass Britain, made this reduced defensive force possible. All at a time when the Allies were presenting stimulating and inviting pre-Overlord targets in the form of troop buildups, concentrations of shipping, and large stockpiles of munitions and other war supplies. Antiaircraft artillery, barrage balloons, ground observer units, and elements of the Royal Navy also became part of the forces available to blunt the expected flying bomb attack. Radar chain stations in southeast England were also to be involved, the assumption being that radar would detect a pilotless V-1 in much the same way it could a piloted plane. Hopefully, differences in track behavior would enable the two to be distinguished. Similarly, it was expected that ground observers would soon learn to recognize the V-1 by both feature and sound identification.

Active preparations for the defense against pilotless aircraft commenced in December 1943, when Leigh-Mallory instructed Hill to draw up plans as to how he would go about destroying missiles that traveled approximately 400 miles per hour at an altitude of 7,500 feet. (Allied assessments of the speed dropped during the ensuing months to about 330 mph at an altitude of about 6,000 feet. In actual fact, the V-1 flew at just under 400 mph at altitudes of between 2,000 and 3,000 feet.) Hill immediately contacted General Sir Frederick Pile, the general officer commanding the Anti-Aircraft Command. As the situation was outlined to Hill, it became

> . . . clear at the outset that to prepare a detailed plan of defence would take several weeks. I therefore decided to submit a preliminary outline plan. I took as my point of departure the fundamental proposition that a pilotless aircraft was still an aircraft, and therefore vulnerable to the same basic methods of attack. Of course, as there was no crew, such an aircraft could not be made to crash by killing the pilot; on the other hand, it would be incapable of retreat or evasion, except, perhaps, to a very limited extent . . . on balance, and considering the uncertainty of our knowledge, it would clearly have been unjustifiable to exclude any of the normal methods of defence which we were accustomed to use against piloted aircraft.

"Normal methods" meant, of course, not only fighters but Pile's antiaircraft guns and searchlights as well as balloons, all so deployed as to reduce to a minimum the danger of mutual interference. (These and other recollections are contained in Hill's "Air Operations by Air Defence of Great Britain and Fighter Command in Connection with the German Flying Bomb and Rocket Offensives," published as a supplement to the *London Gazette*, October 19, 1948.)

Hill and Pile completed their first detailed defense plan toward the end of December and on January 2, 1944, Hill submitted it to Leigh-Mallory. It was soon approved and copies were sent to higher authorities and to the intelligence community for study and comments. Intelligence, for its part, advised that the danger of an immediate flying bomb attack had receded as a result of Allied bombing attacks, while commanders involved in Normandy invasion preparations took note of the possibility that the flying bomb offensive and Overlord might well coincide. This turned out to be the case.

Because of this convergence of the German and Allied offensives, the Hill-Pile plan for flying bomb defense underwent some modification during the winter of 1944 so that by the end of February it came to be called the Overlord Diver Plan—Diver being the code name for the pilotless aircraft (the full title: "Concurrent Air Defence Plan for 'Overlord' and 'Diver'"). In essence, this was the plan Hill would put into effect when the flying bomb campaign finally began three and a half months later:

> ... at every stage the principal object that General Pile and I had in mind was the defence of London, which was the target threatened by the vast majority of the "ski sites." Secondly, we had to provide for the defence of Bristol, which was threatened by a small number of "ski sites" near Cherbourg. Thirdly, we had to bear in mind the possibility that, as a counter-measure to our preparations for the European operations, pilotless aircraft might be used against assembly areas on the south coast, and particularly around the Solent [the strait between the Isle of Wight and Southampton].

The first line of defense was assigned to fighter aircraft:

1. No. 11 Group flying standing patrols by day at 12,000 feet on three lines (30 to 40 aircraft in the air):
   a. 20 miles off coast between Beachy Head and Dover
   b. over coastline between Newhaven and Dover
   c. between Haywards Heath and Ashford
2. After an attack had begun, other aircraft would patrol the same lines at 6,000 feet.
3. At night, fighters would patrol under radar station control, as control from the Sector Operations Room was considered too slow for the mission. (Aids other than radar would be used, also; for example, "solid-power" rockets fired by the Royal Observer Corps and markers set out in the Channel by the Royal Navy.) No standing patrols were assigned to Bristol and the Solent, but fighters would always be ready to attack any oncoming bombs.

Next came the second line, guns and searchlights, though in poor flying weather they were to be considered first line. The availability of sufficient antiaircraft artillery as the spring of 1944 approached was uncertain; the requirements of Overlord were being felt ever more strongly, and earlier plans had to be revised to take this into account. Consequently, Hill and Pile agreed to do the best they could and deploy 192 heavy guns in the hollows of the North Downs (a range of low ridges in southern England), where, it was expected, their radar equipment would enjoy maximum screening from enemy jamming attempts. The heavies were to be accompanied by 246 light AA guns and about 200 searchlights. Bristol would be defended by 96 heavy guns but, in deference to stringent Overlord needs, only 36 light pieces instead of the 216 originally planned. Since the Solent would be well defended by Overlord forces against conventional bombing attacks, no strong reason was seen for installing special Diver defenses.

Behind the guns would be the third line of defense, 480 barrage balloons extending from Limpsfield to Cobham in Kent. The Air Reporting System and the Home Security Civil Defence organizations would prepare themselves for recognition, tracking, and reporting, while bomb disposal units commenced training for the task expected of them. According to Hill:

> It was, then, with the revised plan ready for action that we awaited the beginning of the German attacks. To say that this plan represented a compromise between the requirements of "Overlord" and those of "Diver" would not be strictly true; for the defence of the base [the UK] against "Diver" was itself an essential "Overlord" requirement. But it provided at once the largest appropriation that could be spared for the job, and the smallest that was likely to be effective against the threat which was then foreseen. The number of guns to be deployed, in particular, was not more than a bare minimum. In the circumstances, it was impossible for us to budget for more guns; but we took care to frame the plan in such a way that the numbers could easily be increased if further guns should happen to become available. I took the precaution of pointing out that if the pilotless aircraft should fly between 2,000 and 3,000 feet instead of at the greater altitude expected by the Air Ministry, the guns would have a very awkward task, for between those heights the targets would be too high for the light anti-aircraft guns and too low for the mobile heavy guns which at that time could not be traversed smoothly enough to engage such speedy missiles.

The opening of the flying bomb campaign against the United Kingdom on June 13–14, 1944 was preceded by four hours of cross-Channel artillery bombardment of the towns of Maidstone, Otham, and Folkestone. Its purpose: to create tension and uncertainty as a prelude to the firing of the first V-1s a few minutes later.

When the first V-1 missile flew overhead, one Royal Observer Corps member heard what he described as "a swishing" sound while another talked of "a noise like a model-T Ford going up a hill." The strange robot exploded at Swanscombe near Gravesend at 4:18 in the morning, June 13, 1944. Other missiles came down at Bethnal Green, Cuckfield, and Platt near Sevenoaks. Recalls Hill:

> The attack then ceased for the time being. I came to the conclusion that so small an effort did not justify the major re-disposition of the anti-aircraft defences required by the "Overlord-Diver" Plan. The Chiefs of Staff agreed. . . . In the meantime the existing defences were authorised to engage pilotless aircraft on the same terms as ordinary aircraft.

This "holding" period was to last only a few days, for on June 15 the launch rate of V-1 climbed impressively. The Chiefs of Staff agreed now that Overlord-Diver must be put into full-scale operations, and on June 17 the necessary redeployment of forces commenced under Hill's command. By June 21, all was in readiness. From the point of view of defense, the first phase of the V-1 campaign lasted from June 13 to July 15. It was essentially a learning period for the defenders, who had to adapt conventional arms and tactics to a novel weapon.

Almost from the beginning, problems of interference between ground and air defense forces cropped up. This soon led to the adoption of a plan whereby, during fine weather, the guns would not fire at all (except light AA units that were tied in to the communications link, and only at times when no fighters were visible), thereby giving fighter aircraft complete freedom of kill. On the other hand, in poor weather guns would be relied upon 100 percent. In weather that was neither fine nor poor, but rather middling, the fighters would range from the Channel to the front of the gun belt, entering it only if they were actively chasing a V-1. In most cases, however, the guns would have freedom of operation up to 8,000 feet, considerably above the 2,000 to 3,000 feet selected by the Germans to fly their missiles—an altitude at which the gunners had much difficulty, as Hill and Pile had predicted.

During the first month of the V-1 offensive the Germans managed to get about a hundred missiles across the Channel on an average day, though during the last week it dropped to about seventy. Fighters would bag some of these before the English coastline was reached, and antiaircraft artillery a few more. Then, in the inland zone, fighters would down more—about ten to every four destroyed by AA artillery and one by the barrage balloons. Still, a couple of dozen or more bombs would get through to Greater London every day.

A number of tactics were open to pilots of Tempest V; Spitfire IX, XII, and XIV; Typhoon; and night-fighting Mosquito aircraft: Attempt a stern chase, dive on the V-1 from above, or (most effective) match course with the bomb and, when the missile had drawn level, fire deflection shots into it—care being taken not to approach within 200 yards lest it explode and destroy attacker and attacked alike. Occasionally, they would fly in very close to a V-1 and throw it into a premature dive by tipping it physically (wing-to-wing contact), or they would create an aerodynamic disturbance that would cause it to go out of control.

Fighter pilots would rely heavily on reports from small Channel boats operated by the Royal Navy and from ground radar stations, receiving warnings and running commentaries on the approach and flight paths of the robot planes. At the most, fighters would have only five minutes to intercept from the time an individual V-1 reached the coast to the time it came up against the gun belt. If a pilot elected to pursue his quarry into the belt itself he would have an additional minute to make his kill before the balloon barrage ruled out all possibility of further engagement.

Naturally, many efforts were made to improve the slight speed margin the best British planes had over their small pilotless targets. By stripping off armor, many internal fittings, and outside paint, and by modifying their engines so that they could use a more powerful aviation fuel, up to 30 mph additional velocity could be achieved. This was highly appreciated by pilots as they gave chase to the 400-mile-per-hour robots and sought to destroy them within the relatively few available minutes.

The gunners on the ground, just as the pilots in the air, had to condition themselves to this new enemy weapon. At Anti-Aircraft Command, Pile and his officers all reacted "rather cautiously" at first. "We had been assured from above," Pile complained, "that we should have plenty of warning, perhaps as much as a month, before the enemy began to use his new weapon, and the general opinion was that these few salvoes of bombs were in the nature of a final gesture by the enemy, an attempt to test the possibilities of the flying-bomb before the War ended."

That the Germans meant more became increasingly evident. Soon after the initial deployment, Pile realized that many more guns than had been allocated to the Overlord-Diver plan would be necessary. He set forth to get them and, by the end of June, had managed to deploy 376 heavy pieces and 476 light 40-mm guns, augmented by some American battalions and 560 light units manned by an RAF regiment. At about the same time, it was decided to restrict the firing of the guns located in London; except for a few cases of detonation in the air, all they accomplished was to bring down missiles that if left to themselves would have impacted anyway. By not firing the guns, some of the V-1s would undoubtedly have continued beyond the densely populated central area and crashed in rural areas.

Londoners, who did not understand what was happening, naturally became bitter at the inaction of the gunners. As if this were not enough, the overall performance of AA artillery during the first phase of the V-1 offensive was poor, especially when compared to the bag of bombs falling prey to the fighters. According to Pile in his book *Ack-Ack*;

> As a matter of fact, there were very good security reasons why the gunner's contribution—which, I must confess, was not very impressive—could not be hinted at [publicly]. We were not at all anxious for the enemy to find out what a strain had been put upon our resources, nor did we want it to be realized that the northern vulnerable area now lay open and undefended against attack.

Among the reasons for AA Command's difficulties was the location of the radar equipment in the North Downs. Its placement in the hollows to reduce the danger of enemy jamming also proved detrimental to its own operation. Since the Germans elected not to jam, or were unable to do so because their transmitters had been knocked out as a result of Overlord bombing, Pile wanted to move his guns and equipment to higher ground. But that was, or soon would be, occupied by the balloon barrage. The complex cycle of using fighters in fair weather, guns in inclement weather, and both the rest of the time posed additional complications. "Never before had the guns been so forcibly restrained," he said. "Never before had the rules of defense been so complicated."

From his position as head of the combined defense forces, Hill became only too aware of friction mounting between the various elements. He knew that pilots were not always able to determine when they were entering the gun belt, while gunners were hard pressed to tell when a fighter was legitimately pursuing this or that missile into the belt and when he wasn't. "Charges and counter-charges mounted," he reluctantly had to conclude, "and with deep misgivings I began to sense a rising feeling of mutual distrust between pilots and gunners."

In an effort to find an interim solution, Hill directed that effective July 10 fighters would under no circumstances be allowed to enter the gun belt. Once this was decided, Pile then recommended that virtually all the guns outside the belt be moved inside. This would give the gunners a clear, uninterrupted sphere of action, unperturbed by the danger that friendly fighters would enter it in pursuit of a flying bomb. Until now, however, no one had seriously considered a more drastic alternative: Movement of the entire gun belt to a more advantageous position. "Was there such a better position?" asked Hill. If so, where was it?

To General Pile and to Air Marshal Hill's deputy senior air staff officer, Air Commodore G. H. Ambler, the answer was obvious: the coast. There, the gunners would have an open view of the approaching V-1s, and the radar would no longer be hampered by the undulations of the North Downs. Moreover, proximity fuses could be used with impunity (they could not inland), and most of the missiles destroyed would fall into the sea and not litter up "Bomb Alley." The principal disadvantage to the move, other than logistic, was that it would split the fighters' sphere of operations into over-the-Channel and behind-the-gun-belt zones. Air Vice Marshal Robert H. W. L. Saunders, who was in charge of No. 11 Group, surprisingly did not object to this mode, and soon gave the new plan his endorsement. Hill immediately gave the appropriate executive orders, though not without having weighed carefully the consequences of failing to consult higher authority.

> On the one hand, since the forces which I intended to re-dispose had already been allotted to me for the defence of London against flying bombs, and no move of guns from one defended area to another was involved, I might regard the change as a tactical one and act at once on my own responsibility. On the other hand, bearing in mind that no move involving so many guns had ever been made on purely tactical grounds before, I might adopt a more proscriptive attitude and refer the matter to higher authority first, as I should have done, for example, if I had proposed to move guns from, say, Manchester to the "Diver" belt, or from Birmingham to Bristol.

He felt his decision in favor of the former course was justified in that "the situation had reached such a point that no delay could be accepted." Pile agreed, though he was fully cognizant of what was involved:

> Consider the difficulties at a time when we were in the midst of a battle. Every redeployment of sites meant a complete reorganization of the complicated signals circuits: for instance, the move forward of the light anti-aircraft guns meant that twelve new Gun Communication Rooms had to be provided for them. . . . The changeover of the heavy anti-aircraft equipment meant that 441 static guns had to be uprooted and moved with all their stores, fire-control instruments, and so on from all over England to the Diver defences: 365 of them went by road, the remainder by rail. The portable platforms alone involved a move of 8000 tons of material. . . . Thirteen thousand Anderson shelters and three million sandbags were among the stores to be obtained. . . . Quite apart from the troops of the various supply services, 23,000 men and women had to be transported forward with all their kit and stores: 30,000 tons of stores of all sorts were lifted in one week by 8000 lorries and nearly 9000 . . . men; . . . heavy anti-aircraft ammunitions, weighing an additional 30,000 tons, were moved into the new defence belt. In this one week [beginning July 14] the vehicles of a supposedly static Command travelled 2,750,000 miles.

In actual fact, Pile got his heavy guns to their new positions along the coast (by this time, heavy mobile 3.7 guns were being replaced by static guns equipped by remote control and automatic loading devices), which extended from Beachy Head to Dover. And it was done in only three days. The guns covered a belt some 5,000 yards deep and they fired up to 10,000 yards out to sea.

And so began the second phase of the defense of the United Kingdom against V-1 attackers.

Buzz bomb, doodle bug, flying bomb, robot aircraft, fly, pilotless plane—the names given to the weapon seemed almost endless. And in seemingly endless numbers they poured forth from the French coast, only to come head on with the greatly improved coastal defense of General Pile. Sixty bombs were knocked down in the 24-hour period from July 20 to 21; of these, 23 were by guns, 19 by fighters, 1 by the two together, and 17 by balloons. Part of Pile's improvement was due to the better positioning of the guns, and part was accounted for by the timely arrival of a new American automatic gun-laying radar. Pile's enthusiasm over the new device was so great that he later said that, without it, "... it would have been impossible to defeat the flying-bomb." He also greatly appreciated the twenty American batteries and their 90-mm equipment that General Eisenhower had loaned him. Nor was this all. Before the campaign was over, he added more than seven hundred rocket launches to his defenses that fired surface-to-air solid rockets, along with antiaircraft tanks fitted with Oerlikon guns, and even some 9-inch mortars supplied by the Ministry of Petroleum Warfare.

Since flying bomb attacks could, and did, come at any time, day or night, Pile's men began to feel the strain. "Sleep was difficult," he recounted, "until the troops got used to the fearful noise that went on with bombs crashing and guns firing." On one day, the guns were in action for 24 hours without respite and on another for 23 hours and 20 minutes. Despite this, morale was "colossal," and "... the percentage of those reporting sick was far below normal. The Sussex Constabulary reported that they had never had so many troops in their area and less trouble with them." Above all, their kill rate soared from 17 percent the first week (of all targets entering the compass of the defense) to 24 the second, then 27, 40, and 55 percent the fifth week.

As modified V-1 sites were overrun by Allied armies, the Germans commenced launching farther and farther to the north with the result that the flying bomb entry corridor shifted in the same direction. In fact, as early as July 8, ADGB headquarters had noted that some V-1s were coming in from the east rather than the southeast, reflecting the shift northward of German firing positions. Pile responded to this shift by moving ten heavy and five light batteries toward North Foreland on August 19. Even this new move did not disturb the success of the gunners, who by the last week of the month were knocking out 74 percent of all targets within their range. Most of the V-1s that eluded the AA gunners were brought down by fighters or by balloons; only a handful got through to Greater London. Ground launchings of what Hill had once called "peevish darts" came to an end on September 1, when the last V-1 site was captured.

So, as we saw in Chapter 10, the Germans resorted to air-launching.

The air launching of V-1s surprised the Allies, but they soon learned that He111s and Do217s operating out of Venlo and other Dutch fields were the culprits. Pile responded to this attempt to turn his left flank by establishing a "box" of 208 heavy, 178 44-mm, and 404 20-mm guns east and northeast of London between Chelmsford, Clacton-on-Sea, Whitstable, and Rochester. As the middle of the month of September approached, a lull in this phase of the attack was noted; it was later learned that the Germans had shifted their operation from Holland to bases at Varrelbusch, Aalhorn, Willmundhafen, Münster-Handorf, Hesepe, and Zwischenahn in the westernmost part of Germany. On September 16, however, fire was resumed in some force.

These attacks were not particularly successful, partly because of problems faced by the Germans in maneuvering their planes with the unnatural V-1 payload aboard. Just as they destroyed ground-launched V-1s, so RAF fighters, AA Command gunners, Balloon Command balloons, and occasionally the Royal Navy downed a large bag of air-launched versions. Some missiles did, of course, get through; and, as the Germans started air launching ever further northward out over the North Sea, the problem of defense promised to become more difficult. The evolving situation forced Pile to create, in addition to his main Diver gun belt and the box, a new strip of defenses from Clacton-on-Sea all the way northeast to Great Yarmouth. He did this mainly by removing guns from the main belt. By mid-October, the box and the strip were equipped with nearly five hundred heavy and more than six hundred light antiaircraft artillery pieces.

Gunners on the new strip faced several problems their comrades on the belt did not have to contend with. One had to do with the location of Bomber Command bases and the need for their planes to fly back and forth over the strip. This, of course, restricted fire at certain times of the night to the frustration of the gun crews. Added to this was the fact that air-launched V-1s came in at 1,000 or more feet altitude instead of the customary 2,000–3,000 feet. Low-angle firing with heavy guns was not something the crews relished. Boredom was also a problem, as the enemy was no longer launching with the same frequency in the autumn and winter as he had during July and August.

Hill's fighters also faced new problems, partly because air-launched V-1s almost invariably came across at night and partly because of the low altitudes at which they were being launched. Wave-top flying in itself is dangerous, without having to pursue a tiny target—and to estimate its range in the dark. Fortunately, a range-finder was quickly developed that proved to be of great help in permitting Mosquito and Tempest pilots to bring down their prey.

Efforts to shoot down launching aircraft, either before or after they had dispatched their missiles, were marginally successful. For one thing, tracking radar did not operate at full efficiency at the low altitudes involved, nor could ground control radar stations always locate the wave-grazing enemy aircraft. Still, between mid-September and mid-January 1945, at least sixteen enemy bombers were brought down. Occasionally, Bomber Command and the Eighth Air Force would strike the airfields from which they operated.

Pile's coast defenses were outflanked thoroughly on December 24, when between forty and fifty He111s launched their payloads from over the North Sea directly toward Manchester. The missiles flew along a corridor bounded on the south by Skegness (at the north end of The Wash) and on the north by Bridlington. Although thirty reached the coast, only seven got within a radius of 10 miles of the city center (of which one exploded in the downtown area). Hill observed:

> This was one of the few occasions on which the Germans showed resource in exploiting the capacity of the air-launched flying bomb to outflank the defences. Happily for us they were seldom so enterprising; for however carefully our plans were laid, we could not deploy the defences on every part of the East Coast at once, and if more such attacks from novel directions had been tried, they would inevitably have achieved at least a fleeting success, as on this occasion.

Hill's appreciation was shared by an American observer, Lieutenant Colonel S. McClintic. In a memo to Brigadier General George C. McDonald, director of intelligence, on December 26, 1944, he recounted:

> The interesting part of this attack, which has been anticipated by a few of those studying this subject, is that the Germans are well aware of the fact that the A/A Command are under pressure to release the guns that are being maintained for the defense of London. If these guns are released to the Continent then the effectiveness of the robot bombs on ANTWERP and BRUSSELS will be considerably decreased. The Germans know this and if they can frighten the inhabitants of various parts of the Island, A/A Command cannot release their guns as other areas will demand protection, as the

> Germans must be aware of the impotency of their London attacks.
> The Air Defense of Great Britain is maintaining hundreds of fighter planes to intercept the flying bomb or to intrude into the enemy's coastline so as to arrest the He's [He111s] before the bombs are released. With this new tactic the entire Island is now open to attack, aircraft cannot be released to the Continent, nor can spotters be spared their arduous task. If the past single target attack of London is given up by the Germans then they have complicated the work of ADGB many times and thrown confusion into rather placid minds on where the next attack would come. A truly brilliant move at this time, when the enemy is pushing his offensive "on all fronts."

No further such attacks occurred. On January 14, 1945, the last air-launched V-1 landed at Hornsey at 2:13 AM.

The flying bomb campaign against England was not yet over, however. Between March 3 and March 29, a new series of attacks was unleashed, this time from ramps set up at Ypenburg (close to The Hague), Vlaardingen (west of Rotterdam), and near the Delftsche Canal. To combat these longer-ranged versions, guns were moved once again and fighters redeployed. So successful were the defenses, that of 125 robot aircraft to approach the coast (of 275 launched), 86 were downed by the gunners, 4 by fighters, and 1 by the Royal Navy. Of the rest, only 13 were able to reach Greater London.

The last two missiles to reach the British capital impacted on March 28. Then, on the twenty-ninth the last V-1 of the war that successfully pierced British defenses fell on the village of Datchworth near Hatfield. Damage was insignificant. Later the same day, two other V-1s were bagged by the defenders as the flying bomb campaign came to an end.

While most of the glory for the defeat of the more than nine-month-long flying bomb campaign legitimately goes to the fighter pilots and antiaircraft gunners, two other elements deserve full recognition: the searchlight operators and the barrage balloon troops. In a typical night attack, a fighter would dive down on a bomb, pull off level, prepare his aim, and fire. If he had only the missile's exhaust flame as a guide, his chances of success were at best moderate; but if the bomb was expertly tracked by searchlight, the pilot's chances for a kill were greatly improved. According to Pile, the searchlight contribution amounted to 142 V-1s. While a relatively small number, it nevertheless represented nearly a third of the targets coming into the searchlight area at night.

As for balloons, on the eve of the V-1 offensive, Balloon Command chief Air Vice Marshal W. C. C. Gill possessed some five hundred cable-dangling balloons strung out in a belt across Sussex, Surrey, and Kent. Immediately after the attack opened (in the early morning hours of

June, 1944) he got on with his special deployment, "moving," as a chronicler would later put it, "from the wings to the centre of the stage."

The first auxiliary officer to become an air vice marshal, Gill had an enormous task cut out for himself and for the men under him. First, clearings had to be made and new balloon sites established across the three counties. Mile upon mile of telephone wire had to be laid, new access roads built, and transport arranged so that hydrogen and large quantities of food and other supplies required by the crews could be brought in.

In the past, the barrage balloon had acted more as a deterrent to piloted aircraft incursions than as an active means of warding off attacks. Now, with pilotless missiles involved, the mere deterrent presence of barrage balloons was clearly not enough: New mooring and handling procedures would have to be developed, along with new concepts of mobility. Depending on where the V-1 attacks appeared to be coming from, balloons could be concentrated—and this required mobility, in terms not only of the ability to move balloons themselves from place to place but to change the routes along which trucks brought in hydrogen and other supplies from the Midlands. During the peak week of operations, 18 million cubic feet of this gas were consumed, and by the time the V-1 campaign was over, 40,000 miles of steel cable had been used.

To help ease the manpower shortage, widespread employment was made of Women's Auxiliary Air Force personnel to sew and cook for the troops, assist with the routing of supplies to individual sites, and help accommodate and feed the thousands of transport drivers who continually brought in the supplies. WAAFs also operated the telephone network and plotted the courses of V-1s from radar and visual sighting information fed into the control centers at Biggin Hill and Gravesend.

Although only a relatively few V-1s came through during the initial mid-June deployment, Gill was quite aware that the new weapon was doubtless undergoing a teething period and that heavier attacks could be expected. His Balloon Command set into operation 480 balloons above the North Downs in only five days instead of the 18 that had originally been considered necessary.

As the flying bombs crossed the Channel in ever growing numbers, more and more balloons were deployed, so that soon a full one thousand lazily waiting above the English landscape ready to snare their prey in their dangerous cables. At the height of the German offensive even that impressive number was doubled, forming a spectacular sight as far as the eye could see. Thousands of tons of equipment was needed to sustain the giant operation, not counting transport. Troops and their supplies were often found grouped together in open fields, the men sheltered only by tents. Throughout the entire campaign they enjoyed an excellent morale, which "... rose even higher when their cables began to harvest a steady crop of flying-bombs."

And what was that harvest?

Over an eighty-day period, from June 13 to September 1, 1944, an average of 102 V-1s was launched each day, the maximum number being 295 in one 24-hour period. Within this period, 2,340 reached London and 3,765 were destroyed. Although Balloon Command's total was only 278, it was a healthy number for a completely passive system, one that could not aim at or pursue the enemy. The record kill for a single crew, located at the apex of the V where flying-bomb runs crossed, was three in an eight-day period. To Flight Lieutenant R. F. Delderfield

> it was interesting to note the erratic behaviour of a flying bomb when a cable was struck. No two impacts were the same. Sometimes the bomb exploded in mid-air, sometimes it yawed and crashed or chugged off in another direction, and on other occasions it simply staggered, lost height and fell to earth a considerable distance away. There were a number of casualties, some fatal, among the operators, but losses were seldom inflicted upon the crew claiming the kill.

Master control over the movement of V-1s was maintained in an Operations Room in London, where a huge map was located on which each oncoming bomb's movement was monitored as it penetrated the various zones of defense. Understandably, there was keen competition as to which arm—fighter, ack-ack, or balloon—would make the kill.

Despite the deaths; despite the destruction and damage to homes, factories, and public buildings; and despite the disruption of sleep, work, and recreation, the Londoner learned to live with the buzz bomb or doodle bug—or whatever he chose to call it—just as he had come to terms with the Blitz. He proved that a tough, disciplined, and determined people can maintain their morale in the face of astonishing aerial punishment.*

As the battle against the V-1 raged, intensive efforts were made to determine the feasibility of providing at least some warning of the impending long-range rocket offensive. Expectations that the V-2 could be intercepted and destroyed along its trajectory à la V-1 were few (though means of attempting to do so were carefully investigated), but it was hoped that potentially affected areas could be advised of the rocket

* The problems facing the peoples of Antwerp, Brussels, and other continental cities that underwent missile bombardment were similar, as were defensive measures taken. In the typical period of October 21–November 30, 1944, 851 V-1 missiles were recorded, of which 57 approached Brussels (40 being destroyed by antiaircraft artillery) and 274 approached Antwerp (199 were downed by the gunners).

V-1s Fired from French Coast Destroyed
Between June 13 and September 1, 1944

No. of successfully launched missiles that were destroyed by Allied defenses	Means by which they were destroyed
1,912	fighter aircraft
1,575	antiaircraft artillery
278	passive balloon barrage
Total 3,765	

V-1 Activities During Typical Night (July 8–9, 1944)

Radar estimate of number launched	81
Number crossing English Channel to coast of UK	58
Number entering the London Civil Defence Area	24
Number of impacts	
London area	22
Elsewhere	32
Mean point of impact of strikes	Norbury
Number of V-1s destroyed:	
Day fighters over sea	4
Day fighters over land	29
Night fighters over sea	2
Night fighters over land	3
Antiaircraft artillery	11
Barrage balloons	1
Casualties	
Killed	22
Seriously injured	167
Slightly injured	157
Unclassified	12
Launching area	
Pas de Calais	40%
Dieppe	60%
Estimated number of sites operational	35 to 40
Average rate of fire per hour	
Day	—
Night	10

danger and that launching sites on the Continent could be located so that countermeasures could be taken.

Just as he stood on center stage during the V-1 offensive, so it fell to Air Marshal Hill to coordinate the defense forces being arrayed against the anticipated V-2 assault against southeastern England. His first task was to assess the state of the danger and the means at his disposal to reduce its intensity if not to eliminate it altogether. During the winter and the spring of 1944, most information available to him suggested that a much larger rocket was to be put into action than the V-2 actually turned out to be; but, large or small, Hill and his planning staff reasoned that, at the very least, it should be detectable before it exploded on English soil.

During the summer of 1944, the rocket threat—along with the flying bomb threat—was intimately associated in British minds with the launch sites observed in northern France. Accordingly, five suitably modified radar stations were put into operation between Dover and Ventnor. Their crews were trained to look toward the opposite coast for traces assumed to be characteristic of rocket-powered missiles. Additional stations were added later to the search radar array. At the same time, Royal Artillery teams underwent training in Kent to develop and perfect sound-ranging and flash-spotting techniques it was hoped would enable them to determine the location of V-2 rocket launch positions.

By August 1944, it had become quite plain that the long-range rocket was much smaller than anticipated, and that its warhead weighed but a ton, or perhaps a bit more. It was now known to be a mobile weapon readily transportable by rail and motor vehicle. Hill recalls:

> What we did not know was how (if at all) the rocket was externally controlled once it had left the ground. Misleading evidence on this point led to wasted efforts to forestall, detect, and hamper nonexistent radio transmissions which were expected to be used for this purpose. Not until some time after rocket attacks had begun was the conclusion reached that control of the rocket under operational conditions was entirely internal and automatic, apart from the use of a "beam" to control the line of shoot in certain instances.

(During the V-2 campaign, range control by radio was occasionally employed but without notable success.)

By the end of the month, the rocket threat seemed so near at hand that on August 30 the Air Staff ordered armed fighter reconnaissance to be flown over suspected launching areas. But five days later, reconnaissance flights were abruptly suspended: Higher authority now considered the rocket threat to London to be over. After all, it was reasoned, Allied armies had completed the occupation of the Pas de Calais and adjacent territories where the German missiles were presumed to have been made ready for launch.

Hill and his intelligence staff demurred, pointing out that a 200-mile-range rocket could easily reach England from sites in western Holland. "While recognizing that the Chiefs of Staff were better able than ourselves to foresee the effect of future operations," reflected Hill, "my intelligence officers felt, therefore, that as things stood at the moment we ought to be ready to meet rocket attacks from western Holland within the next ten days." As it turned out, the Vice Chiefs of Staff were quick to see the logic of Hill's concern and ordered preparations to be continued. However, they looked for the Germans to bombard cities in England closer to Holland than the capital. London was out of danger, or so it seemed.

Reflecting on events, Hill recognized that, although his precautionary views were soon to be justified, "there was much that might have been urged on the other side." German rail and road transport in Holland was in a state of disarray, partly because of Allied air raids and partly because of the German retreat into the country from France and Belgium. Moreover, the Germans could not know that the Allies were not at any time going to mount a strong northward push. Would a rocket campaign under such circumstance be warranted? One person at least, Obergruppenführer Hans Kammler, thought it was.

On Saturday morning, September 9, Allied intelligence and air defense authorities all over England knew that two rockets had exploded in the London area early the evening before. One fell on a Chiswick road and the other in an open space in Epping Forest. The sound-ranging microphone lattice in East Kent reported the times: 18:40:52 and 18:41:08, respectively.

The incidents were quickly examined by intelligence and other experts. The Chiswick crater turned out to be from 8 to 9 feet deep and between 31 and 35 feet in its irregular diameter. Gas and water mains under the 12-inch-thick road were ruptured by the explosion, but no fires resulted. Several houses in the immediate vicinity were demolished and others—out to a radius of 300 to 400 yards—were superficially damaged. Three people were killed, nine seriously injured, and another nine moderately hurt. In general, the explosive effect of the Chiswick V-2 resembled that of the V-1 flying bomb. A civilian, interviewed after the incident by an investigating officer,

> said he was in a factory about 3½ miles from the point of impact last evening when the explosion occurred. He stated that approximately 1½ minutes before the sound of the explosion reached him, he and other people in the factory heard a whistling sound similar to a siren but of higher pitch.

Parts of the rocket were immediately gathered up and taken to Farnborough, where cursory comparison with pieces recovered a couple of months earlier from Sweden "led to the reasonable identification of the new parts as portions of a V-2 bomb." Shortly later, Royal Aircraft Establishment engineers could without qualification report that "...the operational rocket was unmistakably the same in its basic design, dimensions, etc, as that which landed in Sweden. Certain apparent differences of construction were consistent with those that would be expected between prototype and production models."

The crater at the Epping Forest site was found to be quite similar to that at Chiswick. Fragments of the rocket were gathered, including a control shaft similar to the drive shaft of one of the propellant pumps recovered from the Swedish incident. Witnesses also reported the whistling sound, and one talked of a "whitish object" coming down at an angle of about 45 degrees to the horizon. Another spoke of "horns" at its upper end that wavered while the shape traveled through the air.

Between September 8 and 13, ten rocket incidents were reported (see the table below). All of these flights were confirmed by secret agent and/or air observations of the launching itself or of the ascent trajectory. Vapor trails were often seen; in fact, the trail-making characteristics of the V-2 proved to be about the same as for Spitfire airplanes. When the humidity was 100 percent, the rocket's vapor trail would form between 30,000 and 50,000 feet, while at 25 percent it would appear from 35,000 to 40,000 feet. One Dutch observer reported the missile to be "long and narrow, about the size of a Spitfire without wings."

First Ten V-2s to Land in England (September 8–13, 1944)

Rocket Number	September	Time	Location
1	8	1840	Chiswick
2	8	1841	Epping Forest (Epping Highlands)
3	10	2129	Near North Fambridge
4	11	0907	Crocken Hill, near Orpington
5	11	0933	Magdalen Laver, Essex
6	12	0614	Kew
7	12	0818	Dagenham
8	12	0851	Biggin Hill
9	12	1755	Paglesham
10	13	1108	Dengie Flats (in sea just offshore)

By the time the first fifteen rockets had reached England, intelligence had determined that nine of them had fallen within a circle of 18½ miles radius centered in Charing Cross. The average range error was estimated to be about 14 miles and line error nearly 7½ miles. From the bits and pieces gathered up, an intelligence report noted:

> There is a very important change in the method of construction of the jet and venturi [combustion chamber] unit. This unit is now entirely

> of steel, the jet face of the unit, which was previously fabricated in light alloy, being built up in steel by a welded construction, thus avoiding by this method the use of complicated light alloy castings, a large amount of light alloy welding and a considerable amount of machining.

The Ministry of Home Security's Research and Experiments Department examined the results of the first week's attack and concluded that the mean aiming error was larger than that of the V-1 during its first week of operation. Moreover, the damage caused was comparable, though the explosive effect of the V-2 seemed perhaps 10 percent greater. The average V-2 crater size was 34½ feet in diameter by 9½ feet deep, the maximum discovered being 50 feet wide and 11 feet deep. (This compared to V-1's average of 17 by 4 feet.)

Home Security was quite aware that some missiles burst in the air; of the first ten reaching England, those coming down near Fambridge and at Paglesham so behaved. Witnesses said that missiles bursting in the air produced a sharp crack followed by a louder and deeper sound. One reported that visually an airburst explosion appeared as a "reddish flash" followed by "a large plume of black smoke."

The first week's casualties produced by the V-2 were 187—22 killed, 68 seriously injured, and 97 slightly injured. The averages per V-2, as compared with the V-1, were quite similar at this early stage:

Average per:	Killed	Seriously injured	Slightly injured	Total
V-2	2.7	8.5	12.1	23.3
V-1	2.7	9.1	9.1	20.9

When the Eighth Big Ben Sub-Committee met at Montagu House Annexe on September 12, a number of new representatives were present, including Colonel B. H. Brock and Dr. C. J. B. Clews of Special Defence Headquarters, Royal Artillery. All clearly realized that the Royal Artillery's sounding and flash detection system was going to be extremely helpful in providing the information needed to pinpoint the V-2 firing sites—or at least, to designate a fairly small area within which they must be located.

Before that subject could be brought up, Professor C. D. Ellis, chairman, suggested that the normal Sub-Committee procedures be modified, and only "those matters arising from the incidents of the previous few days" be considered. He added, however, that "the actual arrival of rockets in no way changed [the Sub-Committee's main task of building up a complete picture of the nature and capabilities of the V-2] but merely opened up a new source of information."

Details on eight incidents, reported up to the time the Sub-Committee gathered, were reviewed. Dr. R. E. Stradling, representing the Research and Experiments Department of the Ministry of Home Security, gave a detailed account of events at Chiswick and Epping. Showing that the weight of the charge was about 1,600 pounds and that it tended to fragment more than the V-1's warhead, he added that "Fusing must have been very rapid."

When it came his turn to talk, Colonel Brock reported that the upper shell wave of the Chiswick rocket was strongly recorded on the Kent sound-ranging lattice up to about 38 miles on the downward part of the trajectory. He added that a vapor trail along the initial upward part of the trajectory had been seen by observers located at Dover and that a Mosquito pilot, on the night of September 10, had watched "a bright light which rose slowly in the sky and accelerated sharply before disappearing." Sub-Committee members would be glad to know, he concluded, that the Colchester sound-ranging lattice soon would be in operation.

Radar also turned out to be effective, with two new stations in addition to the original five already in operation. Mrs. A. Munro of the Interdepartmental Radio Committee said that radar had determined the launch points of the V-2s (that had been fired up to the time of the meeting) to within a mile. "A small rectangular area north of The Hague," she reported, "was the launch area." For the time being, radar could serve to supply only a very short warning time, a couple of minutes at best.

Sir Robert Watson-Watt followed with a summary of his study "Radio and Radar Aids to the Defeat of the Rocket," which had just been prepared for the main Crossbow Committee. Noting that high speed is the principal characteristic of the rocket as a target for radar, he explained that "the most powerful discriminating capacity of the radar stations is in rate of change of slant range." Since V-2 launchings were occurring more than 150 miles from London and thus more than 100 miles from stations in England, he cautioned that at the long ranges involved range radar response unfortunately began to deteriorate. He continued:

> ... many of the rockets will be above the radar coverage before they close to such a range as would, at lower heights, permit observation; and identification of those launched near enough to be observed is dependent on very speedy and skilled interpretation of an unsatisfactory radar characteristic.

The range of the early rockets fired against London was some 190 miles. Flight time was a few seconds over five minutes.

Before the eighth meeting was over, the always intriguing phenomenon of airburst was brought up and duly entered into the minutes:

> The Sub-Committee note that at least two of the impacts had been characterized by reports of double explosions and considerable discussion took place upon their possible significance. It was thought unlikely, in view of the strong evidence of a highly efficient impact fuse, that air bursts were intentional. The more probable explanation was that they were due to some weakness of design, possibly to be found in the fuel tanks, that had caused the Swedish rocket to burst in the air and might also account for similar reports from the experimental grounds in Poland.

By the time of the Ninth Big Ben Sub-Committee meeting later in September, Sir Robert Watson-Watt's Interdepartmental Radio Committee was able to show beyond doubt that the V-2 was "basically automatic in that its trajectory is determined internally, rather than by radio signals." He added that "further speculation [would be] unprofitable until more evidence is available."

On this note, the members adjourned by provisionally agreeing that no date should be fixed for the next meeting "until the outcome of current operations was known." The Big Ben Sub-Committee never again came together in formal session.

From speculation about the rocket and the damage it might cause, the emphasis now shifted to appropriate means of warning, to defense, and finally to countermeasures.

Of the three, the most practical seemed to be the execution of air attacks on targets that reconnaissance, radar, sound- and flash-ranging, or Dutch underground observers might locate, specifically:

(1) the technical batteries (fueling, transport, supply, repairs, test), both in billets and in movement;

(2) the battery supply point, especially at the moment when elements of the firing and technical batteries rendezvoused prior to movement to the actual launch sites;

(3) the possible radio location unit, the position of which—in relation to launching points—it was thought possible to determine;

(4) the launch sites themselves, during periods of missile erection and fueling.

Logical though this approach appeared to be, aerial countermeasures within the launch areas proved to be largely ineffective. On at least one occasion, a great explosion did occur, leading the attacking pilot to believe a missile had been hit. When the smoke cleared away, a large crater was seen. In the long run, though, the day in and day out harassment of rail and motor communications was much more damaging to German field commanders than a lucky odd strike.

There were, of course, occasions when the Dutch resistance could transmit to the Air Ministry such timely information of firings that very rapid reaction on the part of fighter groups was possible. Moreover, Allied planes flying back and forth over Holland would often observe vapor trails and, once in a while, rockets themselves as they started off toward England. One intelligence brief noted:

> A pilot of the 4th US Fighter Group observed a Big Ben act up ready for firing near LOCHEM on the 1st of January [1945]. The pilot dived to take a motion picture of the site and he observed that the rocket was immediately tilted from 85 deg. to 30 deg. We will report on results when obtained.
>
> On a photographic sortie over HAAGSCHE in Holland 12 rockets were found regularly dispersed in horizontal positions along the sides of the road.

The sound-ranging system, of help in locating targets suitable for air attack, involved six lattices of microphones placed on the ground between the launch area and the target. The idea was to detect and record sound during powered flight; but as one officer complained, the system had "... little merit other than the fact that it did provide some data on the rocket's trajectory and that from this data the géneral launching area could be determined." Among the disadvantages he cited were: (1) good communications and an accurate timing system are required—these make the system cumbersome and difficult to operate with success; (2) accurate meteorological data to 50–60 mile heights are essential; (3) jamming of the system by extraneous noises is easily accomplished; (4) data can be obtained from only one V-2 at a time—two rockets passing over the same area at about the same time cause confusion of data; (5) data are inaccurate; and (6) neither a countermeasure during flight nor prediction of point of impact can be accomplished with the system since the speed of the missile is supersonic.

Lieutenant Colonel Norman L. Falcon recalls one instance of "the unwisdom of using strategic intelligence for tactical purposes." During the V-2 onslaught on southern England, he wrote the authors, it was discovered at Medmenham

> that some of our CIU [Central Interpretation Unit] strategic intelligence reports were being used by Fighter Command for tactical bombing. Appreciating the risks involved in using out-of-date, but nevertheless useful in the right hands, strategic reports, we decided to make an interservice party to call on the Senior Intelligence Officer at HQ RAF Fighter Command. A Wing Commander RAF, a Commander RN and a Lt Col (myself) made the visit. The SIO [Senior Intelligence Officer] was somewhat incensed. "Have you come here," he said, "to tell me I do not know my job!" We politely replied that we did disagree with the occasional selection of tactical bombing

> targets at his HQ from our reports, because the information was not up-to-date, and this must in part be his responsibility. We made very little impression I remember. But a week or so later an out-of-date report of V-2 weapons having been seen in a wooded part of The Hague caused a bombing raid that set a large area of The Hague on fire although the target had moved at least a week before the raid.

The sound system was employed for only about a month in the United Kingdom; but during the latter part of the V-2 campaign, when the Germans were aiming principally at Antwerp in Belgium, it was used extensively. This was because in the case of operations against the UK, only a few launch positions along the Dutch coast were feasible and missiles rising from them could more effectively be tracked by radar stations, which were always "looking" in the right direction. When the Germans engaged Antwerp, however, they could launch almost anywhere along an arc of about 125 degrees. This made it difficult for radar, but the sounds of a V-2 taking off could often be picked up right away.

The capability of radar to warn of the approach of an oncoming V-2 was proven during the war, though at first it was thought that the very speed of the missile, its shape, its high-altitude trajectory, and the possible interference of ionospheric echoes would all combine to make radar of marginal value at best. Even under optimum conditions, however, radar could give but a very short warning. Attempts to provide a citywide air raid alert every time a missile crossed the Channel would have been futile, especially since World War II radar was unable to predict the precise point of impact.

One customer did exist for the radar stations: London Transport. Early in the V-2 offensive, authorities feared that a chance hit on the underground tubes passing beneath the Thames would be catastrophic. Consequently they requested, and regularly received, radar warnings so that in the event of an impact being predicted in the general area of the river, the flood gates to the tubes could be shut.

The radar warning system grew steadily during the course of the V-2 campaign, and eventually consisted of:

(1) five RAF Chain Home (CH) stations capable of measuring range up to 230 miles. They were located on the east coast.

(2) standard GL II sets deployed at gun sites in the London area;

(3) modified longer-range GL IIs, deployed north of the V-2 flight corridor at Aldeburgh and Southwold, and south of the corridor at Walmer;

(4) a single SCR-584 positioned at Steenbergen in Holland, some 60,000 yards south of The Hague;

(5) Benjamin, an experimental unit operated for a short time in the UK.

CH stations were capable of detecting V-2s shortly after their takeoff and of following them along the initial leg of their trajectories to an altitude of between 10,000 and 30,000 feet. By combining the point of launch with a number of points along the flight path, the complete trajectory would be constructed. Range accuracy was not, however, at all impressive—under ideal conditions, a mile had to be tolerated, but often up to 5 miles. Partly because of such inaccuracies, and partly because CH stations were subject to jamming and "clutter" from winged aircraft, they were supplemented with the GL II radar, which could "look" at higher elevations and thus was less likely to pick up clutter. GL IIs were placed so as to observe the V-2 at approximately right angles to its Hague–London trajectory. Operators soon learned to distinguish ionospheric scatter echoes from the missile itself.

For a typical missile, the CH would make contact within 30 to 60 seconds of takeoff and thus 4 to 4½ minutes before impact in England. CH would immediately advise GL II, which would refine the trajectory and, if necessary, warn London Transport. From January 8 to March 19, 1945, for example, a total of 516 warnings were given, of which 44.4 percent were false (in that no V-2 fell inside Greater London); 3.5 percent were spurious, i.e., no known incident happened anywhere; and 52.1 percent were followed by an impact within the city's civil defense perimeter. Only eight rockets came down inside this zone without being preceded by a warning.

Toward the end of the V-2 attacks, some thought was given to setting up an advanced detection system that would be practical to provide warnings to the potentially affected population. One radar installation was contemplated for Great Yarmouth and another for the Continent between Ostende and Dunkerque. The fortunes of war were such, however, that the project never left the planning table.

Antwerp's warning system—consisting of five modified GL II radar stations, a number of Ames-type mobile units, and six visual observation posts (to report vapor trails and occasionally exhaust flames)—was used only to alert the sound-recording system. About 75 percent of the missiles that impacted in or near the city were detected; 25 percent escaped notice.

Those charged with the aerial defense of Britain next turned to the question of radio countermeasures. Although it was gradually becoming apparent that most V-2s were guided by a self-contained (inertial) system, the RAE at Farnborough was still confused by the *Wasserfall* radio components found in the Swedish rocket. When the parts were gathered up and studied, they were found to include:

(1) a receiver, Type E 230, used in Hs293 and Fritz X controlled missiles to accept control signals for azimuth correction;

(2) a "Verdoppler" or retransmit unit, a receiver-transmitter permitting the rocket's velocity to be determined;

(3) a control receiver with associated audio and relay units for accepting signals to cut off the rocket's fuel supply;

(4) a transmitter and modulator, thought by the British to be used to transmit experimental data.

This equipment led the investigating engineers to conclude at the time that although radio countermeasures offered little chance of affecting the rocket's course, plans should nevertheless be made to establish not only listening stations but jamming systems as well. Particular emphasis was placed on intercepting possible V-2-directed signals in the air and then reradiating on a VHF link. Ten aircraft equipped with rapidly tunable, high-power transmitters for aerial jamming were available by the time the first V-2s came across the Channel on September 8, though it was realized that without knowledge of operating frequencies, coding sequences, and procedures, not much could be done. At the same time, ground jammers had been installed in the Dover area, and Royal Navy equipment (originally developed to jam the air-to-surface Hs293 weapon) was turned over to the RAF. All to no avail, however, for as one official lamented, the "... use of these sets proved to be futile since the E 230 receiver used in the Swedish V-2 did not appear in later missiles." Peenemünde's engineers had been correct when they forecast that the errant Swedish rocket would bedevil the British for many months to come.

During the latter part of September and on into October, the RAE continued to study remains of V-2s falling on the UK and on the Continent. The only radio controls that turned up at all were occasional Verdoppler receiver-transmitters used for velocity measurement and control receivers used for fuel cutoff. Most rockets, however, did not incorporate the Verdoppler at all, but rather a gyro integrating accelerometer that made ground-emanated control signals unnecessary. An RAE engineer had to confess that "at present, it is not yet clear whether all rockets are now being fitted with the accelerometer only or whether this instrument is being used experimentally either alone or in conjunction with radio." Although prisoner-of-war reports indicated that some rockets fired on Antwerp were using a radio beam for azimuth control, after intensive studies the best the British felt they might possibly accomplish was to prevent the Germans from exercising final corrections. All thought of jamming or taking over control of an in-flight V-2 vanished.

As the V-2 campaign continued throughout the autumn of 1944 and on into the winter, pressure built up for an active defense against the missile. The aerial harassment of launch areas and the logistic chain was not producing the hoped-for results and the application of radio countermeasures had proved futile. The interception of supersonic missiles by fighter aircraft being completely out of the question, what, then, could be done?

Since December 1944, General Pile and his Anti-Aircraft Command headquarters staff at Glenthorn in Stanmore had been debating the feasibility of massing artillery fire at or near the London end of the V-2's entry corridor. Such a barrage, some argued, might result in the destruction of missiles in midair, relieving a weary city of additional tonnage of high explosives.

AA Command planners showed that for barrage fire to have any chance of downing a V-2 it would be necessary to so deploy radar tracking equipment as to lead to the prediction of a rocket's passing through an imaginary square one kilometer on a side. This goal, it was estimated, could be reached for only one out of every ten V-2s coming across the Channel. If four hundred massed antiaircraft guns should fire into the square for 15 seconds, it was at first calculated that the chance of destroying a V-2 would be 1 in 40; however, AA Command later felt it wise to assume an average azimuth error of 5 degrees, in which case the chances of knocking out a missile would drop to 1 in 160. This could be improved to 1 in 80 if the four hundred guns were to fire for 30 seconds (ten rounds per gun for a total of four thousand rounds). "Consequently," an AA Command report pointed out, "the number of 28-lb. shells required per V-2 destroyed would be 400×80=320,000".* The infeasibility of this plan being rather evident, other methods were considered. One report went so far to recommend that "in addition to the possible engagement of the V-2 missile by improved AA fire control methods, all other possible countermeasures should be explored, particularly the use of guided counter-missiles."

These facts were placed before a panel whose members included R. V. Jones, Professor E. C. Bullard, representing Fighter Command; Colonel O. M. Solandt, representing the scientific adviser to the Army Council; and Dr. D. Taylor, representing Sir Robert Watson-Watt's

* Many studies were made by the British on the probable effect of bursting antiaircraft artillery shell fragments on the V-2, with particular attention being focused on the temperatures existing in the warhead and elsewhere in the rocket with the aim of causing premature detonation or other malfunction. The Armament Research Department issued a report on "The Attack of High Explosive by Steel Fragments," in which it was concluded that the detonation of the V-2 amatol explosive would require impacting ½- and ¼-ounce fragments to have 4,200- and 4,700-feet-per-second velocities, respectively. Knowing that the majority of AA shell fragments in that weight range have initial velocities of 3,000 feet per second they concluded that "... the warhead is almost immune from shell splinters which penetrate the casing."

Interdepartmental Radio committee. Called into session at Glenthorn, its chairman, Major General S. Lamplugh, felt constrained to caution that "there is, of course, no certainty that the hits will have any effect at all." Nevertheless, the firing of the guns might have a useful side effect:

> It may be remarked that the firing of the guns will give about three quarters of a minute warning of the arrival of about half the rockets. This warning will extend over Eastern and Central London and will on occasions be audible further West. It is likely that this warning will reduce casualties.

This benefit could be mitigated by the panel's assessment that thirty "blind" shells would hit the ground each week.

It was not until March 21, the first day of spring 1945, that AA Command's final plan was ready. Entitled "Engagement of Long Range Rockets with AA Gunfire," it was formally presented to the Crossbow Committee by General Pile five days later. In his opening remarks, he took a circumspect attitude ("I do not consider that it is possible to make a scientific analysis of the chances of detonating the Rocket in the air, as there are so many imponderables"). Nevertheless, he felt that at least the plan should be tried out. "Such an experiment," he explained, "might enable most valuable information to be obtained on the performance and vulnerability of such targets, and will certainly teach us a very great deal about the problem of engaging them in the future." Realizing that the V-2 offensive must be drawing to a close, he requested permission to fire without delay.

The Minister of Home Security was strongly opposed to the plan for, as he pointed out, not only did it offer but slight chance of success but dozens of "blind" AA shells falling in Greater London inevitably would cause some casualties. After deliberation, the chairman of the Crossbow Committee expressed the opinion that, inasmuch as the proposed action was

> in the nature of an important experiment, it would clearly be necessary to obtain the War Cabinet's approval to the action proposed by AA Command, especially as there are objections from the Ministry of Home Security's point of view. Before the matter can be taken before the War Cabinet, it will be necessary for the Chiefs of Staff to express their views.

Two days later, before any decisions could be made, the V-2 offensive was over.

That offensive lasted about 6½ months and was divided by ADGB planners into six phases:

(1) September 8 to 18—the opening attack on London
(2) September 19 to 25—cessation of fire during regroupment

(3) September 25 to October 12—the attack on Norwich

(4) October 3 to November 18—renewal of operations against London

(5) November 19 to December 31—third series of attacks on London

(6) January 1, 1945 to March 27—final bombardment of London.

During the first phase, detection and warning systems were tuned up and procedures devised with the Dutch resistance whereby the Air Ministry received regular and accurate reports on V-2 launching preparations and the location of structures, vehicles, and troops associated with them. Dr. J. M. J. Kooy, director of the Aeronautical School in The Hague and a well-known expert on ballistics, was one of many technically trained intelligence officers whose reports became invaluable to the Allies. He told the authors:

> I used to watch V-2 launchings through an optical glass. I hid about 700 meters away, protected by a dike. Once, when I was out in the open, a missile exploded nearby so I dove under a German car whose occupants were crouched in a ditch. When the Germans crawled out afterwards, they told me, "You are certainly lucky, because this car is full of explosives." They then drove off, never asking why I happened to be in the area.
>
> It was very dangerous for us to spy on the Germans in the launch zone, for if we had been caught we would have been executed instantly.

Kooy and his associate, Professor Dr. J. W. H. Uytenbogaart, maintained a very complete chart of the successes and failures of rocket firings in The Hague area.

Based on information coming in from the Dutch and from the various detection systems, Bomber Command dropped 172 tons of bombs on suspected V-2 concentrations at Eikenhorst and a smaller tonnage at Raaphorst.

From September 19 to 25, no further rockets were fired; through the Dutch resistance, it was learned that V-2 columns had moved out of The Hague on September 18, firing off a single shot that came down at Lambeth as they left. Hill continued his aerial reconnaissance and intruder sorties over suspected launch and logistics areas, but no bombing was undertaken simply because there were no attractive targets. Then, a little after 7:00 PM on September 25 a rocket came down near Diss in Suffolk, followed by one the next afternoon in East Anglia about 8 miles from Norwich—now obviously the new target. An analysis of reports coming in from Dutch resistance, from vapor trail sightings, and from radar showed that the village of Garderen was the "estimated position" from which the new firings were being made. Then in early October, launchings against Norwich began to be made from positions farther north, apparently along the Zuider Zee beyond the range at which Hill's armed reconnaissance planes could efficiently operate. The movements on the part of the Germans resulted in the Second Tactical Air Force—headquartered in Brussels under Air Marshal Sir Arthur Coningham—assuming the recon function.

The Norwich firings ended the morning of October 12. Meanwhile, beginning on October 3, the attack on London had been renewed from sites in and around The Hague. Added to this, by the middle of the month at least one hundred V-2s had come down on the recently captured Belgian port of Antwerp. This expansion of targets brought the whole question of V-2 countermeasures straight to the attention of Supreme Headquarters, with the result that on October 11 the chief of the Air Defence Division became responsible for the coordination of all missile countermeasures on the Continent. A few days later, the Allied Expeditionary Air Force was formally disbanded—it had been created only to support the invasion of Normandy and the consolidation of the Allies on the European continent. ADGB became Fighter Command again (under Hill), and it and the Second Tactical Air Force under Coningham consequently became independent of Air Chief Marshal Leigh-Mallory's all-embracing authority. This meant that the defense of the UK was completely dependent on a man who had no control over the Second Tactical Air Force. In Hill's words:

> A situation in which I was responsible for defending the country against long-range rockets while responsibility for conducting the only countermeasure open to a fighter force was exercised by another Command, not under my control, was no longer merely inconvenient; it was clearly untenable.

Soon afterward, it was agreed that Fighter Command would assume overall responsibility for the armed reconnaissance of V-2 targets in Holland. To the extent that Hill's fighters could not accomplish the task—and often they could not because of range and poor weather—the Second Tactical Air Force would be called upon.

Although agents' reports and aerial reconnaissance did locate suitable targets at Leiden and at The Hague-Bloemendaal (among them an insane asylum at Ockenburg Klinier where German V-2 troops were quartered) both Bomber Command and low-level attacks with No. 2 Group Mosquitos were ruled out because all aircraft were being utilized in the major campaign against the German heartland. This meant that launching activities in Holland would be disturbed only by armed reconnaissance fighters. This relative respite from attack permitted the scale of daily V-2 firings to rise from between two and three during the first three weeks of October to four by the middle of November and then to six.

On November 17, Hill requested of the Air Ministry that he be given authority to arm his Spitfire aircraft so that they could attack targets that did not appear to risk the lives of Dutch civilians. "What we had to do," he later said, "was to balance the off chance of injury to life and property at The Hague against its certainty in London." Dutch authorities were contacted, agreement was immediately reached, and on November 19 the Air Staff authorized Hill to commence fighter-bomber attacks in coordination with the Second Tactical Air Force's campaign of rail interdiction. Bomber Command support, however, was still ruled out unless the scale of rocket attack should rise precipitously.

Thus, on November 19 a new phase in the V-2 defense program began. During the ensuing weeks, fighter-bombers made precision attacks whenever the weather permitted. Results seemed to justify Hill's proposal; by the middle of December the firing rate of V-2s was off from a high of seven missiles a day toward the end of November to four. A fortnight later, the rate was down to a daily average of over three rockets.

But was Hill's fighter-bomber effort really the cause of the drop-off in V-2 firings? According to Hill:

> At the time I was not altogether prepared to accept [the] conclusion [that it was]. In the light of subsequent experience I feel quite sure that to do so would have been to claim too much for our efforts. The chief factor in limiting the scale of attack was almost certainly the rate at which supplies could be brought to the firing areas; and this in turn must have been mainly determined by the frequency and success of the armed reconnaissance and rail interdiction sorties flown by the Second Tactical Air Force over the enemy's lines of communication.

Hill also speculated that enemy preparations for the Ardennes offensive and the stepped-up frequency of V-2 firings on Antwerp had their effect on the decreased intensity of the attack on London.

The failure of the Ardennes offensive was followed almost immediately by stepped-up launchings against London, bearing out at least part of Hill's surmise. The final phase, lasting from January 1 to March 27, 1945, saw rocket accuracy increase somewhat. Daily incidents in England averaged eight missiles during much of January and about ten in mid-February. Fighter-bomber raids were pressed against suspected liquid-oxygen plants and the Haagsche Bosch assembly and firing site in The Hague. Other targets included Wassenaar, the Hook of Holland, and the vital rail junction at Amersfoort. During the final two weeks in February, 548 fighter-bomber sorties were made, resulting in 108 tons of bombs being dropped on V-2 targets. A particularly intensive effort against the Haagsche Bosch forced the Germans to cease fire for a couple of days while they moved to the racecourse at nearby Duindigt.

When fire was resumed from there, however, it was as intense as ever.

In March, Hill's fighter-bombers dropped three times more tonnage on firing and storage sites than in February; and, in fact, made more sorties than in November, December, January, and February combined. Duindigt alone received 70 tons. The situation became so desperate that the Germans abandoned the site during the middle of the month. Firings continued for a while longer until, on March 27, they ceased altogether. The last rocket to reach England, the 1,115th, impacted at Orpington at 4:45 PM.

The question of accuracy was always of great concern, lest any demonstrable rise presage an increase in civilian casualties. Once the campaign was in full swing, intelligence concluded from statistics that the propellant cutoff point was reached when the rocket was moving at 5,100 feet per second at an angle of inclination to the horizontal of between 33 and 39 degrees. About a quarter of the missiles burst in the air, their components dropping to the ground separately; but, since the warhead still exploded in a normal manner, this did not necessarily mean the mission was a failure. As for range, it was found to average nearly 190 miles, with the maximum being 220 miles. The Ministry of Home Security showed that the vulnerable area over which casualties occurred was about 80 percent larger than the vulnerable area of the V-1. "Since there is little difference in the area of damage caused by the two weapons," they reported, "it must be that the lack of warning of the rockets' approach is the main contributory factor."

In September, the mean range was 189 miles and accuracy was fair. From October 1 to 27, range was erratic with many missiles falling short of their target. From October 28 to December 1, the mean range was 183 miles and it was not erratic. Standard deviation was 7.9 miles. From December 2 to January 12, 1945, mean range dropped to 181.7 miles, while the standard deviation increased to 9.7 miles. From then to February 9, mean range went up to 183.8 miles and standard deviation dropped to 8.3 miles. And so on to the end of the campaign. Range depended on launch site and target, while errors were influenced by such factors as the launch environment, intensity of Allied air attacks, and the reliability of the missile itself.

Taking a typical fortnight period, beginning November 18, 59 percent of the V-2 incidents in the UK were in the London area, the MPI* being in

* Mean point of impact. In figuring accuracy, the term *circular probable error* is used; it is the radius of a circle that will include 50 percent (and therefore exclude the other 50 percent) of the impact points. The MPI, taken as the center of this circle, is shown by the intersection of two lines, one drawn so that half of the impact points lie on one side and half on the other. A second, and similar, line drawn at right angles intersects at the mean point of impact, or MPI.

Barking. The average deviation in line was 4.8 miles and in range 6.3 miles. The following table shows the seriousness of some incidents during the period. The next fortnight showed a marked deterioration, for while the average line error decreased slightly to 3.5 miles, the average range error of 11.4 miles almost doubled.

Typical Serious V-2 Incidents During November 1944

Incident No.	Location	Day of month	No. Casualties	Notes
213	Wandsworth (Hazelhurst Road)	19th	139	direct hit on houses
217	Bromley (Southborough Land)	19th	131	direct hit on public housing
226	Erith	21st	105	exploded on open ground near Erith Oil Works
234	Bethnal Green	22nd	104	demolished houses
235	Bexley/Chislehurst	22nd	122	impact on roadway
248	Poplar (Beale Road)	24th	144	impact on road
253	Deptford (New Cross Road)	25th	391	direct hit on Woolworth store; passenger bus also involved

During the month from mid-November to mid-December, the percentage of London incidents dropped from 59 percent to 40 percent and the MPI shifted eastward about 6 miles. For London, the maximum weekly number of incidents was sixty, established in late February–early March 1945, while for Antwerp more than one hundred were recorded each week during the December 14–January 4 period.

Fortunately for the Allies, the V-2 was the only long-range rocket with which they had to contend during World War II. Yet for several years the specter of an even longer-range missile that might be fired from one or more of the heavy installations in northern France concerned intelligence analysts. As late as January 1945, a report was received from Sweden to the effect that a V-4 "super-rocket bomb" was being developed by the Germans for the purpose of "laying New York in ruins."

The source of the report? The "V-2 scientists," who had somehow communicated with Swedish professors at Uppsala and Lund universities. Not only did the Swedes declare the invention to be "astounding," but they were impressed with "... the genuineness of the German engineers' revelation," which, they added, "cannot be off-written as Nazi propaganda." Among the details given: "Weight of Veefour is approximately fifteen tons. It is fired in seventy-five degree angle to staggering height of 225 miles, practically above the Earth's top air layer. Its velocity is roughly nine thousand miles per hour reaching its maximum altitude in just under two minutes."

When Dr. Theodore von Kármán, Hungarian-born director of the US Army Air Forces Scientific Advisory group, read the cabled report from Stockholm on February 16, he commented:

> There are internal inconsistencies in the report. The maximum range of a missile projected with a velocity of 9,000 miles per hour is about 1,150 miles, neglecting the effect of air resistance. Such a missile would rise to a height of 260 miles. It is obvious, therefore, that the figures given do not justify the assumption that the Germans have a transatlantic missile. However, in the past often such reports were quantitatively inexact; nevertheless some real development was hidden behind the fantastic claims.

Even more fantastic than the purported range of the V-4 was the method by which the range was supposed to be produced. According to Stockholm, it was

> a radical departure from anything so far seen in scientific circles. In tube eleven yards long six inches in diameter is produced electrical tension of ten million volts. . . . High frequency discharges in tube imparts to gas fed into tube from tanks built into Veefour heat of explosive degree propelling bomb in unbelievable number thrusts.

Though he did not rule anything out as impossible, von Kármán remained skeptical:

> No clear picture of the electrical apparatus can be obtained from the descriptions. . . . In which way the propulsive effect is obtained is not intelligible, but it is not inconceivable that such methods may produce a greater momentum than is possible through straight combustion and efflux of the combustion gases in conventional rockets. However, none of our experts seem to be able to sketch a workable method . . .

The origin of the report on the supposed V-4 remains obscure to this day. Certainly, the scientists and engineers at Peenemünde were probing beyond the technology represented by the V-2, and Eugen Sänger and Irene Bredt had written a report on a rocket-propelled "antipodal bomber" capable of reaching New York. But resources and time had all but run out on the Third Reich. Intercontinental missiles were still a decade away.

CHAPTER 12

Effectiveness of the V-Weapon Campaign

As far as one can determine from surviving World War II records and postwar interrogations of the principals involved, slightly more than 25,000 V-weapons were successfully launched against targets in England and in Allied occupied areas of France, Belgium, and even Germany.* "Successfully launched" does not mean that individual V-1s or V-2s hit assigned targets, but only that they took off properly and were observed to proceed toward their destination. Many missiles fell short or wide of their target and, in the case of the V-1, were often destroyed by antiaircraft fire, fighter planes, or barrage balloons.

The table on page 251 records 21,770 successfully launched V-1s and 2,445 launch failures, compared with 3,255 successful V-2 launches. V-2 launch failures are not accurately known, but it is certain that they were significantly reduced by the end of the campaign. Both London and Antwerp were engaged by over 10,000 missiles each, and Liège by over 3,000. In England alone, V-weapons caused more than 30,000 casualties, V-1s being the most damaging because of their greater numbers.

British casualties due to German aerial attacks between September

* A curious report of a V-2 being fired on Leningrad was turned up in an English newspaper. *The Daily Express* stated on December 17, 1944, that a V-2 had somewhat earlier impacted on the Russian city; no details, however, were provided. Interrogations of Georg Rickhey (the former general manager of the Mittelwerk plant at Niedersachswerfen and chairman of its board of directors) also brought up a "report that an A-4 rocket was fired on Leningrad in 1943." There appears to be no validity for either of these reports.

1939 and March 1945 peaked severely when flying bombs were introduced, almost reaching the monthly loss rates of the "night Blitz" period. The number of missiles employed and the number of human casualties they caused far from tell the whole story. In assessing the V-weapon campaign, one must consider what the intentions of the Germans were, how (relatively) successful they were, the extent of the reactions of the Allies, and the burden the mere existence of Hitler's vengeance weapons placed on manpower and matériel resources of the victorious nations before the campaign was over. All of this must be balanced against the manpower and matériel costs to the Germans, coupled with speculations as to what they could have done with these resources had they elected to direct them toward jet airplane construction, air-to-air rocket weapons, more and better bombers, and antiaircraft surface-to-air missiles.

Perhaps the first question that should be asked is: "Was it a sound decision to attempt to develop, produce, and field *two* new, unmanned bombardment weapons?" Might it not have been better to choose one or the other, but not both? Possibly so, but when the decision was made to back the Fi103 (V-1) *and* the A4 (V-2) in May 1943, there was no clear indication which would ultimately prove successful. By supporting both, the Germans hedged their bets, knowing full well that there were decided advantages and disadvantages associated with each of the spectacular—and speculative—missile weapons.

Some of the weaknesses of one weapon were balanced by advantages in the other. If both missiles had been made operational simultaneously and if they had been fielded well before D-day in Normandy, they would doubtless have had a far greater impact than they ultimately did. When the decision was made by the Long-Range Bombardment Commission in 1943 to recommend to Hitler that both missiles be placed into development, Dornberger cautioned, ". . . if at last it was really intended to make practical use of these long-range missiles, there had better be no limit to the strength deployed." If the Allies had had to contend with 50,000 or 100,000 V-1s and V-2s instead of 25,000, and if the weapons had been fielded earlier, the final victory in Europe would have been far more costly and protracted.

One of the most obvious reasons for the German failure to exploit the V-weapons fully was their lack of effective planning of a command structure for field operations. Indeed, it was not until November of 1943 that the decision was reached in the first place to unify under a single command both the Fi103 and the A4. As Lieutenant Colonel M. L. Helfers of the US Army Office of the Chief of Military History observes:

> Six months were thus wasted during which selected senior staff officers and commanders of the V-weapon field command could have

> familiarized themselves with the new weapons and created an organization which would have guaranteed their commitment in the most effective manner. During this period inter- and intra-service rivalries and differences of personalities could have been resolved. As it was, both the commanding general and the chief of staff of LXV Corps were thrown into the midst of developments which had been going on for several years as independent projects of the Army and the Air Force. Perhaps even worse was the last minute selection of an Army general [General Metz], totally ignorant of previous V-2 developments and not entirely in sympathy with the program, to command the V-2 field units, while the general who had developed the weapon was assigned a less glorious role in the ZI [zone of interior, e.g., inside of Germany]. Had the Army and the Air Force gotten together from May 1943 on and organized a firm V-weapon development and field command structure, the commitment of both weapons might have come about simultaneously, possibly even months before the V-1 was committed piecemeal in June 1944 to be followed by the V-2 in September. Certainly a united Army–Air Force front would have hindered unqualified SS personnel from seizing control first of the V-2 and then of the V-1.

Nevertheless, despite all the difficulties, inefficiencies, and rivalries, both weapons did get into the field and, albeit after the Normandy invasion had begun, did go into operation against London and other targets.

As a result of what General Günther Rüdel, inspector of Antiaircraft Artillery, called the London "field test," some points of operational usefulness emerged. First, he felt that the V-weapons could be fully effective only when fired "against suitable decisive targets as part of the over-all operation," and "in close coordination with all other weapons." He found them "particularly suitable for use against large-area targets at long range, such as industrial centers, supply centers, seaports, and air terminals ... [as well as] troop assembly areas." It was clear to him that the "greater the concentration on one or few targets, the greater the effect." Supply facilities, he recognized, were particularly vulnerable; to solve the problem wide dispersal was advisable, but that also served to exacerbate the supply problem.

Before attempting to determine what benefits the V-weapons gave to the German war effort beyond killing thousands of people, mostly civilians, it is instructive to take a look at the relative importance of the V-weapon program within the German military establishment and just how much it ended up costing. Postwar analysis leads to the conclusion that the Germans spent more on rockets and guided missiles during the war than on any other type of ordnance. As for their importance, War

Production Minister Albert Speer, in an interview on May 21, 1945, noted that the V-2 especially was

> protected with particular love by the Army Armament Office, and it was in contrast to all other items that the Army got a special quota in order that nothing could happen to it. One could almost have gotten the impression that we did not have any other special interest.

It is estimated that the German missile program cost in the neighborhood of $3 billion wartime United States dollars. If one should work out the cost effectiveness of the program only in terms of missiles launched—disregarding enemy casualties, anxiety among the populace, loss of manpower efficiency, property damage and, most important perhaps, the Allied efforts directed at protection and countermeasures—one would come up with the figure of some $120,000 per successful missile (averaging V-1s and V-2s), each of which carried about a ton of explosive. Even assuming that all 25,000 V-weapons impacted at or near their targets, only some 25,000 tons of explosives would have been delivered. By way of comparison, the Allies dumped more than 36,000 tons of bombs on V-weapon launching sites in the December 1943–D-day (June 1944) period alone, losing 154 aircraft and 771 men in the process. The German $3-billion V-weapon effort was more costly than the US $2-billion Manhattan program that produced the atomic bomb.

Opinion differs among Germans as to the worthwhileness of the V-1 and V-2 programs. Hitler once said that missiles spared his men and his planes. He even suggested that fuel would be conserved as there was no need for the return trip to Germany. Speer, in his May 1945 interview, was convinced that

> from the point of view of their technical production the rockets were a very expensive affair for us, and their effect compared to the cost of their output was negligible. In consequence, we had no particular interest in developing the affair on a bigger scale. In this case, the person who kept urging it was Himmler. He gave one Obergruppenführer Kammler the task of firing off these rockets over England. In Army circles they were too expensive; and in Air Force circles the opinion was the same, since for the equivalent of one rocket one could almost build a fighter. It is quite clear that it would have been much better for us if we had not gone in for this nonsense.

Dornberger, however, recorded that in mid-1943 Speer had confided his "... full confidence ... in the success of the scheme [the V-2]." Moreover, the general felt that the fighter program was then moving along unhindered by the V-weapon program, "right up to the closing days of the war." Dornberger insisted, "We lacked not fighters but gasoline. Our vital artery, fuel, had run dry."

On the cost issue, he stated, that "every A4 in mass production cost thousands of marks less than a torpedo and less than a thirtieth of the price of a twin-engined bomber," and questioned "... how often, after 1941, could a German bomber fly to England before being shot down?"

In making this partially justified comparison, Dornberger was thinking only of the production-line price tag of a V-2, not the entire physical and human structure that made its operation possible. Furthermore, he showed that "the spread [deviation] of the V-2 in relation to its range was always less than that of bombs and big guns." Recognizing that it was idle to speculate on its possible effects on the war, he concluded that "only one thing can be said with absolute certainty: The use of the V-2 may be aptly summed up in the two words 'too late.' Lack of foresight in high places and failure to understand the scientific background were to blame."

This conclusion was echoed by General Georg Thomas, Chief of the Economic and Production Office of the Armed Forces High Command from 1934 to 1943: "The low priority allotted to the development of rockets ... has without doubt caused the loss of much precious time, which in 1944 was evident in a most disturbing way." Nevertheless, in an interview in July 1945, he admitted that "no one conscious of his responsibilities believed that this weapon would ever decide the outcome of the war."

The low-priority status suffered by the V-weapons at various stages of their histories affected not only their development as reliable weapons but the very numbers that could be built. Even after the war, it was impossible to construct an absolutely accurate account of all missiles launched in combat. However, as the table on page 251 indicates, the number was not far from 25,000.

A successful launch did not mean a successful impact on target. Of slightly more than 1,400 V-2s that took off from the launch positions in Holland and started on the 5-minute trajectory toward the United Kingdom, only 1,115 got through.

September 8–December 31, 1944	418
January 1945	229
February	239
March	229
Total	1,115

Continental launchings of the missiles were not as accurately recorded, leading to discrepancies between many Allied and German sources. Listings under "other continental" in the table refer to firings against Brussels, Ghent, Maastricht, and Lille, principally from Trier,

Montaburg, Hachenburg, Eskirchen, and Heek in Germany. German records suggest that some 4 percent of the missiles successfully launched aborted shortly after takeoff, while another 6 percent underwent airburst on the upward trajectory. Of the total number of rockets placed on the launch pad, about 15 percent were judged unfit for firing because of loss of liquid oxygen, icing up of valves, or other problems.

The table shows nearly 22,000 V-1s to have been fired against targets in the United Kingdom and on the Continent. The table is essentially accurate except for the number of air launches, which was possibly higher. Basil Collier, in his *The Defence of the United Kingdom*, estimates the number may have reached 1,600, though German sources record no more than 890 He111 and Do217 sorties. Whatever the total of all types launched, 7,488 V-1s were observed by the British to cross into their defensive network. Of these, 3,957 were destroyed, 3,531 eluded the defenses, and 2,419 reached the London Civil Defence Region.

Of course, neither missile was accurate. V-2s produced some 4.8 fatalities per round landing in Greater London, versus 2.2 for flying bombs. Outside of London, rates were much lower; for example, the V-1 produced only 0.13 deaths and 0.44 injuries per impact. Probably the least effective of all was the air-launched V-1, which, aside from its nuisance value, did very little harm. It could not be fired in poor weather, required the costly use of manned airplanes, and was only a fourth as accurate as the ramp-fired V-1. Only 6.5 percent of the air-launched V-1s reached London and 55 percent other parts of Britain; 38.5 percent were destroyed. To get the 58 hits registered, 890 aircraft had to be dispatched, meaning that 130 pounds of bomb payload was delivered per effective sortie. Though the percentage of ramp-launched V-1s destroyed was higher (48 vs. 38.5), so was the percentage reaching the target (29 vs. 6.5). This simply meant that by outflanking defenses, the Germans could land more missiles *somewhere* in England, but accuracy was so low as to have made the effort rather futile except insofar as it caused the British additional strain and anxiety.

The V-1 campaign was broken not only by Allied air attacks on, and later occupation of, launch positions, but a greatly improved defense system. Thus, during the first week of the flying bomb offensive, 35 percent got through to London, 32 percent hit other parts of southeast England, and 33 percent were destroyed. The figures for the final week (August 25–September 1, 1944) were 9, 19, and 72 percent, respectively.

If time and resources had permitted the Germans to improve their weapon, they might have offset the Allied defenses. When they used the standard 138-foot launch ramp, for example, the accident rate was about 8 percent. This dropped to 2 percent on experimental 170-foot ramps, greatly boosting the morale of the troops.

"Although the accuracy of the individual [V-1] bomb was not of a high order," a January 1947 Aircraft Division Industry Report published by the U.S. Strategic Bombing Survey stated, "the statistical average of hits was good evidence of the fact that no district in London was spared." Other studies also took note of the strong psychological effect flying bombs had on a populace never certain when an attack would take place. "The robot shelling of London greatly affected war production," a British report admitted, "subjecting the populace to unending strain and thus the physical damage [was] enhanced by the psychological effect."

The September 1944 through March 1945 casualty figures of 10,626 killed or seriously injured were largely due to the V-2 missile. About 2,700 deaths and 6,500 serious injuries made up the V-2 total of 9,277. The remaining casualties were caused by air-launched and Holland-based ramp V-1 firings plus an occasional bomber raid. If one added the 15,000 persons slightly hurt, the total casualty figure would rise nearly to 25,000 persons. On the V-1 side, total casualties were approximately 60,000, including 24,165 in the killed or badly injured categories (compared to 112,932 by conventional bombing, 9,277 by V-2 rockets, and 402 by cross-Channel artillery). While serious, in terms of the effort mounted and the costs involved, these figures do not seem impressive. What the casualties might have been if the V-weapons had been introduced six months or a year earlier and in the quantities originally planned is a matter of speculation.

When the Ministry of Home Security in London was estimating casualties and damage from warheads weighing 5, 10, and even 15 tons, with aluminized explosive filling, the figures were chilling. Using known flying-bomb casualty rates at the time of the study (2.4 persons killed, 7.6 seriously injured, and 11.1 slightly injured, for a total of 21.1), Home Security multiplied the 21.1 by 6.1 to determine the effects a 5-ton Big Ben warhead, by 12.2 for a 10-ton warhead, and by 18.3 for a 15-ton warhead. Their standardized casualty rate was defined as the number of casualties expected to be caused by a given warhead when falling in an area with a population density of one person per 1,000 square feet (or 44 persons per acre).

Even before the V-weapon campaign was over, military economists in Britain went to work to assess what they called "the economic balance." The idea was to attempt to determine what the V-1 cost the Germans in development, production, and operational terms, and then to work out the economic effects their pending and actual existence had on the British.

British Air Ministry studies published on November 4, 1944, two months after the V-1 campaign from France was over, showed that, as regards labor costs, it ran "greatly in the enemy's favour." In fact, as the following table shows, the estimated ratio of Allied to German costs was about 4 to 1.

Principal military countermeasures	Cost to the Allies as a fraction of total German cost
I. Damage and loss of production in the United Kingdom	1.46
II. Bombing attacks by Allied Expeditionary Air Forces, including fighter interception patrols	0.34
All other bombing attacks on V-1 sites	1.54
Defense fighter interception	0.30
Antiaircraft protection	0.09
Passive balloon defense	0.07
Total	3.80

Enemy costs were measured by the survey in terms of the industrial labor needed to produce the weapon (about 800 man-hours per unit, not including the warhead) and its equipment, service personnel diverted from other parts of the war effort, and civilian labor required for special constructional work. Not considered, however, were the cost of research and development at Peenemünde and elsewhere, nor the construction of the large servicing and launch installations on the French coast. (Interrogations conducted during a survey of the Pas de Calais area yielded estimates that these installations required some 15,000 laborers during the year-long construction and repair period; antiaircraft protection cost an additional 200,000 man-months.)

On the Allied side, costs were divided into two major categories: (1) direct damage as well as loss of production in terms of industrial labor diverted from other war activities; and (2) active and passive military countermeasures. Here, too, some costs do not appear. Examples: the cost of fighter escort for bombers attacking the launch sites, photoreconnaisance and interpretation, preparation of target materials, and high-level planning and assessment.

Each of the countermeasure actions had to be considered in terms of loss of weapons and equipment, loss of trained personnel and service manpower, and the need for special construction.

As far as loss of equipment was concerned, costs were calculated for 93,000 tons of bombs at 88,000 man-months and fuel at 350 man-hours or 1.8 man-months per aircraft sortie. Repairs to damaged housing ran 260,000 man-months up to early November 1944, at which time the Ministry of Works estimated that 100,000 laborers would be needed to repair the remaining damage.

Up to early November, the man-months expended by the Allies in response to the V-1 danger were:

	Man-months
(1) Damage and loss of production	
a. repairs to damaged houses and building	260,000
b. loss of production	680,000
(2) Military countermeasures	
a. bombing attacks	1,209,602
b. fighter interceptions	195,461
c. antiaircraft artillery	54,100
d. balloon barrage	43,616
Total	2,442,779

As R.V. Jones points out in his *Most Secret War*, the economics of the V-1 campaign were balanced in Germany's favor because of these and other factors. But, he adds, "The balance on which judgement must be passed is not between British and German expenditure but between our [British] expenditure in countermeasures and the damage that would have ensued in lives, material and morale if those countermeasures had not been undertaken."

Opinions in Germany varied as to the effectiveness of the V-weapons, whatever the cost might have been. At first, there was optimism in some quarters, almost none in others. Hitler blew hot, then cold, then hot again. To what extent he was convinced of the real military merit of the weapons as opposed to their propagandistic value is hard to assess, but by January 1945 he apparently knew they could not affect the war situation. The following dialogue is excerpted from a Situation Report dated January 9, 1945:

> JODL: I have one more thing to mention, that is to say, a report from an agent in Antwerp who claims that on December 17, 1944 a V-2 hit the Rex movie theater during a very crowded performance. There were 1,100 casualties, including 700 soldiers.
>
> HITLER: Well, that would finally be the first real hit. But it is so much like a fairy tale that I, skeptical as I am, don't believe a word of it. Who is this agent? Is he getting paid by the men who launch the V-2s?
>
> JODL: This agent goes under the strange name of Whisky.
>
> HITLER: That's not exactly a star witness.

This turned out to be one of the few tactical successes of the V-2. As Hugh M. Cole pointed out in his "Ardennes: Battle of the Bulge" in *The*

United States Army in World War II, "The military casualties inflicted by this V-weapon attack [on Antwerp and Liège during the German Ardennes offensive] were slight, except for one strike on 16 December which destroyed an Antwerp cinema, killing 296 British soldiers and wounding 192." "Whisky" was a day off on the date and exaggerated the casualties. Incidentally, the best hit on a supply target occurred the next day in Liège where 400,000 gallons of gasoline were destroyed.

Many Allied and German military experts felt that the propagandistic effect of the V-weapons outweighed their military value. Propaganda Minister Goebbels seized on the "vengeance weapons" to the fullest, with the undisputed result that the mere existence of the V-1 and V-2, coupled to the inflated claims made for their efficacy, contributed to German defensive stubbornness. As Speer recorded in mid-September 1944, "A belief that new, decisive weapons will soon be employed is generally prevalent among the troops.... Even high-ranking officers seriously share this belief." When the weapons failed to appear, the collapse of Nazi Germany became inevitable.

Although Speer realized that the V-weapons could not possibly lead to victory or even a negotiated peace, he did expect the V-weapon offensive "... to make the British population tired of the war, for now and then we received reports that their morale wasn't as high as before. We expected more of a political reaction to the weapon." As for von Rundstedt's opinion, his biographer, General Günther Blumentritt, writes that the Field Marshal "took a very cautious view of these weapons. He believed that they would certainly damage the enemy, but he saw in them no decisive effect upon the war."

General Dwight D. Eisenhower, Supreme Commander of the Allied Expeditionary Forces, was not so certain. In his *Crusade in Europe*, he wrote:

> It seemed likely that if the Germans had succeeded in perfecting and using these [V-1 and V-2] new weapons six months earlier than they did, our invasion of Europe would have proved exceedingly difficult, perhaps impossible. I feel sure that if they had succeeded in using these weapons over a six-month period, and particularly if they had made the Portsmouth-Southampton area one of their principal targets, "Overlord" might have been written off.

Churchill disagreed.

> This is an overstatement. The average error of both these weapons was over ten miles. Even if the Germans had been able to maintain a rate of fire of a hundred and twenty a day and if none whatever had been shot down, the effect would have been the equivalent of only two or three one-ton bombs to a square mile per week.

Speer, looking back at events from the perspective of two and a half decades, hardly changed his mind. In his *Inside the Third Reich* he wrote:

> The whole notion [of producing 900 V-2s a month] was absurd. The fleets of enemy bombers in 1944 were dropping an average of three thousand tons of bombs a day over a span of several months. And Hitler wanted to retaliate with thirty rockets that would have carried twenty-four tons of explosives to England daily. That was equivalent to the bomb load of only twelve Flying Fortresses.
>
> I not only went along with this decision on Hitler's part but also supported it. This was probably one of my most serious mistakes.

In retrospect, Speer realized that Germany should have put all her efforts into attempting to defeat the strategic bombing offensive. Had this been the major objective, and had Germany pressed forward with its promising jet aircraft and the *Wasserfall* surface-to-air missile program, the war in Europe might well have lasted until the advent of the atomic bomb. "To this day," muses Speer, "I think that [using] the *Wasserfall* in conjunction with jet fighters would have beaten back the Western Allies' air offensive against our industry from the spring of 1944 on." Speer continued:

> Instead, gigantic effort and expense went into developing and manufacturing long-range rockets which proved to be, when they were at last ready for use in the autumn of 1944, an almost total failure. Our most expensive project was also our most foolish one. Those rockets, which were our pride and for a time my favorite armaments project, proved to be nothing but a mistaken investment. On top of that, they were one of the reasons we lost the defensive war in the air.

Ironically, on August 15, 1944, at about the time the V-2 was entering into large-scale production at Mittelwerk, Churchill's scientific adviser, Lord Cherwell, confided to the Prime Minister that it seemed inconceivable the Germans would seek to develop such a complex and costly device to deliver a mere ton of explosives on London. The V-1, he said, could do the job almost as effectively and much more cheaply. Hitler would "... be justified in sending to a concentration camp whoever advised him to persist in such a project," he added.

These judgments may be too harsh, but one is always tempted to search for the alternative "ifs." What if the *Wasserfall* had been accorded the priority the A4 received? What if other surface-to-surface missiles, and jet fighters, and air-to-air missiles had been pushed? Certainly the war would have lasted longer, the Allied losses would have been greater ... and Germany would have suffered not only a more prolonged conventional bombardment but the onslaught of atomic bombs.

Supposing the efforts spent on the heavy installations had been devoted to strengthening the Atlantic wall and building additional submarine pens? Suppose Mittelwerk had produced only jet fighter planes? Or synthetic oil? Or tanks? Speer himself said that if the 30,000 V-1s and the 6,000 V-2s estimated to have been built had not been manufactured, 24,000 fighters could have been (one fighter equaling five V-1s and one-third of a V-2). Or suppose that the more than 300 heavy antiaircraft guns used to defend the ski sites and heavy installations and the various supply depots had been installed at the great oil production plants at Brux and Politz, defended by only 52 and 76 guns, respectively—or at Hamburg (which had 232) or Bremen (with 188). (The shortage of explosives became so acute that antiaircraft fire at all these sites had to be restricted.)

The "ifs" could go on for ever. Whether the V-weapons were or were not successful depends almost exclusively on what might have happened if the human energy behind them had been devoted to something else. What, in effect, would the 200,000 persons associated with the A4 program have done had there been no von Braun? No Dornberger? No rocket? Improved tanks? Would they have probed into the atomic field? Developed better airplanes or submarines?

Dornberger's simple statement that the V-weapons were too late to be effective during the Second World War against a vastly superior enemy is, in the final analysis, the best that can be concluded. They, along with the US atomic bomb, British radar, and jet planes developed by all three powers, were the crowning technical achievements of the epoch. Not only did they change fundamental concepts of warfare but they laid the foundations for stunning scientific and technological achievements in the decades to come.

Without the atomic bomb research, one of man's greatest energy resources would still remain in the pages of science fiction. Without radar and jet engines, the rapid, complex—and safe—world travel by airplane would be an impossibility. And without the advances in military missilery, space travel would still be a dream.

Such are the more beneficial dividends of war.

A summary of the V-Weapon Campaign

Target	Successful V-1 launches[1]	Unsuccessful V-1 launches	Launch Area	Successful V-2 launches	Unsuccessful V-2 launches	Launch Area	Total successful launches
UNITED KINGDOM							
London	8,839[2]	1,043[3]	France, Holland	1,359	169	Holland	10,198
Ipswich	—	—	—	1	0	Holland	1
Norwich	—	—	—	43	?	Holland	43
Southampton	53	9	France	—	—	—	53
Miscellaneous	890	not known	air[5]	—	—	—	890
BELGIUM							
Antwerp	8,696	1,009	Holland and Germany	1,610	?	Holland and Germany	10,306
Brussels	151	18		under "other" below	—		under "other" below [151 V-2s]
Liège	3,141	366		86	?		3,227
FRANCE							
Paris	—	—	—	19	0	Germany	19
OTHER CONTINENTAL	—	—	—	137 (including 11 at Remagen)	?	Holland and Germany	137
Subtotal	21,770	2,445		3,225[6]	(169?)		25,025

[1] From point of view of German launch crew. Not all successful launches resulted in successful impacts on target. Many missiles fell short or wide, or (in the case of the V-1) were shot down en route.
[2] 8,564 from France; 275 from Holland
[3] 1,006 from France; 37 from Holland
[4] Itemization of targets not available; missiles principally flown against London with secondary targets Southampton and Bristol. Covers period September 1–December 13, 1944 only. Some reports say that nearly 75 more air-launched V-1s were flown in January and February 1945, of which 45 were directed at London.
[5] Off the Dutch and Belgian coasts
[6] Of the 1,403 successful V-2 launches against the UK, 1,115 missiles actually got through to create an "incident." Of the 1,852 successful Continental launches, it is not known accurately how many actually created "incidents," but the number 1,775 has been estimated. See page 79.
N.B. Data from report prepared by the staff of Armee Korps z.v. on April 8 1945, approximately a week after the V-weapon campaign ended. It is based on data provided by Division z.v. and the Fifth Flak Division.

Casualties and Direct Damage Caused by V-Weapon Campaign[1]

Casualties	Killed	Seriously wounded
Civilian	9,768	24,494
Military	2,917	1,939
Civilian property[2]		

Housing/Buildings	Destroyed	Damaged
Greater London	23,000	100,000
Greater Antwerp	6,400	60,000
Greater Liège	4,300	44,000

Military property	
No. of aircraft lost	498
Damage to military and port installations	modest[3]

[1] Thompson, Royce L., *Military Impact of the German V-Weapon.*
[2] Damage to other targets roughly in proportion to number of missiles engaging them.
[3] Damage rapidly and rather easily repaired; caused no major slowdown of the Allied war effort.

Estimated Cost of V-Weapon Program (US Wartime Dollars)

V-2 Development Program		
Kummersdorf, 1932–1937		300,000
Peenemünde, 1937–1945		1,900,700,000
	Subtotal	2,000,000,000[1]
V-1 Development Program		
Peenemünde		200,000,000[2]
V-1 and V-2 Construction[3]		
Organization Todt		273,000,000
Other Construction[4]		
Lehesten, Raderach, etc.		100,000,000
Miscellaneous Costs[5]		
Military salaries for maximum of 20,000 men, transportation, intelligence, antiaircraft protection, etc.		500,000,000
	Total	$3,073,000,000

[1] Estimated from Johns Hopkins University—Operations Research Office Study ORO-R-3, Part IV, Appendix C, January 21, 1950.
[2] During interview with Minister Albert Speer on May 21, 1945, it was estimated that the V-1 cost about 1/20th the V-2 or $100 million, but Dornberger feels that 1/10th is more realistic, or $200 million.
[3] Estimate of Herr Dorsch of Organization Todt.
[4] Combustion chamber test and other sites.
[5] Lt. Colonel M. C. Helfers, *The Employment of V-Weapons by the Germans During World War II.*

CHAPTER 13

Exodus from Peenemünde

> In January 1945, I found myself confronted by several conflicting orders—from the local defense commander, the *Gauleiter* of Pomerania, the Ministry of Armaments, and the Army Ordnance Department in Berlin, as well as several lesser organizations. All orders dealt with the problem of how to cope with the Soviet army then advancing on Peenemünde from the east. Some directed the Peenemünde management to stay put and help the *Volkssturm* [home guard troops] "to defend the holy soil of Pomerania." Others directed immediate evacuation of the center so the invaluable development team could continue to make its contributions to the still-expected "ultimate victory."

With only slight hyperbole, von Braun could later say to the authors of the situation confronting him at that time, "I had ten orders on my desk. Five promised death by firing squad if we moved, and five said I'd be shot if we didn't move."

That he personally would have to take the initiative to save the team had occurred to him as early as October 1944. During a visit to his father at the family home in Silesia, von Braun senior told his son of the decision made at Yalta by the Allied leaders. Germany was to be partitioned after the war. He had heard the news on radio broadcasts from the Swiss, French, and British. It was then that von Braun junior decided to throw his fortunes in with the West.

When in mid-January 1945 the rumor that the Russians were at Eberswalde to the south of Peenemünde reached von Braun, he knew that he must take some action in view of his conflicting orders. He met secretly with a few close and trusted friends. The group gathered around the iron stove in the parlor of the Inselhof Hotel in Zinnowitz. It consisted of von Braun, Steinhoff, Rees, Stuhlinger, engineer Werner Gengelbach, and von Braun's secretary, Dorette Kersten. He put it to them bluntly and asked for a vote. The result was a unanimous decision to leave Peenemünde and move to a region likely to be occupied by the American army.

The responsibility for the relocation of *Elektromechanische Werke, Karlshagen*, the at-the-time designation of the Peenemünde establishment, fell to von Braun rather than Dornberger for the reason that the latter, who had been for so long in command at Peenemünde, was no longer so. As part of the hysteria of the final days of the war, Dornberger found himself in a curious position. On January 12, 1945, he was named to head *Arbeitsstab Dornberger* (Working Staff Dornberger), a department within Speer's Ministry of Armanents that replaced the impotent and largely inactive Long-Range Bombardment Commission under the ailing Prof. Waldemar Petersen. Shortly after Dornberger's appointment, Obergruppenführer Kammler prevailed upon Reichsminister Göring to make him Special Commissioner for Breaking the Air Terror. Once so named, Kammler had *Arbeitsstab Dornberger* assigned as his own technical staff. The transfer gave Dornberger advantages he never possessed as a mere major general in the German army responsible for developing its guided missiles.

Wearing his Ministry of Armaments hat, he could direct civilian authorities and industries to do his bidding. Wearing his army hat, but never the skull-and-crossbones insignia of the SS, he could deal similarly with the military, using the implicit clout of the SS, with which he was associated. It was potentially an extremely influential position for developing all V-weapons and antiaircraft weapons as well, with the exception of conventional guns. However, his rise to prominence occurred when the war had less than sixteen weeks to run.

On January 31 von Braun received two more mutually contradictory orders, and they forced his hand. He had two courses of action open to him, and he had to pick one or the other of them. Kammler had decreed that von Braun remove his personnel to the Thuringian area around the V-1 and V-2 production plant near Nordhausen. On the other hand, the army commander charged with the defense of Pomerania, which included Peenemünde, had ordered him to have his men form into *Volkssturm* units and turn back the Russians at Peenemünde. As ridiculous as this latter proposition was, since December 12, 1944, the civilians at Peenemünde had spent what little free time they had on Sunday morning receiving instruction in bayonet drill, rifle marksmanship, and even hand-to-hand combat for just that purpose.

Thus, von Braun could stay and face the prospect of having the Russians capture the team en masse, as well as valuable documentation and hardware; or he could pack up and move south, buying at least a little time in which to plan further on what ultimately to do.

As von Braun pondered his plan of action, business—more or less as usual—went on at Peenemünde. The last of the test A4s was launched on February 19. In view of the circumstances, he decided that Kammler's order fitted in nicely with the decision he and his friends had made covertly several weeks earlier. They would pack it in and head south for Thuringia.

Planning for the move got under way immediately. With characteristic energy, von Braun took personal charge of the complex logistic arrangements. The motor vehicle transportation was in the charge of Erich Nimwegen, a mysterious individual who appeared for work one day, no one knew from where, and became an indispensable member of the Peenemünde team. He always operated just within the law, especially in obtaining a variety of scarce items demanded by the engineers and technicians. Nimwegen was not above fobbing himself off as a member of Himmler's staff and calling the admiral in charge of naval activities at nearby Swinemünde to get fuel and food that would have otherwise been unobtainable. For the journey south, he also "organized" twenty *Gulash Kanonen* (goulash cannon) or army field kitchens.

On February 3, von Braun held his final meeting for consolidating the movement.

"We will go as an organization," he said firmly. "This is important. We will carry our administration and structure straight across Germany. This will not be a rout."

In terms of personnel alone, the task facing von Braun was staggering. No fewer than 4,325 people were currently employed at Peenemünde. The largest number, 1,940, even at that late date, were assigned to A4 development and modification. Another 270 individuals were at work on the winged version of the A4, called the A4b, and 1,220 were devoting their efforts to *Wasserfall*. *Taifun*, the small but potentially very effective antiaircraft rocket, by then engaged only 135 people. Supporting the efforts of the technical staff were 760 administrative and logistical personnel. It was decided that 30 percent of the total number, most of them reluctant to leave for various personal reasons, could be left behind without seriously impeding the work to be done in the new location.

When someone pointed out that the trains allotted for the move would hardly be sufficient to haul all the matériel to Sangerhausen, their proposed destination, von Braun informed them that a number of barges also had been made available as a complementary means of transportation. Once loaded, the barges would be towed across the Baltic to Lübeck, down the Trave-Elbe canal, and thence down to Schönebeck, near Magdeburg, while the trains proceeded as best they might.

The movement plan conceived by von Braun was to have families and personnel, and some less bulky equipment, moved out by train. The heavier equipment and records would follow by barge. Finally there would be motor vehicle convoys to evacuate remaining personnel, files, and equipment from Peenemünde.

The pace of activity quickened daily. Dieter Huzel, von Braun's assistant, later described the hectic time in his book *Peenemünde to Canaveral*:

> We went to work with a vengeance. Virtually all the coordination came through von Braun's staff, and this kept us busy night and day. Such simple things as the procurement of boxes for packing was in itself a large task. Some of the technical sections needed hundreds of them. We devised a color-coding system for ready identification of each box on arrival at our new headquarters: white for administration, green for design and development, blue for manufacturing, red for test, and so forth. All this time each department was frantically trying to determine which thirty percent of their people were not to be taken along, and how many of the seventy percent who were going had families, and how many people were in these families.

Somehow, things got sorted out; and the first train left Peenemünde on February 17 with 525 people, most of them in freight cars. Whole families were crowded into them and slept on straw. For the babies in the party, several cows were taken along to provide milk. The cows, in turn, ate the straw, solving at least one logistic problem. At the same time Dornberger and his staff left Schwedt-an-der-Oder for Bad Sachsa, a resort-village 20 kilometers north of Bleicherode, which was to be his new headquarters. Von Braun had flown ahead to prepare a place for them in the Harz Mountains, in and around Bleicherode.

At this point, a note of almost classical farce entered the proceedings.

Earlier, in a move to associate *Elektromechanische Werke, Karlshagen* visually with the SS, letterhead stationery had been ordered. It was to have read *BZBV Heer* (the designation of Dornberger's new SS organization). Somehow the printer had scrambled the initials to *VABV*, a meaningless jumble of letters that made no sense whatsoever.

The imaginative Nimwegen turned the printer's error into what later proved to be the salvation of the Peenemünde team. Huzel explained how:

> He made *VABV* a top secret agency [*Vorhaben zur besonderen Verwendung* or Project for Special Dispositions, a name equally as

> meaningless as the random array of initials], not to be interfered with by anyone save Himmler himself. Soon *VABV* signs began to appear in letters several feet high on boxes, lorries, and wagons. Indeed, it provided complete protection against any interference; thus, despite the growing confusion inside Germany, all of our essential personnel and a good portion of our equipment did arrive at the new location.

Nimwegen's guile and bluff paid off almost at once.

Scarcely had the convoy got under way than it was stopped by an SS unit on the lookout for soldiers who might be absent without leave or civilians who were failing to take up arms in defense of the homeland. Flashing the letterhead stationery and waving to lorries and packing crates emblazoned *VABV*, Nimwegen overwhelmed unsure SS officers who wanted no part in crossing Himmler even at that late date.

Von Braun made his last trip to Peenemünde on February 27. He flew back there from Bleicherode to explain to those left at the rapidly disintegrating center the new organization in the Harz Mountains.

"What has been set up," he said, "is *Entwicklungsgemeinschaft Mittelbau*, a sort of central cooperative developmental structure. We will be only one part of this organization, which will also include Henschel Aircraft, the Ruhr Steel Corporation, Aircraft Components, the Kreiselgeräte Company, Dornier Aircraft, and the Walter Company, as well as a number of smaller organizations."

Avoiding the eyes of his engineers as much as possible, he went on hurriedly "For *our* effort, several test areas will be established. There will be a smaller area set up for missiles X4, X7, *Taifun*, and *Schmetterling* missiles, and larger setups for the A4 and *Wasserfall*."

The incredulity of his colleagues is best summed up by the thought that ran through Huzel's mind at the time, which he later recalled in his book: "Test stands for the A4, just by snapping your fingers?"

But von Braun was playing the game. He knew that every move he made was being closely watched by the Gestapo. Only as long as he kept talking about continuing to develop the weapons that were deemed so crucial to winning the "ultimate victory," could he muster the massive logistic support from other agencies that was required for the evacuation of Peenemünde. The slightest doubt of the merit of continued technical efforts by the team not only would have cut off this support but also would have cast the suspicion of defeatism on the move.

"Launching from Peenemünde will definitely halt," said von Braun. "Dr. Debus, your mobile launching convoy will be formed and dispatched to the general area of Cuxhaven."

Accordingly, Kurt Debus suspended operations and readied his convoy, which he headed westward through Mecklenburg. However, it soon became obvious that the British would reach Cuxhaven before he could arrive and establish a launching facility, so he turned south to join the others in Thuringia.

Debus and his group continued to an area just north of Garmisch-Partenkirchen. There they remained in absolute idleness until they were taken into custody by the American army.

On the orders of Major General Erwin Rossmann, who had become commander of Peenemünde in 1944, a systematic destruction was begun of the larger facilities by placing explosive charges in them. Sergeant Heinrich Schulze, who had trained men in the preparation, checkout, and launching of the V-2, destroyed his training material and literature and then made a final trip down to his old training field, Test Stand 10. There was still a V-2 on a launcher. Carefully, he placed amatol charges around and in it. Then, with a sigh but no tears, he set off the detonator and left Peenemünde, though not for good as he thought at the time.

In preparation for the move south, the V-2 guidance and control personnel stationed at Lubmin, some 14 kilometers to the west of Peenemünde, dismantled their huge *Wurzburg Riese* radar antenna, and made ready to leave their cozy quarters in the resort hotel atop which it was located. As much equipment as could be accommodated was loaded into trucks for the motor convoy. Women and children were packed into buses. From somewhere there appeared fifteen mules, which were harnessed to wagons loaded with delicate electronic parts for the radar. Quite literally, additional horsepower became available as well. A horse-breeder with friends and relatives at Peenemünde stopped off in Lubmin with a herd of fifty thoroughbreds as she and her equine charges fled westward from East Prussia. Her show horses joined their hybrid cousins in pulling for Peenemünde.

The *Wasserfall* guidance and control group, located at Bansin and Pudalga, 27 kilometers to the southeast of Peenemünde, had just two hours' notice to be ready to move out by train from the station at Zinnowitz on February 22. In the limited time, Dr. Ernst Geissler, former chief of a unit performing flight dynamics and control systems studies on the *Wasserfall*, somewhat irrationally settled for an armload of books, a few potatoes, some underclothes, and a painting by a favorite uncle.

The departure of the train was delayed at Zinnowitz because the local commander of the *Volkssturm* insisted all men detrain and take up arms—if they could find any—in defense of the area. After much discussion with senior army officers on board, the train was permitted to leave.

The personnel of Hans Lindenmayr's valve laboratory near Friedland packed up not only a supply of materials and machine tools but also two wooden barracks, which they disassembled and put into their train headed south. They set up in a castle near the village of Leutenberg, some

10 kilometers south of Saalfeld near the Bavarian border. The two barracks were erected in the middle of the village football field to serve as a shop and storehouse for the machine tools and supplies.

On March 19, Hitler issued an order that all research facilities and their important documentation were to be destroyed, as a part of his "scorched earth" policy for what remained of the Third Reich.

The order was honored more in the breach than the observance by the rocket and guided missile segments of *Entwicklungsgemeinschaft Mittelbau*.

Its liquid-oxygen plant in the Redl-Zipf brewery ignored the order. The key personnel simply took off for the surrounding woods to escape their SS overlords. Remaining behind in the underground plant were a thousand or so inmates of a nearby concentration camp, who provided the labor for operations. These men were largely Spaniards who had fought in the losing war against Franco during the Spanish Civil War in 1936–39 and subsequently escaped to France, where they were interned. With the German occupation of France in 1940, they were transferred to prisons within Germany. Some of them were employed in a carefully screened-off and guarded section of the former brewery and worked on a curious project. Under German direction, they were producing extremely accurate counterfeit English banknotes with which the Germans were planning to wreck their enemy's economy by releasing the money through outlets in various countries.

Dr. Rudolf Hermann, director of Peenemünde's supersonic wind tunnel, then at Kochel, simply disregarded the order. He had the more trusted of his two hundred associates round up the most critical data and reports and bury them nearby. The precious wind tunnel and ancillary facilities were not destroyed. Thus, when the Americans captured the facility a month later, their first task, under the direction of Dr. Fritz Zwicky, California Institute of Technology, a consultant to the US Army Air Corps, was to finish the reports on which they had been working when the war ended.

Similarly, the static firing test stands at Lehesten for the V-2 engines were not demolished either. They were in operating order when the Americans took them; and, indeed, several engines were test-fired for them after the war ended. As impressed as the Americans were, they insisted on leaving the stands intact for their Soviet allies, who would and did take them over in a few weeks.

Throughout March, several engineers returned to Peenemünde for various reasons.

Heinrich Schulze went back on a train to dismantle several wooden barracks for use at the new location at Bleicherode. Helmut Hölzer returned on a motor bicycle to look for portions of his PhD dissertation that had gone astray in the movement to the south. In a ludicrous note, engineer Wernher Dahm was ordered back to Peenemünde to serve a jail term to which he had earlier been sentenced. While still a member of the *VKN*, he had overstayed a leave and had subsequently been condemned to spend a week in close confinement. After the removal to Bleicherode, a sharp-eyed clerk noted that the sentence had never been served. Regulations were regulations. They might be bent at times, but they could never be broken. Dahm was sent back to the nearly deserted facility to spend his week in the stockade.

By that time, Peenemünde was a lonely place. There was no activity at all, and the only people about were SS troopers preparing for a last-ditch stand against the Russians expected momentarily from the east. They had a long wait. It was not until May 5 that Major Anatoli Vavilov, commanding a unit of the Second White Russian Army, actually captured the facility. He found it deserted except for a number of maintenance men. There were few casualties among the former personnel who had remained behind. Oberstleutnant Richard Rumschöttel, former adjutant to General Zanssen, was living in the village of Bansin when the Soviet troops took over. Several drunken Russian soldiers broke into his apartment and attempted to rape his wife. When he attempted to fight them off, the soldiers killed both him and her.

Later on the evening of Easter Day, April 1, von Braun instructed Huzel to find a hiding place for the critical documentation on all former Peenemünde projects, a collection of reports and drawings that represented the result of thirteen years of research and engineering development. He had reason; on that day, American tanks had been reported at Mülhausen, only 20 kilometers south of Bleicherode. He also suggested that Huzel might make use of fellow engineer Bernhard Tessmann.

The value of these documents in advancing the state of rocket technology in the West cannot be overestimated. They were a technological treasure trove and as such were priceless. Huzel later said of them:

> These documents were of inestimable value. Whoever inherited them would be able to start in rocketry at that point at which we had left off, with the benefit not only of our accomplishments, but of our mistakes as well—the real ingredient of experience. They represented years of intensive effort in a brand new technology, one which, all of us were still convinced, would play a profound role in the future course of human events.

In the midst of his planning, Huzel had a call from von Braun:

> "How's it going?"
> "As well as can be expected, I suppose. Everything is so scattered."

"Yes, it's pretty messed up. ... You stick to what you're doing, though. We can't afford to slip up on this."

Von Braun then gave Huzel a letter that probably would cow the most dedicated SS officer and surely would intimidate the most dedicated *Volkssturm* leader he would encounter. It said, basically, that Huzel's mission was top secret, that it had to be completed, and that Huzel was to be assisted in every way and in no way delayed.

Von Braun then suggested that a mine or cave might make a good hiding place, an obvious solution in a region honeycombed with both. He suggested that Huzel contact the mining authorities at Clausthal-Zellerfeld, a village 40 kilometers northwest of Bleicherode. He also added that a corporal and eight or ten enlisted men from the *VKN* should be assigned as laborers.

Huzel and Tessmann went to work immediately; as Huzel later recalled:

> The central gathering place was a room in the potassium mine administration building and by mid-morning Tuesday it was nearly full. It was obvious that three trucks were not going to be enough. Together Tessmann and I went down to the car pool and managed to shake loose two 2½-ton trailers, but not without considerable haranguing. Rain added to our troubles. By noon, it was pouring steadily. Finally, on Wednesday morning, everything was in hand—except the contributions from the laggards in the crucial base-section and rudder design group. They had dutifully packed everything up, and then simply had taken off, leaving a one-ton chest of documents for whoever could lift it.

The task fell to a cursing and sweating lot of *VKN* soldiers, with Huzel and Tessmann lending a shoulder as well.

The heavily loaded convoy left Bleicherode and slowly headed north, taking secondary roads to avoid the streams of refugees clogging the main ones. The going was slow because of the great weight of the vehicles, but the trip was relatively uneventful except for a narrow escape from a flight of American P-47s, called *Jabos* by the Germans. Since any convoy acted as a magnet for such aircraft, of which the skies at the time seemed constantly full, Huzel and Tessmann were unwelcome guests in the small villages along their route. They were constantly being urged to speed through such hamlets by fearful inhabitants.

Nevertheless, Huzel parked the convoy in the small village of Lerbach, the hilly, winding, and wooden roads of which provided both concealment and cover for the motor vehicles. He left Tessmann in charge and then went on ahead to Clausthal-Zellerfeld for conference with the local mining officials about a suitable hiding place. What Huzel wanted was a remote, almost inaccessible cave or, perhaps, an abandoned mine shaft or tunnel. He was disheartened to learn that there simply was no such place in the neighborhood. The *Bergrat*, a local official with whom he discussed the problem, sent him on to Goslar, 20 kilometers farther to the northeast, where there was a suboffice of the Supreme State Mining Authority. There, he was put on to just what he sought: an abandoned mine near Dörnten, about 12 kilometers farther north.

Herr Cornelius, the local official, led him to it.

"This is it," he said. "Shaft *Georg Friedrich*. That vertical shaft hasn't been used for several years. . . . There is another horizontal shaft in that shallow hill behind the tower, discontinued some years ago."

Accompanied by Herr Cornelius and Herr Nebelung, the caretaker of *Georg Friedrich*, Huzel explored the proposed vault. It was perfect. Some 300 or 400 meters into the horizontal shaft, he was shown a room that was sealed by an iron door. It was especially dry. Huzel was delighted; he excitedly returned to Lerbach, which he reached too late to permit a return to Dörnten. On the following morning, he explained his plan to Tessmann and the *VKN* soldiers:

> We will drive the trucks to an old quarry close to Goslar. It will make an excellent center of operations. It's located in a narrow part of the valley and should be safe from *Jabos*. When nightfall comes, I'll drive the first truck and trailer myself to the hiding place. With the exception of Tessmann, all of the rest of you will be locked inside the truck. That way, you will be able to say with complete honesty that you have no idea where we have hidden the documents. Two of you will remain behind, one for each of the remaining trucks, just in case. . . . When we've unloaded the first truck, I'll come back and pick up the next, and then the third.

So went the plan; the *VKN* troopers, still cursing and sweating, loaded the boxes onto small flatcars provided by Herr Nebelung. Once full, they were pulled by a small mine locomotive to their destination within the mountain. Thus continued the operation until daybreak, when Huzel arrived with the last truck and the two men who had been left behind. By noon the job was finished. Nebelung promised to finish off the job by sealing the corridor and iron door with charges of dynamite that would collapse the rock in the ceiling.

The following day, April 7, Huzel and Tessmann returned to Dörnten to check on the demolition work of Herr Nebelung. They found it less than satisfactory. A great deal of rock was indeed piled before the iron door, but not enough. Men could easily crawl over it. Herr Nebelung promised he would finish off the job properly that night. Huzel instructed Tessmann and the exhausted *VKN* soldiers to return to Bleicherode with the trucks, being sure to keep one of them because "we've got to pick up some of the professor's [von Braun's] stuff." He kept a car for his own transport and decided to stay overnight to ensure that Herr Nebelung kept his word.

On the evening of the next day he received news that made him forgo his inspection of Nebelung's handiwork with the dynamite. The Americans were at Goslar. He decided to return to Bleicherode immediately, trusting to the integrity of Nebelung to finish the job—which he did with efficiency, as the Americans later found out.

On April 9, Huzel returned to Bleicherode after having stopped off in Nordhausen to inform Karl Otto Fleischer, business manager at Niedersachswerfen, of the exact location of the documents at Dörnten. He then sought out his colleagues in the Bleicherode area, only to find them for the most part missing.

Those he did meet were suffering deprivations of various kinds, including hunger. However, when food became critically short, Nimwegen commandeered a truck and a local butcher and searched the countryside for stray cattle, which were slaughtered on the spot and dressed. The blood was also collected and made into *Blutwurst*, a nutritious sausage that sustained many of the former Peenemündians for weeks.

The reason why Huzel found few familiar faces in Bleicherode was that on the night of April 1, Herbert Axster, Dornberger's chief of staff, had received a call from Kammler's adjutant. He had an order from the *Obergruppenführer*, said the adjutant, continuing:

> You have a list of four hundred of the top men. You, as the chief of staff, are responsible to gather these people together and are to move them to Oberammergau at the earliest possible time. There, at the Messerschmitt installation, will be a part of Kammler's staff. They will take over. You know that this is an order from Kammler.

Axster remonstrated that it would be impossible to move men and their equipment into the Bavarian Alps at short notice. The men as well as the equipment were scattered, and—

"I will be expecting to hear from you in twenty-four hours."

It was early in the morning, still dark, when Axster rousted von Braun from his bed and went over the story. Von Braun groaned, not so much from the pain in his broken arm, as from the thought of another move for his equally weary and downhearted engineers and scientists. Axster sympathized but added, "I must follow orders or otherwise I hang tomorrow!"

The next morning the men were hastily assembled. There was neither time nor space to collect and pack personal belongings—a few under clothes and that was it. They were piled into a truck convoy and aboard a train and sent off at once. It was the same train that had served as dormitory and quarters during the V-2 firings in Poland and had served the same purpose while parked in the railyard at Nordhausen. The train continued to be called the "Vengeance Express" by its cynical passengers.

Not all members of the team made it aboard, despite a search by frantic supervisors and armed SS troopers. Rudolf Schlidt was more concerned with locating his recent bride, Dorette, the former Fräulein Kersten, who had gotten separated in the chaos. He found her, after the train departed, and the two newlyweds simply sat tight in Bleicherode, awaiting the Americans. After they appeared, Schlidt was soon thrown together with Major Robert B. Staver, an energetic young ordnance officer who was to play a pivotal role in Project Paperclip, a story that unfolds in the following chapter. Also among those left behind in Bleicherode, for some unknown reason, was the invaluable Nimwegen.

Because of the continuous air attacks on all railways and roads, the journey to Oberammergau took six days. Frequent rerouting found the Vengeance Express meandering through Czechoslovakia, back to Augsburg, and then to Munich as it slowly wound south.

Among those not traveling by train were engineer Heinz Millinger and three colleagues who went by automobile. Having arrived in Oberammergau, the wary Millinger reported their arrival to the SS but decided not to move in with the others. The four men stayed in their car for several days, in case circumstances dictated a hasty flight. Intuitively these men felt, as did their fellow Peenemündians, that they could become either hostages or sacrificial pawns in a last-minute deal Kammler might try to make with the Americans.

Von Braun, also, was not a passenger on the Vengeance Express. He drove to Oberammergau by car so that he would arrive well before his rocket team. His mission was to look over half of the *Jägerkaserne*, a cantonment for mountain infantry troops in earlier days, now surrounded by barbed wire and guarded by soldiers. The other half was already occupied by designers of the Messerschmitt aircraft company, who had previously been relocated.

Kammler had arrived earlier and had set up his headquarters in the Haus Alois Lang, an inn owned by Alois Lang who for more than a decade had played the part of Christ in the famous passion play. Indeed, his inn was known to the local citizens somewhat irreverently as "Jesus' house." (After the war, an Allied investigative board determined that with the exception of the man who played the part of Judas Iscariot, Lang and all other members of the cast were members of the Nazi Party.)

About a week after von Braun arrived in Oberammergau, he was summoned to Haus Alois Lang by Kammler. As von Braun entered Kammler's office, he saw that the latter was in good humor and solicitously asked after von Braun's broken arm. Kammler told the technical director of *Elektromechanische Werke* that his own new duties as head of the General Commission for Turbojet Fighters (yet another of Kammler's titles accruing as the war ended) required his presence elsewhere; and therefore he had to leave. Without further explanation, Kammler dismissed von Braun, who immediately felt better. It was the last time the two were ever to meet.

With Kammler gone, von Braun set about getting his team out from under the guns of the SS. He devised a plan with the help of Steinhoff. Together, they approached Major Kummer, the officer appointed by Kammler as chief in his absence, with the proposal that the men, now under one roof in the *Jägerkaserne* for the sake of safety, should be scattered about in villages around Oberammergau, as they had been at Bleicherode. The argument was that if the buildings were to be bombed by the Americans, the group would be destroyed in one fell swoop, and Germany would lose its only hope of turning the tide of the war. Kammler probably would have seen through the ploy immediately; luckily for von Braun, Kummer was not as shrewd as his superior.

While Kummer was debating the plan with himself, a flight of *Jabos* streaked low over Oberammergau. They could not have arrived at a time more propitious for von Braun and Steinhoff. Still, Kummer felt that the plan could not work because of the transportation problems involved. He simply did not have enough fuel for the few vehicles available to be moving such a large number of people about.

Steinhoff had a ready answer: There was plenty of alcohol on hand and it would provide a fuel capable of doing the job. (Practically all the automobiles and trucks of the unit were using alcohol at the time.) At that moment, the *Jabos* screamed back over the village again.

Kummer was convinced.

As a result, 400 key men of Peenemünde were dispersed in 25 small hamlets and villages about Oberammergau. They were billeted in varying sized groups, under guard; but von Braun was happy. At least all of Kammler's eggs were not in one basket.

Von Braun, his brother Magnus, Kurt Lindner, and von Braun's secretary Hannelore Bannasch, found quarters in Weilheim, 32 kilometers south of Oberammergau. Once installed, von Braun came close to total collapse. For the past several weeks, he had been traveling almost constantly, getting little sleep or rest, and working too long. His broken arm was giving him trouble, and he knew he must tend to it or run the risk of having it amputated.

There was a hospital at Sonthofen, 70 kilometers to the southwest across the rugged Allgau Mountains that had an excellent orthopaedic surgeon; and so von Braun made yet another tortuous journey by car. The surgeon removed the cast and reset the bones in the arm, without the benefit of anaesthesia since it was in short supply and used only for what the doctors felt were the most serious cases. Von Braun was rigged into traction with the injunction not to move at all. He was in this condition when a flight of *Jabos* began bombing and strafing Sonthofen. Several bombs fell near the hospital, and the most serious cases were rushed to the safety of the basement. They did not include von Braun, who lay trussed up, trying not to move as bombs rained about him. Fortunately, the hospital was spared during the raid.

For several days, he remained in bed, getting the rest he so badly needed. But he also had time to worry about what was happening to his teammates at Oberammergau as well as his own fate. Lying helpless in a hospital, von Braun could be scooped up at any moment by the SS and hauled away as hostage or disposed of so that he could not be captured by the French said to be closing in on Sonthofen at the very moment.

On April 6, *Arbeitsstab Dornberger* had moved from Bad Sachsa into Bavaria, as had a few other members of the *Elektromechanische Werke* staff. Dornberger and some hundred of his military personnel settled in Haus Ingeburg, a ski resort inn at Oberjoch. On hearing that von Braun was in the hospital at Sonthofen and that the town was being bombed, Dornberger sent an ambulance and driver to fetch him to Oberjoch before he could return to Weilheim.

Thus, the move to Oberammergau ended any possibility of additional rocket development.

The design office moved into the village of Kinsau, and its members sat in the summer sun speculating on when and what would happen when the *Amis* (Americans) arrived. Logically, they began speculating about what they could do in Germany, a nation certainly without the need for rocket engineers. Someone suggested that there was a great quantity of aluminum on hand from V-2s as well as from German planes that had never made it into the air and Allied planes shot down in fields and cities (the metal of which actually reappeared as V-2s). There would be an obvious market for cooking utensils after the war.

Willi Mrazek and his fellow engineers in the hamlet of Hönfurch, near Schongau, were not so industrious or forward-looking. They engaged in day-long games of skat, flipping the cards without real enthusiasm as they waited for the inevitable arrival of the *Amis*.

One member of von Braun's team seemed unable to accept the idea of joining the *Amis*, or at least working anywhere except in and for Germany. Dr. Hermann Steuding was the brilliant mathematician and chief of the flight dynamics division of Steinhoff's guidance, control, and telemetry department at Peenemünde. In those early days of guided missiles research, before such tools as high-speed electronic digital computers were available, the mathematical analysis of control systems was a tedious and time-consuming process; it was Steuding who headed this effort on the V-2 and *Wasserfall*. After a week or so of idleness in the hamlet of Böbing, Steuding became depressed and morose. He had apparently been brooding over his nation's future. A bachelor in his early sixties, he had been born in Russia. However, because of his high idealism (which had led him to join the Nazi Party, though he later became disillusioned with its excesses), he could not accept the prospect of being invited or forced to continue his work in another country, in either the East or the West. Such an action to him would be treasonable.

One day, he gave away his most prized possession, his slide rule, and distributed his other personal effects to friends. Then he packed a kit bag with what little he had brought from Bleicherode. His close friend and colleague Dr. Rudolf Hölker, also a flight dynamicist, walked with him for an hour or so among the firs and pines. The two said practically nothing. Suddenly, Steuding stopped and took his kit bag from Hölker, saying, "Now, I must say good-bye." He went on alone.

No one ever saw or heard from him again. While his friends assume that he committed suicide, they had no conclusive evidence. The man had simply disappeared.

With the end of the war at midnight on May 8, Wernher von Braun was in the Allied detention camp at Garmisch-Partenkirchen. Other members of the Peenemünde team were scattered about the country, and several found themselves denounced by their fellow countrymen.

Hermann Weidner, who had been in charge of the liquid-oxygen plant at Redl-Zipf and had remained there, was pointed out by an inmate of the concentration camp which had supplied laborers for the plant as a "war criminal." He was quickly taken in tow by the US Army and sent to an interrogation camp. While awaiting the worst, an inmate of the camp, who had worked for Weidner, came in.

"Hermann, what are you doing here?" he inquired.

"The man's a Nazi war criminal," a soldier replied.

"Nonsense! That's Hermann Weidner; I worked for him. He's a good fellow," Weidner's former employee responded.

Weidner was then cleared and permitted to rejoin his colleagues.

At Leutenberg, Walter Wiesman received rough treatment at the hands of the American soldiers who captured the village and its castle-cum-rocket-valve-laboratory. With Walter Riedel, who had been deputy to von Braun as chief of research and development at *Elektromechanische Werke*, both were pointed out as developers of a bacteriological bomb by two German movie actors, who had taken refuge in the village when the war forced them to leave Berlin. To back up their accusation, the actors produced a drawing of the bomb, which in reality was a drawing of a *Wasserfall* valve. To a nontechnical and highly imaginative eye, it could well have resembled a bacteriological bomb, even though none existed at the time.

The two men were taken to the castle from their room in the village for interrogation by soldiers who knew no more about such bombs than the two Germans. Riedel became short-tempered, and it cost him a couple of clouts. The Americans were in no mood to listen to cock-and-bull stories about valves for a rocket that would shoot down aircraft. They knew a bacteriological bomb when they saw one, or at least a drawing of one. All afternoon, the questioning continued, and the two began to fear for their lives. Finally, wiser heads arrived, and Wiesman was released. Riedel was removed to a jail in the nearby town of Saalfeld.

Perhaps one of the more sorely put upon of the team was Werner Gengelbach, an electrical engineer in charge of ground control electronics at Peenemünde. He was detained by US Counter-Intelligence Corps agents who put him in the small jail at Bleicherode, explaining that only people with political records were there.

The next day he was removed to a jail in Bad Sachsa, where he was incarcerated. The American officer in charge removed Gengelbach's sunglasses, pointing out he would not be needing them in the cellar room that night. On the following day, he was taken to the former concentration camp near Nordhausen for the political and criminal prisoners, who were now the jailers for inmates largely made up of high-ranking Nazi Party members and SS officers. The SS officers were still in their uniforms but lacking their highly polished black boots, which were on the feet of their guards and erstwhile charges. The officers, in turn, wore the wooden clogs that formerly had shod their present guards. Gengelbach spent some five days or so here, then was sent on to yet another prison camp at Naumburg, some 35 kilometers southwest of Leipzig.

It contained between thirty and forty thousand prisoners. In the center of the camp was a camp within a camp. Behind a barbed-wire compound milled a group of some eight thousand "political prisoners." Gengelbach later explained to the authors:

> There was no shelter—nothing. Whatever we had on provided our only protection against the rays of the sun, rain, and the cool breezes at night. We were fed only once a day but were issued no eating utensils. Within a day, however, we all had managed to come by something that would hold food. We had to keep moving constantly at night in order to stay warm. . . .

After several miserable days Gengelbach was packed into a truck that was one of several in a convoy. On the first curve in the road away from the camp, the body of one truck gave way because of the excess mass of the tightly packed prisoners. They were thrown into the road and off it as well, killing several. The remainder of the convoy continued to Counter-Intelligence Camp No. 95 in Ziegelheim, fifty kilometers south of Leipzig.

As they were processed into the camp, the men were told to check in their money, watches, rings, cigarette lighters, etc. Fearing—correctly, as later events proved—that such items would not be returned, Gengelbach slipped his wedding ring from his finger into his mouth and thus saved it. In Camp No. 95, he found himself crammed into a barrack with eighteen other prisoners. A democratically elected leader was in charge of maintaining discipline and distributing food (four loaves of bread for twenty men, soup made from dehydrated vegetables and supplemented by weeds gathered in the compound.) Escape, had there been any motivation for it, was impossible, as Gengelbach later explained to the authors:

> There was no way to establish an escape route from this camp. It was surrounded by barbed wire, watch towers, searchlights; and the guards would immediately open fire—even if members of a family approached the fence to talk to a prisoner.

Since he was an electrical engineer, Gengelbach earned cigarettes and chocolate for repairing faulty radios of his guards. He also made a deal with a sympathetic American mess sergeant to allow him to visit the guards' mess hall and gather leftover food, which he took back to the barracks in his pockets and distributed to his fellow prisoners.

One day, the public address system blared out that any prisoner who had worked on "secret projects" during the war should make himself known. Gengelbach was one of the few who approached the guard on the gate. He was taken to a Lieutenant Langer, who spoke perfect German, having been born and brought up in Germany before emigrating to the United States several years before the war. The rocket engineer told him his story and was returned to his barracks. A fortnight later, Gengelbach was released from the camp. With no more explanation of the charges against him than Franz Kafka's Joseph K. received in his novel *The Trial*, Gengelbach joined his Peenemünde teammates at Garmisch-Partenkirchen.

For the most part, the end for the team came surprisingly peacefully. The US Army had decided to bypass Oberammergau rather than capture it. When the war ended, it found the men of Peenemünde sitting leisurely in the shadow of the beautiful mountain Zugspitze, with their faith not so much in the Lord as in Wernher von Braun.

The Heidelager camp, Blizna, Poland. Messhall, left; latrines, right.

As part of an Allied mission to inspect German secret weapon test installations in Poland, a team looks at a recovered German V-2 combustion chamber along the bank of the river Bug. Standing: British Captain S. Masterson, Colonel T. R. B. Sanders bending to the right with trousers rolled up.

Lineup of Meilerwagen V-2 rocket trailers in the Haagsche Bosch, The Hague, Holland, one of the few pieces of evidence of operational launch sites.

Residents of Antwerp work at clearing away the destruction caused by the explosion of a German V-2 on November 27, 1944, at Lange Nieux and Eikensh Straats, a residential section.

The scene at one of the balloon defense sites where a flying bomb was brought down by a balloon cable. The bomb fell just behind the small farm building on the right.

SS General Hans Kammler arriving at the V-1 site at Saleux on August 10, 1944.

Air Chief Marshal Sir Roderic Hill, commander of the defense of Great Britain from 1943 to 1945 and commander-in-chief of Fighter Command.

A V-1 in flight over England.

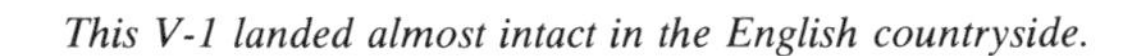

This V-1 landed almost intact in the English countryside.

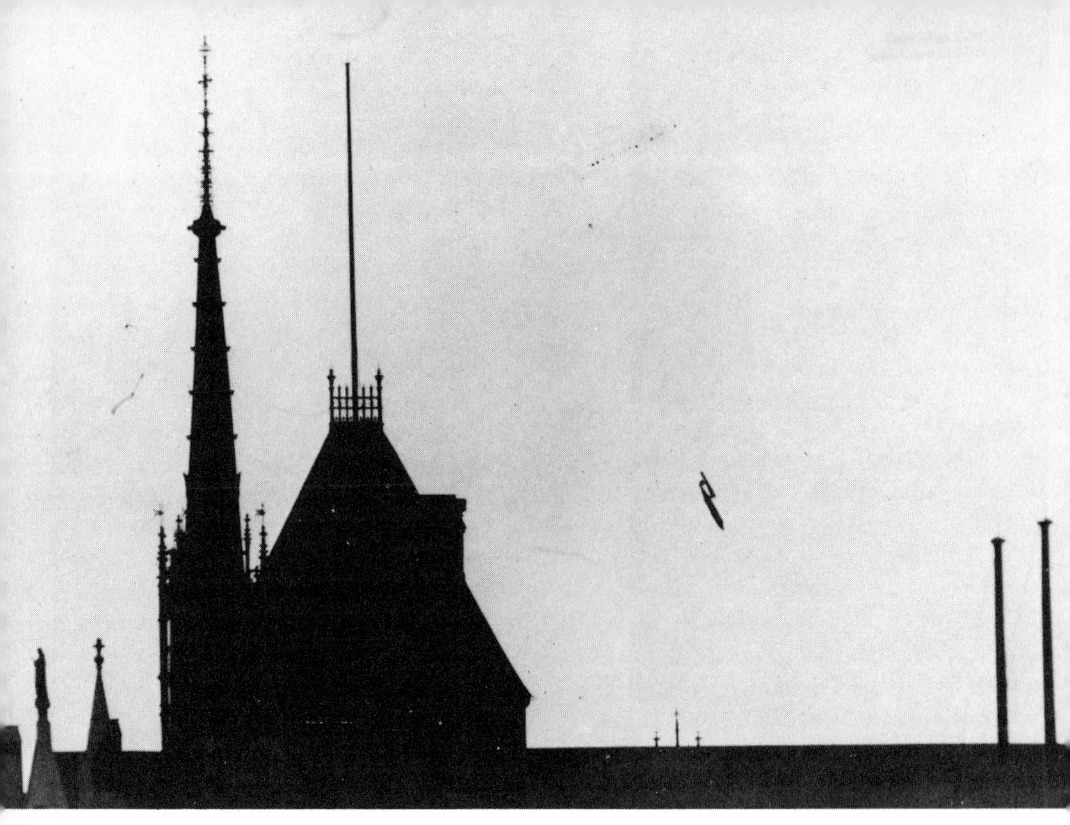

V-1 just before impacting on London.

V-2 rockets shot up by RAF flyers and later captured by advancing Allied forces (April 16, 1945).

Entrance to mine at Dörnten, where secret Peenemünde archives were hidden.

A grim-faced Hitler inspects bomb damage in Germany in 1944. Almost to the end he pinned exaggerated hopes on his V-weapons to retaliate for this aerial punishment.

A V-2 removed from the Mittelwerk en route through Brussels to Antwerp (May 31, 1945), where it was to be shipped to the United States.

Winston Churchill talking to General Eisenhower at 16th American Corps Headquarters, Germany. March 25, 1945. Both were highly relieved as the V-weapon campaign drew to a close.

Dependents' housing for Paperclip personnel at Camp Overcast in Landshut, Germany, 1945.

CHAPTER 14

Roundup in Bavaria

As World War II approached its end, the greater number of the Peenemünde team waited, some apathetically, some apprehensively, in the hamlets and villages around Oberammergau for the arrival of the Americans. Von Braun, of course, was not with them since he was then still in Haus Ingeburg, the ski resort outside Oberjoch. The first Americans arrived in Oberammergau on May 1, driving the ubiquitous jeeps. If the Germans were ill at ease, they were no more so than the Americans. Practically to a man, these were civilian employees of the US Navy clad in the green fatigue uniform of the US Army! The men were members of the Naval Technical Mission in Europe under the command of Commodore H. A. Schade. They had gingerly set foot on the beaches of Normandy following the first waves of American infantry on D-day, June 6, 1944, and they had dogged their heels ever since.

Curiously enough, the Navy men were not concerned with the Peenemünde rocket engineers. But they were interested in Professor Dr. Herbert Wagner, chief designer of guided missiles for the Henschel Aircraft Company, who was also rumored to be in the neighborhood, as indeed, he was. The navy was very keen to locate him and his engineering team because it was especially intrigued with the Hs293 missile Henschel had developed. The US Navy felt that it could be used to good advantage against the Japanese in the war in the Pacific.

The questioning of the team members at Garmisch-Partenkirchen, later in the first week of May, began logically enough with Dornberger and von Braun. It was to a large degree organized and conducted by Dr.

Richard W. Porter, of the General Electric Company, which was under contract to the US Army Ordnance Department to research and develop long-range guided missiles under a project known as Project Hermes. However, Porter was not concerned solely with the Peenemünde group. For most of April, he had been interrogating the faculty of the University at Heidelberg and the Technical University at Darmstadt concerning research in the general field of guided missiles. For the interrogation of the Germans at hand, Porter said to his assistant, Dr. Herman A. Liebhafsky:

> Let's start with General Dornberger. He had overall responsibility for the German army rocket program and can give us a broad base and reference for the more detailed interrogations which will follow. I'll handle the preliminary discussions with Dr. von Braun. Also, I think von Braun is preparing a general summary and several technical papers for our review and to help expedite the overall interrogation.

The paper that von Braun was preparing had a formidable title, even in German: *"Übersicht" über die bisherige Entwicklung der Flüssigskeitsrakete in Deutschland und deren Zukunftsaussichten* ("Survey" of Previous Liquid Rocket Development in Germany and Future Prospects). This thesis began:

> We consider the A4 stratospheric rocket developed by us (known to the public as V-2) as an intermediate solution conditioned by this war, a solution which has certain inherent shortcomings, and which compares with the future possibilities in this field roughly in the same way as a bomber plane of the last war [i.e., World War I] compares with a modern bomber or large passenger plane. We are convinced that a complete mastery of the art of rockets will change conditions in the world in much the same way as did the mastery of aeronautics, and that this change will apply both to the civilian and military aspects of their use. We know, on the other hand, from our past experience that a complete mastery of the art is only possible if large sums of money are expended on its development and that setbacks and sacrifices will occur, such as was the case in the development of aircraft.

The survey went on to describe the various rockets developed at Peenemünde as well as those planned. It ended on a note of prophecy:

> When the art of rockets is developed further, it will be possible to go to other planets, first of all to the Moon. The scientific importance of such trips is obvious. In this connection, we see possibilities in the combination of the work done all over the world in connection with the harnessing of atomic energy together with the development of

> rockets, the consequence of which cannot yet be predicted. . . . A prophecy regarding the development of aviation, made in 1895, and covering the following 50 years, with the corresponding actual events, would have appeared at least as fantastic then as does the present forecast of the possibilities of rocket development. . . .

With exception of the Project Hermes team, the questioning was done for the most part by people with little or no appreciation of what their captives really knew. The Germans were amazed that only a few of the so-called experts were technically capable of asking intelligent or relevant questions. Porter's men, the US Army Ordnance Department representatives on Combined Intelligence Operations Subcommittee (CIOS) Team 183, were clearly among the most technically qualified even though their knowledge was not nearly as deep as that of the Germans. The CIOS had been established in the summer of the preceding year by the British-American Chiefs of Staff to plan and administer in an organized way the exploitation of German scientific targets. Several teams of Allied military technicians and civilian scientists were formed to investigate special areas such as artillery, rockets, ammunition, explosives, chemicals, etc. The degree to which CIOS succeeded is reflected by the fact that its acronym soon became known colloquially as CHAOS.

It also soon became apparent to von Braun and Dornberger that, despite Porter's attempts to approach the interrogation in an orderly fashion, no one person was actually orchestrating it. On May 22, for example, Dr. Fritz Zwicky, an internationally renowned astronomer and physicist of the California Institute of Technology and US Army Air Forces representative on CIOS Team 183, questioned von Braun for several hours on various jet motors, the V-2 oxygen vent valve, and the *Taifun* rocket. This interview took place only a week or so after practically an identical quizzing by Lieutenant Colonel G. J. Collin and Flight Lieutenant H. M. Stokes, British members of the CIOS Team. During the first week in June, von Braun was again interviewed by Porter on pretty much the same subjects.

This disorganization was emphasized by Zwicky in an official report he made to the US Army Air Corps five months later: "There were too many technical teams, both British and American, the members of which conducted interviews without any coordination with others and with little regard to what had previously been done." CHAOS had prevailed and CIOS had failed.

The situation played into the hands of the Germans, who had already agreed among themselves not to be too chatty about what they knew. They had decided to hold back until they got a reasonable offer to continue their work before they told all.

Some of the interrogation verged on the absurd.

Bernhard Tessmann, who a month earlier had helped hide the Peenemünde archives, was questioned by an American major who sought to impress him by waving a loaded .45 caliber automatic pistol in his face. However, Tessmann refused to be intimidated, and simply said he would not speak until the pistol was put away. Seeing what was going on, a friendly chap named Carlson, in the uniform of a lieutenant colonel, came up and spoke to Tessmann in impeccable German, apologizing for the major, whom he then dressed down in flawless American army idiom.

Other questions were as naive as the replies were ironic or cynical.

> "Did you ever try to kill Hitler?"
> "Did you ever try to kill Roosevelt?"
>
> "Suppose you were working for us, and we had another war with Germany."
> "Not likely in my lifetime."
> "Suppose it happened, though."
> "Then I am still a German."
> "You'll never make it to America talking like that!"

Not all of the Germans were enthusiastic about going to the USA or even for working with their former enemies in Europe. One engineer pointed out to Porter: "We despise the French; we are mortally afraid of the Soviets; we do not believe the British can afford us, so that leaves the Americans."

By the end of May, almost half of those at Garmisch-Partenkirchen were released to return to their homes. Preliminary negotiations on contracts for work in the US also started about the same time. These, however, got nowhere.

Von Braun later summarized the situation to the authors: "Our primary concern was 'stability' and continuity. We were interested in continuing our work, *not* just being squeezed like a lemon and then discarded. A three-year contract, we felt, would reflect long-range intentions on the part of the US, but such a contract could not be attained. Higher-ups obviously wanted to see how things would work out with the Germans in the US."

The team members at Garmisch-Partenkirchen had been told that their families were in Witzenhausen and Eschwege. To check the veracity of the report, they nominated several members to sneak out of the compound and go to those villages. However, the men so designated simply picked up their families and disappeared. For the second attempt, the team chose a bachelor who returned with the news that the families were indeed so located. Relations with the Americans improved considerably after that.

To pass the time away, the team at Garmisch-Partenkirchen organized and attended informal lectures on everything from "The Fundamentals of Gyroscopes" to "How to Play Chess."

To relieve some of the tedium, Magnus von Braun turned impresario. With the help of a former Peenemünde soldier, who had been a professional composer and orchestra conductor before the war, Magnus adapted Wilde's *The Importance of Being Earnest* to a musical show. In addition to providing the lyrics, Magnus took the leading male role. Some three or four performances were given to an audience that included British and American military personnel as well as the Peenemünde team. After one performance, a British officer paid Magnus a compliment by saying that "Oscar would have enjoyed it." The success of the play led Magnus and his collaborator to follow it with another show.

That the first Americans to arrive in Oberammergau had been sailors rather than soldiers is illustrative of the confusion and lack of organization about what should be done with the likes of von Braun and Wagner. Detailed planning, even broad policy, on what to do with the enemy's scientific research and development facilities as well as its industrial base simply did not occupy the British and the Americans to the extent it did the Soviets. The American attitude was that reparations should be made, of course, but within reason. President Harry S. Truman felt that Germany should be left without industrial potential for war but with enough of an overall industrial base so that the nation would not wind up on the dole, as it had following World War I. Truman clearly foresaw from where the charity for such a Germany would have to come.

Earlier, at Yalta, Truman's predecessor, Franklin D. Roosevelt, had dealt with an adamant Stalin, who demanded nothing less than $20 billion in the way of reparations. It would take the form of forced labor, industrial tooling and equipment, and stockpiles of current production. Half of this sum was to go to the USSR. Roosevelt agreed to the sum only as a talking point. When the Allied leaders met in Potsdam, only Stalin was among the original of the Big Three left. He had to deal then with Truman and Clement Attlee. And again, he insisted on the figure of $20 billion. Truman then made his own position clear.

As the argument continued, it became obvious that the Russians were not going to yield and that they had indeed already started their own reparations. In June, following their occupation of Berlin less than a month earlier, an order from Moscow to the head of Soviet forces in that city commanded him to "prepare" the western zone of Berlin for occupation by the British, French, and Americans. Immediately a shock troop of engineers and technicians of the *Vsesoyuznyi institut aviatsionnykh materialov* (All-union Institute of Aviation Materials), dressed in Soviet Army uniforms, descended on the Kaiser Wilhelm Institute in Dahlem.

A Soviet member of the VIAM who took part in the "preparation" later wrote:

> It is difficult to portray the scene presented by that world-renowned institute. First occupied by the German population, then by General Konev's troops, which were quartered on the premises, the institute looked like a field devastated by locusts. There were piles of rubbish, broken windows, and empty drawers pulled out of the tables and dressers all over the floor. As in other places which had been dismantled, valuable instruments had been thrown together in a heap with broken and smashed equipment. . . . What remained of the equipment was promptly thrown on trucks and transported to Adlershof [an airfield], which was in the sector assigned to the Russians, where it was carted up by Germans and repatriates. . . . The three Soviet scientific research institutes . . . which took part in the dismantling of these two objects [i.e., the institute as a whole] received not more than 25 percent of the equipment in good order, and even that was incomplete. The remainder was either ruined or reduced to scrap.

On July 26, Edwin Pauley, US Commissioner on the Allied War Reparations Commission, protested that the Russians were stripping plants that had nothing to do with the manufacture of war matériel.

"What we saw amounts to organized vandalism," Pauley complained to US Secretary of State James F. Byrnes, "directed not alone against Germany, but against the US Forces of occupation. . . . In the area which we captured and turned over to the Russians, we made no removals except for a few samples of unique equipment."

In his authoritative study *Project Paperclip: German Scientists and the Cold War*, Dr. Clarence G. Lasby points out such looting was not restricted to the Russians:

> Several days later Pauley learned of the evacuation of virtually all of the equipment from the German property owned by such companies as Anaconda Copper, IBM, American Radiator, Gillette Safety Razor, Ford Motor, National Cash Register, General Electric, and Paramount Pictures.

Late in 1944, British and American military organizations had collaborated on Operation Eclipse, a plan for implementing Project Safehaven, conceived by the US Department of State as a means of controlling German technical personnel who might attempt, after the war, to revive Germany's war potential by subversive activities in foreign nations. However, these were largely abstract plans and had no counterpart at SHAEF, which on May 15, 1945, sent a secret cable to the War Department in Washington asking for policy on controlling German science and research personnel and their facilities. Predictably, the cable provoked a "staff study of the problem," a proven means in the army for delaying decisions. While the War Department studied the problem of what to do with the facilities, an idea of what should be done with the personnel occurred to officers of the US Army and the US Army Air Forces in Europe.

Major General Hugh J. Knerr, Deputy Commanding General for Administration, Headquarters US Strategic Air Forces, on June 1, 1945, recommended to Lieutenant General Carl Spaatz, the commanding general, that the US Army Air Forces make full use of captured German Air Force research facilities and their personnel. Knerr also had earlier discussed the subject with Robert A. Lovett, US Assistant Secretary of War (for Air), when he visited the European Theater of Operations in April.

Knerr's plan was at once imaginative and bold, thus suspect by higher command. It consisted of nothing less than transporting enemy scientists, along with their families, to the USA, where they would be kept under strict surveillance and guard. A close scrutiny would determine eventually who of the men were most professionally qualified, hence desirable, and who were obviously no political threat. Those who passed scrutiny on all counts would be offered citizenship, while the rest would be returned to Germany. In the process, of course, all candidates were to be exploited to the fullest.

Early in May, just after the war ended, other military elements in the US became interested in the idea of exploiting German scientists and engineers. On May 14, the Director of Intelligence for the US Army Service Forces sent a memorandum to the Assistant Chief of Staff, G-2 [Intelligence], of the War Department General Staff. In it, he pointed out: "General Somervell and several of the Technical Services are desirous of transporting selected German scientists to the United States to assist in research and development agencies of the Army Service Forces in perfecting types of equipment to be employed in the war against Japan." A list of names of specific Germans desired was appended to the memorandum. Several days later, Somervell sent a similar letter to the Chief of Staff of the US Army.

While this activity was going on within the US Army Air Forces in Europe and the US Army Service Forces in the US, a brash but dedicated young ordnance officer appeared on the scene in the European theater of Operations. He was Major Robert B. Staver, who had arrived in February under orders from Colonel Gervais Trichel, chief of the Rocket Branch in the Ordnance Department at the Pentagon. Staver was assigned as chief of the Jet Propulsion Section of the Research and Intelligence Branch of the US Army Ordnance Technical Division in London.

One of the tasks given him by Trichel was to direct the Project Hermes team under Porter to the most promising rocket research targets in Germany. He also compiled a "black list" of important German rocket engineers and scientists, based largely upon intelligence supplied him by the British. At the head of the list he put the name Professor Dr. Wernher von Braun.

Obviously a figure rather low in the Ordnance hierarchy, Staver found himself at a further disadvantage. Colonel Horace B. Quinn, his immediate superior, did not share Staver's enthusiasm for rockets: "I don't care if the Russians get all of those Krauts. I say good riddance." He reassigned Staver's officers—a captain and three lieutenants—to tasks he considered of more importance than V-2s and V-1s.

Shortly after the word of the capture of *Mittelwerk* reached Paris on April 11, Colonel Holger N. Toftoy, who was chief of US Army Ordnance Technical Intelligence in Europe, organized an ad hoc unit called Special Mission V-2. Its objective sounded simple enough: "Go to Niedersachswerfen [*Mittelwerk*] and get up enough V-2 components to make up 100 complete rounds. Ship them to the US." Less than a month previously, Toftoy had received a request from Trichel for one hundred *operational* V-2s for firing at the newly established White Sands Proving Ground in New Mexico, an experience he felt would further the objective of Project Hermes. However, completely assembled and ready-to-fire V-2s did not exist in such quantities. The missile was made to be launched as soon as possible after being manufactured. There were no magazines or warehouses of stored missiles. Hence, the order had to be filled with components that could be put together in New Mexico. Staver was made available by Trichel to assist in the project, particularly in locating documents and personnel.

Two weeks later, to Toftoy's surprise and dismay, he learned that Niedersachswerfen and Nordhausen were to be turned over to the Soviet Army at an unspecified but quite early date (soon set for June 1). The loss of this valuable facility was assured. In November 1944, the European Advisory Commission, consisting of representatives of the US, Great Britain, and the USSR, had issued a decree that was to become effective upon the formal surrender of Germany. It stated, in part, that "all factories, plants, shops, research institutions, laboratories, testing stations, technical data, plans, drawings, and inventions" would be held intact and in good condition until the arrival of Allied personnel designated to dispose of them.

Niedersachswerfen certainly was such a plant, and there was no doubt that the V-2 was an invention.

And yet another stumbling block lay in the way.

There was a gentlemen's agreement in the European Theater of Operations between the naive Americans and the clever British. It concerned the disposition of captured enemy weapons of a new type, such as the V-2. If two were so obtained, one would go to the British and one would go to the Americans. However, if only one were picked up, it would go to the British, who would study it and share their findings with the Americans. Further to frustrate the plan was the fact that an order from SHAEF stated that no captured matériel would be removed by one ally out of a zone destined to be occupied by another after the war.

There had been for many years an axiom in the US Army, and presumably in those of other nations as well, that in any operation 10 percent of those involved never get the word. Thus, officers at the level of Toftoy were unaware of the various agreements existing at higher military and diplomatic levels, especially those concerning the exploitation of scientific personnel and the disposition of war matériel of an advanced technological nature. Toftoy, being a West Pointer, was an officer who would never have considered "bucking the system." He acted only within the framework of orders, guidelines, and policy. In the absence of these, he acted decisively on his own, as he had been trained to do. Special Mission V-2, then, was his solution to a problem in which he had to act quickly without sufficient direction from "above."

Special Mission V-2 was commanded by Major William Bromley, a close friend and college classmate of Staver's. Bromley, for some reason, had been ordered to report to Toftoy through Major James P. Hamill, designated by Toftoy to coordinate the transportation of V-2 components from Nordhausen to Antwerp, the port from where they would be shipped to New Orleans.

To accomplish its task, Special Mission V-2 had at its disposition two Army units: one, the 144th Ordnance Motor Vehicle Assembly Company, then in Cherbourg, 770 miles distant, had to move from there to Nordhausen post haste; Company B, 47th Armored Infantry Battalion, 5th Armored Infantry Division, was already on the scene and surrounded the plant, preventing entry to anyone not cleared by Bromley.

Because of the orders and policies mentioned above, Bromley could not rely upon the other technical services of the army. The Transportation Corps, for example, refused to have anything to do with an operation that could perpetrate an "international incident" among the Allies. Thus, he had to rely largely on his colleagues in the Ordnance Department. "For this reason," Hamill later remembered, "the Ordnance Department was required to run its own railroad during this period."

Special Mission V-2 operated under the constant pressure of the Russian takeover on June 1 (as things transpired it was July before they arrived).

The group's greatest deterrent was the lack of a parts list for an assembled V-2. None of the German personnel available knew, in detail, what was required. Thus, it was largely a matter of picking up one hundred of everything in sight. A zealous second lieutenant, M. S. Hochmuth, rounded up mountains of fiberglass (the insulation for the liquid-oxygen tank of the V-2 rocket), which was crated and sent to America. The Germans who later inventoried these components in New Mexico were mystified by the enormous amount of a substance so easily obtainable in nearby El Paso and the almost complete lack of the crucial carbon jet vanes. There was a trainload of some fifty V-2s in the rail marshalling yard at Nordhausen that had been severely damaged by American bombers. Nevertheless, Hochmuth managed to remove some undamaged parts of the critical guidance sections and rudder actuators, leaving the burnt remains for the Russians.

Somehow, enough components were collected to build the hundred V-2s. The 144th Ordnance Company got them together and moved them to the marshaling yard at Nordhausen, where there were more than enough German freight cars in perfect condition to carry them. However, to Hamill's dismay, he found that the Army Air Force had bombed all of the railway bridges leading into and out of Nordhausen except one. As a result, freight cars tended to congregate since there was no other place for them to go. Then, on May 19, near disaster struck from an unexpected quarter. The US Army Transportation Corps decided to impound all German freight cars and move them into the safety of the newly established American zone.

Hamill was ready to throw in the sponge. He explained the apparently hopeless situation to his highly resourceful Technician Fifth Class Bob Payne and went to bed in despair. On the following morning, Transportation Corps officers arrived to remove the cars only to find that the remaining railway bridge had been destroyed by explosive charges during the night. Counter-intelligence Corps agents were willing to believe that the demolition was the work of the vaunted Werewolves (nonexistent German guerrillas), but Hamill never was convinced that the providential destruction was not technically a case of sabotage by a soldier of the US Army. The bridge was not totally destroyed, only partially so. It did not take the 186th Combat Engineer Battalion long to repair the damage, as well as to place still another bridge into usable shape. In the interim, however, Hamill managed to convince the Transportation Corps of his urgent need for the cars, and they were left at his disposal.

Meanwhile, Staver and Hochmuth searched the neighborhood for fugitive engineers and scientists of the Niedersachswerfen plant and hidden treasures as well. In a network of artificial caves near Wofleberg, two miles northwest of *Mittelwerk*, Staver found an almost complete *Schmetterling* antiaircraft missile and the guidance and control unit of an

Hs298 missile. At Friedrichslohra, hidden in a pigsty, Hochmuth discovered a trove of phototheodolites, kinetheodolites, and other precision optical instruments for tracking guided missiles. Additionally, he turned up several pieces of V-2 equipment in a forest not far away. Together with Staver, he uncovered a complete guidance unit for a *Wasserfall* missile cached in a barn.

Being a lowly lieutenant had disadvantages; Hochmuth found himself rousted out of bed at two o'clock one morning by a weary RAF wing commander from SHAEF, who demanded that he have Hochmuth's bed! Hochmuth remonstrated, but to no avail; rank had its privileges, as every officer knew. On the next day, he escorted the wing commander on a tour of *Mittelwerk*, but refused to allow him to remove any matériel. Having seen what technological treasures were in store at Niedersachswerfen, the wing commander left for Frankfurt full of ideas for "securing" them for the Allies, i.e., the British.

On May 12, Staver, following an intelligence tip, located Karl Otto Fleischer, former general business manager of *Elektromechanische Werke*, who was in a mood to cooperate with the Americans. His most valuable knowledge was the location of the Peenemünde documents, buried earlier by Huzel and Tessmann. However, he did not immediately reveal this information. Fleischer did, though, introduce Staver to a man who was probably the second most important of the Peenemünde team. He was Eberhard Rees, former director of Prototype Production. Several days later, the busy Staver found another member of the team who would prove to be of great assistance with Fleischer and Rees. It was Walter Riedel, who had earlier been jailed at Saalfeld as a developer of a bacteriological bomb. Staver secured his release and removed him to Nordhausen to help in rounding up still other members of the team.

On May 18, confusion increased considerably.

Dr. Howard P. Robertson, a civilian assigned to the Field Intelligence Agency, Technical (FIAT) (yet another of the many Allied intelligence units running amok in the theater at the time), arrived in Nordhausen to inform Staver that he was there to pick up Rees and Fleischer, as well as others, and take them to Garmisch-Partenkirchen. On the same day, the 144th Ordnance Motor Vehicle Assembly Company clattered into town after a long, tedious motor journey from Cherbourg. Staver managed to convince Robertson that it was out of the question for Rees and Fleischer to be removed to Garmisch-Partenkirchen for interrogation; indeed, he desperately needed von Braun *from* Garmisch-Partenkirchen.

Somehow, he found time to sit down and write a letter, much as General Knerr had done earlier, to Paris, suggesting that the Peenemünde personnel he was turning up in the area be put under contract and sent to the United States for utilization in the war against Japan.

On May 22, the first train of V-2 components left Nordhausen guarded by soldiers of Company B and driven by German railway personnel. It went as far as Erfurt, where personnel of the US Army Military Railway System took over for the remainder of the trip to Antwerp. Each day for the ensuing nine, a train averaging forty cars moved over the same route. However, the story did not end at the Belgian port. British intelligence agents there spotted the V-2 matériel on the docks and immediately sent word to London of what was afoot. A spirited attempt was made to stop the shipment. The strongest complaints were made to Eisenhower through British officers attached to his staff in Frankfurt; but before any action could be taken, the ships were at sea.

On the same day that the first ship left, Staver learned the exact location of the documents hidden by Huzel and Tessmann on April 3. He did so by a clever ruse. A couple of days earlier, talking with Fleischer in Bleicherode, he casually pulled a notebook from his pocket and read an imaginary entry:

> Von Braun, Steinhoff, and all the others who fled south have been interned at Garmisch. Our intelligence officers have talked to von Poletz [Kammler's intelligence officer], General Dornberger, General Rossmann, and General Kammler. They told us that many of your important drawings and documents were buried underground in a mine somewhere around here, and that you, Fleischer, could help us find them.

As he finished, he watched the German closely. Fleischer was upset by what he had heard and reacted visibly. Staver, not wanting to appear overly eager to find the cache of documents, merely told him to think things over and that he would see him the next day.

On the following day, Staver returned to Bleicherode to pick up Fleischer, only to find he was missing. Staver was told that Fleischer was waiting for him in the nearby hamlet of Haynrode, where, indeed, he was. After a night of soul searching, Fleischer had decided to reveal the location of the Peenemünde archives to Staver, using the face-saving excuse that his superiors had already revealed their existence to the Americans at Garmisch-Partenkirchen. Fleischer then suggested that he and Rees search out the exact location of the documents. Staver was reluctant to allow his two prize Germans to take off on their own, but he agreed because they had been completely cooperative up to that time.

Fleischer and Rees finally located the mine, but the loyal Herr Nebelung refused to admit that it held the priceless documents. It took loud arguments and threats before Fleischer convinced Nebelung that he was the manager of the company that owned the documents. Then they returned to Nordhausen to inform Staver, arriving there at 1:30 AM on May 22.

Now that he knew exactly where the buried treasure lay and the way in which it was sealed into the mine, Staver could make further plans. At 3:00 AM, he tumbled a weary Hochmuth out of bed and set him about making the arrangements to disinter the papers at Dörnten. Staver then drove to the airfield at Kassel and approached the first pilot he saw, who happened to have a P-38 ready to fly to Paris, exactly where Staver wanted to go.

There was only one problem, as the pilot pointed out. The P-38 was a single-seater. By no means daunted, Staver said that he would double up in a cramped space behind the pilot's seat for the two-hour flight and so flew to Paris for an important conference with Toftoy.

At Dörnten, Hochmuth found things well in hand. Before he left, Fleischer had hired three shifts of local miners and put them to work around the clock removing the rubble that filled the shaft. Work came to a halt on the following day, however. Col. William J. R. Cook, British commander of CIOS Team No. 163, dropped in to see what all the activity was about. The British were to take over the Dörnten area on May 28, and they were getting acquainted with the neighborhood—or so they said. Hochmuth led the British to believe that his people were engaged in a survey of natural resources. To convince them, he spent a day having his miners round up samples of low-grade iron ore and put them in the boxes meant for the documents.

In Paris, Staver had two things to accomplish in the plush offices of the Ordnance Office in the Plaza Athénée Hotel. The most pressing one was to secure assistance in removing the Peenemünde documents from Dörnten to a safe place in the US Zone of Occupation. Colonel Joel G. Holmes lent a hand. He assigned two 10-ton semitrailers and ordered them to rendezvous with Staver at Dörnten. Staver's other task was to brief Colonels Quinn, Holmes, and Toftoy on the situation as it was shaping up in the area around Nordhausen and Bleicherode.

He convinced Holmes that the Germans should be utilized in the American war effort and drafted a cable for Holmes' approval to the Chief of Ordnance. The cable, later approved, bore Eisenhower's signature and it read in part:

> Have in custody over 400 top research development personnel of Peenemünde. Developed V-2. Latest development named *Wasserfall*, a 2000 kg flak rocket with 280–300-pound explosives. ... Believe this development would be important for the Pacific war. The research directors believe if the group were taken to the US that after

> one month of adjustment and reorganization and three months of hard work they could reproduce complete drawings of *Wasserfall*.... They are anxious to carry on research in whatever country will give them the opportunity, preferably US, second England, third, France. The thinking of the scientific directors of this group is 25 years ahead of US. ... Recommend that 100 of very best men of this research organization be evacuated to US immediately for purposes of reconstructing complete drawings of *Wasserfall*. Also recommend evacuation of all material, drawings, and documents belonging to this group to aid their work in the US. Immediate action is recommended to prevent loss of whole or part of this group to other interested agencies. Suggest policy and procedure be established for evacuation of other German personnel whose future scientific importance outweighs their present war guilt. Urgently request reply early as possible.

Staver was under no illusion that the *Wasserfall* drawings could be reconstructed within the three months he suggested. He also felt, shrewdly enough, that none of the Germans to whom he had talked had any perceptible "war guilt." However, his primary task in the cable was to sell the project to a group of Stateside officers who were concerned with the war still going on against Japan and who had the preconceived, somewhat naive notion in their minds that all Germans would be guilt-ridden and repentant in the fortnight following the end of the war in Europe.

Early on the morning of May 24, Staver left Paris to return to Nordhausen. Two days later, he learned through Toftoy that "Paris and Washington are both working on the problem of evacuating the German technicians and their families. In the meantime, it is requested that you remove the German technicians and their families to an area under US control."

Thus, Staver had Washington's approval, but the problem of implementation was dropped back into his lap.

He immediately sent a letter off to Quinn asking for his four officers back and for von Braun and other key members of his team to be sent to Nordhausen. Then, he turned his attention to the problem rapidly shaping up at Dörnten. Only one semitrailer had appeared; the other had broken down en route. He obtained six 2½-ton trucks from the 71st Ordnance Heavy Maintenance Company in Nordhausen and started off full speed over the Harz Mountains for the mine where the Germans were still frantically removing rubble and, finally, documents. The vehicles were loaded immediately and turned around for Nordhausen, crossing over into the US Zone of Occupation only two hours before the British put up their roadblocks.

Further to complicate Staver's problems, a party of Soviet Army officers showed up on May 26 to tour *Mittelwerk*. There was nothing to do but let them in, and so he did. Their presence in the area added to his impatience in getting the Peenemünde team out of the area as soon as possible.

On May 31, Bromley was able to report that the objective of Special Mission V-2, shipping the components for one hundred V-2s to the US, had been accomplished. A grateful Toftoy saw to it that Bromley received the Bronze Star medal. However, the pressure did not let up on Staver, whose efforts did not end with those of Special Mission V-2, since he worked for Trichel in Washington. He faced the increasingly stressful task of looking for and sorting out the more important members of the Peenemünde team scattered over Thuringia. While Fleischer, Riedel, and Rees were extremely helpful, Staver needed still more assistance. On June 1, he received an order from Trichel (though signed by Eisenhower), through an apathetic Colonel Quinn, who could not understand all the fuss and bother over what he considered "this Buck Rogers stuff." It asked Staver to select five of his German technicians and send them to the US.

The cable was ambiguous at best. "Note that technicians are required. Colonel Trichel wants people with actual working experience rather than theoretical. Problems of theoretical group not yet answered. Conditions of evacuation of group to United States are cooperation and willingness to assist. Need for approximately six months and amount of pay will probably be reasonable. Believe they will be subject to only nominal amount of restraint. Permission might possibly be obtained to move dependents to American Zone of occupation since better work may result under such circumstances."

On June 2, Staver received another cable from Trichel. It advised him that Dr. Richard Porter and the Project Hermes team, in Garmisch-Partenkirchen, would be available to assist in selecting Germans to be sent to the US.

With these men joining Rees, Fleischer, and Riedel, the pace of identifying and locating Peenemünde people picked up considerably. Despite his awkward plaster cast and painful arm, von Braun pitched in to round up his fellow team members after he arrived in Bleicherode on June 19.

Some three hundred motor vehicles, of all types, were assembled. Each was provided with one man who knew exactly the persons to be contacted and generally where they were. Steinhoff located Dr. Ernst Stuhlinger, his former chief in charge of developing measuring instruments, with twenty of his people as far away as Weimar. In each case, the message was the same to those located: "You have a quarter of an hour to make up your mind—stay here and face the Russians or pack up what you can and go with us." Each was allotted 100 pounds of luggage. The search went on around the clock.

By noon of June 20, only 24 hours before the Russians were scheduled to occupy Thuringia, almost one thousand Peenemündians had gathered in the railyard for transportation west. Some fifty freight cars were waiting, but there was no sign of a locomotive. First Lieutenant George Gross, one of Staver's officers, began to share the mounting apprehensions of the nervously milling Germans: "... I was close to being a mental case waiting at the station. Every time a German would say 'Russki,' I would jump ten feet!" Just as Gross had made the decision to move out in a motor convoy, the locomotive arrived; and the train was on its way to Witzenhausen, some 40 miles to the southwest and just across the river Werra that separated the Soviet and American zones.

With the Peenemünde archives, nearly one thousand Peenemünde engineers and their families, and the components of one hundred V-2s under American control, it would seem that little was left for the British and Soviets. One thing did remain, however. It was a collection of official documents, weighing some 260 pounds, belonging to Dornberger, which he had buried at Bad Sachsa. With less than 12 hours before the Russians were to take over there, Porter and Staver organized a last-minute hunt for them, finally finding them with the aid of a military land-mine detector. The five crates were loaded aboard a truck and sent off at top speed to Kassel.

Ironically, after all the haste in evacuating personnel and papers, there turned out to be little need for it. The American army did not vacate the areas to be turned over to the Soviets until July 1. The Soviets were planning a big parade in Moscow which required the presence of Marshal Zhukov and other top-ranking military leaders in Germany—thus, the postponement.

While Staver and his associates (American and German alike) were so busy in Thuringia, events of some magnitude were taking place in the United States that would later affect the future of the Peenemünde team.

On May 28, Undersecretary of War Robert P. Patterson wrote to the Secretary of the War Department General Staff:

> I strongly favor doing everything possible to utilize fully in the prosecution of the war against Japan all information that can be obtained from Germany or any other source.... It is also my feeling that the information from these men should be obtained to the maximum possible within Germany and only those should be brought here whose particular work requires their presence here. It is assumed that such men will be under strict surveillance while here and that they will be returned to Germany as soon as possible.

On June 30, Staver's entreaties for the evacuation of the Peenemünde team members to the US were granted. On that day Major General Gladeon M. Barnes, chief of the Technical Division, Office of the Chief of Ordnance, in Washington, sent forward for approval a letter to Major General Clayton Bissell, assistant chief of staff, G2 [Intelligence], War Department General Staff. In it he laid out a detailed plan for the transfer of select members of the Peenemünde team to America. It went into matters such as housing, medical care, security, food, and even religious freedom—an item that puzzled the Germans, unfamiliar with this typical American attitude that seemed to rank religious freedom as equal important as food and housing. On July 6, G2 approved the Ordnance plan; and it was further approved by the US and British Joint Chiefs of Staff. On the following day, Bissell informed Eisenhower that "the Chief of Ordnance has directed Col. Toftoy to represent him in the selection of and negotiations with these Germans, of whom there are estimated to be about eighty."

On July 19, 1945, Operation Overcast was officially established by a secret memorandum from the Joint Chiefs of Staff:

> OVERCAST—Project of exploiting German civilian scientists, and its establishment under the Chief, Military Intelligence Service, on an island in Boston Harbor at a camp formerly known as Fort Standish [i.e., Fort Strong].

The point was stressed that the "purpose . . . should be understood to be temporary military exploitation." Thus, Operation Overcast was from the beginning conceived as a purely short-term arrangement. A total of 350 German specialists (only 100 rocket men for Ordnance) asked for by name would be brought to the United States under contract for a period of six months. The contracts could be renewed for an additional half year, and then the men would be returned to Germany. At no point was it planned for their families to join them. Those US military agencies participating in the operation included the US Army Ground and Service Forces and the US Navy.

In practice, Overcast turned out to be anything but a well-organized and coordinated military operation. No attempt was made to produce an overall plan. Various and sundry intelligence teams ran hither and thither tracing down and gathering in "targets" wanted by their sponsors. The Ordnance Department responded quickly. On July 24, Toftoy, who had been appointed to succeed Trichel and had been temporarily in the United States, had himself posted back to Europe to carry out his part in Overcast.

In the first week of August, he journeyed from Paris to Witzenhausen to meet Staver. His purpose was to recruit among the Germans who had recently been evacuated there from the Nordhausen-Bleicherode and Garmisch-Partenkirchen areas. A large group was billeted in Witzenhausen's pre–World War I *Kolonial-Schule*, where Germans preparing to move to that nation's colonies in Africa studied tropical and subtropical agriculture. Its members were crowded into the building, as Toftoy recalled in 1959 for TV news reporter Vincent Hackett:

> It was a two-storey, small, European-type country schoolhouse, having a very small kitchen with one two-burner stove, one small hot-water heater, and a number of ordinary-size schoolrooms. The schoolhouse was arranged as a huge dormitory for some eighty of the Germans, as I recall. Beds were lined up from one wall to the other. Families went to bed one after the other according to the time they usually went to bed. That is, the youngsters and aged dependents first. There was no space between the beds, so that there was in effect one long, continuous bed across the room sleeping as many as five to eight people. A small amount of personal property for the various families was stacked in the corridors.

The Germans were not under extremely tight guard, allowing Hannes Lührsen, von Braun's former fellow prisoner of the Gestapo in Stettin, to sneak away on a mission of mercy. In the scramble to evacuate people from Bleicherode, Hans Palaoro, a colleague, had of necessity left behind his infant son with the boy's grandparents. The boy was ill at the time and could not be moved. Lührsen volunteered to cross back into the Soviet Zone and bring the boy to Witzenhausen. To do so, he concocted a story of traveling across the line of demarcation to retrieve a very important German scientist. The Americans responded, somewhat reluctantly, by providing him with a jeep and a guard. After several hair-raising days behind the "enemy lines," Lührsen returned with the youngster.

At Witzenhausen, however, Toftoy had a difficult task as he, Staver, and Porter sat round the stove in the kitchen of the schoolhouse.

He was trying to get the team to sign contracts that held very little for them. The Germans at first wanted a three-year contract, which was standard for engineers in their country. However, the best that Toftoy could offer the men was one for only six months. Then he had to tell them that their families could not go to the US, at least at that time. Negotiations became protracted because Toftoy was dealing with civilians and not prisoners of war, who presumably could be sent summarily to the US, as were hundreds of thousands during the war. The Germans were not interested so much in the money as in other things, as Toftoy later recalled to Hackett:

> The Germans indicated that [if they stayed in Germany] they could repair bicycles, radios, cut wood, and do other tasks which would provide eggs and other food, or fuel, on a barter basis. Therefore, they did not feel they could leave their dependents behind under such circumstances.

The issue of the families soon became paramount. Steinhoff and several other top men flatly rejected the first contract because no mention was made of dependents. Finally, Toftoy decided:

> The only solution I saw was to set up a camp for dependents in Europe, at which they would be guaranteed by the government the necessities of life; not under charity, but as a part of their pay . . . my team and I reconnoitered the countryside, found three places which might be used as a dependent camp. We then went to General Bull's headquarters. General Bull was in command of the American forces in that area. I was required to take the matter up with his staff at first, and his staff reported back to me that General Bull had stated that he would not be able to set up a dependent camp for the Germans. I persisted, however, and demanded to personally present the case to him and after considerable effort was permitted to outline the facts in the case and request one of three areas, top priority being a former German cavalry barracks [at Landshut]. General Bull, on hearing the story, said, "Oh, that's different. I thought they were just displaced persons." The request was granted; American doctors would check them medically; American commissaries would sell the necessary food; and the dependent camp was set up at Landshut.

Having been assured of care for dependents, the Germans then began haggling on lesser matters. Would they be permitted to mail packages back to Germany? What would be their status in the US? Would they eventually be permitted to apply for citizenship, if they desired? Toftoy told them frankly that he simply did not have answers. The Germans admired his honesty as well as his compassion for the plight of their families, traits which he never relinquished in his dealings with them both in Germany and later in the US.

The salary scale had been set by the War Department and was not negotiable. However, the Germans found it satisfactory. The top pay (i.e., von Braun's) was 3,000 marks ($750) per month. A doctor-professor grade individual received 2,500 marks ($625), while a senior doctor of engineering got 1,400 marks ($350). A junior doctor of engineering or diploma engineer made 1,100 marks ($275), while a bachelor of engineering received only 800 marks ($200). Skilled technicians without degrees (of whom there were very few) made 650 marks ($162). The salaries were to be paid to the families at Landshut in marks. The men in America were to subsist on a special $6 per diem.

The contract negotiations went on for some time, and Staver later reported to the Pentagon: "Contrary to what some Ordnance officers expected, these Germans were not at first anxious to come as a group, and many of them were undecided for several weeks." (The first ones did not actually sign until September 12, just before leaving for the US.) Toftoy took it upon himself to employ 115 rather than the specified 100 because the larger number was necessary for a balanced team.

Two members of the team, not at Witzenhausen, were most eager to sign contracts when offered them. Whatever the terms, they were far better than their current position. Eberhard Spohn and J. G. Tschinkel were members of the *VKN* who were still in uniform when the war ended. Thus they went into a POW camp. The two rocket engineers were finally located working at a coal mine in Belgium in the company of civilian criminals.

The Americans were not the only bidders for the Germans in Witzenhausen and nearby Eschwege.

Just across the river Werra, in the Soviet Zone, others were also interested in the rocket technicians and scientists so tantalizingly close by. On August 15, an engineer name Elmi reported to the American authorities:

> I had been for several days in Russian occupied zone around Bleicherode to pick up my baggage, which had been left there. At this occasion I spoke to an old collaborator, whose name I give for internal use only, in order to prevent him from personal difficulties by the Russians. He told me that the Russians intend to develop a big rocket for a normal range of 3,000 miles and that they are needing specialists with knowledge of the theory of flight mechanics and control equipment. He told me that the Russians set big prices for getting over to Russian area Prof. V. Braun and Dr. Steinhoff.

Former Lieutenant Hochmuth later recalled to the authors: "The Russians at this time were not working directly in the area, but were broadcasting daily and continuously from Radio Leipzig. They were going to collect all Germans and equipment, and everyone connected with Peenemünde should come over to them. They would be well fed; their personal safety would be guaranteed; they would be well paid, etc."

But Hochmuth was mistaken; the Russians *were* working directly in the area.

A secretary who had formerly worked at Peenemünde appeared one day in Witzenhausen after having spent several weeks in Bleicherode under Russian occupation. She warned her colleagues that the Russians had put prices on the heads of von Braun and Steinhoff, and she advised the Germans in the area never to go out alone—the Soviet zone lay just across the river Werra. As a result, the Germans always went about in groups of four or five and even then stayed within shouting distance of one other.

Her warning was not without foundation.

One day, a group of men in American army uniforms entered the schoolhouse in Witzenhausen. They began a friendly conversation with several members of the team and suggested they all go into the village for a few drinks. However, the Germans were suspicious of the English they spoke—it was neither "English" English nor "American" English. The Russians left without captives.

In early September, von Braun, Rees, Steinhoff, and Herbert Axster were flown from Kassel to London for interrogation by the British; Dornberger had arrived earlier from Cuxhaven. In London, they were billeted in a temporary prisoner-of-war camp in Wimbledon and driven daily to Shell-Mex House for interviews by Sir Alwyn Crow, Comptroller of Projectile Development in the Ministry of Supply. For the best part of a week, the Germans were kept busy answering Crow's questions, which were largely technical, concerning guided missiles. However, he wanted also to know if the Germans selected at Garmisch-Partenkirchen for participating in Operation Backfire, the British project to launch V-2s from Cuxhaven in September or October, were the best available and could do the job adequately. After being assured that they were, Crow then asked the Germans to develop a plan for a rocket and guided missile development center. It was to include personnel, organization, requisite laboratories, and testing stands. He also was interested in radar and optical tracking, telemetry, and launching site requirements for such a proposed center. Finally, he asked von Braun and Dornberger to recommend, among the Germans not contracted to the Americans, the best men to form a similar team for England. However, Crow was told by von Braun and Dornberger that they felt these were matters left to individual negotiation between the British and the Germans involved. Sir Alwyn said that he understood and respected this position.

From Witzenhausen and Eschwege, in the latter part of September, the German dependents were moved to Landshut, some 20 miles southeast of Munich. It consisted, eventually, of two areas: the *Kaserne* and the *Piflas*. The former was an old cavalry compound of twenty-eight buildings with a total of 303 rooms; the latter was a group of thirty-two residential buildings with 273 rooms. In all, 570 people were accommodated. An additional 365 lived outside the area but were considered a part of it for administrative purposes. However, they did not receive supplementary rations as did those in Landshut. There were also ninety-four dependents still inside zones other than the American. Regardless of where they were, each dependent had a "letter of protection." It stated that the bearer was exempt from arrest and from seizure of his property.

Such was to be their way of life for the next two years.

The buildings in Landshut were divided into apartments, each of which had a cooking stove and at least one heating stove. Initially, there were problems, such as shortages of furniture and cooking utensils. The quarters were comfortable, if a bit crowded, the wives reported to their husbands in America. During the winter of 1945, the facility was expanded. A combination recreation hall and chapel was built, and a school for the young children was opened.

While things might have seemed luxurious for the dependents in Landshut, a problem soon developed for the American civil affairs officers in charge of local government. The German populace outside the compound became almost belligerent toward their well-nourished countrymen living in comparative luxury at Landshut, known locally as Camp Overcast (much to the dismay of the US military, which still considered Overcast a secret operation).

As a result of this indiscretion on the part of the dependents, Operation Overcast on March 16, 1946 was officially renamed Operation Paperclip.

Typical of the causes of friction between the Overcast Germans and their less fortunate neighbors was the fact that those living inside could have two pairs of shoes per family resoled per month, while the others not so privileged could have only one pair of shoes per family per year resoled. The basis of the friction might seem trivial today, but the difference loomed very large in those early postwar days. While the shoe-repair situation caused hard feelings enough, there was still a greater injustice in the eyes of the burghers of Landshut who had no husbands or relatives in America. Each family in the compound received 165 pounds of wood and 165 pounds of coal bricquets per month. Those beyond the American pale received only a meager ration of wood.

Despite the enmity with their neighbors, the women of Landshut spent their time as profitably as they could. They gathered mushrooms in the nearby forests and, above all, studied English. Especially, they attended many films supplied by the American army from which they learned a great deal of "American" English—and gained a completely warped view of life in America.

One problem was not resolved for well over a year at Landshut: mail—both that sent from America to Piflaserweg 22a V, Landshut, and that sent from Landshut to 13B Munich 2, Postschliessfach 1 (the address for the men in the US). Because of an absurd system of double censorship, letters mailed by husbands took as long as ten weeks to reach Landshut. As a result, morale deteriorated rapidly both among husbands in Texas and families in Germany. It was a situation that was not satisfactorily resolved until 1947.

Thus, comparatively comfortable in a desolate Germany, several hundred women and children at Landshut waited for a year or two before they could join their husbands and fathers in America.

CHAPTER 15

Backfire at Cuxhaven

While Special Mission V-2 was collecting V-2 components at Nordhausen, the British 21st Army Group was capturing rockets abandoned by artillery batteries as the Germans withdrew eastward from firing sites in the Hellendoorn-Zwolle-Enschede region of northeastern Netherlands and the Hachenberg-Montaburg region northeast of Coblenz. The British also began capturing the Germans who knew best how to prepare and launch the V-2, while the Americans were rounding up the Germans who had designed and built it.

With both men and rockets available, it was only a matter of time before someone thought of putting the two together for the edification of the Allies. The idea first occurred to Junior Commander Joan C. C. Bernard, Army Territorial Service, an aide to Major General Alexander M. Cameron, chief of the Air Defense Division (ADD), Supreme Headquarters, Allied Expeditionary Force (SHAEF). On May 1, 1945, she proposed a demonstration firing that would be fully documented and photographed. The task of selling the idea to the British Imperial General Staff and War Office fell to Colonel W. S. J. Carter, assistant chief of the ADD, who went to London and briefed such influential individuals as Dr. Charles D. Ellis, adviser to the Army Council; Sir Alwyn Crow, Comptroller of Projectile Development; and Lieutenant General Sir Ronald Weeks, deputy chief of the Imperial General Staff. Ellis was especially enthusiastic about the proposal. He felt that it should be got under way as soon as possible before the Germans lost their skills. The project would also provide the Allies with knowledge and skills in launching the large rockets that obviously would be the weapons of the future.

After Carter had returned to SHAEF, several events took place that furthered the idea. On May 4, SHAEF had issued an order that no captured V-2 matériel would be removed from the theater until further orders. Five days later, on the day the war ended in Europe, the last V-2 firing unit was captured. An extensive series of interrogations of German prisoners of war had resulted in a much better knowledge of the ground support equipment that would be required for such an undertaking. Troops especially skilled in preparing and launching the rocket were segregated in a POW camp outside Brussels. By May 20, the interrogations had produced a list of approximately thirty vehicles that would be necessary to form a firing unit. The list, with a suggestion that thirty V-2s be made available for such a project, was sent to G-2 (Technical), SHAEF.

Intelligence reports from Allied tactical units and information from the Secret Intelligence Service and the POW interrogations indicated there were four likely sources for the matériel needed. These included the vast underground assembly plant for V-2s near Nordhausen as well as various supply points, maintenance facilities, and depots within the area of both the 12th Army Group, under General Omar N. Bradley, and the 21st Army Group, commanded by Field Marshal Bernard Law Montgomery. Additionally, there were factories and smaller plants in all parts of Germany as well as some in France, Denmark, and Belgium that had produced components for the rocket and its complex ground support equipment.

On May 6, CIOS Team 163, with Col. William J. R. Cook, visited the factory near Nordhausen and found some V-2 matériel which they sent to London. About the middle of the month, it was decided that the factory should be exploited as soon as possible. There were two compelling reasons: first, it had become known that Nordhausen lay in the zone of Germany that would be turned over to the Russians in the near future; second, much of the V-2 matériel already was badly deteriorated, having been designed for immediate use rather than open storage in a war zone.

The Supreme Commander, SHAEF, was convinced of the desirability of the proposed plan, soon to be code-named Backfire by Colonel Carter. The responsibility for its organization and prosecution was delegated to Major General Cameron and the ADD.

On May 26, the 307th Infantry Brigade Headquarters, commanded by Brigadier L. K. Lockhart, was designated to provide logistic support for the project. Specifically, Brigadier Lockhart was charged with reconnoitering and selecting a site for the operation, construction of requisite facilities, reception and storage of matériel, and security for the site.

Selecting a location proved not to be as difficult as one would suppose. Cameron and Carter agreed that it would have to be in the 21st Army Group zone and that it would be in a region that permitted radar tracking from sites stationed parallel to the flight path of the rockets. The Schleswig peninsula was an obvious choice. Lockhart suggested the former Krupp naval gun range located some five miles south of Cuxhaven on the coast of the North Sea. The missiles could be launched in a northern direction while being tracked by radar continuously from the river Elbe to the Danish border. In addition, the Krupp range already had rail sidings and buildings that would serve adequately for anticipated operations and roads that could be used to good advantage in transporting heavy equipment.

Also on May 26, a discouraging report came in from the field. Practically every V-2 located was missing its critical guidance control equipment. There were several reasons for this. Allied fighters had damaged some of the rockets being transported by rail, but these components also were the parts the Germans first destroyed or mutilated when capture of the rockets was imminent. In addition, such items were in short supply generally because of damage to manufacturers and because of their inability to obtain raw materials as the war drew to an end. To further complicate the search for V-2 parts, some factories, such as Wumag Abt. Maschinenbau, which had built the turbopumps, were in the hands of the Soviets.

As May ended, ADD made what was in reality a pro forma request for a senior US officer to be assigned to the project to help organize it. However, the Americans were in no hurry to appoint such an individual. They knew that the demonstration was to be a British operation, and they also knew the US Army was preparing for a similar but more extensive one to be held in the deserts of New Mexico beginning in the following year. It was July 4 before Colonel W. I. Wilson appeared at Cuxhaven. Several other officers accompanied him, but they were content for the most part to observe activity rather than to participate in it.

Meanwhile, ADD search parties were ranging over Germany, France, the Netherlands, and Belgium looking for V-2 matériel. Each party generally consisted of a British officer, from ADD or detailed from an outside unit; a German officer or noncommissioned officer who knew what the group was searching for; an interpreter if needed; a British NCO; and six enlisted men. These parties covered approximately 288,000 miles in about six weeks. The booty they turned up filled some two hundred trucks and four hundred freight cars. In addition, some seventy aircraft loads of special tools and equipment were flown in from England.

Even so, the acquisition of usable ground support equipment was expecially troublesome.

Since the V-2 firing batteries in the area of The Hague and Enschede had withdrawn into positions north of Hanover, these latter positions were the first to receive the attention of a search party. Much equipment was found, but it was all in poor condition or partially destroyed. However, in an area around Nordhausen one party turned up some serviceable equipment as well as some unusable rocket matériel. Unfortunately the latter did not lend itself to "cannibalization"—the destruction of one piece of equipment to obtain parts for rebuilding another. In order to locate additional ground equipment, the British turned again to interrogating POWs. Those who recalled "burying" things were sent along with search teams to point out "graves." In one case, it was necessary to dredge a river to find a critical item.

Searches often proved frustrating. A party would drive a hundred miles or more only to find that an abandoned peroxide truck or air compressor was still in place but that it was utterly useless because of the ravages of weather or looters. In some cases needed components lay tantalizingly just inside the Soviet zone, and the party had to turn back empty-handed. However, there also were occasions on which the search teams actually crossed into the Soviet zone and made off with materiel, sometimes pursued by Soviet troops.

In addition to the V-2 parts and ground support equipment streaming in from the search parties, some 640 tons of special tools were supplied from the Nordhausen plant. They arrived in five trains at Cuxhaven on June 20.

Eisenhower issued SHAEF Backfire Instruction No. 1 on June 22, stating in broad terms the objectives of the operation:

> The primary object of this operation is to ascertain the German technique of launching long-range rockets and to prove it by actual launch. ... In addition to the primary object, the operation will therefore provide opportunities to study certain subsidiary matters such as the preparation of the rocket and ancillary equipment, the handling of fuel, and controls in flight. [SHAEF was unaware of the problem of obtaining guidance control parts for the V-2.]

Dieter Huzel, who had hidden the Peenemünde documents only a month previously in a mine near Dörnten and who only grudgingly took part in Operation Backfire, later defined the objective of the project more succinctly, saying the British wanted "to become familiar with the other end of the trajectory."

While the search parties were rounding up equipment, Allied interrogations of the former scientific and technical personnel of Peenemünde continued at Garmisch-Partenkirchen. On June 20, Flight Lieutenant H. M. Stokes and Captain W. N. Ismay (of ADD), members of CIOS Team 183, questioned General Dornberger about the dangers that launch crews would run in Operation Backfire. Cooperating fully, Dornberger stressed the safety precautions for handling V-2 propellants. He also pointed out that the weapons should be fired as soon as possible after assembly and checkout. Additionally, he supplied a list of thirty men at Garmisch-Partenkirchen who would be of assistance in Backfire.

By July, the German military contingent had grown to 137 officers and enlisted personnel, all billeted in a special camp near the village of Altenwalde. These men were to form the firing battery. An additional 274 POWs were assigned as a provisional labor company, but they were not to take part in assembly or firing of the V-2s. All in all, some 1,000 German nationals and 2,500 British military personnel and civilians ultimately were to be involved.

Despite the extant Krupp facilities, some construction work had to be done, mainly by two thousand engineers of the Royal Canadian Army. They performed such feats as constructing a rocket assembly shed almost 300 feet long, with a 10-ton overhead crane down its length, in only three weeks. Four V-2s could be worked on simultaneously within it. In only two weeks, the Canadians succeeded in erecting a vertical checkout stand for the rocket, using sections of the military Bailey bridge. They also built new roads, laid additional railroad spurs, and cast heavy concrete bunkers for the launching crew. Among the special facilities constructed was a museum that featured a sectionalized V-2 and many of its subassemblies. The museum was used in training the nontechnical British personnel and later in briefing visitors to the site.

The head of the German military contingent that would fire the V-2s, Lieutenant Colonel Wolfgang Weber, was a man of considerable experience. Formerly commander of the experimental firing battery established in 1944 at Peenemünde to learn the V-2 and formulate firing drills, he had recently commanded Tactical Group South, consisting of the 836th Artillery Detachment (motorized) and the 444th Battery. Weber's unit had launched their last V-2 on March 16, only four months previously, so procedures were still fresh in mind. As commander of *AVKO (Altenwalde Versuchs-Kommando),* the German effort organized for Backfire, Weber was in charge of 591 military and civilian personnel.

By the beginning of July, the basic plan for Backfire had been revised drastically: Search parties had not been able to recover enough batteries, gyroscopes, and control amplifiers to assemble more than eight rockets. Thus, one of the most important of the technical objectives of the program could not be met—firing the V-2 under inertial guidance control used by the Germans, The V-2s would have to be fired solely as ballistic missiles.

With the dissolution of SHAEF on July 14, an administrative reorganization of Operation Backfire took place. However, it did not impede the progress or the schedule. The British component of SHAEF (ADD) was simply redesignated by the British War Office as Headquarters, Special Projectile Operations Group (SPOG) and moved from Frankfurt to Cuxhaven.

Meanwhile interrogation of the Peenemünde Germans continued at Garmisch-Partenkirchen. A group of eighty-five of them was selected by the British, with American concurrence, to participate in Backfire. These men were gathered for a briefing on July 21 by Captain M. W. S. Meyer, SPOG security officer and interpreter. Speaking circumspectly about the operation and even more so about their duties, he would not divulge even the location of the operation. With considerable misgivings, the Germans left Garmisch-Partenkirchen on July 23 in a convoy of six British army trucks—all, that is, except Huzel, who hid in his room.

Meyer had originally decided to allow his charges to stop in Witzenhausen for a brief reunion with their families billeted there. However, probably because of Huzel's defection, he changed his mind and stopped the convoy outside town, permitting the Germans only to write letters to their families instead of visiting them. Then, he told them to choose one of their number to be the postman. Heinz Scharnowski, a new father who had never seen his child, was selected.

During the trip, the Germans first became acquainted with a druidical ceremony (to be observed at Cuxhaven) that has long puzzled all non-Britishers. At 3 PM each day, the convoy would slow down perceptibly as the driver of the lead truck sought a shady forest or a spot more congenial and pleasant than the dreary succession of bombed and burned-out villages. It was, of course, tea time.

The perpetually hungry Germans soon began looking forward to these daily bonuses of tea, biscuits, and milk. At night, the convoy stopped at a farm or a village where there were enough houses left standing to accommodate it. When there were not, the British slept inside and the Germans shifted for themselves.

The Germans were divided into two groups of about forty men each by Meyer, and separated on arrival at the Backfire site. One group, in charge of Dr. Kurt Debus, was sent to Camp C at Altenwalde, while the other was billeted at Camp A in nearby Brockeswalde. Dornberger, who had been brought to the site separately by Major R. T. H. Redpath, SPOG senior intelligence officer, was housed with Lieutenant Colonel Wilhelm Zippelius, the former logistics officer on Dornberger's staff in charge of supplying propellants to V-2 units. Even though he was in Altenwalde,

Dornberger was held incommunicado, excepting his roommate Zippelius, from his fellow Germans there because the British felt that he would have too much influence over them. He, though, had been questioned previously about the capabilities of his former colleagues and the best means of organizing them for the job at hand.

The British had divided the Germans into two groups in order to use one as a check against the other, thus ensuring that truth would prevail during Backfire. The group at Altenwalde assisted actively in the preparation, checkout, and launching of the V-2s, while the unit at Brockeswalde was available to confirm procedures. However, in actuality it did nothing constructive. Zippelius, for example, was asked to write a report on his specialty. He dragged his feet, dawdled, asked for a typist, and generally showed himself to be in no hurry to oblige. Huzel, after belatedly joining his teammates at Brockeswalde, was asked to write a lengthy report on Test-stand 7 at Peenemünde, an assignment that could hardly have contributed to the success of Backfire. So Huzel and his fellow inmates in the Schutzenhaus restaurant at Camp C absorbed themselves in playing skat and chess, attending the recently opened library and movie theater in Cuxhaven, luxuriously smoking English cigarettes, and swimming from a pleasant beach about half a mile away. In between times, they worked diligently—if unhurriedly—on their lengthy "memoirs."

Paradoxically, then, in the midst of a devastated and prostrate Germany in the summer of 1945, a small but fortunate group of the Peenemünde team serenely enjoyed what amounted to paid vacations, first in the Tyrolean Alps and then on the shores of the North Sea.

Their situation also had a humorous aspect.

Some of the Germans on arrival were issued brand new Nazi Party uniforms, but without insignia—the only clothing for them the British quartermasters could find. Thus clad, they milled around behind the barbed wire of Brockeswalde and Altenwalde, fences easily penetrated for friendly visits. Sympathetic villagers assumed that the men were high-ranking party members who had been corralled by the British for summary extermination. Offers of help by the villagers to escape were made, but mystifyingly refused.

On their first night, the men were asked if they would prepare V-2 rockets for launching and assist in firing them. The Germans, taken unaware, refused. As a result, they were required to sleep on mattresses without bedsteads. The following day, however, the team members received a lengthy briefing in German on the project, and attitudes changed.

Part of the reluctance of the Germans stemmed from their having become used to the largesse of the more affluent American army, particularly its commissary department. At Garmisch-Partenkirchen, they had first been given American C rations, which they thought were splendid—a view not shared by their captors, who had subsisted on them for some two years in Europe. Later, they had been supplied with a relative abundance of fresh food, further spoiling them.

The British commissary was nowhere near as varied and well stocked, even for their own troops. The Germans, unreasonably, failed to see the problem. At first, they were issued a smaller ration than the British soldiers with whom they worked. They staged what amounted to a two-day strike or slowdown until the British agreed to put them on the same ration as their soldiers.

Additionally, the Germans chafed at the excessive formality of the British. They had become accustomed to the informality and, in some cases, overt friendliness of the Americans. However, they understood the situation.

"At that time it was an offense for British troops to fraternize with the Germans. This does not mean that we did not have a very good working relationship with them, but it was strictly business, and you must take account of the atmosphere of that time. Many of us had seen German internment camps," Colonel Carter later recalled to the authors.

Morale did not become a serious problem among the Germans as they adjusted to the British way of doing things. Indeed, there was a good deal of intramural horseplay among them. Generally, it was perpetrated by Karl Heimburg and his crony Helmut Hölzer. At first, Heimburg shared a room with Arthur Rudolph, formerly of the V-2 plant near Nordhausen and now in charge of the production department for Backfire. Colonel Weber, the commander of the German military contingent, demanded that Rudolph be given a private room because of his position. The British agreed and furnished him one. His erstwhile roommate retaliated by sneaking into the room and nailing Rudolph's furniture to the floor upside down.

In a later prank, Heimburg hit upon a means of breaking up the snoring of Dr. Kirschstein, a specialist in infrared sensors, who roomed next door. Heimburg and Hölzer rigged a purloined electrical bell underneath his bed in a hidden position. The wires ran into a switch in Heimburg's room. Each time Kirschstein began snoring, Heimburg pressed the button.

By the beginning of August, all rocket equipment for Backfire was on hand. To the amazement of Rudolph and his colleague Hans Palaoro there was an almost complete set of V-2 manufacturing drawings, with changes made at Peenemünde which had not yet reached them before the war was over! The two Germans marveled at the efficiency of the British secret intelligence service and its coordination with the underground in wartime Germany.

One of the most important logistic problems facing Major C. W. Lloyd, general staff officer 2 (technical), who was responsible for providing propellants, was getting liquid oxygen. Because of the lack of proper storage facilities and liquid-oxygen trailers, he knew that a supply relatively close to Altenwalde was needed. He found such a plant at Fassberg, some 125 miles away. The plant, inoperative since early spring, was capable of supplying 2 tons of liquid oxygen per day. It also had storage facilities for 56 tons. However, it could not operate at full capacity, and an additional 13 tons were to be supplied from a plant at Brunswick, approximately 120 miles to the southeast.

To assist with the liquid-oxygen production, Cameron called upon a dour little Scot, Lieutenant Colonel J. S. Brown, a former engineer with British Oxygen Company.

Brown had been told by Germans in the town that the plant at Fassberg was booby-trapped and that it would blow up if he attempted to get it back into production. Not believing them, he poked around the plumbing a bit, cleared the plant of all personnel except himself, and turned it on. It worked almost perfectly, and by August 7 it was "on stream."

The needed fuel supply was found near Nordhausen and sent immediately to Cuxhaven. Some 70 tons of 93 percent pure ethyl alcohol were thus obtained. At Altenwalde, it was diluted by the addition of 10 percent distilled water to produce 80 tons of alcohol with the required specific gravity of 0.86. Since each V-2 required only 4 tons of fuel, more than enough was available.

There was, indeed, enough for nonpropulsive purposes. Some mechanically skilled members of the Peenemünde team manufactured metal flasks in the machine shop and fitted them cunningly into the bottoms of their omnipresent attaché cases. Thus, some of the alcohol was smuggled into the camps at Brockeswalde and Altenwalde because security was lax. Zippelius, who was as familiar with chemistry of fuels as oxidizers, set up an illicit distillery for converting the raw alcohol into schnapps.

"The first batch was so strong that it dissolved all the mucous membranes in our throats, so we diluted it six to one, and wound up with seven batches of drinkable stuff," he recalled years later to the authors.

When British soldiers smelled the process, Zippelius invited his recent enemies in for a friendly drink. After several, former enmity gave way to something approaching deep *Kameradschaft.* The British were taught, phonetically, the words to an old German song and joined in enthusiastically if not melodiously.

At the height of the merriment, a British officer came by. Hearing some decidedly English overtones in the erratic harmony, he decided to investigate. Upon doing so, he found twelve of his soldiers with glasses brimming and voices raised in "*Wir fahren gegen England!*" ("We March Against England!").

The other V-2 propellants provided no problems, logistic or otherwise, both being unpotable, indeed highly toxic.

Hydrogen peroxide to provide steam for turbopumps was obtained from the Walther factory at Kiel, then under the control of the Royal Navy; and sodium permanganate, to catalyze the hydrogen peroxide into steam, was located in sufficient quantities near Nordhausen.

By mid-August, Operation Backfire had crystallized into two distinct, sequential phases: the assembly and checkout of complete rockets and the recording of all technical details in so doing; and the procedures for the launching of the rocket and the recording of its performance.

In mid-August also, Major Redpath escorted Dornberger to London for what was alleged to be an interrogation. The general was taken to a POW camp for high-ranking German officers on the estate of a British tea king of the nineteenth century named Sassoon, near London. There he was given a light brown uniform with no insignia of rank but the prominent letters PW emblazoned in white upon the back of the tunic. He found himself among familiar faces: former Field Marshals von Rundstedt and von Brauchitsch.

For interrogation, he was taken to the "London Cage," headquarters of the British War Crimes Investigation Unit. Its commander was Lieutenant Colonel Andrew P. Scotland, who informed Dornberger in a friendly way that since the British could not find Obergruppenführer Kammler for trial at Nuremberg, they were willing to settle on him as a surrogate lamb for sacrifice. Not unreasonably, Dornberger protested that he had had nothing directly to do with launching V-2s against London and, in lesser numbers, Norwich. He asserted that if he were to be tried for such an act of war, then the scientists and military leaders of all countries (i.e., the Allies) should be given similar treatment.

Unimpressed by his prisoner's argument, Scotland remained detached. He said that Dornberger's fate and future would be up to the Cabinet and Sir Hartley Shawcross, chief British prosecutor at the war crimes trials at Nuremberg. In the meanwhile, he suggested that Dornberger pass his time in writing an aide-mémoire on the V-2. Petulantly, Dornberger refused. He was subsequently transferred to a detention center for high-ranking German officers at Bridgend in South Wales. There he spent two years, presumably contemplating his technologically bellicose sins against Great Britain.

On August 21 Operation Backfire received an unexpected and most welcome boon in the form of a dozen V-2s in excellent condition. They were found near Leese, west of Hanover, near Duren. One rocket was almost in mint condition. The first phase of the project thus benefited greatly and stayed on schedule. Finally, eight rockets were assembled and checked out for firing.

As the month progressed, the US authorities at Frankfurt requested the return of twenty-six of the German engineers and scientists in Camp A and Camp C. The official reason given was that they were urgently needed for use in the war against Japan. The British replied that their loss would jeopardize the success of Backfire. While the negotiations were going on, the war against Japan formally ended on September 2, effectively negating the American position. Nevertheless, the British compromised by returning fourteen of the men. In return, the Americans replaced them with an additional twenty-five from Garmisch-Partenkirchen.

Throughout August, the Backfire search parties continued their quest for vital V-2 components and ground support equipment.

The Germans, military and civilian, were used to having everything needed to do the job. They found it difficult if not impossible to adapt to the expediency of improvisation urged by the British. The German insistence on having special tooling often necessitated trips to the plants of manufacturers, only to find their factories were nothing but masses of bombed-out rubble. (Another complication, which will not recur should history ever repeat itself, lay in the fact that all German V-2 tools were in metric sizes, while those available from the British were in a variety of English sizes.)

Mid-September found all necessary ground support equipment for Backfire on hand and in reasonable working order. This milestone was no mean accomplishment. Much of the critical matériel had arrived at Cuxhaven in a wholly unserviceable condition and had to be completely rebuilt and tested by the German technicians and their British supervisors. As the end of the month approached, British and Germans alike were eagerly awaiting October 2, the launch date for the first missile. One might think that the Germans would be rather blasé about an event so commonplace as the launching of a V-2, but as the countdown began they, too, were caught up in the same exhilaration they had felt on June 13, 1942, when they first attempted to launch the V-2 at Peenemünde.

From the outset of Operation Backfire, the decision had been made not to fire the V-2s to the maximum range of 200 miles. The reason was that they would impact too far from the Danish coast for proper tracking by cameras and radar. (The only tangible or "hardware" American contribution to Operation Backfire was five antiaircraft radars loaned to the British.) The range of the V-2s was set at 150 miles. The launching direction was 336°51' North. The aiming point in the North Sea was some 46 miles southwest of Ringkoebing, Denmark.

All shipping was warned away from an area that included a 30-mile radius of the aiming point. Aircraft over the area were also notified of the time of firing, flight, and impact and warned accordingly.

The launching site itself was not the tactical one to which Weber's troops were used. A concrete base or pad was built for the launcher. Trees in the area were cleared so that cameras would have an unobstructed view. Fire control was exercised from a concrete bunker 1,000 feet from the launcher, rather than the *Fürleitpanzer* or super armor-plated tank used by the launching platoon of the tactical V-2 firing battery. With everyone in place and the V-2 on its launcher, the engine ignition command was given.

Nothing happened.

The British were disappointed, but the statistically minded Germans merely shrugged. This was not the first failure by a long shot. The ground power supply plug had failed to eject, and the main stage propulsion system did not ignite even though the pyrotechnic device had lighted and flames appeared in the combustion chamber. It was put out by means of fire extinguishers, while the Germans pondered the problem.

A technician in a raincoat was sent up into the chamber amidst a drizzle of ethyl alcohol to make some repairs. (The British press later reported that the man became intoxicated by the fumes, but this was denied by Cameron in his official report to the War Office.) A second attempt to launch failed when the pyrotechnic device lit but was expelled from the engine before the propellants could ignite. Because of the lateness of the hour, no further attempt was made to launch on October 2. So it was decided to set up and check another missile.

The following day was a different story. The ignition signal was given at 2:43 PM, and the V-2 rose slowly off its launcher and into a cloudless sky. The observers in the bunker were deafened by the pulsating roar. In 4 minutes and 50 seconds, the rocket sped from the launcher and impacted only one-half mile to the left and one mile short of the aiming point.

In the control bunker a German engineer named Zimmermann, noting the presence of Junior Commander Joan Bernard, who had precipitated Backfire, announced with solemnity to the British that so far as he knew only one other woman had ever been present when a V-2 was launched: Eva Braun!

The second launching took place at 2:15 PM on October 4. It was the recalcitrant rocket that had failed two days earlier. Demonstrating the inherent perversity of complex electromechanical objects, it fell only 14.97 miles from the launching site, its engine having operated for a truncated 35 seconds.

The third and final launch, designated Operation Clitterhouse, provided a grand finale for an elite group of observers from the British, French, American, and Soviet armed forces and the press.

It took place at 3:06 PM on October 15, in ghastly conditions, with low clouds and a steady ground wind of 40 feet per second. Despite the inclement weather, the primary objective of putting on a show for the visiting brass, many of whom had earlier experienced the lethal end of the V-2 trajectory, was achieved. All present were impressed with the noise, smoke, and flame. The flight was a qualified success, falling short of the target by 11.2 miles and to the right of it by 3.2 miles.

Most fretful among the visitors were the Russians. Present officially were Colonel Yuri A. Pobedonostsev and Colonel Valentin P. Glushko. The former was currently leader of the Special Technical Commission (rocket) in Berlin. The latter was directing the test firing of V-2 engines at Lehesten. (Both men had considerable experience in liquid propellant rocketry, dating back to the mid-1930s when they built and test-fired such motors in Moscow. They later would become key figures in that nation's program to place man into space.) With them was another officer, identified only as General Sokolov.

Additionally two more Russians showed up unannounced. One of them was Colonel Sergei P. Korolev, disguised as a captain, who was deputy to General Gaidukov, chief of the Soviet Special Commission at the recently reopened V-2 facility at Nordhausen. Korolev at the time was active in extending the capabilities of the V-2. (In the mid-1950s, he became the mysterious "Chief Designer of Spacecraft," who developed the Vostok, Voskhod, and Soyuz manned spacecraft for the USSR.) He was accompanied by an unknown colleague.

Korolev and his fellow officer were not permitted into the launching despite a great deal of arm waving and shouting. Cameron stood fast, and the two were eventually escorted just outside the gates, from which place they saw the launching.

Following the launching, the two Russians were again incensed at not being allowed to tour the assembly and checkout area and launching bunker, especially after having been urged to do so by an overly helpful American colonel with the suspect name of Shedersky.

The Russians as well as the other observers had been restricted to a fenced-off area during the launch. They could have binoculars but not cameras. During the prelaunching activities, Lieutenant Hochmuth, one of the observers, engaged Pobedonostsev in conversation. Later he recalled to the authors:

> He knew my name and that I had been there [*Mittelwerk*]. He told me the stuff [the matériel removed by Hamill] was going to White Sands (this was supposed to be a secret). We began to discuss engineering. I asked him how things were at Nordhausen, and he said he was having a hell of a time because we had cleaned the place out.

He was a very technical guy and said if they were able to see White Sands, we could see Peenemünde.

Hochmuth dutifully reported this offer to his superiors, who immediately turned it down. What a chance the US Army missed! The Russians would have seen hundreds of square miles of barren desert and a few dilapidated wooden buildings. In turn, the Americans would have seen the remains of their late enemy's most sophisticated rocket research center.

On October 21, the British began processing the German civilians before releasing them. Luggage was inspected, identity cards were issued, and letters of commendation were given to those who had done especially good jobs. The group at Brockeswalde was transferred to Altenwalde. Individuals who intended to remain in Germany began leaving on October 23, and two days later those already under contract to the Americans departed.

Earlier in the project, most of the Germans had been interviewed at length and individually on the prospect of their going to England to continue their rocketry. Some twenty who had not been offered contracts by the Americans took them up. By and large, however, the British had little to offer. The Germans were nothing if not shrewd. They knew that economically Britain was up against it and that there would be no funds available for large rocket development for at least a decade.

Operation Backfire had accomplished its purpose. The project produced the only detailed set of documents describing the procedures for assembling and firing a V-2 rocket. While the radars did not work as well as could have been desired, they and the tracking cameras turned in an acceptable job of trajectory recording. The five-volume report submitted by Cameron to the War Office in January 1947 is a completely illustrated manual for readying a V-2 and launching it.

The Germans during the war produced no equivalent manuals. Security regulations forbade it. Each soldier was taught his function and station only. If he were captured he could reveal only what he knew. Even manuals for officers included a less-than-complete description of the complete V-2 system.

Operation Backfire accomplished in five months what would take the Americans in the desert of New Mexico considerably longer a year later, even with the assistance of the Backfire report and 127 top men from Peenemünde.

Among the Americans at the final launch were several with a vested technological interest rather than mere curiosity. One man with a hearing aid in his left ear was wearing the crumpled uniform of a full colonel, but somehow did not look the part. Indeed, he was not. Dr. Theodore von Kármán, the aerodynamicist, formulator of the principles of supersonic flight, and at the time director of the Jet Propulsion Laboratory of the California Institute of Technology, was more in costume than uniform—as were his Soviet neighbors. Standing not far from him was Lieutenant Commander Grayson P. Merrill, a legitimate US naval officer, with experience in designing pilotless aircraft and later editor of the 10-volume series *Principles of Guided Missiles*.

Not far from these two stood Dr. Howard S. Seifert and Dr. William H. Pickering, both of the California Institute of Technology. Seifert was on leave from the Liquid Rocket Section of the Jet Propulsion Laboratory, and was currently a technical adviser to the military attaché in the American Embassy in London. New Zealand–born Pickering, an assistant professor of electrical engineering at California Institute of Technology, would later become director of its Jet Propulsion Laboratory.

Following the launching, von Kármán and Merrill discussed what they had seen and its immediate potential for the still amorphous US military rocket program. As a result of this conversation, Merrill wrote into his report on Backfire that V-2s could be launched from the decks of ships or submarines on the surface of the sea. He further went on to propose that several V-2s be obtained from the British for launching from ships to prove the feasibility of the technique, not knowing that at the time components for one hundred of the rockets were already stacked in the desert near Las Cruces, New Mexico, while Wernher von Braun, in El Paso, Texas, was waiting for some of the Germans at Cuxhaven to join him in assembling and launching them. Despite Merrill's suggestion, it was not until 1957 that a V-2 was launched from the deck of the American aircraft carrier USS *Midway*.

Merrill, in later years, recalled to the authors:

> In looking back, I feel the Cuxhaven launchings had a definite bearing on the genesis of the now famous Polaris program. For example, it impelled me to work with Rear Admiral Calvin Bolster about 1951 to set up a program for adapting the Navy's Viking to submarines.

Thus Operation Backfire had ramifications that SPOG could not have foreseen. What Merrill had observed stirred interest in the launching of large rockets from surface ships and submarines. (He became technical director of the Polaris missile in 1956.) Some of these rockets later would be adopted by a financially ailing postwar Britain for her own defense: the Polaris, Poseidon, and Trident.

The United States obviously received nothing of value from Operation Backfire except the detailed report that Cameron made. However, one observer at Clitterhouse later looked back on the firing and saw the potential the V-2 held for America. Richard S. Lewis, a reporter for the US Army newspaper *Stars & Stripes* at the time, wrote in his book *Appointment on the Moon* (1968):

> That demonstration marked a transition point in the development of rocket technology in the West. That was the last V-2 fired in Germany. The engineering science which the weapon represented was carried off by the victors as spoils of war. In the United States, it was to evolve into an interplanetary spaceship technology.

CHAPTER 16

The New Beginning

Seven dejected members of the Peenemünde team hunched miserably along the hard wooden benches in the back of the US Army $\frac{3}{4}$-ton truck as it bounced and jolted westward from Kaiserlautern toward Saarbrucken. They already ached from the drive southward from Frankfurt, which had begun several hours earlier on September 12, 1945; and they still faced another 300 miles before they reached their destination: Le Grand Chesnay, an estate near Paris that had been converted into a POW camp for high-ranking German officers.

As the truck clattered across the bridge over the river Saar into France, Wernher von Braun said to his colleagues, "Well, take a good look at Germany, fellows. You may not see it for a long time to come."

Puzzled, August Schulze, a former systems engineer in the Test Laboratory of *Elektromechanische Werke*, took issue with him: "What do you mean? You know we only have a six-month contract with the Americans."

"We may have only a six-month contract now; but I still don't think we will be back for a long time to come," von Braun went on apocalyptically.

More or less sharing von Braun's views were Eberhard Rees; Erich "Maxi" Neubert, recent production liaison manager for the Electronics, Guidance, and Telemetry Laboratory; Theodor Poppel, sometime chief of the ground support equipment branch of the Test Laboratory; Walter Schwidetzky, previously an instrumentation engineer in the Electronics, Guidance, and Telemetry Laboratory; and Wilhelm Jungert, a highly skilled technician who had worked for Rees.

As a prognosticator, von Braun produced mixed results.

Jungert returned after only a year. However, von Braun was correct in the case of Neubert: It was 13 years before he was to return, as an American citizen visiting Germany on official business for the US Army. Rees did not return for seven years; when he did so, it was to recruit scientists and engineers for the US Army Ordnance Missile Laboratories, in Huntsville, Alabama.

The truck stopped for the night at a US Army camp near Reims. On the following morning the travelers got off to a very early start and arrived in the suburbs of Paris around 5:00 AM on September 13. The American lieutenant escorting the group then proceeded to drive around for several hours seeking a hospital in which his brother was a patient. Having got them properly lost, the lieutenant refused von Braun's directions to their destination on the grounds that a German could not be expected to know much about Parisian geography. Von Braun's explanation that he had visited the city several times during school holidays did nothing to dissuade the lieutenant. Instead, he approached a local citizen and asked him to show them the way to Le Grand Chesnay, some 9 miles west of the city not far from the military academy at St. Cyr. The Frenchman, with Gallic charm and graciousness, proceeded to take them over a good deal of the city and its outskirts before admitting he had no idea where the place was.

When they finally reached Le Grand Chesnay, von Braun and his associates found themselves in a sylvan setting more appropriate to a romantic novel of the eighteenth century than a POW camp. Many of the inmates were engaged in assisting US Army historians to process captured documents for shipment to the United States.

At Le Grand Chesnay, the von Braun group met several fellow German scientists who were also awaiting transportation to the US. Drs. Muller and Meyers, designers of propellers and turbines, were under contract to the US Navy and had been at the estate since April. Others were assigned to the US Army Air Corps. Among them were Drs. Gerhard Braun, Theodor Zobel, Wolfgang Nöggerath, and Rudolf Edse, formerly with the renowned Hermann Göring Aeronautical Research Institute at Brunswick.

During the five days they spent at Le Grand Chesnay, the small Peenemünde team had little to do except wander along the wooded paths of the estate and discuss their unclear future.

On the evening of September 18, a trim and efficient young officer of the US Women's Army Corps collected sixteen of the Germans and drove them to the US Officers' Club at Orly Field. There the men dined, confusing, mystifying, and at last angering the French waitress, who had assumed she had heard the last German spoken in her country. After dinner, the Germans were driven to a remote C54 cargo plane at the far end of a runway and hustled aboard it with a few US soldiers returning home. Their escort officer turned smartly and marched back to the bus without wave or good-bye.

There was little conversation on the plane as it flew first to the Azores and then to Newfoundland for refueling stops before landing at Newcastle Army Air Base, in Wilmington, Delaware. There, the group was kept together and segregated from American personnel as far as possible. Still, they received many suspicious glances. The loose, double-breasted jackets (a sartorial embellishment forbidden at the time in the United States in order to save cloth) and longish hair styles immediately marked them as European to returning US soldiers.

As soon as it was serviced, the men were hastened aboard a chartered DC-3 and were flown to Logan Field in Boston, where they landed on September 20. US Army sedans immediately drove them to Boston Harbor, where they were transferred to a small boat for the five-mile trip out to Long Island and Fort Strong, a dreary old fortress erected in 1898 to protect the city from invasion by sea; it was serving as a post of the US Army Intelligence Service.

For two weeks, the team members sat around at the fort, doing little after having been photographed, fingerprinted, and answering what to them were inconsequential or irrelevant questions or repeating answers to questions asked several months previously at Garmisch-Partenkirchen. The damp, chill wind sweeping over the island made von Braun acutely uncomfortable. His recently broken arm ached constantly because of the miserable weather. To add to his discomfort, he contracted hepatitis.

For recreation, the seven team members engaged in marathon games of Monopoly. With Teutonic vigor, they modified the already complicated rules to include provisions for forming holding companies and cartels. As Schwidetzky fondly remembered an all-night session: "The last three 'tycoons' were involved in intense trilateral negotiations and each needed an assistant to administer the complex contracts invented on the spot." The game, with its rapidly accruing rules and regulations, accompanied the group first to Aberdeen Proving Ground, in Maryland, and later to the deserts of Texas and New Mexico.

When the excitement of vicarious high capitalism palled, the men walked upon the beach, not particularly cheered by the mournful clanging of the buoy in Boston Harbor. Luckier, for the moment, than the rocket scientists were their four aeronautical colleagues—Braun, Nöggerath, Zobel, and Edse. They spent only four days at Fort Strong before flying on to Wright Field in Dayton, Ohio.

On October 1, Major James Hamill, testy after a recent illness, arrived at Fort Strong and dutifully signed a US Army Custody Document in triplicate for the group. While perusing the form, he noted an entry: "Probable date of return to Fort Strong." He left it blank; and, indeed, von Braun and the others never returned there. The fort was closed in 1948. On the following day, the group, less von Braun, entrained for the Aberdeen Proving Ground, between Washington and Baltimore. Having seen them safely off in charge of Private First Class Eric M. Wormser, Hamill and von Braun proceeded to Washington by car. They checked into the Bachelor Officers Quarters at Fort Myers, and for several days von Braun was closeted in the Pentagon with Major General Gladeon M. Barnes, Chief of Technical Division, Ordnance, and Colonel Leslie E. Simon, his chief of Research and Development. The two told him of the V-2s that had been shipped to White Sands Proving Ground and that they wanted von Braun there as soon as possible. Barnes and Simon also spent a great deal of time sizing him up, being suspicious of a man who had developed the V-2 and was now eager to help them develop its successors.

The discussions over, Hamill took von Braun to Washington's Union Station, where they boarded the train for El Paso, Texas. It was to be a memorable journey. Hamill was under orders from Colonel Toftoy to "stick to von Braun like glue." In particular, von Braun was *not* to engage in conversation with other passengers.

Von Braun tried to cooperate, but it was not always easy. As the train approached Texarkana, Arkansas, Hamill saw that he was in an animated conversation with a man sharing his compartment. After the man had pumped von Braun's hand, slapped him on the back, and gotten off at Texarkana, Hamill's charge recapitulated what had happened.

"Where are you from?" his companion had asked.

"Switzerland," von Braun replied.

"What business are you in?"

"Steel."

As things turned out, the man himself was in steel and knew Switzerland from border to border and apparently a great number of its population. Just as the conversation was beginning to strain von Braun's talents of improvisation, the train pulled into Texarkana.

In leaving, the steel expert said, "If it wasn't for the help you Swiss gave us, there's no telling who would have won the war."

It was with great relief that Hamill arrived in El Paso. Scarcely had he put down his luggage at Fort Bliss than he wrote to Toftoy, "Since I came back from Europe, instead of being allowed to go and see my wife, I've been made to honeymoon with von Braun."

The group that remained at Aberdeen Proving Ground after Hamill and von Braun had gone to Texas was faced with a complicated task. The

14 tons of material, including 3,500 reports and 510,010 engineering drawings, that Dieter Huzel and Bernhard Tessman had cached in the mine near Dörnten on April 4 as well as documents from Dornberger's files from Bad Sachsa that had arrived at the installation several months earlier. Initially, Colonel James G. Bain, of the Rocket Development Division, Ordnance Research and Development Service, had assigned Private First Class Wormser to sort them out and prepare abstracts of them. Wormser had been born and partially educated in Germany and knew the language fluently. He also had a Master of Science degree in mechanical engineering. These qualifications notwithstanding, Wormser soon found that the nature of the material was alien and far beyond him both in German and in engineering expertise. The Germans who had generated much of the documentation and who had assisted in cataloging it in Germany seemed to Bain a logical choice to assist Wormser. Thus, Wormser had been sent to Boston to get them.

Upon arrival at the proving ground, Neubert and his colleagues found the crates had been opened and the contents hopelessly jumbled by Wormser and a team of hastily assembled German-speaking helpers. They had to sort it out by project (i.e., V-2, *Wasserfall*, *Taifun*, etc.) and by subject matter (development, design, testing, etc.). Since their knowledge of engineering English was extremely limited, Bain made arrangements for the actual translating tasks to be done by volunteers among some 130 volunteers in a group of German naval officers of a POW camp at nearby Fort Hunt, in Virginia. Until the end of the war, these men had been the most aggressive of all German prisoners in escaping and generally harassing their American captors. The partial story of Fort Hunt and these prisoners is described in John H. Moore's *The Faustball Tunnel, German POWs in America and Their Great Escape*.

All was not work at the proving ground, however. On weekends, Wormser and his fellow soldiers would accompany the Germans to New York, Washington, Baltimore, and Philadelphia. En route to the latter city, the group sometimes stopped off in the "Pennsylvania Dutch" region of the state which had been first settled in 1683 by German emigrants seeking religious freedom following the persecution of the new Protestant sects in their native principalities. Only with difficulty could they communicate with those descendants of their common ancestors who spoke seventeenth-century German in mid-twentieth-century America. In New York, on the other hand, language was no barrier; their lodgings were provided by relatives of Walter Schwidetzky, and they spoke contemporary German.

One of the soldiers who accompanied the Germans was Technical Sergeant John A. Reinemund, now Chief, Office of International Geology of the U.S. Geological Survey. He recalled to the authors that "for travel within the Aberdeen Proving Ground and to nearby cities, an extra-long-wheelbase military-type bus was assigned, which was promptly dubbed the dachshund 'dachel' by the Germans and which became a familiar and distinctive indicator of the presence of the German team in and around Aberdeen. It also had a tendency to scrape bottom, because of the long, low undercarriage, and on many occasions the Germans and their enlisted companions could be seen scrambling out of the 'dachel' to give it added clearance over rough or uneven roads."

By the end of December, the group had done as much as they could to get the paperwork in order; and engineers from Dr. Richard Porter's Project Hermes team began arriving to go through the recently abstracted material looking for documents of value to them in fulfilling their contract with the Ordnance Department. As the Germans were preparing to leave Aberdeen, they ran into an old friend. Major Robert Staver, who had collected many of them in Bavaria, was at the post for demobilization. They were dismayed to find that he would not be with them in Texas; but his position had been taken by Major Hamill who stood in greater favor with Colonel Toftoy.

On January 11, the team members boarded a train for El Paso. Its members also were under strict orders not to speak to fellow passengers and to be as inconspicuous as possible. However, circumstances intervened, and they were not to remain incommunicado during the two-day trip. The train was delayed for a day in Kansas City, and the army intelligence personnel accompanying the Germans let them wander about in the railway station, explaining to curious travelers that the aliens among them were members of a Hungarian gypsy band, none of whom spoke English. In Kansas City the Peenemündians were not likely to be addressed in Hungarian.

Throughout the trip, they gazed curiously at the country through which they passed. They were amazed to see "acres and acres of airfields loaded with planes," a sight that brought home to them for the first time the fact that the American bombers that had decimated their hometowns a year previously had been only a token of the military might of the United States. Additionally, they were fascinated by the architecture—something that always mystifies Europeans. Every house, it seemed, was made flimsily of wood! They asked their escorts if such domiciles were temporary structures, for so they seemed to Germans used to solid stone and mortar.

While they had studied the Mississippi River cursorily in their elementary geography classes, they had always assumed that it was merely the Rhine of the United States. They were incredulous when they saw its almost mile-wide breadth as they crossed it near New Orleans, 100 miles above its mouth.

The second group of the team, those who had been left at Le Grand Chesnay, boarded the *Argentina* at Le Havre on November 9, accompanied by a group of five US Army Intelligence officers. After a trip marked by rough seas, salt-water baths, and strict isolation below decks, the Germans arrived in something less than a joyous mood at Pier 90 in New York on November 16.

In addition to its members, there was a small group of aeronautical engineers, including Dr. Bernard H. Gothert, former chief of the Department of High Speed Aerodynamics at the Hermann Göring Aeronautical Research Institute in Brunswick, who were assigned to the US Army Air Force at Wright Field. The group was bundled into a bus for a quick drive through the city to Grand Central Station, where they were put aboard a train for Boston.

Once at Fort Strong, they, too, spent monotonous hours filling out personal history and experience forms and answering a bewildering number of apparently irrelevant questions: "Do you know who George Washington was?" "Where is Chicago?" "Who was Robert E. Lee?" "Who was Ulysses S. Grant?" "Who won the Civil War?" The queries were put to them by German Jewish refugees serving in the American army as interpreters. Behind the apparently pointless questions, intelligence specialists were looking for answers that indicated the degree to which the Germans were interested in the United States prior to the war and hence would probably be motivated to work for it subsequently.

The group took up Monopoly as had its predecessors and further passed its time studying English. The more athletically inclined engaged in soccer games with the German POWs at Fort Strong, who endlessly questioned them about conditions in Germany. Further mystifying the rocket specialists were the endless films they were shown on duck and pheasant shooting, a sport with which they were reasonably familiar. Other films were of a more general nature, dealing with the variety of birdlife in America as a whole. After several such shows, the men began to wonder if they were to be employed as rocket technicians or ornithologists. The simple truth was that the films were the only ones readily available.

The third group left Landshut on November 21 and three days later was in Le Havre. These men left aboard the *Florence Nightingale* on November 28 and arrived in New York on December 6. After spending a bleak Christmas at Fort Strong, they left for El Paso on January 10, 1946.

The fourth segment of the team left Landshut on a train January 11. Their escort was First Lieutenant Enno Hobbing, a strapping and friendly American officer who spoke perfect German. In Munich several well-known scientists slated for duty with the Army Air Corps at Wright Field joined the group. They included Dr. Alexander Lippisch, designer of the

Me163 jet fighter, and Dr. T. W. Knacke, a pioneer in the design of parachutes.

In Munich the men spent the night at Hotel America, an army transient camp filled with homeward-bound US soldiers. On the following day, Hobbing took them on through Paris and to Le Havre, where he placed them in another transient center, Camp Home Run. With characteristic romanticism, the Germans assumed the name came from the fact that the happy American soldiers made their run for home from it. Their imagination was rudely punctured by Hobbing's explanation that it was, as were other transient camps, named after brands of American cigarettes, and in the case of Home Run not one of the more popular ones.

On January 2, the men trudged aboard the liberty ship *Central Falls* before it was loaded with the soldiers. After a leisurely twelve days, the *Central Falls* docked in New York on February 3, but the Germans were kept below decks several hours until all the troops had disembarked. Only then were they furtively trotted down the gangway into a warehouse and aboard a waiting commercial bus. Instead of proceeding to Fort Strong, it took them to Washington, pausing on the way to stop at a puzzling structure Hobbing told them was a *hamburger stand*. In it, they dined on a most curious form of sandwich, which would become a staple in their diets over the succeeding years. Once in Washington, they transferred to army vehicles and were driven to Fort Hunt, which besides being a POW camp housed a small intelligence post.

There they went through the same procedure that their colleagues had earlier undergone at Fort Strong. For three weeks they were fingerprinted, questioned, and filled in lengthy forms. They also wandered along the tree-lined Potomac River, studied English with the assistance of their amiable German-speaking inquisitors—and played Monopoly, German style.

As they waited, they received sad news that their friend and colleague Hans Lindenberg had died at Fort Bliss. On February 21, they finally entrained in a special railway car for their new homes in the deserts of New Mexico and Texas. Two days later they arrived in El Paso, and the Peenemünde team was complete in Texas.

CHAPTER 17

V-2s on the Steppes

As negotiations between the US Army and members of the team at Witzenhausen were in progress, one former Peenemündian decided to cast his fortunes with the East. He was Helmut Gröttrup, erstwhile assistant to Dr. Ernst Steinhoff, director of the Guidance, Control, and Telemetry Laboratory, and ex-cellmate of the von Braun brothers in the Gestapo prison at Stettin.

Gröttrup had graduated "with distinction" from the *Technische Hochschule* in Berlin in 1939 with a degree in physics. He had paid his own way through the university and even sent money home to his parents by serving as a technical consultant to various companies in Berlin. Following his graduation, he had spent six months in the laboratory of the renowned physicist Professor Manfred von Ardenne, an electron microscopist, before accepting a position at Peenemünde.

Gröttrup's reasons for going with the Russians were complex and varied. For one thing, he was unhappy with the contract offered by Colonel Holger N. Toftoy. His wife, Irmgard, later wrote in her book, *Rocket Wife*:

> When hard facts are involved, civilization is thrown to the winds—first come, first served is the rule. The Americans were acting on that principle when, after ceding Thuringia—and with it Peenemünde, which had been evacuated in the path of the advancing Russians—they grabbed Wernher von Braun, Hüter, Schilling, Steinhoff, Gröttrup and other leading rocket experts. We were housed at Witzenhausen and interrogated. After weeks had passed,

> Helmut was handed a contract offering him a transfer to the USA without his family, a contract terminable by one signatory only: the US Army. Since we wanted to remain in Germany, we moved back to the Russian Zone...

One should not infer that Gröttrup chose the Soviets because of political inclinations. Upon being interrogated by British and American intelligence personnel in 1954, Gröttrup struck his interviewers as being cooperative and having no discernible influences of Marxist philosophy. Later still, with two decades of reflection, his former colleagues at Peenemünde, then in America, drew an ambiguous portrait from memory: "an intelligent person, with a slight intellectual flair ... Gröttrup's 'defection' puzzled many people; probably, he saw too many competitors in the von Braun group ... he was a liberal intellectual ... he was a Communist ... indispensable to the electrical side...."

In all events, it is unlikely that a Communist could have managed to rise to the position Gröttrup occupied in such a secret organization as Peenemünde. As incomplete a portrait as it is, perhaps the best of Gröttrup is that of Albert E. Parry, in his book *Russia's Rockets and Missiles*: "Complex personal grievances and ambitions rather than political convictions played their role here."

As early as May 16, 1945, Gröttrup had told Major Staver in Bleicherode that he did not want to go to America. He also got into a violent political argument with Lieutenant Hochmuth at the same time. Nevertheless, he and his family were evacuated to Witzenhausen along with the other personnel of *Mittelwerk*. Once there, he again stated his position and turned down the US contract. Having refused the American offer, Gröttrup set himself up in the village as a dealer in scrap metal; such a mundane occupation simply was not for him, however.

At this time, he was approached by the *Burgomeister* of Bleicherode, who crossed the river Werra into the US zone to bring him an offer from a Major Chertok, a member of the Soviet Special Technical Commission and current officer in charge of reconstituting the plant at *Mittelwerk*. Gröttrup made two clandestine trips back into the Soviet zone to negotiate with the Soviets. Finally, in mid-September, he took his wife and two young children and returned to Bleicherode for good—or so he thought at the time.

Until he actually left for the last time, he tried unsuccessfully to do some local recruiting. In the nearby village of Ermschwerd, he attempted to win over Dr. Ernst Geissler. He explained to Geissler that the Russians were not the "bad guys" they had been painted and that he would be remaining in Germany while continuing his career. Geissler was not persuaded.

Once back in Bleicherode, where the Soviets furnished him with a fine home by simply moving its affluent merchant owner out, Gröttrup found out that the Russians intended to live up to their promises. His salary was set at 5,000 marks ($1,250) per month, although there was little he could buy with it in the Eastern Zone. Food was excellent; he and his family received the best rations available to the Soviet armed forces, even though most of it came from the local economy. His German employees were similarly well looked after: A typist made 300 marks ($75), and a working-level engineer earned upwards of 1,400 marks ($350).

Following their occupation of the Nordhausen area on July 5, the Soviets immediately set about putting things into order. There was some debate over whether the equipment and matériel in the region ought not be collected and shipped to the USSR. However, men such as Colonel Grigori A. Tokady, chief rocket scientist for the Soviet Air Force, later recalled the circumstances in his book *Stalin Means War* (1951):

> The problem we face today is this: we have no leading V-2 experts in our zone; we have no complete projects or materials of the V-2; we have captured no fully operational V-2s which could be launched right away. But we have lots of bits and pieces of information and projects which may be very useful to us. We have the free or compelled co-operation of hundreds of German workers, technicians, and second-rate scientists, whose experience could be of value to us. In the circumstances, I think the best thing to do is to organize all these into a group, in Peenemünde, to give it a set task, and to find out what it can do for us here in Germany.

The logic of this view prevailed, for a while at least.

The Soviets set two immediate goals at Niedersachswerfen: reconstruction of a full set of production drawings for the V-2 and reestablishment of a pilot production line for the missile. To accomplish these aims, the *Institut Rabe—Raketenbau und Entwicklung* (rocket manufacture and development)—was formed. It was under the command of the Soviet Special Commission (Nordhausen), in Berlin, under a General Kuznetsov initially and later a General Gaidukov. A group of Germans was also at work in the old *Mittelwerk*, largely engaged in writing manufacturing procedures and putting its facilities in order. The test stands at Lehesten were in perfect shape, since they had been left untouched by the Americans for the Soviets. The Gema plant in Berlin, an organization primarily concerned with missile control systems and the development of antiaircraft rockets, was also made a part of *Institut Rabe*.

In early 1946, four additional elements were added: *Zentralwerke*; *Nordhausen*, *Werk II*; Sommerda; and Sonderhausen. *Zentralwerke* consisted of a pilot production line in the old V-2 repair depot at Klein

Bodungen. *Nordhausen, Werk II*, also known as *Montania*, assembled V-2 engines and propulsion equipment. In Sommerda, some 80 miles east of Leipzig (in the former Rheinmetall Borsig plant), a large design office and laboratory was set up. A factory for manufacturing all electrical equipment for the V-2 was set up in Sonderhausen, 12 miles south of Nordhausen.

In addition to reconstructing the V-2 and its manufacturing line, Gröttrup had the additional duty of developing two special trains given the deliberately misleading nomenclature of *Fahrbarbare Meteorologische Station* FMS (Mobile Meteorological Station). Each had between eighty and one hundred cars and provided full facilities, men, and matériel needed to prepare and launch V-2s. There were sleeping and dining cars, laboratory cars, as well as freight cars for the V-2s and their launchers.

Gröttrup, in the spring of 1946, was placed in charge of *Zentralwerke* and was asked by General Gaidukov to take over responsibility for all guided missile development being done in the Soviet zone by Germans. In less than a year, his organization grew from the thirty-man *Büro Grö* (his initial office) to the five-thousand-man *Zentralwerke*. Between September 1945 and September 1946, he accomplished all that the Soviets had asked of him. The drawings had been redone and a pilot production line was turning out flightworthy V-2s. By the beginning of September 1946, some thirty rockets were ready. When parts could not be obtained in the Soviet zone, representatives of *Zentralwerke* contracted for them in the British and American zones. More often than not, they were bartered for food, drink, and tobacco rather than paid for with money. The completed components were smuggled into the Soviet zone by bands of "adventurers" paid in a similar fashion.

In lieu of launching, the V-2s from *Zentralwerke* were static-fired at Lehesten, where a stand that would take the complete missile had been constructed. This facility, under the direction of Valentin P. Glushko, fired its first V-2 propulsion system on September 6, 1945. Firings were conducted at first by Dr. Joachim Umpfenbach. Later, Soviet crews under Glushko performed tests as they became qualified through the tutelage of the Germans.

In the summer of 1946 the Russians asked Gröttrup's team to suggest technical improvements in the V-2 by mid-September. Some 150 such proposals were made, most of them based on concepts that had originated at Peenemünde. The Soviets accepted about half of them and asked that the other half be studied in greater detail before being resubmitted. Among the improvements the Russians accepted were pressurized propellant tanks; the relocation of all control equipment behind the propellant tanks; and turbopumps for the propellant driven by exhaust gases tapped off the V-2 engine.

Also, by mid-1946, Sergei P. Korolev, General Gaidukov's deputy on the Soviet Special Commission, was busy with calculations on an improved V-2, known to British intelligence as K1. Korolev himself was more interested in the airframe and propulsion systems than the guidance and control systems, with which he had little experience. The new missile was essentially a "stretched" V-2. By lengthening the propellant tanks of the German missile by some 9 feet and increasing the thrust of its engine from 25 tons to 32 tons, Korolev produced a design that had an estimated range of 400 miles, or about twice that of the V-2. His one difficulty lay in the fact that the V-2 turbopump was not designed for the higher flow rates he required. He chose to remedy the problem by adding a second pump in series and ignoring German advice for the "bootstrap" of exhaust gases from the engine back into the turbopump. He did like their idea of a warhead compartment separated by explosive bolts from the rocket body, when its engine had shut off, to increase its speed.

Indeed, Korolev, who spoke and understood German, made as little use as he could of the Germans in *Zentralwerke*. Thus, at that early date, it became apparent the Russians were clearly on the road to independence in missile development.

The production drawings for the K1 were all in Russian and were sent to the manufacturing facility at Sommerda from the Special Commission headquarters in Berlin. The drawings, which were different from those used to manufacture the V-2, apparently had been made in the USSR. All components for the K1, including standard V-2 components made in other facilities of *Zentralwerke*, were crated and sent to the USSR. Some of them went to *Zavod 88* (Factory 88), a rocket plant formerly producing oil-well drilling equipment until the machinery was evacuated to the Urals in 1942. *Zavod 88* was located in Podlipki (later renamed Kaliningrad), some 10 miles northeast of Moscow. Other components were shipped to *Zavod 456* in Khimki, 4 miles northwest of Moscow, a plant where jet engines for aircraft and rocket engines for missiles were being built. None of the K1 missiles were destined to be launched or test-fired in Germany.

In the autumn of 1946, Gröttrup was asked to produce a rough design for a missile with a range of 1,500 miles, with no specification as to warhead weight or accuracy. Such a sketch was completed in a few days and handed over to Korolev for transmittal to Berlin and thence to Moscow. Gröttrup had based it on the A9/A10 concept developed a few years earlier at Peenemünde.

Relationships between the Germans and the Russians were proper and in some cases even friendly. There was a brief period of initial distrust on the part of both, however. The Russians were suspicious of the true intentions of the Germans to put forth their best effort in reconstructing the V-2 or making improvements on it. Some even feared that the Germans would actively sabotage such efforts. The Germans, on the other hand, worried about the permanence of their jobs and had the nagging fear of being sent to the USSR.

Gröttrup in particular got along well with Korolev, being very impressed by Korolev's professionalism and the respect shown him by the other Soviet engineers. Korolev had a sympathetic ear for the personal problems of the Germans at *Institut Rabe*, and on one occasion used his authority and energy to rebuff an MVD (secret service) attempt to harass Gröttrup's secretary.

Things were going well for the Germans in Bleicherode and the vicinity as autumn of 1946 approached. A meeting had been scheduled for early October to discuss improvements to the V-2, but was postponed until October 21. The conference was a spirited one and presided over by General Gaidukov. Proposed plans for increasing the range and accuracy of the rocket were discussed, and Gaidukov seemed to be in agreement and at his amiable best. Following the meeting, he insisted that Gröttrup and his managers join him in a party. Gröttrup knew that he could forgo any plans for getting home at an early hour. He had been a guest at Gaidukov's parties on the latter's previous visits to Bleicherode.

The party began with the inevitable toasts to comradeship between the Russians and the Germans, and they were made with the best vodka, shipped in seemingly endless quantities from Moscow. The Soviet toasts were returned by the Germans as protocol demanded. These, in turn, prompted another counter round of toasts by the Russians. Thus the party continued until 4:00 AM on October 22.

An hour or so before the party broke up, Irmgard Gröttrup was awakened by a telephone call from the wife of one of her husband's engineers. The distraught woman asked if Irmgard was going to Moscow with the other Germans of *Zentralwerke*!

"For Heaven's sake! What a time for bad jokes! You must be drunk," she said as she hung up the telephone.

It was the first of a series of calls that continued for an hour or so. Later, Irmgard recalled:

> Then there was the sound of powerful engines . . . Cars stopping at the door. How many? Beginning to realise what was happening, I jumped out of bed. From every window I could see Russians; the house was surrounded by soldiers with tommy-guns. Outside were cars and lorries, nose to tail.
>
> Someone pressed the front door bell and kept his finger on it—fists hammered on the door—the noise echoed through the house.

The same scene was being played all over the Soviet zone. At each of the homes of the some six thousand German citizens employed by the USSR, a young army officer pounded upon the door and upon its opening immediately began reading:

> As the works in which you are employed are being transferred to the USSR, you and your entire family will have to be ready to leave for the USSR. You and your family will entrain in passenger coaches. The freight car is available for your household chattels. Soldiers will assist you in the loading. You will receive a new contract after your arrival in the USSR. Conditions under the contract will be the same as apply to skilled workers in the USSR. For the time being, your contract will be to work in the Soviet Union for five years. You will be provided with food and clothing for the journey which you must expect to last three or four weeks.

The notice was the culmination of a plan, the last details of which had been worked out several days previously by Colonel General Ivan A. Serov, deputy special commissioner of the Soviet Military Administration in Germany and first deputy of the MVD. His operation between October 12 and 16 resulted in the removal of some twenty thousand Germans to the Soviet Union. They were transported there in ninety-two trains that had been previously positioned at railway stations throughout the Soviet zone of Germany.

Irmgard, once the initial shock had worn off, called Gröttrup. His voice held nothing of reassurance: "Do be sensible! General Gaidukov is with me, the room is full of officers—You understand, don't you? There is nothing I can do. I may come home, on the other hand I may not see you until we get on the train. I may have to fly over in advance or follow afterwards."

As events transpired, the Gröttrups were to travel to the USSR in three compartments of the sleeping cars parked on the siding at Klein Bodungen. However, it was 3:00 PM on the following day before the train departed. Several members of the group had yet to be rounded up. A Dr. Ronger appeared with a nearly incredible story: The Russians had orders to evacuate him and his wife. Ronger's wife had died three days previously, and the Russians urged him seriously to take any woman he wanted—he could get married in Moscow.

The train had scarcely got under way before Gröttrup was dictating to his secretary. Entitled "Official and Formal Protest Against the Deportation of Central Works Employees to Russia," the lengthy document opened with the preamble: "In the numerous discussions which went on right up to the aforementioned dated of 23.10.46, the delegates of the Special Commission, and in particular Colonel Korolev and General Gaidukov, repeatedly stressed that a removal of the works to the USSR or any part thereof would only be considered within the next few years."

When the protocol was presented to the Russians, who suddenly seemed much more distant and far less friendly than they had at *Zentralwerke*, the result was predictable. *Nyet!*

The train arrived in Moscow on October 28, as Irmgard recorded in her diary:

> We were neither gassed nor received at the Kremlin. It all happened like this: General Professor Kolsianovich [in fact, Pobedonostsev] was waiting with a few officers to welcome us on the platform. I had known him and some of his officers in Germany so it was quite a pleasant reunion. He advanced towards me with outstretched hands as if I were his personal guest: "Welcome to Moscow! What have you brought for me? A cheerful mood, I hope!"

On the arrival in the USSR, the Germans from *Zentralwerke* were split into two approximately equal groups. One was sent on to the island of Gorodomlya in Lake Seliger, some 150 miles northwest of Moscow. Conditions on the island were incredibly primitive, and the island was in a region that had seen some of the bitterest fighting on the Russian front only four years earlier, in January 1942. As a result, the *Nemtsy* (Germans) were received with outright hostility by the local Russians. The other group, the Gröttrups among them, was settled in the northeastern section of Moscow, near Datschen, in relative comfort.

The Gröttrups rated a six-room villa with an entrance hall and two anterooms, the home of a former council minister. The other members of his staff did not fare so well. They were quartered in great mansions of tsarist days: one room to a family of three, two rooms to a family of four, with university graduates having an additional room.

On November 4, the Gröttrups' BMW automobile arrived from Bleicherode, but they could not drive it since they had no Soviet driver's licenses and the Russians did not recognize international permits. However, their obliging hosts provided them with a chauffeur, Ivan Ivanovich, who soon proved to be a great help to Irmgard, protecting her from rapacious and conniving Moscow pickpockets.

The Germans worried about when if ever they would be permitted to return to Germany. The salary situation also was a critical issue, odd as this may seem under the circumstances. The Germans arrived in Moscow with no contracts, and the Soviets seemed in no hurry to draw them up. Negotiations proved useless. After attempting to reason with the Russians for several months, Gröttrup simply went on strike on April 30, 1947. He refused to go to his office in *Nauchnii isledovatelskii instituit Nii-88* (Scientific Research Institute 88). He presented a list of ten or so complaints to General Leo Gonor, the commander, who ignored them completely. However, on May 2 salaries were set. In general, they were much lower than the previous monthly stipend. Gröttrup's wages were set at 8,500 rubles monthly (he had been receiving 10,000). Even so, this sum was four times that paid to his Soviet counterpart in industry.

Relatively few rocket specialists such as Gröttrup were transported to the USSR—only two hundred men and their families were taken from among the five thousand employed by the *Zentralwerke*. The remainder were individuals currently working in production and research facilities in every field of technology and science. In each case, their fate was the same. It is described by Robert A. Kilmarx in his book *A History of Soviet Air Power*:

> The pattern of exploitation ran along these lines. A German design team in Russia would be given specifications on which to base an independent developmental project. In most cases original German ideas were eagerly sought. Russian engineers and technicians were attached to the German groups and learned all they could from them before being replaced by another group of trainees. At a certain stage, an independent Russian design-research team would be established. The team would siphon off German advances and incorporate them into its own work. From time to time progress among the Germans would be investigated by the Russians and the project might even be canceled—if the state of Russian progress or priorities so dictated. By the early 1950s, the usefulness of most groups of Germans had passed and, after a cooling-off period, they were allowed to return to Germany. Only in certain critical areas like missile guidance were the captured [sic] Germans retained longer.

The first six months in Moscow continued to be chaotic. While most of Gröttrup's men stayed with him or went to Gorodomlya Island, several key members of the former *Zentralwerke* were sent to the Communications Ministry, the Air Ministry, and other organizations. Furthermore, engineers and scientists from other institutions or companies evacuated at the same time as the *Zentralwerke* were gratuitously assigned to Gröttrup. In none of the personnel changes was he consulted. It was clear that the Soviets had no idea of using the Germans as a team, in Moscow at least. Those on Gorodomlya Island, however, were so organized and utilized.

Conditions at *Nii-88* were later described by Irmgard:

> Here the men have to make do with badly equipped laboratories and take to work such tools and apparatus as they still happen to have among their private possessions.... What's more one of the men even took the kitchen alarm clock to pieces because the clockwork had the very steel spring he needed so urgently! To his wife's exclamation: "Surely this joke is going too far! Aren't there any steel

springs in this strength to be had in the whole of Russia?" He replied: "Not in stock, and it would take a year for the head of the buying department to get them."

One of the first tasks for the people at *Nii-88* was to assist the Soviets in setting up a pilot production line for the V-2 in *Zavod 88.* The guidance and control experts on Gorodomlya went to work on building a *Bahnmodell*, a simulator for rocket trajectories, which was completed in about a month and sent to *Nii-88*. Work in both locations was hampered by the fact that the documentation from *Zentralwerke* had not accompanied the personnel in October. Indeed, it did not arrive until the summer of the following year. Thus, on many projects, engineers had first to reconstruct the drawings and documents that were lost somewhere in the Soviet bureaucracy. A team of some twenty Germans under a Dr. Putze had been sent to *Zavod 456* to work with its new director, Valentin P. Glushko. However, Glushko did not like the Germans and largely ignored them; he made no effort to stop Gröttrup's request to have them transferred to *Nii-88.* Still other propulsion specialists, such as Dr. Ferdinand Brandner, wound up in Kuibyshev under the direction of A. G. Kostikov, who had developed the solid-propellant *Katyusha* rocket of World War II.

During Gröttrup's strike, Professor Colonel Yuri Pobedonostsev, chief engineer at *Nii-88* to whom he reported, gave Gröttrup a report to read "during the holidays," as the Russian put it. It was "*Über Einen Raketenantriebe Fernbomber*," dated August 1944, and prepared by Dr. Eugen Sänger and Dr. Irene Bredt of the *Deutsche Luftfahrtforschung* in Ainbring, only a hundred copies of which had been printed. In it, Sänger and Bredt proposed an "antipodal" bomber—one that would be boosted into space by a large rocket, circle the world by alternately dipping into the upper atmosphere and pulling up until it reached the target, drop its bomb, and by the same maneuver return to its base.

The Soviets were at the time showing great interest in the report. On April 14, there was a meeting at the Kremlin in the office of M. A. Voznesensky, deputy chairman of the Council of Ministers, chairman of the State Planning Commission, and member of the Politburo. Present were G. M. Malenkov (who would later succeed Stalin), deputy chairman of the Council of Ministers, and second secretary of the Central Committee; D. F. Ustinov, minister of armaments; Air Marshal K. A. Vershinin; A. S. Yakolev, aircraft designer; A. I. Mikoyan, aircraft designer; Lieutenant General T. F. Kutzevalov, Soviet Air Forces; Colonel G. A. Tokady, chief rocket scientist of the Soviet Air Force; and M. V. Khrunichev, minister of aircraft production.

They were there to discuss the technical feasibility of the Sänger project and were looking to Tokady for an expert opinion. The colonel (as described in his book *Stalin Means War* and articles in *Spaceflight* magazine) explained that he had been given only the briefest time to examine the very detailed and highly technical report. Nevertheless, the group pressed him for an answer, especially Voznesensky. After some hesitation, Tokady replied that he "was inclined to think that Sänger was a gifted scientist but rather academic and lacking in practical experience." He also thought that Sänger's "physics and mathematics should be carefully checked." A day later found Tokady in the presence of Josef Stalin and the Politburo Council of Ministers. Stalin's mind was probably made up before he listened to the colonel's opinions of the Sänger proposal. Still the latter expressed his opinions rather freely. He mentioned that the Germans, for example, were at least ten years ahead of the Russians in jet- and rocket-propelled aircraft and missiles. Stalin, then, replied, "In other words, we shall have to learn from the *Nazis*—is that it?" Tokady was hesitant to answer the question; when he was pressed by Stalin, he said, "It is hard for me to say, Comrade Stalin; perhaps the Germans have given more thought to war than we have. Perhaps in their history militarism occupied a more fundamental place."

Stalin permitted the colonel to speak for some three-quarters of an hour. Later, Tokady recalled:

> Sänger had produced a number of stimulating notions, but he was not an engineer. I suspected some of his equations: my own calculations so far suggested that he was wrong in putting the thrust of his machine at 100 tons. In any case, we had no experience in the field of rockets, we lacked the men, the research institutes.... Our metallurgists had not yet produced metals sufficiently heat-resisting to use for the combustion chamber....

At the time, Tokady's choice of words worked to his disadvantage. To those in the room, he seemed to be implying that the USSR thought of war and espoused militarism, both of which were counter to the current "party line." He was reprimanded for his words after the meeting by Serov.

The result of the conference with Stalin was the formation of Special Commission No. 2 for investigating the Sänger project and reporting directly to the Council of Ministers. (Curiously enough, it had not only a title similar to that of the special commission headed by Hans Kammler in the latter days of the war in Germany but also had a similar purpose.) It was directed by Serov, with Tokady as the deputy. Also on the committee was Major General Vasily I. Stalin, the son of the premier. The commission went off immediately to Berlin to pursue its task of coming up with a final report by August 1, 1947. However, like so many bureaucratic undertakings in the USSR at the time, the commission failed to provide any useful information, and the project came to naught.

Later in 1947, the Sänger report was sent to the Germans on Gorodomlya for further study and comment. Apparently, someone in Moscow still held hopes for the concept. However, they received little encouragement from the Germans, who doubted the calculations of a mass ratio of 0.1 postulated by Sänger. (The mass ratio of a rocket vehicle is the empty weight divided by the weight at launch. The value proposed was about that realized some two decades later with the American Saturn 5 rocket.) Additionally, they pointed out that the "skip" flight path foreseen for the winged craft would impose severe loads on the wing root structure and that Sänger did not say how they were to be accommodated. Gröttrup himself did not believe that an exhaust velocity of 9,900 to 16,500 feet per second and a chamber pressure in the engine of 1,470 pounds per square inch was attainable with the current Soviet technology. He also felt that Sänger had inadequately dealt with the problems of reentry. The method of launching, from a 1.8-mile-long ramp, also seemed to have no advantage over a vertical lift-off. All in all, the Germans corroborated the criticism of Tokady, and there probably was no subsequent attempt to build the "antipodal bomber," once the Soviets appreciated the advances in the state of technology it required and their own shortcomings in fields such as high-temperature metals.

The Germans also were asked to design a new rocket, the R10. Actually, they had started limited work on the project while still in Germany, and the rocket was then labeled G1. By early autumn 1947, the documentation and drawings for the R10 had reached an advanced stage, even though the Germans did not have access to their previous drawings done at *Zentralwerke*. Both the Germans at *Nii-88* and at Gorodomlya were occupied on the project.

R10 was one of the few projects on which Gröttrup's men would work together or, in modern industrial terminology, "apply the systems engineering approach." The rocket design that emerged had features that would also appear in the first generation American missiles designed by Wernher von Braun's team in the United States.

The R10 was to be 46.5 feet in length and 5.3 feet in diameter, making it rather similar in external dimensions to the wartime V-2 rocket. The newer missile would have a gross weight of nearly 41,000 pounds compared to the V-2's more than 28,000 pounds. The respective empty or unfueled weights were 4,250 and 8,750 pounds. A major advance in the R10 was its separable warhead—that of the V-2 remained attached to the empty rocket body all during flight. The Von Braun team in the

United States followed the same approach with their Redstone derivative of the V-2.

The R10 would feature a modified V-2 propulsion system, producing 70,400 pounds of thrust using liquid oxygen and ethyl alcohol. The V-2 produced only 55,000 pounds. The additional thrust was to be obtained by increasing the flow rates of the propellants to the engine, which raised the combustion chamber pressure to 294 pounds per square inch, compared to the 227 pounds per square inch of the V-2. The interior surfaces of the chamber were to be cooled, as was that of the V-2, by emitting a film of alcohol through small holes in its wall.

While the R10 would bear an external and superficial resemblance to the V-2, its structure was innovative. Gone were the internal aircraft-type propellant tanks surrounded by an aerodynamic shell. The R10 would have featured a *monocoque* structure, in which the pressurized propellant tanks formed an integral, load-bearing part of the rocket. (This weight-saving technique was also utilized in the Atlas, America's first intercontinental ballistic missile, and Britain's short-lived counterpart, the Blue Streak.) To lighten weight, and hence increase range, the propellant tanks were to be of very thin aluminum or steel.

These tanks were pressurized to two atmospheres. The liquid-oxygen tank was pressurized by the natural "boil off" of the liquid oxygen. The fuel tank was pressurized by the cooled gas tapped from the turbopump.

The high-test peroxide generator of the V-2, used to provide steam to turn the turbopumps, was done away with in favor of utilizing gases from the combustion chamber piped back to them. Initially, the pumps were brought up to speed on the launching pad by means of compressed nitrogen from a source on the ground. A weight savings of some 400 pounds was thus achieved.

The control section for the missile was moved to the aft end from its forward position in the V-2.

At the point in the trajectory where the propellants were cut off, the warhead was separated from the rocket body with no loss in momentum by two small retro-rockets on the body, fired at the moment the engine shut off, and they, in effect, backed the body away from the warhead. The warhead was covered with plywood, which in charring upon reentry into the atmosphere protected the explosive from the heat so generated. A steel warhead case was also designed at the request of the Soviets.

The R10 was to be controlled by means of jet vanes, as was the V-2. However, while there were small fins on the aft of the missile, they had no movable aerodynamic surfaces as did those of the German rocket. An improved servo system for the vanes obviated the need for these surfaces on the fins.

Guidance for the R10 consisted of "beam riding," in which four antennas on the ground established the very narrow beam path in vertical and horizontal planes. The missile lifted off vertically and was programmed to pitch over into the guide beam. Velocity was measured by an on-board Doppler transponder. Once the proper velocity for a given range was reached, the engine shutoff signal was sent to the missile. The range of the missile was considered to be about 570 miles and its accuracy would theoretically have 25 percent of all missiles fired falling into a square 0.6 mile on a side.

Despite this quantum leap in rocket technology, there is no indication that the Soviets went into serial production with the R10. However, they did adopt many of its features, especially the guidance scheme, for the wholly Soviet-designed missiles that appeared in the early 1950s.

In the autumn of 1947, Gröttrup took a break from his duties at *Nii-88*. It was not, however, a holiday.

On August 26, he was ordered onto the FMS-1 train along with several other Germans, such as Dr. Kurt Magnus, the gyroscope expert; Dr. Johannes Hoch, chief of Guidance and Control at Gorodomlya; Karl Munnich, an electronics and radar expert from the Ministry of Communications; and two propulsion test engineers from *Zavod 456*, Alfred Klippel and Otto Meier. He left in such haste he was not permitted to telephone his wife. The train wound southward for almost a week before turning eastward into the steppes beyond Stalingrad (now Volgograd). It halted in a siding of the Ryazano–Uralskaya Railway in the village of Kapustin Yar, some 75 miles from Stalingrad. The FMS-2 train was already there, apparently with a full military complement aboard.

Gröttrup and his colleagues were at the railhead of the Soviet Union's first long-range rocket range, upon which construction had begun earlier in the summer. It was still a primitive place, though. It had been built by some eight thousand military engineers who were still at the site living in American army tents. They were there to assist in launching the first V-2s.

Among the Soviet dignitaries on hand was Colonel Spiridonov, the affable commander of Branch No. 1 of *Nii-88*, as the Germans on Gorodomlya were denominated; he spent most of his time in the club car of the FMS-1 train, living up to his sobriquet of "Rumbarrel" bestowed on him by the Germans. Marshal Nikolai N. Voronov, commander of the artillery that effected break-through of the Mannerheim Line in Finland in 1939 and also artillery commander at Stalingrad during World War II, was there. With him was Col. Tyulin, a leading ballistician from the Academy of Artillery Science, the commander of which in 1950 would be Voronov. Also from *Nii-88* were Kurilo, head of the assembly shop, and Ginsburg, an electrical engineer, both of whom were constantly busy readying the V-2s.

Next to the railhead was an equipment dump, largely in the open. Gröttrup recognized some machinery from *Zentralwerke* and Lehesten rusting away. Some 2 miles to the north was a dirt airfield. And about 9 miles to the northeast lay the range head proper. It consisted of a horizontal test stand for checkout of the V-2s and a rail siding for the shop cars of the FMS-2 train. A mile or so to the east was a vertical static-firing test stand on the cliff of a dried river bed also with a rail siding for other cars of the FMS-2. About a mile and a half to the northwest of the static firing stand was the launching point itself. It was served by FMS Car No. 28, which had the same function as the *Fürleitpanzer* did in the mobile V-2 battery. Nearby were the German *Messina* and *Hawaii* telemetry receiving antennas. (The Soviets also attempted to measure the attenuation of radar signals passing through the highly ionized exhaust stream of a rocket motor. Curiously enough, the experiment was made with an American SCR-548 radar, which had tracked British-launched V-2s in Operation Backfire.)

As Gröttrup worked at Kapustin Yar, living in one of the sleeping cars of FMS-1, Irmgard had no word from him and could find none at *Nii-88*. Her continual harassment of the officials, however, paid off.

On October 19, she was permitted to fly to Kapustin Yar to be with him. She was destined to be one of the few females at the range. (Another, in an official capacity, was Mme. Katya Chernikova, chief of the laboratory at Gorodomlya that was concerned with electrical measurement instrumentation.) Irmgard's diary later offered a revealing glimpse of living conditions on the steppes:

> The testing ground is some way off: a launching base and testing ground for carrying out tests on the automatic and electric controls of the unfuelled rocket, and a test-bed for the firing of fully loaded rockets.... We also have a mobile army bath unit. The instructions for us are to enter the carriage, knock on the wall and wait until the *Moujik* [peasant] has pumped water into the water container. Then turn on shower. I go in and knock and proceed to soap myself well all over. I turn on the shower but nothing happens. The *Moujik* has fallen asleep.

While Irmgard Gröttrup contented herself walking in nearby Kapustin Yar, where there were more camels than cars, and enjoying the incredibly sweet melons grown on the collective farm nearby, her husband was extremely busy. Kapustin Yar was no Peenemünde by any means, although practically all of the equipment there had come either from Peenemünde or Lehesten. There were green US army tents instead of the sturdy masonry barracks that had housed the soldiers of the *VKN* at Peenemünde. Gröttrup lived and worked in a train rather than the modern *BSM Haus* as he had done at the center on the Baltic.

On October 27, as the last details were being taken care of for the launch of the first V-2 only two days away, Gröttrup was talking with a group of visitors from *Nii-88* when suddenly a Russian worker fell 65 feet from a scaffolding and cracked his skull. Gröttrup turned pale, but his Russian visitors did not miss a word of conversation as the late worker was dragged away by his fellows. A day later, a beam came loose and crashed to the ground, killing a leader of a welding group sent in from a factory in Stalingrad to assist in getting the range ready. Again, there was no discernible reaction among the Russians. Clearly, industrial safety, which had been closely checked at Peenemünde, was of no concern at Kapustin Yar.

Irmgard was on hand for the launching of the first missile on October 30. She later recorded her impressions:

> Last night, probably the most exciting and memorable of all nights in Russia, no one can have slept. Differences in rank simply ceased to exist; there were no more professors, ministers, or officers, just one enormous wildly excited family.... It was just like Peenemünde when we made our first experiments and I went to Helmut's room only to find that it was occupied by an exhausted professor from Dresden University.... The queen bee of our hive is the rocket!
>
> A beautiful clear morning dawns over the steppes of Kazakhstan.... The tension has become so acute that I could scream.... We are off! "Zero minus 10 ... zero minus 9 ... zero minus 8 ... zero minus 7 ... zero minus 6 ... zero minus 5 ..." Suddenly the launching platform collapses sideways and with it the fully loaded rocket. One leg of the platform has given way....
>
> We make a dash for the bunker, while workmen run toward the rocket and, with absolutely no sign of fear, winch the whole thing back into position, platform, rocket and all, and prop it up with girders. There's Russia for you!

The countdown was resumed, and the rocket was launched. Then there followed a few emotionally charged minutes as everyone waited for the word from the Askania cinetheodolite stations, the superb monitoring equipment which came from Peenemünde, lining the route of the missile as it sped almost due east. Suddenly, Minister of Defense Armaments Ustinov grabbed Korolev in a bear hug and danced him about. Korolev, in turn, did the same with Gröttrup. There was almost pandemonium at the launch area as word was announced that the rocket had flown almost 175 miles and landed reasonably near the target.

Gröttrup, however, soon overcame his emotion. The fact that the first launch was so successful was simply a statistical occurrence that could easily have gone the other way, as, indeed, it had done at Peenemünde when the first V-2 was launched.

"We can't play around here like naked savages," he said to his wife. "Besides, there is more to come tomorrow. The next missile has to be got ready!"

On the next day, the second V-2 behaved just as badly as had the first one at Peenemünde. It began rolling faster and faster about its longitudinal axis until the fins sheared off and the rocket tumbled into the ground from an altitude of about 500 feet. While Ustinov began muttering to himself about the possibility of sabotage, Gröttrup and his launching chief, Fritz Viebach, merely shrugged. Korolev sympathized with his German associates. Being the engineer that he was, he knew that the complexity of the V-2 made perfect take-offs each time an impossibility.

In all, some twenty V-2s were launched from Kapustin Yar before the Gröttrups were permitted to return to Moscow, which they did on December 1, 1947, traveling on the FMS-1 train, that had been parked in a siding near *Nii-88*.

A variety of warheads or payloads was fired during this series. Some were merely instrumented ballasts to determine heats of reentry, etc. Others were the standard, high-explosive warhead used with the tactical V-2. Yet others were scientific instruments to measure the cosmic ray flux above the sensible atmosphere. These instruments were supplied by the Academy of Sciences of the USSR, several members of which were present for the firings.

The twenty-odd V-2s were launched alternately by Viebach's all-German crew and an all-Russian crew under the direction of L. A. Voskresensky, who at *Zentralwerke* had been Soviet manager for the development of the FMS trains. Upon removal to *Nii-88*, Voskresensky rose rapidly. He became a member of the *Nauchnii Tekhnicheskii Soviet* (Scientific and Technical Council) NTS of *Nii-88* and an assistant to Korolev. (In later years, he would become Korolev's deputy as a space vehicle systems designer and play a key role in the early manned spaceflight programs of the USSR. He died on December 15, 1965, only a month before Korolev.)

Scarcely had Gröttrup settled down to the routine at *Nü-88* than he was called upon to attend a meeting of the NTS. This group was a policymaking body for *Nü-88*. Some eighteen or twenty members were associated with the establishment while others were not. Gonor, Korolev, Pobedonostsev were all members. Glushko from *Zavod 456* at Khimki was a member, as was a Professor Frankel, an Austrian aerodynamicist who had emigrated to the USSR in the early 1930s and engineer Mikhail K. Tikhonravov was there, too. He had worked with Korolev and

Glushko a decade earlier in Moscow as a member of the Group for Studying Reaction Propulsion, which built and static-fired liquid propellant rockets.

Meeting with the Russians across the red-draped tables on the ground-floor conference room of *Nii-88*, next door to the technical library, were most of the department heads from Branch I of *Nii-88*. Among those present were Dr. Waldemar Wolff, chief of ballistics; Dr. Werner Albring, chief of aerodynamics; Dr. Joachim Umpfenbach, chief of propulsion; Engineer Josef Blass, chief of design; Dr. Johannes Hoch, chief of guidance and control; and Dr. Franz Mathes, chief of propellant chemistry.

The occasion was similar to an academic event in the West where a candidate for the PhD degree must defend his dissertation against his faculty. The Germans earlier had been asked to propose a design for the R10 rocket. Their first attempt had not pleased the Russians of the NTS, who demanded more details. The Germans now had returned with more than two hundred detailed drawings and several hundred pages of engineering reports, including the involved mathematics that so pleased the Russians. Impressed, the committee approved the project and allocated the funds necessary to undertake it provisionally.

However, the Germans were to be disappointed. For all their hard work, they were assigned very few tasks on the program. They were permitted to do some structural testing of the missile body and to assist in the development of the gas-bleeding process for driving the propellant turbopumps. They also worked some on the Doppler radio system for the R10 guidance and control. Other than these areas, they were not called upon. The Germans soon lost interest when it became apparent that they were not to be permitted to develop the overall rocket.

With the beginning of 1948, the Russians began removing Germans from *Nii-88* and other locations in Moscow to Gorodomlya. Gröttrup was among them; he was not particularly downcast to leave *Nii-88*, but his pay was cut to 5,000 rubles. Umpfenbach took his place at *Nii-88*. Gröttrup's position and authority had gradually been eroded, both by the Russians and by several ambitious Germans who worked for him. On February 20, 1948, Gröttrup and his daughter Ulli left for the island, while his wife and their son Peter stayed behind for a few months because of the boy's illness.

To add to the low morale of the Germans at *Nii-88*, Korolev had decided to push ahead with his pet project, which was the improved V-2, begun at *Zentralwerke* and improved upon by features he had gleaned from the documentation produced by the Germans for the R10. He had the ear of General Gonor, and he had his own fabrication shop adjacent to *Nii-88*, but he lacked the technical expertise of the Germans. These latter had no desire to work for Korolev on his project because it simply was no professional challenge. "Stretching" the V-2 to get another 75 or so miles range did not interest them. Also, if they went to work in his shop, their integrity as a team—or what was left of one—would be compromised. Thus, there was another motivation for not staying in Moscow.

"If we let Korolev have his own way, he'll grab all the best people for himself and we shall be turned into mere drudges, or, at best, walking technical handbooks. . . . I proved to him ages ago that his plan of simply building a longer V-2 and providing it with a more powerful thrust won't materially increase the range," said the perpetually gloomy Umpfenbach.

The lot of the Germans on Gorodomlya had improved considerably by the time the Gröttrups arrived. Earlier, in October 1946, when the initial group had stepped from the ferry that brought them across the lake to Oshtakov, they were met by the genial, bald, and rotund Colonel Spiridonov, an affable party hack and great friend of Ustinov, who was chief engineer on the island under its manager, Suchomlinov.

What the Germans found ashore were filthy and verminous huts, water that had to be hauled from the lake, and no sewage or public sanitation by Western standards. All of the "pioneers" assured the Gröttrups that things were immeasurably better than when they had arrived. Irmgard was hard to convince. Used to the luxury of a villa near Moscow and a chauffeur, she found it hard to adapt to her new quarters. Her diary furnishes evidence of her frustrations and privations:

> *15.11.1948.* At nine o'clock I took a hatchet to the shed and hacked off a piece of frozen meat. By eleven o'clock when it was almost defrosted, I decided it was almost bones. So it had to be cabbage soup again. My stove, bearing the proud inscription "Made in Gorodomlya," is incredibly primitive, constructed of clay and stones. . . . I'd much rather have windows that shut properly, I'd like some decent shoes for the children. . . . "What, cabbage again?" was the general outcry at lunch. Helmut was busy making calculations on the edge of the table. I pointed out to him that he was not setting a good example, but all I got in reply was an absent-minded glance and a remark about a fully automatic test stand. . . .

During the remainder of 1948 and 1949, the Germans on Gorodomlya did some useful work. They were given selective design or consulting jobs on several rockets being considered by the Soviets. From the summer of 1948 until the spring of the following year, the group worked on the R12, a multistage rocket that would send a 2,200-pound warhead to a distance of 1,500 miles. However, the engineering problems associated with stage separation and particularly second-stage ignition were so formidable that nothing came of this effort. In midsummer 1949, the Germans were consulted on the R13 rocket being studied at *Nii-88*. Curiously, it seemed to be a technical retrogression. It was to have a 2,200-pound warhead and a range of only 75 miles.

On April 4, 1949, Ustinov visited Gorodomlya with a proposal. He enlisted the assistance of the German group on the design characteristics of the R14, a rocket that would send a 6,600-pound warhead to a distance of 1,800 miles. He gave them three months to have preliminary designs ready but imposed no other limitations upon them.

The project in some ways breathed new life into Gröttrup's people on Gorodomlya. The fact that a man as high up as Ustinov brought them the project seemed to indicate that it would be an important one, one in which they "would have a piece of the action."

Several alternative approaches were considered. Some wanted to use as yet unproved high-energy propellants to boost performance. Others wanted to use multi-engines. Some thought that a two-stage or three-stage rocket would be required. Methodically, each one of these concepts was discussed and rejected or modified in some degree. The R14 finally proposed by the Germans was certainly no "uprated" V-2. It was a new departure in rocket design. Indeed, at the time, it was considerably in advance of anything being proposed or thought of by von Braun and his team in the United States. Branch 1 proposed a single-stage, conically shaped rocket 77.6 feet long and 9 feet in diameter. It had a relatively low initial acceleration, which meant that the structure of the rocket would not have to be excessively heavy. There were to be no fins or other aerodynamic control surfaces, and the rocket on the pad would weigh some 156,200 pounds. With an empty weight of about 15,400 pounds, including a 6,600-pound warhead, the Germans were proposing a vehicle with a mass ratio of 0.1. The *monocoque* structure was to be of pressurized stainless steel tanks for the propellants and the warhead section, although the plywood ablation technique for cooling the warhead was maintained.

This version of the R14 utilized alcohol and liquid oxygen. (The Germans seemed to have a technological fixation on this combination, both in the USSR and the US, simply because they were familiar with it.) The engine was to have a thrust of 220,000 pounds, with the combustion chamber regeneratively cooled by circulating alcohol through its walls. The turbopumps for the propellants were to be driven by hot gases "bootstrapped" from the combustion chamber of the main engine, as had been proposed with the R10. A novel feature for roll control, or the ability to keep the missile from rotating about its longitudinal axis, was the use of the exhaust gases from the turbopumps vented through a nozzle which could be swiveled to counterreact rolling. (This same feature would be introduced as a secret feature on the Jupiter intermediate-range missile produced by von Braun's team at Redstone Arsenal in the United States a few years later—another transfer of technology to both the US and the USSR due largely to the engineering foresight and imagination of Ludwig Roth's advance planners at Peenemünde three or four years earlier.) Path control was accomplished by mounting the engine on a double knife edge or alternately on a ball and socket and moving it with pneumatic or hydraulic actuators.

The warhead compartment was to be separated from the rocket body by explosive bolts, a significant saving in weight over the retro-rockets proposed for the same purpose on the R10. Also, the warhead was to be encased in plywood as a means of insulating it from the heat of reentry into the atmosphere.

While mobile ground launching equipment was proposed for the R14, Gröttrup personally was convinced that it was unnecessary. The advantage that could be gained in moving the missile a few hundred miles from the point where it was manufactured to its launch site was negligible in terms of intercontinental targets. He proposed quite simply, and logically, launching the missile from its manufacturing plant!

Accordingly, Heinz Jaffke and his associate Anton Närr designed an underground factory, similar to *Mittelwerk*, from which the R14 was to be built and launched. (Jaffke was a talented designer of such structures. He assisted in the design of the first large rocket static-test facility of the Russians outside Zagorsk, some 36 miles northeast of Moscow. In time it included a 220,000-pound thrust test stand, a structure that came in handy when the USSR began planning space missions not too far in the future.) The facility Jaffke proposed could not only manufacture the R14, but it could also extract the oxygen needed from the air and store sufficient quantities of ethyl alcohol for the rockets that were to be launched from an underground silo adjacent to the plant!

On October 1, 1949, the NTS arrived on the island from *Nii-88*. Along with Gonor and Pobedonostsev was Korolev. They were briefed on the missile design and took away with them all the drawings and reports that had been generated by Gröttrup's task force. Then the Germans heard nothing more of the R14, except that they were asked in the following month to redesign the warhead section to include its kinetic energy in the destructive effect on the target. Work on that project continued through February of the following year. The other task given them was to redesign the rocket body to utilize aluminum rather than steel.

Following these instructions, the Germans received no bonuses, but they also received no criticisms. It is possible that by then the decision had been made to return the Germans to their homeland and that the Soviets simply wanted to gain as much technical data from them as was possible.

An alternate approach to developing a missile with a warhead weight of 6,600 pounds and a range of 1,800 miles was followed by a somewhat smaller group on Gorodomlya led by Albring. It was designated R15 and was an unmanned version of the Sänger-Bredt antipodal bomber. With it, an R10 or V-2 booster would send a small, pilotless bomber several miles into the air. There its ramjet engine would ignite and send the craft up to a height of 8 miles, from which it would dive on to the target. Drawings and reports were submitted to the NTS, but nothing more was heard of the project. In 1951, Dr. Joachim Umpfenbach received two tasks that may have been related to developmental work on the R15 or a similar vehicle. One was to design a statoscope, or instrument to maintain a constant height for a vehicle in flight by means of a barometric device, such as had been used on the V-1. The other task was the design of a gyroscope using the concept of an oscillating pendulum instead of a rotating mass.

There were other rocket projects on which the Germans worked piecemeal. The Soviets showed interest in an antiaircraft rocket similar to the *Wasserfall.* A small group at *Nii-88*, under the direction of a Soviet engineer named Silnishnikov, was employed on such a weapon. There was also a test stand for the *Wasserfall* engine located at *Nii-88*. Karl Harnish, a former Peenemündian with a master's degree in engineering, was involved briefly in the static firing of the engines at *Nii-88.* Additionally, Albring and Hoch, at Gorodomlya, were asked to produce *Taifun* drawings, and a few engines were made in the shops at *Nii-88* and static-fired there, too. Also at *Nii-88* a Soviet engineer named Rashkov had a small group that was studying the *Schmetterling* antiaircraft missile. Two Germans, Dr. W. Quessel and Karl Falkenmeyer, assisted him for a short time.

In the summer of 1949, Gröttrup was asked by Spiridonov if the German group could undertake the design of an antimissile missile. Considering the time being spent on the R14, Gröttrup said no, and added that he felt such a rocket was too far beyond the technology of the day. Apparently, it was a passing fancy at the time; Spiridonov never brought the subject up again.

Following the completion of the studies on the R10, R14, and R15, life on Gorodomlya became stultifying at best. Some ideas of conditions there can be gleaned from Irmgard's diary:

> 28.7.1949. We live surrounded by water; there is even an inland lake, but we have no drinking water. The water from the tap comes

> out dirty, full of seaweed and tiny animals. . . . The atmosphere is very tense and unpleasant. . . .
>
> 5.8.1949. . . . for most houses here have raw floorboards with wide cracks. The advantage of these is that the dirt can be brushed straight into the cracks; on the other hand, you can't get rid of the bugs.
>
> 13.4.1950. The days seem to drag on. These long nights painfully increase my longing for home and freedom. . . .
>
> 20.4.1950. Mrs. R. [Rebitzki] had a nervous breakdown and was taken to the clinic at Kalinin. She has been forced to have her hair cut, sit around with fifty insane women and to eat miserable food out of tin bowls.

That petty squabbling and marital infidelity occurred among the Germans was predictable, if, indeed, not understandable under the circumstances. That many of the talented and frustrated men turned to vodka once the stimulus of constructive and challenging work had been denied them was equally predictable. For some, infidelity and vodka were not strong enough measures. Mrs. Engelhardt Rebitzki hanged herself, and an engineer named Möller wandered out in the snow to accomplish the same end, only to be found in time; he lost both hands to frostbite. Similar things happened in other German collectives. By the summer of 1954, when Dr. Brunolf Baade, former chief engineer of the Junkers Aircraft Company, returned to Germany with the remainder of his original team of eight hundred men and their families, twenty-five had died in the USSR, five had committed suicide, and two had gone insane.

On December 21, 1950, Gröttrup was relieved as chief of the German collective. Previously, it had been agreed among the Germans that they would do no more work for the Soviets. Taking a leaf from the texts of their masters, the Germans were going to show strength through solidarity. On that day, Ustinov sent his deputy to Gorodomlya with a new set of tasks. Seeking a way of refusing to comply, the Germans hit upon the idea of refusing on grounds of health hazards. The new project required the use of nitric acid rather than oxygen as a propellant. It would give off exhaust fumes that were poisonous.

However, a small faction of the Germans decided to break ranks, defecting to the Soviets. Once they did, the deputy minister announced that Gröttrup was no longer in charge. Dr. Johannes Hoch was appointed to fill his position and moved to Moscow, where he worked for five years until his death there.

In contrast to the primitive domestic accommodations that the Germans had found on Gorodomlya in 1946 were the island's engineering facilities. There was a Mach 5 wind tunnel and a fully instrumented static test stand for liquid propellant engines with thrusts up to 17,600 pounds.

Additionally, there was an excellently equipped high-frequency laboratory for the electronics and communications specialists. The Soviets also provided the best of scientific instruments for their German tutors—which was, in effect, what they were to become. It was increasingly apparent to Gröttrup, after his arrival in 1948, that the primary function of Branch No. 1 of *Nii-88* was pedagogical.

With the beginning of 1951, there was an influx of young Russian engineers just out of the technical universities. They were eager to learn from the Germans; and they were polite, friendly, and appreciative as well. However, the Germans had long since lost their motivation and enthusiasm. While a few tasks related to the R14 continued to come from *Nii-88*, they received only half-hearted attention. Most of the men had given up hope of ever seeing Germany again.

Then, with the sly perversity shown on so many occasions, the Russians announced that the first group of twenty Germans would be leaving for their homeland on March 21, 1951! The people in this group were all technicians and thus most easily dispensed with.

To his further humiliation and discomfort, in September Gröttrup was required to give up his small house, the only one on the island with an indoor bath, and to move into a smaller apartment. In February of the following year, he fell seriously ill. For a fortnight he was comatose, and his temperature rose alarmingly. It took two and a half days before the local *feldsher* (medical technician) appeared. She made him comfortable, gave him some caffeine, glucose, and penicillin. Apparently the last drug turned the tide. Gröttrup recovered, but was not in really good health for well over a year.

By mid-1952, the young Soviet engineers were still arriving on the island. They began taking over more and more of the work of the Germans, utilizing them only as a sounding board for their own ideas and for technical approaches to problem solving. Concomitantly, the tasks assigned to the Germans grew fewer and increasingly unrelated to rocketry. Gröttrup was asked by the university in Leningrad to design a computer for controlling a lens-grinding machine. Colleagues who had worked so dedicatedly on the R10, R14, and R15 found themselves designing a vehicle that could skim across the surface of ice-covered bodies of water. By then, the Germans knew that they were being replaced with native talent.

But what would be their fate?

On June 15, a commission arrived from Moscow. After a long conference with the men of the island, the news was released to the women and children. All but twenty of Branch No. 1 of *Nii-88* would be returned to Germany on June 21. Gröttrup was among the twenty designated to stay.

With the death of Stalin on March 5, 1953, the Russians on the island all seemed in agreement that the remainder of the Germans would soon be on their way home too. Yet when the men reported for work on that day, expecting a national day of mourning, they were told: "Comrades, let's honor the dead by working harder than ever!" Disheartened and gloomy, they turned to their dull tasks.

But rumors kept cropping up that they would be repatriated. On November 15, the commission from Moscow returned. Their statement, read to the German engineers, was explicit: "All German specialists with the exception of twelve are to return to their country on November 22, 1953. They must leave within two days of that date. We take this opportunity to express our thanks for the work done."

Gröttrup's name was *not* among the twelve to stay.

The dozen who stayed were all specialists in rocket guidance. Under a Dr. Faulstich, they were removed to Moscow and given five-year contracts with good salaries and excellent accommodation. Thus, by the end of 1953, the Russians felt confident in their own competence in the areas of propulsion and structures; but they still were not so confident of their expertise in guidance and control. On November 28, 1953, Helmut and Irmgard Gröttrup, with Ulli and Peter, crossed the river Oder into Frankfurt-an-der-Oder, "the river which to us is the frontier between two worlds, between past and future," as Irmgard later put it.

After seven years, the majority of the men of *Institut Rabe* and *Zentralwerke* were home. In five more years, the remainder would also return.

What had the Russians gained from them?

In the final analysis, they certainly had obtained some talented men from Peenemünde, although by no means in the numbers that the Americans had Scientists and engineers such as Gröttrup, Magnus, Umpfenbach, and Viebach were without doubt of assistance to them in specialized areas. However, such professional talent must be cultivated and motivated if it is to be optimally productive. Clearly, the Soviets did not realize this fact or were consciously unwilling to do so.

As a result, the Russians deprived themselves of many valuable and creative contributions that these men could have made to Soviet rocketry at a very crucial point in its postwar development. Fundamentally, the Russians failed to realize that the team approach or "systems engineering" technique was what had produced the V-2, while their own engineers could produce only the small, solid-propellant *katyusha*. In compartmentalizing the Germans and fragmenting their efforts and skills, the Russians wasted a great engineering potential.

Despite this treatment of the Germans, the Russians took what they had learned from them and quickly put it to use. Men such as Sergei P.

Korolev, Valentin P. Glushko, and Aleksei M. Isayev (a designer of rocket engines) who had nothing to gain from the Germans in theory quickly absorbed their engineering and managerial techniques and then went on to form teams to build the rocket boosters that sent the first unmanned and, later, manned spacecraft into orbit about Earth and scientific probes to the Moon, Venus and Mars.

CHAPTER 18

V-2s in the Desert

As Wernher von Braun and some of his Peenemünde team were being interrogated at Garmisch-Partenkirchen, a lanky American ordnance officer, Lieutenant Colonel Harold R. Turner—though he did not know it at the time—was preparing a new home for them in the deserts of New Mexico. Turner had recently returned to the United States from London (where he had just missed death by V-2 when one fell in Hyde Park, not far from the Great Cumberland Hotel in which he was billeted). There, he had been in charge of modifying American fighter aircraft to accept the army's 3.5-inch rocket.

In February 1944, Major General Gladeon M. Barnes, chief of the Technical Division, Office of Chief of Ordnance, in Washington, had sent teams of Ordnance Department, Corps of Engineers, and War Department civilian engineers across the country to search out a site for a long-range proving ground for rockets. The urgency was dictated because the army by then had under contract several weapons that would need such a range.

Barnes's instructions to that group obviously directed its search to the spacious western portion of the nation. He told its members to look for a place with "extraordinarily clear weather throughout the preponderance of the year; large amounts of open, uninhabited terrain; accessibility to rail and power facilities; and . . . proximity to communities to provide for the cultural needs of personnel to be employed at such an installation."

These specified characteristics, with the equivocal exception of the last one, were met in an expanse of desolation some 80 miles north of El Paso,

Texas, and some 30 miles east of Las Cruces, New Mexico. There, in an area about 60 miles wide and 90 miles long, lies the Tularosa Basin, a flat sea of sand between the Sacramento Mountains on the east and the San Andreas Mountains on the west.

It is a region rich in American history. The area around Las Cruces is associated with Billy the Kid, the infamous outlaw of the late 1870s. Sheriff Pat Garrett, the man who gunned Billy down in 1881, was shot to death seventeen years later at nearby Organ. Adjacent to the eastern boundary of the range is Dog Canyon, where the Apache Indian Geronimo ambushed "pony soldiers" from Fort Bliss.

The region is also historically connected with modern American rocketry. Only 120 miles to the northeast lies the Mescalero Ranch, near Roswell, New Mexico, where Dr. Robert H. Goddard conducted his research during the 1930s and early 1940s.

In the annals of the history of science and technology, the range has a more ominous page. It was and still is indelibly described in the sand some 80 miles north of Turner's desert headquarters at White Sands Proving Ground. The place had received a top-secret place name in September 1944: Trinity Site, where the world's first atomic bomb was detonated on July 16, 1945.

"I had arrived in Las Cruces two days before the explosion. When the bomb went off, I was asleep in the Amador Hotel in Las Cruces," Turner later recalled to the authors. "The first I knew of it was on reading the morning newspaper although it was on property I was supposed to command. It was the army's best kept secret."

After reconnoitering his new domain, Turner quickly realized that his first priority should be drilling for water. "Our first day, I felt like Moses," he said. "Plain nothing was in front of me, and I went around putting a stick here and there, and ordering 'Let there be water here and let there be water there.'"

In creating the new establishment, Turner had to overcome a variety of problems seldom faced by a military commander. He solved them by constantly shouting "Damn it, do it!" In doing so, he had to improvise and take chances in the face of the inertia of the American military machine, which was getting back to its peacetime norm. Indeed, his zeal almost had him court-martialed by Major General John L. Homer, commanding general of Fort Bliss, nearby in El Paso, Texas.

It seems that friends in the Corps of Engineers had told Turner that it would be permissible to tear down some old cavalry buildings at Fort Bliss in order to obtain lumber he critically needed in the desert. With his troops and a fleet of trucks, he attacked the buildings like termites. A speechless General Homer could not believe what was happening to his real estate. Turner's friends had not gotten around to putting through the proper papers informing Homer that the razing of the buildings was authorized. In Washington, Colonel Toftoy took up the problem of the court-martial with Colonel Trichel, who diplomatically settled the case out of court.

In his rush to finish the base, Turner received news that would have shattered a less dedicated and resourceful man. In mid-August, Barnes called to tell him that three hundred freight cars of V-2 components were on their way to him. The Santa Fe Railroad promised to spot ten cars per day in Las Cruces. While Turner was pondering the problem of unloading them, he was informed that every siding from El Paso to Belen, New Mexico, a distance of 210 miles, was full of his cars. Washington, in turn, was adamant that he not incur any demurrage (a fee paid by a shipper or recipient if a freight car is not unloaded and made available to the railway company by a certain date).

"We hired practically every flatbed truck in Dona Ana County," he later remembered, "and moved the mess in twenty days."

The first group of Germans arrived at White Sands Proving Ground in October. Billeted initially in Building H, a huge wooden airplane hangar to be used for assembling V-2s and other missiles, they wondered if they had been hasty in agreeing to come to America. By Christmas their sense of isolation and loneliness became especially acute. To brighten spirits, the imaginative Magnus von Braun again turned impresario as he had done earlier at Garmisch-Partenkirchen. He produced a kind of Teutonic "Christmas Carol." There were no ghosts of Christmas past, present, or future. However, Magnus described an event that would take place in the distant year 2000: *man's first flight into space.* Observing the hustle and bustle of men and machines around a huge rocket there in the desert was a curious figure that most people seemed to ignore. Constantly in the way was an old man, bent with age and supporting himself shakily on a cane. It was eighty-eight-year-old Wernher von Braun, who had lived to see his dream come true.

The first days at White Sands and Fort Bliss were tedious ones for the Germans. Not only was there practically no recreation available to them, but also army policy prohibited social intercourse between officers and enlisted men and their former enemies. They had to look to each other for diversions, which inevitably included horseplay and practical jokes.

Hans Palaoro had spent many weeks completing a meticulous drawing of a proposed ramjet engine. Once he had put the finishing touches to it, he went off to find Dr Gerhard Heller, his boss. When the two returned, there was an 8-inch rusty spike driven through the center of the drawing and apparently through the board. Palaoro turned white and then began screaming in rage. He grabbed the spike and tugged on it. It came out so easily he fell backward onto the floor. A tiny needle had been soldered to half of a sawed-off spike. The drawing was undamaged aside from an almost microscopic hole.

Life in their barracks also gave rise to tricks. At night, to relieve the boredom of being POPs ("*prisoners of peace*," as von Braun called the team), the Germans often staged battles between barracks. For weapons, they used pillows, sandbags, and fire extinguishers. In one of these forays, von Braun led a group onto the roof of an enemy barracks and began pelting its inhabitants from that vantage point. As the battle developed, the leader of the opposing forces shouted up to him that he might as well surrender because of the greater number of the forces against him.

"That may be," called down von Braun, more the physicist than the military tactician, "but you forget our higher potential energy; we're *up* here throwing *down* there!"

While von Braun was cautious and diplomatic in dealing with the officials at Fort Bliss, his associate Dr. Ernst Geissler did not mind making a few waves. He demanded from General Homer the use of the Fort Bliss swimming pool under the threat of returning to Germany upon refusal. It was a canny ultimatum. Geissler figured that if the demand were rejected it meant the Germans had no future in the United States. Homer capitulated, permitting its exclusive use by the Germans one afternoon a week. He also made the bowling alley similarly available.

Once a month, in groups of four and with an army enlisted man as escort, the Germans went into El Paso to shop, have a meal in a good restaurant, or simply to sit on the benches in the Alligator Plaza in the center of town. When such excursions took place immediately after payday, the restaurateurs were always taken aback by four men paying for a $1.50 dinner with $100 bills. The army always paid the team its per diem with a $100 bill and several $1 bills.

There were other forms of recreation as well. Excursions were made to the Carlsbad Caverns, and in the spring and summer some went mountain climbing in the Organs, which they found a small-scale substitute for the Bavarian Alps. Helmut Schmidt, a former Olympic ski champion, looked forward to winter when he could take to the nearby mountains.

Despite the recreational opportunities, living accommodations and working facilities were something less than desirable, and in some cases inadequate. Displaying the same resourcefulness and skill at improvisation they had shown at Garmisch-Partenkirchen immediately after the war, the team made the best of a bad lot. While its members were given the equivalent of junior officers' quarters at Fort Bliss, such officers in the US Army did not live in anything approaching luxury. In addition, there was a shortage of furniture, and the quartermaster seemed under no compulsion to find any. Using scrap lumber, wooden boxes and crates, and other such material as they could "organize," the Germans made their own.

Hans Grüne explained fruitlessly to the supply people that he had to have a linoleum floor on the room in which he assembled gyroscopes because the sand kept sifting through the cracks in boards and getting into his precision instruments for the V-2 guidance unit. For one reason or another, the linoleum was not forthcoming. Grüne then resorted to the old army practice of "moonlight requisitioning." He simply went out at night and pilfered enough of the material to cover his floor and installed it himself.

In the winter of 1945, the team applied the same skill in converting a set of old noncommissioned officer quarters to a social club. It consisted of a reading room (least crowded), game room (always crowded), and a bar (generally packed to the point of spontaneous combustion). Gerd de Beek, former chief illustrator at Peenemünde, added a touch that remained a popular feature until the wives started arriving at Fort Bliss. He drew the silhouette of a voluptuous Rhine maiden on the window shade of the bar. When the shade was drawn, the bar attracted an unusual number of customers—not that business was ever slack.

"A shot and a half of 'red eye' cost twenty-two cents," Walter "Junior" Wiesman (who had shortened his name from the former Wiesemann) later recalled fondly. "For a buck, you could be in orbit."

To inaugurate the club, Wiesman collaborated with Magnus von Braun on a spoof of the good old days at Peenemünde. Typical dialogue included that between a harassed V-2 manager and Paul Figge, the key procurement figure at Mittelwerk.

> V-2 MANAGER: "Sir, we need high-speed drills desperately!"
> FIGGE: "Now, let me see . . . the last time I saw the Bishop of Andorra, I traded him some good cognac for the metal we need for those drills."

Time was spent in more constructive ways as well.

All the team members sought to overcome the difficulties of a new language. Their patient editor of technical reports, Hoffman Birney, described their problems—and his—at the time:

> When it comes to editing their writing, that's another story. They write in German, then translate with the help of a dictionary, and the result is something like the interlinear translations of Virgil and Ovid that helped you pass freshman Latin. In German it's quite grammatical to say: "My old grandmother a red cow with a bell around its neck has"—and their English is likewise, I'm sure. Add to that the fact that their stuff is scientific, largely theoretical, and that it bristles with equations.

Little by little, their English continued to improve. Their instructional means were informal for the most part. For some, learning meant spending hours in sundry stores, listening to Americans order food, buy clothing, or haggle for automobiles. For Wiesman and Magnus von Braun, it meant sitting in an El Paso movie from the time it opened on a Saturday morning until the last show of the evening, repeating, to the annoyance of the other patrons, line after line of generally insipid dialogue to each other. When Wiesman was not in the theater, he was working part-time in Schneider's Grocery with a Mexican-American clerk named Choppo. Thus, his English took on an odd Germano-Hispanic accent. To further his familiarity with the American idiom, Wiesman also listened to comedian Jack Benny on the radio. The scholarly Dr. Carl Wagner broadened his knowledge by reading professional journals of theoretical mathematics in English.

Despite their growing fluency in English, certain things confused them. One shop in town that mystified them was The Venetian Blind Man. They simply could not imagine what type of work a blind Italian from Venice could be doing in El Paso. An engineer wanting to buy some flints for his cigarette lighter was delighted to see a Firestone store. Proud of his familiarity with English and knowing that the German word for lighter flints was *Feuerstein* or firestone, he went in and ordered a dozen.

English was also spoken as much as possible on the job, at von Braun's insistence. However, when things required technical discussions, the team lapsed into German. They also looked about carefully to see that Herbert Karsch, the American technical director, was not around before using the phrase *Amerikanische Dummkopfe* (stupid Americans). Several members had been caught by the German-speaking Karsch and called down about their language.

Any ideas the Germans had of advancing their work at Peenemünde quickly vanished after they had been at Fort Bliss and White Sands for a while. Their mission was largely one of technical exploitation and assistance to others. They were there for three purposes: to serve as consultants to American industry and research institutions involved in guided missiles research; to assist in the assembly, checkout, and launching of the V-2s sent over from *Mittelwerk*; and to conduct studies and propose new guided missile projects. Under this last phase, they became involved in the design of a ramjet engine under the code name Hermes 2.

Major General Barnes visited Fort Bliss on January 11, 1946, and von Braun made a presentation to him on a scheme for using the V-2 as a first stage for a ramjet-propelled second stage. The concept had been discussed as an intellectual exercise during the period the Germans were whiling away their time in Garmisch-Partenkirchen. At Fort Bliss, the idea became attractive because there were no really large rocket engines available or being planned for the near future. Von Braun and his colleagues thought that they could sell the plan to the army because it would not require expensive outlays for a lot of new rocket components or massive test facilities.

Hermes 2 was not to be a tactical weapon; rather, it was to be the prototype of one that would demonstrate the feasibility of a two-stage missile with a supersonic ramjet propulsion system. As proposed, the V-2 stage would boost the ramjet stage to an altitude of some 12 miles. The second stage would then ignite and add its velocity to that of the V-2 to obtain a speed of 2,100 miles per hour.

Why the Germans should have been encouraged in the project is not easily explained. Clearly, they had no experience in that form of propulsion. In addition, the Applied Physics Laboratory of Johns Hopkins University, under Project Bumblebee, was developing a ramjet weapon called Cobra. However, none of the reports on Cobra and its progress were permitted to the Germans for security reasons.

For a while, the group sought to impress its new employer by designing a gigantic rocket motor with a thrust of 300,000 pounds. Such industry and thinking on a large scale probably prompted the army to construct what is still called the White Elephant at White Sands. It is a static test stand for motors with thrusts up to 500,000 pounds. Cut into the side of a small mountain, it was later used to test the Redstone and Nike Hercules rocket motors, neither of which produced anywhere near 500,000 pounds.

Not only did the men of Peenemünde serve American industry well during this period, but also many of their reports of previous experimentation at the former German research center found their way into the nascent American missile industry. Several such documents were sent from White Sands to the Consolidated Vultee Aircraft Corporation at Downey, California, for its use in designing a research rocket known as the MX-774. This rocket was then being developed for the US Air Force, and it had far-reaching results. (The MX-774, on paper, went through three designs. Design A was a subsonic missile called Teetotaler, because it did not use alcohol as did the V-2. Design B was named Old Fashion because it externally resembled the V-2. Design C was dubbed Manhattan because it was being designed to carry a nuclear warhead, the allusion being to the wartime Manhattan Project to develop the atomic bomb.) It was the forebear of Atlas, the first US intercontinental ballistic missile.

Not all members of the group were assigned to work on the V-2 effort in the desert, however.

Klaus Scheuffelen, Herbert Dobrick, Friedrich Dohm, and Kurt Neuhoffer were engaged in another rocket project. It was, in fact, an extension of work done earlier at Peenemünde on the *Taifun* antiaircraft rocket. Its American counterpart was dubbed Loki and was to be developed by the Bendix Corporation's Pioneer Division, in Teterboro, New Jersey.

These team members, in October 1947, became the first to live in the local economy. First Lieutenant M. S. Hochmuth, who had been with some of them in Germany, was their keeper and found apartments for them near the Bendix plant; but they were under surveillance by intelligence personnel from the US First Army. The sergeant placed in charge of them proved to be a soldier with a long memory. He made life as miserable for them as he could until Hochmuth found out about the situation and had him transferred to other duties.

Professional relationships between the Germans and their new American colleagues were for the most part amiable. There was a certain reluctance on the part of the American engineers in both industry and the civil service to admit that the Germans were "smarter than they were" in rocketry. They also were puzzled by the German insistence upon a rigid organizational structure with many fanciful titles and positions. One early American member assigned to the group described it as having "a president and 124 vice-presidents." The protocol also agitated Major Hamill, who was not the easiest boss for whom to work, being alternately and unpredictably happy and cooperative or short-tempered and uncommunicative. Often von Braun felt that his authority was being diluted by Hamill's way of doing business. There were several times when he verbally or graphically turned in his resignation as leader of the team; on one such occasion he suggested Dr. Ludwig Roth as his replacement. With each submittal, Hamill simply threw away the memorandum or ignored the phone call.

Engineers and scientists from almost every discipline associated with rocket theory and technology found their way to the unique facility in the western desert between 1946 and 1950. Typical of such consultations was a session von Braun had on June 24, 1946, in which the War Department General Staff G-2 (intelligence) received his estimate that the German rocket team then working for the Russians could bring the A10 weapon proposed at Peenemünde into production by 1949.

Acknowledged experts in the group were in almost constant demand. Men such as Walter Riedel and Gerhard Heller spent several months at the plant of North American Aviation in California, where they assisted fledgling American propulsion engineers in the fundamentals of rocket engine design and fabrication. A little more than a decade later, engines made by that company sent the first men to the Moon. Similarly, Ernst Steinhoff was sought for his knowledge of electronics. The US Air Force borrowed him to set up an integrated optical and radar tracking and measuring system for its nearby facility at Holloman Air Force Base.

One day in the summer of 1946, Hamill asked Sergeant Erich Wormser, who had been with the first group of Germans at Aberdeen Proving Ground, if he would interpret during an interview of von Braun and others for a US naval officer, who was visiting Fort Bliss. The officer wanted to know only one thing: "Did you people ever consider launching V-2s from submarines?" Von Braun explained that such a project had gotten to a stage of serious consideration just as the war ended. He then told the story of *Prufstand XII,* the floating launch pad under development at Peenemünde when the war ended.

Activities such as these made up for the hours that the Germans felt were largely wasted in assembling and firing aging V-2s or solving technical problems associated with such launchings. Standing before a blackboard covered with equations and talking to their American peers more than compensated for the hours spent in tracing down a faulty circuit or locating a balky valve in a rusty V-2.

It was not long before von Braun and his colleagues seriously began to wonder if they had not made a mistake in accepting Toftoy's contract. The operation at White Sands and their function at Fort Bliss were certainly of a shoestring type. Funds simply did not exist for the basic tools and materials they needed to do the job the army had prescribed.

"Frankly, we were disappointed with what we found in this country during the first year or so. At Peenemünde, we had been coddled. Here they were counting pennies. The armed forces were being demobilized and everybody wanted military expenditures curtailed," von Braun was later to remark to the authors.

Typical of the way in which they had to operate was the manner in which they acquired casters for a mobile test stand they were building. Money was not available to purchase commercial casters, and none were available in the local army warehouses. So the Germans and their military assistants removed the axles from a Meilerwagen missile transporter and machined them into the bearings and casters needed.

When Arthur Rudolph suggested that work would speed up on the assembly of the V-2s if General Electric, under contract to support the Army, would put four men on the job instead of one, he was told that there was only one metric wrench available! Rudolph then suggested that GE go into the hardware store in Las Cruces and buy several one-inch wrenches. He would then have the machine shop file them to the desired metric size. However, GE balked at spending the money. Finally, in desperation, Rudolph drew up the design for the wrench required, and the soldier machinists of Battery B, 1st Guided Missile Regiment, at Fort Bliss made them for him.

Rudolph did not know that at the same time, half a world away, an identical complaint was being voiced in the USSR by Helmut Gröttrup. "How are we to work if there is nothing to work with? No materials, no tools—not even tables! Test stands, dismantled in Germany, can be seen for miles along the railway lines. The A and B containers [V-2 propellant tanks] are bent; girders are corroding in the snow—soon they'll be nothing but scrap metal," he complained to Minister of Armaments Ustinov.

At White Sands, Erich Kaschig once needed a length of copper tubing to assemble a V-2 component upon which he was working. His GE counterpart took him to the salvage yard for wrecked motor vehicles.

On hearing that there was a shortage of tools, a sergeant wishing to help said he knew where there was a freight car full of them. The POPs, the GIs, and the GEs all hurried to the railway yard and literally tore open the door of the car. It was full of shovels, picks, and axes.

On March 22, 1946, Toftoy made one of his frequent trips to see how the program was shaping up at White Sands. Hamill told him that only thirty-five missiles could be assembled from the components available. Toftoy detailed Hochmuth to track down as many parts as he could in the United States and then return to Europe if necessary to look for more. Hochmuth's job was not an easy one. The V-2 had been designed to be launched—hopefully—within a week of manufacture. It simply was not meant to be a munition stored for an indefinite time in a magazine or warehouse; thus, there were no stockpiles of spare parts produced.

Having been unable to find any V-2 parts in the States, Hochmuth returned to Europe in May. He had letters from the Chief of Ordnance directing all US Army ordnance personnel to assist him. He even went to the British and begged for parts left over from Operation Backfire, but they refused him. Hochmuth also had letters from the Germans at White Sands written to companies in Germany that had manufactured V-2 components. By and large, this approach yielded little because postwar German industry was still in a shambles. Even more frustrating was the fact that factories that could have assisted were forbidden to do so by the Allied military government: no German plant could produce war matériel.

He had trouble placing an order for the critical carbon vanes for the control system with a French company that had manufactured them for the Germans during the war because the French did not want to admit that they had done so.

In the end, many parts had to be manufactured by GE and the Douglas Aircraft Company, as well as many other American industries.

During the six years between April 16, 1946, and September 19, 1952, a total of sixty-four V-2s was launched from White Sands Proving

Ground. Additionally, one was launched from an aircraft carrier at sea; and two were fired as part of the Bumper program from the Long Range Proving Ground in Florida, which later became the US Air Force Eastern Test Range and the Kennedy Space Center.

The V-2 gave American scientists for the first time the means for probing the upper fringes of the earth's atmosphere or—put another way—the lower fringes of interplanetary space. The rocket could place a payload of scientific instruments weighing about a ton briefly to a point well over a hundred miles above the earth or five times as high as current sounding balloons. A special V-2 Upper Atmosphere Panel with members from industry, universities, and the armed forces coordinated the instruments for each missile fired.

Writer Daniel Lang, of *The New Yorker* magazine, in a visit to the White Sands Proving Ground, interviewed Dr. Charles F. Green, a consultant to GE. The significance of the V-2 to scientific research was succinctly put by Dr. Green; "Right now, we know as much about upstairs as a fish does about land." That fish, he went on to elaborate his metaphor, knew what a fisherman looked like but it could not explain him. The V-2 permitted man to explain things about the upper atmosphere that previously he could only guess at. By using sensors in a V-2, scientists had learned, among other things, that the average separation in distance between molecules there was 370 inches, while the distance on the surface of the earth is only about one millionth of an inch.

Typical of the investigations made were studies in solar spectroscopy; solar radiation; meteorites; temperatures; winds; ambient pressures at very high altitudes, and photography of the earth. Despite the often erratic behavior of the parachutes used to recover instruments and cameras, many valuable data in all these fields accumulated at an unprecedented rate.

The first White Sands V-2 was launched on April 16, 1946, and it failed. The rocket reached an altitude of some 3½ miles when a fin came off, and it had to be destroyed by shutting off the propellants. Happily for Turner, the next firing on May 10 was a success. It had been laid on to impress a group of generals and admirals from the Pentagon, a number of journalists, and top scientists and industrialists. With an ear-splitting roar, it lumbered off the launcher and rose to an altitude of 71 miles. While it was ultimately not to be a record, Turner gratefully presented small wooden models of the V-2 to the members of his staff and the team who had helped make the show a success.

Despite the relative crudity of range safety equipment in those early days, wayward missiles and consequent damage were rare. When one did go astray, it did so with a certain technical élan. On May 29, 1947, at 7:35 PM, a V-2, prophetically numbered Missile O, took off normally enough, but a gyroscope in the guidance unit malfunctioned. The missile arched backward and headed south for El Paso. It continued over the city, crossed the Mexican border, and impacted a mile and a half south of the city of Juárez, which was thronged for a fiesta. As it ploughed into a rocky hillside, just outside Tepeyac Cemetery, the missile barely missed a building where the construction companies of Juárez stored their dynamite and blasting powder. Incredibly, no one was killed or injured and no damage was done to buildings or homes. The missile, of course, had no warhead; but it blasted a crater 30 feet deep and 50 feet in diameter through its kinetic energy alone.

"In a few minutes, everything broke loose," Colonel Turner later recalled. "I checked with my friend the commanding general of the State of Chihuahua. He assured me that there was no damage and that he would clear the unfortunate event with the authorities in Mexico City."

Scarcely had he put down that phone when he found himself on another, explaining what had happened to General Dwight D. Eisenhower, chief of staff of the US Army in Washington. Having satisfied Ike, the harassed commander of White Sands a few minutes later found himself talking to Secretary of State George C. Marshall.

The gravity of the situation had completely escaped one enthusiastic young lieutenant in Turner's command. He proudly boasted that he belonged to the first American rocket unit to fire a guided missile against a foreign country!

Turner later explained what had happened: "The accident was a result of mistakes both mechanical and human. A faulty gyroscope caused the rocket to head south instead of north. A civilian worker neglected to push the button that would have cut off the rocket's fuel supply and have made it fall within the limits of the target range. He misjudged the direction of the missile. It looked to him that it went straight up."

However, Ernst Steinhoff, who was the "civilian worker," had a different story: "I was the range safety officer, and I had to tell a sailor, who came from the Naval Research Laboratories, to push the button, which would cut off the fuel to the engine. However, I knew from experience what happens when a V-2 with residual propellants impacts. It starts a nasty fire. So I told the sailor not to push the button. I wanted to let the propellants all burn out and have the missile travel as far as possible, hoping it would clear both El Paso and Juárez."

A board of investigation after the incident saw the merit of his action, although one naval officer remarked, "You know, I've never seen such a cold-blooded bastard."

Steinhoff later recalled that within ten minutes of the impact, food stalls in the vicinity of the cemetery were selling still warm souvenirs of the V-2.

He added, with only slight hyperbole, that altogether the Mexicans sold at least 10 to 15 tons of material (much resembling tin cans) from a rocket that weighed only 4 tons.

Diplomatic repercussions from the mishap were few when it became known that the US Army would make financial restitution. Among the claims was an inquiry from a Mexican gentleman who stated that his wife had been so frightened by the blast that she had become sexually frigid. When it was explained to him that he would have to file a *written* statement of all circumstances, the claimant stated that rather than do that he would gladly settle for $250 in cash.

The launch on December 17, 1946, provided a mystery for the Germans as well as their American colleagues. At midnight, a V-2 with a very important scientific payload took off. In the warhead compartment were several army rifle grenades filled with steel pellets. At peak altitude, these were to be fired from the V-2, producing what would have been the first man-made meteorites. The rocket climbed to a height of 100 miles, but no trace was ever found of the warhead compartment, which was to have been recovered by parachute. Neither were any meteorites seen. The question rather irrationally arose as to whether the compartment or some of the pellets might not have gone into orbit about the Earth. While the former was clearly impossible, several "lucky" pellets might have achieved the necessary velocity to do so.

Other V-2 activity of note during the desert years included the launching of six Bumper missiles from White Sands. Bumpers, the first two-stage missiles in the United States, consisted of V-2 first stages and Wac Corporal second stages. One launched on February 24, 1949, set an altitude record for the time of 244 miles. Such experiments provided much-needed data for the design of multistage rockets.

Later, in 1948 and 1949, Project Albert launched some of the first animal payloads into space. Two monkeys named Albert were instrumented to determine the effects of high acceleration on a primate. Unfortunately in both launches the parachute for the monkey compartments failed to deploy, and the animals were killed when they crashed into the desert. However, both monkeys had been instrumented to telemeter data concerning heart rate and respiration rate. Data from lift-off to crash indicated that they successfully survived the rigors of the flight, including a lift-off acceleration of about 5.5 Gs, in which their usual 9 pounds of weight increased to 49.5 pounds.

Even though they were busy preparing and launching V-2s and carrying out their other assignments, morale was low among the Germans during their first year at White Sands. Mail from their families at Landshut was irregular and sometimes impossibly slow. Alfred Finzel and

Ludwig Roth had no idea whether they were fathers or still expectant fathers. Frantically, Hamill sought Toftoy's help again. As usual, the always warmhearted and resourceful colonel came through, arranging a special mail delivery outside the usual army mail channel. For some reason, the Department of State had written into the Paperclip agreement that none of the Germans brought to the United States would be permitted to send packages through military channels back to Germany. With their wives and children in want for clothing and drugs, the Germans in El Paso found themselves moving wistfully through the plentifully stocked department stores and drugstores.

On December 4, Hamill found himself with a particularly heavy cross to bear. The War Department had decided to permit newspaper interviews with the Paperclip personnel for the first time. The El Paso *Herald-Post* sent out reporter Virginia Strom (who later became Mrs. Harold R. Turner and editor of the newspaper) to talk to the von Brauns, Carl Wagner, Martin Schilling, Ernst Steinhoff, Bernhard Tessman, and others. But she was especially interested in some opinions held by Walter Riedel, a man noted among his colleagues for speaking up in no uncertain terms on subjects about which he felt strongly. Her headline for his story read: "American Cooking 'Tasteless,' Says German Rocket Scientist; Dislikes 'Rubberized Chicken.'"

"Everything is fried," he told Miss Strom. "The bread when you cut it looks like cotton. Then, you serve what we call 'rubberized chicken' fried to a crisp."

Riedel had trod on ground sacred to middle-class America, especially in Texas. To hold American fried chicken up invidiously to any other national culinary treatment was cause enough to reopen the war or send the German lot at Fort Bliss packing back to Germany so far as the local citizens were concerned. They expressed themselves sternly and forcefully on the subject in letters to the editor of the El Paso *Herald-Post*.

Then there were problems with the US Internal Revenue Service. For almost two years, no income tax was deducted from the pay of Paperclip personnel through an oversight on the part of those who had drawn up the initial contracts. It was not until July 1, 1947, that such taxes began being withheld. The Internal Revenue Service demanded back payment of all monies owed on both the salary paid in German marks and the $6 per diem paid in the US—in addition to 6 percent annual interest on the total. Charitably, for the Internal Revenue Service, it did not impose a penalty payment as well.

Dieter Huzel, who was to be one of the first of the group to leave it for employment in the budding American aerospace industry, later described the financial troubles in his book *Peenemünde to Canaveral:*

> Naturally, as good future US citizens, we objected to the double taxation: a legal battle, with attorneys' fees, ensued. By the time the final decision was handed down that we had to pay the tax (plus 6% interest per year), the currency reform had become effective in Germany, reducing all monetary values to 5% of their original rate. Thus, the taxes paid turned out to be higher than the amount received.

(Almost unbelievably, at the same time, their former colleagues at Peenemünde then working for the Russians at the *Centralwerke* near Nordhausen faced exactly the same problem. However, their Russian employer was more beneficent than the regulation-bound US Army. The Soviets solved the problem pragmatically by merely raising the wages of the Germans to cover the taxes imposed by the renascent German revenue service.)

Suspicion of the ultimate intentions of the Germans lingered in the minds of some American officials for several years. In the spring of 1947, the loyalty of the Germans was apparently still in doubt. Despite the fact that they had for the most part been given contracts with the same rights as American civil service employees (except that they were forbidden to leave the continental US), military intelligence at Fort Bliss darkly noted from "reliable sources" that "the Germans from Fort Bliss are establishing relations with persons in El Paso who were born in Germany or are of German descent and are using Schneider's Grocery as a gathering place." The intelligence report further noted that "several German scientists and technicians have been known to cross the border into Juárez, Mexico." The late hours in the El Paso *Biergarten* often ended in a trolley ride to Juárez for more fleshly delights across the border.

The problem of security so far as the Germans were concerned remained an ambiguous one for several years. There was no question in the minds of Hamill and Toftoy of their loyalty; not so, however, among the faceless bureaucrats in Washington.

The awesome and omnipotent force of J. Edgar Hoover, director of the Federal Bureau of Investigation, was felt by the Paperclip Germans on September 13, 1947. Hoover wrote to Lieutenant General Stephen J. Chamberlain, director of intelligence of the US Army, that he felt no classified technical information should be given to the personnel at Fort Bliss, a restriction that would seriously impair their effectiveness. He based his recommendation on a letter from a minor functionary in GE, who clearly saw another rung on the ladder of advancement within his grasp.

Hoover's influence then, as it remained until his death in 1973, was considerable. Even the US Army dared not counter him. Every effort was made by the army to assuage him, but Hoover was not a man whose mind was easily or quickly changed. To his credit, however, it must be said that once Hoover was convinced that no danger to the nation existed, he did change his mind.

On May 11, 1948, Chamberlain made a presentation to him on the worth and value of the Germans to the defense of the United States. He stressed their contributions in the fields of antiaircraft rockets, fire control instruments, and the cold extrusion of precision metal parts. Chamberlain cleverly wove a Communist ogre into it to impress Hoover, and impressed and convinced he was. He agreed to expedite the visas of the Germans and to cut what red tape he could within the Department of Justice.

By the following month, the FBI director found the Germans in El Paso were so clean from the detailed background investigations that in the future he would accept Hamill's word on loyalty.

The visas were necessary because the Germans had entered the United States illegally. The army had simply scooped them up and flown them in without passports or visas. As far as the State Department was concerned, the group was not in the country at all. However, the problem was solved by an imaginative scheme. All those wishing to do so clambered upon streetcars in El Paso bound for Juárez, Mexico. They got off at the American consulate and were issued visas to enter the United States legally. Thus, their immigration papers bore the curious statement that the port of embarkation was Ciudad Juárez, Mexico, and the port of arrival was El Paso, Texas—both ports being 700 miles from the nearest body of navigable water.

While no spies were found among the Germans, FBI informant PT-1 on September 3, 1948 reported to the El Paso office that one of the barbers in the post exchange at Fort Bliss was a member of the Communist Party of Mexico and that he had been infiltrated to obtain photographs and information on US missiles that would be sent to the Soviet Embassy in Mexico City.

The exodus of wives and children from Landshut began in November 1946, when quarters for them became available at Fort Bliss. In preparation for their arrival, Hamill and Toftoy had convinced the Surgeon General of the US Army to give up the William Beaumont Army Hospital Annex, adjacent to Fort Bliss. The hospital was ideal as a billeting area for the growing German colony. Not only was it completely fenced in, but also it had a swimming pool and was isolated from the rest of the huge army post.

To ensure that he would be sole owner of the former hospital, Hamill, on October 26, had begun moving offices in as quickly as they were vacated by the medical personnel and their dwindling number of patients, who were being transferred to other military hospitals in the United

States. Sometimes his occupation was swifter than the Medical Department's internal communications system.

One day Hamill was sitting in his new office when the phone rang.

"Who is it?" shouted an agitated female voice.

Hamill identified himself by his rank.

"This is Captain Clark. Come over to Ward 13 immediately," the voice continued in desperation.

"I'm sorry—" Hamill began to say.

"You'll have to come over immediately," the voice broke in with near hysteria.

"Well, you don't understand. I'm Major Hamill and our job is developing rockets," he replied.

"Oh, my God, I'm developing a baby over here and in a hell of a rush and obviously you're not going to be of any help to me," replied a distraught army nurse.

The Beaumont Annex wards had been redone by the Fort Bliss post engineers into apartments. In anticipation of his coming marriage, von Braun had been assigned an apartment with two bedrooms, a combination living room and dining room, kitchen, and bath. It was next door to that of Krafft Ehricke, ex-tank commander on the Russian front and V-2 propulsion engineer at Peenemünde. Both were in what had been the former mental ward of the hospital.

On December 8, the Neuberts, Poppels, and Schulzes were the first to arrive, and domesticity began to descend upon the German commune at a US Army camp in the western deserts of Texas.

In the spring of 1947, Wernher von Braun joined his colleagues in ultimate domesticity. He wed Maria von Quistorp, after having mailed his proposal of marriage to the beautiful eighteen-year-old blonde. Von Braun was permitted to return to Landshut, where they were married in the local Lutheran church on March 1.

Having acquired a wife, von Braun quickly adopted the American custom of also acquiring a new car. While his colleagues were generally purchasing used vehicles, von Braun indulged in a new 1947 Nash Ambassador. It had a feature that appealed to both his German sense of thrift and appreciation for technical innovation: The seats converted into a bed; thus, costly stopovers at motels could be avoided.

On December 9, 1948, the first American von Braun made her appearance at the Fort Bliss Army Hospital. Daughter Iris was born.

Despite the happy homes and humdrum chores of designing military rockets, the team maintained its old dream of space travel. While the army had absolutely no plans for artificial Earth satellites or space flight, von Braun took every opportunity to bring up the topic. During an interview with journalist Virginia Strom, he told her that it would be possible within a decade to have a space station in orbit around Earth at a distance of 5,000 miles. Its proposed purpose: a refueling station for rockets bound for the Moon.

"Technically and theoretically it is possible. It is only a question of efforts which can be made in that direction," von Braun said.

Toftoy, earlier on May 8, had made an even more remarkably prophetic statement on space travel. He said, "It is possible this generation will see huge rocket ships carrying passengers, that can circle the Moon and return to Earth safely. If work could begin on such a project immediately, and enough money to finance it in the interests of pure science, it could be done and witnessed by persons who are alive today."

In addition, von Braun also evangelized within the army itself. On February 7–18, 1949, he lectured on rocket weapons to the Senior Officers Indoctrination Course No. 4 at Fort Bliss. Brigadier General Julius Klein and a group of fellow generals listened, some in awe and some in patent disbelief, as their recent enemy talked about satellites of the Earth and traveling to the Moon in rockets. To demonstrate how a satellite could be placed in orbit around Earth, von Braun used a three-stage model of a rocket.

"On his desk were about eighteen inches of books with all the calculations to back up his statements regarding the setting up of a satellite," General Klein later recalled to the authors.

In the previous year, von Braun had approached no fewer than eighteen publishers with the manuscript of a book he titled *Das Marsprojekt.* He had started it in 1947, largely as a means of combating the monumental boredom of life in semicaptivity with a lack of challenging work. The high, thin clouds above the light sandy soil of the desert had turned his receptive imagination to thoughts of Mars. He formulated a plot and a story of seventy passengers and their journey to the red planet. (It was not until 1950 that the book was printed. Even then only a portion of it appeared. The German publisher Otto Bechtle visited von Braun in Huntsville, Alabama. After looking over the manuscript as a book expert rather than a space enthusiast, he agreed to publish the appendix only, a mathematical proof of the feasibility of the proposed interplanetary flight. The novel, *qua* novel, proved beyond doubt that its author was an imaginative scientist but an execrable manufacturer of plot and dialogue.)

After five years in the desert von Braun decided to go public in pressing for space exploration. One day, while walking among the sagebrush with his associate Dr. Adolf K. Thiel, he suddenly turned to him and said with bluntness and facility in the idiom of his newly polished English, "We can dream about rockets and the Moon until Hell freezes over. Unless the *people* understand it and the man who pays the bill is behind it, no dice. You worry about your damned calculations, and I'll talk to the people."

With this manifesto there began a barrage of talks and articles on every aspect of astronautics that was to fascinate Rotarians, Kiwanians and, for a few critical years, senators and congressmen, as well as readers of the popular press.

Were the years in the desert worth it?

They decidedly were to the US Army and a large segment of American industry that would later provide the nation's guided weapons and space vehicles. The use of the Germans paid off in both time and money. Indeed, the army later announced that the practice of sending American engineers and scientists to confer with the Germans gained ten years in directing United States research into channels that would ultimately be the most rewarding. In Germany, the von Braun team of governmental, industrial, and educational investigators had explored many areas of every known aspect of rocket technology. Many of them proved to be unfruitful; thus these men were able to steer their American counterparts away from blind alleys. In terms of money, the army estimated that this "brain picking" had saved the nation at least $750 million in basic rocket research alone.

In 1947, the newly created US Air Force reported the approximate savings "to present and future" developments in that branch of service at $30 million. By 1960, Air Force Lieutenant General Donald L. Putt, recently retired Deputy Chief of Staff for Development, with the advantage of hindsight, placed the total ultimate savings from Project Paperclip at some $2 billion. The US Navy has refused to place a dollar value on its savings through Paperclip. However, in a report issued in 1949, it stated: "It is probable that no program has ever paid more rich dividends. It is not only the direct savings in time and money . . . it is also the acquisition for this country of some of the finest technical brains in the world—invaluable additions to the nation's resources."

For the German prisoners of peace, the desert years certainly offered an opportunity for respite, for collecting their thoughts after six grim war years, and for familiarization with their new country. Professionally, however, for rocket development engineers used to ample resources and facilities, they were years of frustration and disappointment. Little did they know that they were to idle along for a while longer among the green hills of Alabama before the missile and space frenzy of the late 1950s and 1960s.

CHAPTER 19

Thrust into Space

AUF WIEDERSEHEN!

Somehow, the banner above the entrance to the boxy new glass courthouse in the center of Huntsville seemed incongruous. February 25, 1970, was a cold, gray, rainy day, but the occasion was one of civic warmth, enthusiasm, and pride. One of the city's best-known citizens was leaving. He had arrived on April 15, 1950, in the sleepy little rural town of 16,000 in northern Alabama where the foothills of the Smoky Mountains taper off along the banks of the Tennessee River into broad fields of cotton.

Now, more than 5,000 of the city's 140,000 citizens turned out for a ceremony in honor of Wernher von Braun, who was leaving his home of twenty years for Washington. Refusing to believe the move was permanent, one banner read: DR. WERNHER VON BRAUN—HUNTSVILLE'S FIRST CITIZEN—ON LOAN TO WASHINGTON. There were many who felt that he someday would return to "The Space Capital of the World," as the billboards on the edges of town proclaimed Huntsville.

Von Braun's arrival with key members of his rocket team two decades earlier certainly was not celebrated with such *élan* by the local citizenry, who, at the time, proudly advertised their city as "The Watercress Capital of the World." Most of its inhabitants were either unaware of his arrival or unconcerned. A few were quite naturally antagonistic, having lost relatives in a war still vivid in their memories.

"So soon after the war, we had expected to find considerable animosity, but here, there was none," said Karl Heimburg, overly charitably. "It was very surprising. There was an understandable reticence on the part of the townspeople. It took a few weeks, but then they began seeing the person, not the nationality."

Southern hospitality and generosity soon manifested themselves. The Chamber of Commerce provided advice on housing and community life. A special welcoming committee of townspeople provided blankets and other needed items. The First National Bank made it a policy to provide a $400 loan to each German without collateral or a financial statement.

"We did that consistently and never lost a dime. They were the best credit risks in town," bank president M. B. Spragins later recalled.

At first, there was a language barrier as there had been in Texas, even though most of the Germans had been learning and speaking English for five years. Most of the stores in town closed just as the men were getting off work and returning home. As a result, some of their shopping had to be done very rapidly. One German raced to the door of a shop just as its owner was closing it. The owner shook his head and said firmly that it was closing time, then was astonished and angered by the newcomer's seemingly insolent reply: "Will you shut up now?"

The southern idiom also confused the Germans, who had grown accustomed to the western dialect. After making a purchase and turning toward the door, a former Peenemündian heard the salesman call to him in a friendly voice, "Y'all hurry back, now." Obediently, the German turned about and walked rapidly back to the counter and waited expectantly. After a few seconds of puzzled silence he made his exit, none the wiser for having complied with the command of his fellow Huntsvillian.

The Germans were also shocked at the number of Americans who suggested that the newly arrived citizens send *gift* packages to their relatives in Germany. In German, the word *Gift* means "poison."

If Wernher von Braun and the US Army were coming to town to build rockets, the majority of Huntsvillians was not wholly overjoyed at the prospect: the army had been in town once before. In 1941, the Chemical Warfare Service acquired 22,890 acres of fertile cotton fields and useless swamps and sloughs a few miles south of town on the bank of the Tennessee River. Huntsville Arsenal was built, under the command of an officer with the seemingly fictional name of Colonel Rollo Ditto. Its lethal products included mustard gas, Lewisite, tear gas, phosgene, white phosphorus, and carbonyl iron. Four years later, the US Army Air Corps requested the arsenal to test liquid propellants and methods for using them to launch the JB-2 or Loon, an early American guided missile copied from the German V-1.

The wages offered at the newly opened arsenal were well above those paid to sharecroppers, cotton pickers, field hands, and factory workers.

Predictably, workers in the city, Madison County, and towns in nearby Tennessee rushed to the new jobs. Why should a man, his wife, and older children pick cotton for $1 per 100 pounds or work as "lint heads" in one of the three local cotton mills for 28 cents an hour when they could make as much as $5 a day at even the most menial labor at Huntsville Arsenal? The cotton fields, ginning mills, and watercress ponds were almost deserted overnight. By the end of 1942, 40 percent of the employees of the ordnance plant were women.

The deadly output of the Huntsville Arsenal was loaded into a variety of artillery shells, aircraft bombs, rifle and hand grenades, and other munitions, including small rockets. However, that activity was performed at the smaller Redstone Arsenal of the US Army Ordnance Department, located next to Huntsville Arsenal. It had also been opened in 1941.

With the end of the war in 1945, the two arsenals closed. Their thousands of temporarily affluent workers returned to the cotton fields or competed for jobs in the local cotton mills, and a recently opened shoe factory, at greatly reduced wages. By 1950, the economy had somewhat stabilized. Still, few people in town were enthralled with the idea of a new army activity. Once bitten, twice shy.

There was an additional factor that weighed against the community's becoming involved with the army again. The more heavily funded US Air Force was then looking for an engineering center for its nascent rocket and jet aircraft programs. Senator John Sparkman, a native of Huntsville, had been using his influence in Congress to have the facility stationed at the old army property in his hometown. If Huntsville had again to depend upon the armed forces for its prosperity, it would be better to go with a richer branch of the military service. Unfortunately, Senator Sparkman's efforts failed. The plum went to Tullahoma, a much smaller town some 75 miles to the north in Tennessee.

Whatever the reservations of Huntsville's citizens, the newcomers were more than happy to be in Huntsville. "As Germans, we felt much more at home here," later recalled Hannes Lührsen. "Everything was so green and beautiful. I remember smelling the fresh-mown hay at Redstone. And the trees! At Fort Bliss, we had to drive two hundred miles to see five trees together!"

Within a few months, native Huntsvillians became accustomed to seeing families of flaxen-haired parents and children hiking over the wooded Monte Sano and Green mountains. Some of them even wore *Lederhosen,* perhaps pretending they were back in the *Schwarzwald.* Monte Sano became home for many of them as they began building in later years. The varied flora of the mountain soon attracted the attention of Dr. Ernst Stuhlinger, a physicist who had become von Braun's expert in electric rocket propulsion. Botany was his hobby; and in his spare time and weekends, he produced a comprehensive catalog of the plant life on Monte Sano.

As the leaves were turning color and beginning to fall on Monte Sano, Green Mountain, and Garth Ridge, in the fall of 1950, the last of the German families arrived in Huntsville. Their immediate goal was to obtain not merely places to live, but homes. For the past five years, they had existed in apartments made out of hastily converted army barracks. For a year or so prior to that, they had been cooped up in the cramped quarters of the former German army *Kaserne* at Landshut. Wives and children were adamant in demanding homes—no more apartments!

However, several factors stood in the way of immediate houses. For one thing, not that many were available. More important, the Germans did not possess the money for down payments. While the First National Bank was charitable with small personal loans, its lending officers were hard-nosed businessmen when it came to home loans. Initially, many of the families had to rent apartments or rooms.

The housing problem was solved with typical German organization, industry, and thoroughness. Fifteen families managed to come up with $7,200 and purchased thirty-seven acres on the crest of Monte Sano Mountain. The property was subdivided into thirty lots, and families built upon them as they were able to. Eventually, they were joined by another twenty families who had first settled elsewhere, outside the original German enclave. Others, including von Braun, preferring to live in the valley and hence closer to Redstone Arsenal, bought lots in an area across the street from Maple Hill Cemetery.

Karl Heimburg's imagination stood him in good stead when he personally faced the housing problem. After four months in Huntsville, Heimburg still did not have the financial wherewithal to build a house, and his wife was pressing him in the matter. He went to see George Gesman, a local real estate dealer. Gesman pointed out to Heimburg that he needed money to build a house. Heimburg pointed out that he did not have any. Gesman then said that if Heimburg had a bank account with $2,500 in it, a house could be started. That was all he needed. Heimburg made the rounds of his affluent German friends and opened a bank account the following day for $3,000.

What followed is best explained by Heimburg himself.

"With this same $3,000, switching it from account to account, four of us went to this same man in a four-day period, and we built our houses," he said. Heimburg, as did the others, later felt that Gesman knew exactly what was going on, but chose to look the other way. Indeed, he was known among them as "Gorgeous George," after a well-publicized wrestler of the day.

* * * * *

Hans Hüter, left, and von Braun, right, prepare to board an American C-47 at Munich airport in July 1945.

Lining up gyroscopes of V-2 with theodolite at Operation Backfire, October 1945.

Left to right: Lieutenant Colonel W. S. J. Carter, Major General A. M. Cameron, Brigadier L. K. Lockhart, at Operation Backfire.

A tense moment at Operation Backfire as five Russian officers show up instead of the announced three. The British stood firm and admitted only the three with credentials: Glushko, Pobedonostsev, and Sokolov.

White Sands Proving Ground, New Mexico, as it appeared when the von Braun group arrived in 1945.

Von Braun group at White Sands Proving Ground, 1946.

Members of the Army Ballistic Missile Agency team reveal their Explorer Satellite in 1957. Standing left to right: William Mrazek, Walter Häussermann, and Ernst Stuhlinger. Seated left to right: Eberhard Rees, Major General John B. Medaris, and von Braun.

Press Conference on January 31, 1958, at the National Academy of Science following the launch of the Explorer I satellite by the Jupiter-C. Left to right; Dr. William H. Pickering, Director, Jet Propulsion Laboratory; Dr. James A. Van Allen, Chairman, Physics Department, State University of Iowa; Dr. Wernher von Braun, Director of Developmental Operations, Army Ballistic Missile Agency; Dr. Richard W. Porter, General Electric.

In the early days of Redstone Arsenal, Major General Holgar N. Toftoy and Wernher von Braun were concerned with producing such missiles and rockets as, left to right: the Nike Ajax, Honest John, Corporal, and Redstone.

A tense moment in the blockhouse at Cape Canaveral. Dr. Kurt Debus, center, must decide whether to launch Pioneer 4 on March 3, 1959, or not. Left to right, in the foreground: von Braun; Debus; Walter Häussermann (slightly to the rear); Brigadier General John Barclay, Commander of the Army Ballistic Missile Agency; Karl Sendler, Debus' longtime associate.

A heated discussion of the best method of going to the Moon occurred just after this photograph was made. Von Braun, second from left, and Dr. Jerome Wiesner, second from right, argued with each other for some minutes until President Kennedy, left, intervened. Vice President Lyndon B. Johnson, fourth from right, appears lost at this point in the tour of the space facility at Huntsville, Alabama. Others shown are: NASA Administrator James Webb, third from left; US Secretary of Defense Robert McNamara, third from right.

In the early 1960s, von Braun, backed up by Dr. Robert Seamans, was a familiar figure before congressional committees that determined the amount of money that would be available to NASA. Always conscious of the value of "props" in a performance, showman von Braun brought along models of his "hardware."

The second Saturn Apollo space vehicle assembled at Kennedy Space Center emerges from the Vehicle Assembly Building on its way to the launch pad. February 6, 1968.

As homes were being completed, the Germans also began a subtle alteration of the city's cultural life.

In *Collier's* magazine, Hodding Carter reported that "when the Germans came to town, they picked up their library cards before they had their water meters turned on." He may have exaggerated, but not by much, according to Mrs. Bessie Russell, a staff member of Carnegie Library. "I don't know if that's true, but it seemed they came here the minute they hit town," she recalled.

Almost before the members of the Junior Chamber of Commerce knew it, in 1952, Walter Wiesman, a wheeler-dealer ex-sergeant of the Luftwaffe and not yet an American citizen, was their president. His election was all the more noteworthy in view of the fact that of the hundred or so members, seventy were veterans of World War II. ("So am I," said Wiesman, when the question was raised.) A fellow member of the Jaycees said, only half humorously, "I can just see him next becoming a member of the American Legion." Similarly, Walter Burose soon became a dedicated member of the Lions Club and later commander of the Huntsville Yacht Club. (Burose had a fatal heart attack while collecting funds in a door-to-door drive for the Lions Club.)

Von Braun's first contact with a local civic organization was not an overwhelming success. Speaking before the Rotary Club and clearly ill at ease, he began rather tritely by saying how much the country around Huntsville reminded him of his native Silesia. This caused some confusion among the members of his audience, several of whom muttered that they thought this fellow was from Germany. His subject, among other things, covered sending men to the Moon, a feat that could be done within 15 years, if the money were made available. Following a round of polite applause, von Braun walked toward the door and on his way overheard a man, obviously a cotton farmer, say quite audibly, "That damned fool is wanting to spend a lot of money that ought to go toward boosting our cotton crop."

A decade later, while addressing a district Rotary Club meeting, von Braun drew greater applause as well as laughter when he said, "Personally, I feel like an old-timer in Huntsville, for I have been a resident longer than eighty-seven percent of the present population of 124,000." Then, he added, in his slightly Germanic accent, "You can distinguish us old-time Huntsville residents from the newcomers by our southern drawl."

Within a few weeks after arriving in Huntsville, the German families had helped to organize St. Mark's Evangelical Lutheran Church, since no such denomination existed among the town's Protestant churches. Pastor George Hart, who specialized in getting new churches organized and built, arrived from Jacksonville, Florida, to be its minister. Even he was surprised at the industry of his new flock. A former funeral home was purchased; and when Pastor Hart remarked that it somehow did not look like a church, Wilhelm Angele and Hans Friedrich designed and built a steeple. Other members of the congregation, such as Martin Schilling, Johann Tschinkel, Hans Fichtner, Herbert Fuhrmann, and Otto Eisenhardt, also added their manual labor to refurbishing the interior of the new edifice. Pastor Hart spurred them on by attempting to communicate in German, a tongue he remembered imperfectly from childhood. Once completed, there was only one thing lacking: a bell for the steeple. It was supplied by the Southern Railway Company, which served Huntsville and donated one from a locomotive no longer in use.

Within a year or so of founding, the church's Boy Scout Troop 13 had more eagle scouts than any other in town. They included Eberhard Ball, Rainer Heller, Bernd Hellebrand, and Harmuth Schilling as well as Pastor Hart's son David.

The willingness of the Germans to contribute muscle as well as money also extended to cultural projects in the growing town. When the Rocket City Astronomical Association was founded, the decision was made to build an astronomical observatory on top of Monte Sano. Teenagers helped clear the land, and their parents laid the brick, ran the plumbing, and worked on its dome. In that phase, Wernher von Braun, Angele, Stuhlinger, Gerhard Heller, and Helmut Zoike joined forces with such native Huntsvillians or recent newcomers as Billy Isbell, Ben Teeter, Eugene Mechtly, Conrad Swanson, George Ferrell, Quincy Love, Bob Bradspies, Charles "Ted" Paludin, and Clarence Ellis. Angele designed and built a polishing machine which was installed in Ellis' home. The latter spent 1,600 hours, sometimes minutes only at a session, polishing a 53-centimeter mirror that, when completed, made the association's telescope one of the best in the South at the time.

Another area of the cultural scene in which the Germans entered almost at once surprised their fellow citizens. Classical music appreciation in the town during the early 1950s was—to put it as delicately as possible—limited. There was a small but dedicated group interested in chamber music, and it occasionally met for an evening's mutual entertainment. With the arrival of the Germans, this group was amazed to find that many of the scientists and engineers who designed the V-2 missile were also very talented musicians.

The day after arriving in Huntsville, Werner Kürs received a phone call from Alvin Dreger, a cellist and local music teacher. He had heard that Kürs was a violinist and the group could use one.

"I was very astonished, " Kürs later said. "Mr. Dreger soon started to arrange playing sessions for us in homes and churches. We were introduced into quite a number of very friendly families interested in cultural activities and education. I experienced a welcome in this city that I had never experienced before anywhere."

From the nucleus of the chamber music group grew the Huntsville Civic Orchestra. The orchestra held its first concert on December 13, 1955, in the auditorium of the Huntsville High School. (Concerts of the Huntsville Symphony Orchestra in the late 1970s would be held in the concert hall of the Von Braun Civic Center.) Conductor Arthur Fraser mounted the podium, glanced to his left where sat concertmaster Kürs, and raised his baton to lead the orchestra in a spirited rendition of Corelli's Concerto Grosso in G Minor.

The orchestra was a cultural admixture of German and American scientists, engineers, military personnel, and talented Huntsvillians not connected with the recently formed rocket center. For example, in the first violin section, under the eye of Kürs, sat J. E. Gunn, an employee of the local US Post Office; and George Detko, an engineer at the arsenal. Mrs. Charles A. Lundquist, wife of Private First Class (PhD) Lundquist (later assistant director of the Smithsonian Astrophysical Observatory and director of the George C. Marshall Space Flight Center's Space Sciences Laboratory), played viola. The trombone section was led by William Hay, who worked at Redstone Arsenal. With him was Thomas Stogner, Jr., who painted signs for the local Coca-Cola Company. Lieutenant Eugene Mechtly played oboe.

While he was not a member of the symphony, von Braun played both the cello and piano, and his brother Magnus was a skilled accordionist. Several years later Wernher surprised his guide at the Mormon Tabernacle in Salt Lake City by playing creditably if not professionally on the temple's famed organ.

Within a few years after moving to Huntsville, the older German children began graduating from high school. In 1953, Magda de Beek, daughter of Gerd de Beek, who had been head of the illustrating department at Peenemünde and held the same position at the newly instituted Ordnance Guided Missile Center, was named Good Citizenship Girl by the Daughters of the American Revolution.

The curtain that von Braun always drew around his personal life and family was raised briefly, much to the delight of his fellow citizens in Huntsville, when an essay written by his daughter Margrit for an English class appeared in the Huntsville Times (in 1963);

> For 11 years Wernher von Braun has been my father.... His favorite dishes are spaghetti, steak, fish and Chinese food.... The only thing wrong is that he usually counts the calories after he has eaten. He has fairly good manners at home, but sometimes Mother has to remind him to keep his elbows off the table....
>
> Daddy always tries to be the big fix-it-man around the house or

> yard. He usually succeeds but very often has to go all round town looking for a certain piece and then picks up the wrong one....
>
> When my 14-year-old sister misbehaves he gets very mad at her and has a long talk which usually helps. When my sister and I have fights, boy does he get mad. He usually shouts, "I come home from the office to have some peace and quiet and talk to your mother!"

In addition to such domestic glimpses, Margrit also revealed her father's unsuspected skills as a handicapper of the ponies:

> The family was going to the Kentucky Derby in Louisville. The night before the derby, Daddy and Mom went to a party. Every man that was willing deposited five dollars and drew out the name of one horse that was to race the next day. Daddy drew out the name Chateaugay.... Chateaugay stayed fifth most of the time and then in the backstretch took a big jump and came in first. Daddy had bet on Chateaugay to come in first and he won twenty dollars and eighty-five cents. Then from the pool game at the party he won $45. In a race before the derby he bet on a horse to come in first that did also....

After finishing high school, most of the children attended colleges and universities within the state, since the tuition was cheaper than at out-of-state institutions. And, of course, there were marriages, but strangely, none of the German children of the original group married each other. Curiously, too, few of the children decided to follow in their fathers' footsteps. By 1978, only two had become civil service engineers at the Marshall Space Flight Center (MSFC). They were Uwe Hüter and Axel Roth.

In July 1950, the Office of Chief of Ordnance directed Redstone Arsenal to study a surface-to-surface missile with a range of 500 miles. Two months later, the Hermes C1 missile project, which had been under study by the General Electric Co. since 1946, was transferred to the arsenal. Early in 1951, the military requirements of the missile were changed by increasing its payload weight, thus decreasing its range. The missile slowly emerging on the drawing boards was known variously as Hermes C1, Major, Ursa, XSSM-G-14, and XSSM-A-14. On July 10, 1951, the development of the rocket was formally assigned to Redstone Arsenal (the name Redstone would be officially assigned to the rocket in April 1952).

The Redstone was one of the first American weapons to combine the two radically new technologies that emerged from World War II: the guided missile and the atomic bomb. Nevertheless, as a technical challenge, the Redstone rocket seemed to offer little since it was really nothing more than a second-generation V-2, a ballistic missile with a range, like the V-2, of only 200 miles. The only real opportunity it offered was to put into "hardware" several ideas that had earlier been proposed by Ludwig Roth's advanced planners at Peenemünde almost a decade earlier. The warhead compartment of the Redstone was to separate from the rocket proper and reenter the atmosphere by itself. Since the burnt-out rocket need not survive the reentry, it could be made of aluminum rather than steel. The structure of the Redstone, therefore, was of the *monocoque* type, in which the propellant tanks formed the outer surfaces of the rocket itself. The warhead compartment was guided to the target by a novel control apparatus: an inertial system that relied only on instruments aboard the reentry vehicle. This had been in the testing stage when Peenemünde was evacuated in 1945.

With less than a year in their new center, the team faced the first of many visits by very important people from Washington. On February 21, 1951, K. T. Keller, director of guided missiles for Secretary of Defense Charles E. Wilson, landed at the Redstone Arsenal airfield in the company of Major General Elbert H. Ford, chief of ordnance; Brigadier General Leslie E. Simon, chief of research and development for the Ordnance Corps; and Colonel Holger N. Toftoy, now chief of the Rocket Branch.

Colonel Carroll D. Hudson, commander of the arsenal, was most anxious that the visit go off smoothly because he wanted his arsenal and its team of German designers to impress Keller with the facility and its professional competence. Hudson told his staff to "pull out all the stops" to get things ready.

To Captain Henry E. Attaya, he gave the order to have a stairway made to accommodate Heller's Super Constellation aircraft. Attaya, in turn, passed the word on to Hans Maus, in charge of the Manufacturing Division of the Guided Missile Development Center. Hurried telephone calls to Birmingham, where the Super Connies regularly landed, produced the height above the runway of the door of the plane. With that information available, it was a matter of simple trigonometry to design and build the required stairway, both of which were done with dispatch.

As the Super Connie stopped and shut down its engines, the stairway, produced by the men who had developed the V-2 and *Wasserfall*, was rolled up to its door. Keller, standing in the doorway, found himself looking through the top several steps, the uppermost being about a foot above the top of the plane's door. Someone had placed the hypotenuse on the wrong leg of the triangle! The problem was solved by placing the stairway sideways and having the guests climb over the rail to descend with something less than the dignity called for by the occasion.

If the Germans faced a lack of money for building their homes, they were only a little better off on the job. The army had allocated few funds for converting the former chemical and ordnance arsenals into a research and development center for large missiles and rockets. With a skill born of necessity and experience, they all pitched in as they had at White Sands Proving Ground and Fort Bliss to help build things themselves. Hannes Lührsen and Hans Grüne joined their colleagues in converting a former army hospital—or part of it—into a Guidance and Control Laboratory for Dr. Theodor Buchhold. With only $250,000 budget for the job, the thrifty Germans decided to use it for materials while they supplied the labor gratis.

Aghast at hearing a civilian contractor wanted $75,000 to build a static firing stand for rocket motors, Karl Heimburg and his testing personnel began rounding up scrap steel and other cast-off equipment. For an observation and control center, they utilized three abandoned railroad tank cars, removed their wheels, cut windows in their sides, and covered them with earth. The facility was made for $1,000, excluding the cost of the conscripted labor.

Keller, apparently having forgiven the earlier mistake in trigonometry, approved a vehicle flight program soon after returning to Washington. The Redstone rockets were to be manufactured in the Fabrication Laboratory of the Ordnance Guided Missile Center, which received its first motor from North American Aviation Co. in 1952. In June 1953, the Chrysler Corporation received a contract to assist in the manufacture of the early missiles.

The marriage of the Germans of Redstone Arsenal and their recently selected counterparts from Detroit, with their vast experience in the automotive industry, was an interesting one. The latter came to Huntsville, for the most part skeptically, to learn from the Germans. But learn they did. The men of Peenemünde demanded and accepted no compromise in the areas of quality control and reliability, areas in which the builders of automobiles had often looked the other way.

Problems inevitably cropped up just as they had with V-2 and *Wasserfall.*

Development of the inertial guidance system for the Redstone lagged behind that of the other major components of the missile. As a result, the unit was not available when the remainder of the rocket was ready for flight-testing. Since the early launches were designed to test the structure of the rocket and its engine, only a simple autopilot was required for its guidance. The thrifty Germans retrieved the drawings of the LEV-3 autopilot used for the same purpose in the early flights of the V-2 and turned them over to the Ford Instrument Company on Long Island; Ford Instrument produced nine sets for use in the early Redstone flights.

The first flight, made on August 20, 1953, was a failure. However, as they always had, its engineers learned more from a failure than from a random success.

During the early flights from Cape Canaveral, someone suggested that the fourth Redstone be decorated, as had been the early V-2s flown from Peenemünde. Each missile had a cartoon or symbol painted on its aft end with the number of the missile. Gerd de Beek, who had done the job at Peenemünde, found it would be too difficult for his American illustrators to paint directly on the skin of the Redstone. Thus, the good luck symbol—a very voluptuous blonde nude—was painted on cardboard and taped to the aft end of the "bird."

The Redstone lifted off and ascended as scheduled. However, its vibration loosened the drawing, which fell away and fluttered back to Earth. For some reason, unexplained by engineers, the tracking radar became attracted to the nude as it fluttered into the Atlantic, leaving the rocket to go where it would. Flight engineers were not amused. And that, one of them reported, "was our last *Playboy* Redstone."

During this period the team faced the first stress in its hitherto absolute solidarity. For the first time since arriving in the United States, several members expressed dissatisfaction and began making plans to leave for jobs with industry or other government agencies. Their reasons were various. Some were tired of working for the military; others felt they had no professional challenge; and personality conflicts had inevitably arisen.

Krafft Ehricke left to join former Major General Walter Dornberger at the Bell Aircraft Co., in Buffalo, New York. Ehricke, a highly imaginative and restless individual much like von Braun temperamentally but without the latter's patience, could foresee no immediate trips to the Moon or Mars or even men orbiting Earth in satellites. After a restless year or so at Bell, he moved to California and the Consolidated Vultee Corporation, which had a contract with the US Air Force to develop the Atlas intercontinental ballistic missile; Ehricke soon proved, on paper at least, that it could be made into an Earth satellite.

The same corporation attracted the talents of Dr. Hans R. Friedrich, who contributed greatly to the development of the Atlas in the field of guidance and control. He died in 1953, only two weeks before Atlas 10B, better known as Project Score, was launched, broadcasting a taped Christmas message to the world from President Eisenhower. Walter Schwidetsky, who had accompanied von Braun to the US in September, 1945, also joined Friedrich and Ehricke on the Atlas project. Dr. Adolf Thiel left Huntsville to become project engineer on the Thor missile and later vice-president of TRW, Inc. Dr. Martin Schilling joined the Raytheon Corporation as a vice-president.

Unable to meet increasingly heavy medical bills on his civil service pay and equally unable to get a raise, despite Colonel Toftoy's efforts, Dr. Theodor Buchhold reluctantly left his team members for General

Electric. Dr. William Raithel also joined the same company, specializing in problems of nose-cone reentry. Magnus von Braun joined the Chrysler Corporation in Detroit, and remained with its automotive division until he retired.

Another reason that some of the Germans left was that the army's budget for guided missiles got smaller rather than larger, and there was little hope for work in space research. The more faint of heart detected the beginning of a trend in which the US Air Force would push the army out of the missile picture. In 1953, after fighting ended in Korea, Toftoy was on the telephone almost constantly with the Office of the Chief of Ordnance, pleading for dollars and people. He was refused both. Funds for Redstone Arsenal were reduced, and all work on the Hermes B2 ramjet, begun at Fort Bliss, was canceled.

In a telephone call to Major General Simon on February 15, 1954, Toftoy came close to giving up altogether:

> I'll tell you trying to operate the way we are this year is frustrating. I haven't been complaining so far because I felt that there was nothing much that could be done, but we have now come to the end of our rope. This is a heck of a way to have to run a big enterprise like this. It certainly dissipates a lot of time of my key people, continually trying to hurdle these unnecessary administrative blocks. I feel that we simply cannot continue major projects efficiently on a delayed funding basis like this. Actually, it is a disgrace to the government.

Despite its funding woes, work on Redstone missile continued. In time, the highly adaptable rocket appeared in modified form as the Jupiter A, a test vehicle; as Jupiter C (with three solid-rocket upper stages), also a test vehicle, mainly for nose-cone materials; and as Juno 1 (with four solid-rocket upper stages), a space booster.

Jupiter A was a standard Redstone changed internally to accommodate and flight-test components being developed for the Jupiter intermediate-range ballistic missile. The Jupiter C was a three-stage rocket based on a modified, higher performance Redstone as the first stage, with clusters of small, solid propellant rockets as the upper two stages.

In the midst of the funding and personnel problems at Redstone Arsenal, on June 23, 1954, von Braun received a phone call from a friend that quickened his pulse considerably. It was tangible proof that someone in the United States was seriously interested in space travel. The call came from Frederick C. Durant III, former president of the American Rocket Society and current president of the International Astronautical Federation.

"I just had a very interesting talk with a man in the Office of Naval

Research; he wants to get rolling on a space vehicle," Durant said, and then added, after a pause, "Do you want to meet him?"

The question was purposefully cadenced by Durant's sense of humor; he knew full well how much von Braun was chafing to get into space, while seemingly bogged down in redeveloping the V-2, as the Redstone was viewed by the majority of the Germans.

Two days later, von Braun made a trip to Washington.

There, Durant introduced the man from the Office of Naval Research to whom he had alluded earlier. He was Commander George W. Hoover. Others were present, too. Their stature in the scientific community clearly showed that the meeting was no mere gathering of space enthusiasts. The group included Dr. S. Fred Singer, a physicist at the University of Maryland; Dr. Fred L. Whipple, a world-renowned astronomer at Harvard University; Dr. David Young, of the Aerojet-General Corporation; and Alexander Satin, chief engineer of the Air Branch of the Office of Naval Research.

Hoover came right to the point.

"Everybody talks about a space satellite, but nobody does anything about it," he said. "I'd like to get the ball rolling with a small satellite carried into orbit by a combination of existing rockets."

The agenda also clearly indicated Hoover's seriousness:

1. Status of satellite projects
2. Criteria for Earth satellites
3. Discussion of desirable features of satellites
4. Discussion of status of propulsion
5. Discussion of instrumentation for satellites
6. Discussion of possible programs

It was soon established that nowhere in the country was there an Earth satellite program seriously under development, although several theoretical studies were being sponsored by the various armed forces. The question logically arose: "Do we have the means for orbiting a set of lightweight scientific instruments?" Von Braun's answer was yes. He briefed the group on a study that had been made by his team in Huntsville. It showed such a launch vehicle could be made from the Redstone missile and upper stages consisting of clusters of small, solid propellant rockets used with the Loki antiaircraft rocket. Von Braun said that a satellite of five to seven pounds, depending upon the number of rockets used in the upper stages, could be orbited.

Everyone agreed that a scientific satellite of useful value, even though it weighed only seven pounds, was worth the effort. Their findings were sent on to the Chief of Naval Research, and he agreed. More significantly, he authorized further discussions with the US Army at Redstone Arsenal.

On August 3, the navy personnel visited Toftoy and von Braun. Toftoy gave his own tentative approval for the proposal, but stated rather cautiously that he would have to have the concurrence, in turn, of General Simon in Washington. He made a hurried trip to the capital and received Simon's qualified approval—only so long as it did not interfere with missile development, i.e., the Redstone and several other missiles currently being developed at Redstone Arsenal. The Ordnance Corps at that time was very hesitant about undertaking any rocket project that was not clearly related to the needs of the Army Ground Forces, the only customer for its wares. Considering the army's miserly attitude towards funding the Redstone, the satellite project could hardly be considered an impediment.

On September 15, von Braun wrote a secret report, entitled "A Minimum Satellite Vehicle Based on Components Available from Developments of the Army Ordnance Corps." It was basically a summary of what he had said at a meeting at the Office of Naval Research a month earlier. In it, he asked for only $100,000, a modest sum indeed for the man who had spent millions of marks to develop the V-2 and who would later spend several billions of dollars to produce the Saturn 5. In this report, he again displayed his uncanny gift for technological prophecy:

> The establishment of a man-made satellite, no matter how humble, e.g., five pounds, would be a scientific achievement of tremendous impact. Since it is a project that could be realized within a few years with rocket and guided missile experience available *now*, it is only logical to assume that other countries could do the same. *It would be a blow to U.S. prestige if we did not do it first.*

On January 20, 1955, a proposal based upon this report was made to the assistant secretary of defense for research and development, a preparatory step in getting it first to the secretary of defense and ultimately to the President. And in April, the Office of Naval Research began planning an expedition to the Gilbert Islands to survey a launch site on Abemama atoll. Unfortunately for Project Orbiter (as the satellite was to be called), it attracted the attention of quite a few people in Washington. A member of the International Geophysical Year Committee of the National Academy of Sciences stated in effect that Orbiter was too inelegant. There were others who carped that the first American satellite ought to be launched by Americans, not by the German team then in Huntsville.

The daily problems of Redstone and the frustrated desires for getting into space were put aside on April 15, 1955. Forty members of the German team with their wives and children gathered in the auditorium of the Huntsville High School. Two federal district judges administered to them the oath of citizenship. Some twelve hundred fellow townsmen turned out, and Mayor R. B. "Spec" Searcy officially declared the occasion to be New Citizens Day.

In the following month, the new American citizens in Huntsville received news that bolstered their long hope for getting into space. The Department of Defense established the Ad Hoc Committee on Special Capabilities under the chairmanship of Dr. Homer J. Stewart, a physicist of the California Institute of Technology. The name of the committee was made purposefully obscure. Its task was to study proposals by the army, navy, and air force to orbit an artificial satellite around Earth. It was then to recommend the most meritorious. The committee had been reluctantly appointed by Secretary of Defense Charles E. Wilson, whose attitudes toward scientific research were well known: "Who cares why the grass is green?" and his belief that "What's good for General Motors is good for America."

In the spring of the year, the Central Intelligence Agency had informed the National Security Council that the Soviets were well along in developing the capability for orbiting a satellite. This revelation probably prompted President Eisenhower to order his secretary of defense to come up with a US satellite proposal.

The committee considered three plans. The army proposed Project Orbiter, with von Braun's assurance that a 15-pound payload could be circling Earth by mid-1956. The navy presented a proposal by Milton Rosen, of the Naval Research Laboratories, for building a new rocket launching vehicle, Vanguard. It would consist of a first stage, upgraded from the Viking sounding rocket; a second stage of NRL's modified Aerobee-Hi sounding rocket; and a third stage to be designed and built from scratch. Rosen, stating that such a vehicle could orbit a 40-pound satellite, assured the committee that such a vehicle could be flying by the same time as the army's Juno 1, the launch vehicle proposed by the army. The air force proposed an ambitious, very heavy satellite to be launched with its Atlas ICBM. However, with the Atlas currently in technical troubles, little consideration was given to placing national prestige on such a shaky launch vehicle.

When the votes were tallied in August, the hopes raised in May at Huntsville vanished. The majority of the committee voted for the navy proposal.

Following the decision to go with Project Vanguard, Stewart, who had voted for the army, visited Huntsville and urged von Braun to keep his Juno 1 as an ace in the hole. He and his associates did so in spite of some very severe restrictions imposed by the Department of the Army after losing its bid. The Jupiter Cs left over from the Jupiter nose-cone developmental flights were stored at Cape Canaveral. However, suspecting von Braun or others might "accidentally" launch a satellite, the

Department of the Army sent insepectors to the Cape to ensure that there were no live fourth stages, which could be used to launch satellites.

Some of the disappointment of losing Project Orbiter was ameliorated by a decision of the Technological Capabilities Panel of the Secretary of Defense: It recommended that concurrent with the development of missiles that could travel between the US and USSR, an intermediate-range (1,500 miles) missile also be developed.

The group at Redstone Arsenal in July 1955 proposed a single-stage rocket, named the Jupiter, with a 1,500-mile range and a payload capability of 2,500 pounds. It would feature a newly developed 150,000-pound thrust engine burning kerosene and liquid oxygen. The rocket would draw on experience gained with the Redstone and, like that missile, would be capable of being launched from a mobile unit in the field as had been the V-2. At the same time, the air force began development of a similar rocket that had to be fired from a fixed base. It was called the Thor.

Why two such expensive weapons, almost identical in performance, were to be undertaken by competing services was a question soon to be raised by both the Congress and the American public. Nevertheless, Wilson decided that both would be developed. The army's Jupiter, ultimately to be built by the Chrysler Corporation, was also to have the capability of being launched from the deck of a ship, thus satisfying the US Navy's requirement for a fleet ballistic missile.

In July 1955, another member of the team from Peenemünde joined his colleagues in Huntsville. Hermann Oberth, the last surviving member of the trinity of pioneers in astronautics consisting of the American Goddard and the Russian Tsiolkovsky, at von Braun's urging had taken a job with him. Oberth, independently of Tsiolkovsky and Goddard, had formulated and elaborated much of the theoretical and mathematical foundation for space travel. For a brief time during World War II, Oberth had worked in an advanced planning group at the famous rocket development center on the Baltic Sea. After several lean years in postwar Germany, Oberth was only too happy to join his former student and boss in America.

Once in Huntsville, Oberth found himself in an environment that would have appealed to the imagination of Franz Kafka. He went to work for the army performing advanced space studies. The results of his work were often classified secret. Since he was not a citizen of the United States, he could not be granted security clearance for military secrets. Thus, he was unable on occasion to have access to his own work, and so on ad infinitum.

Oberth soon became a familiar figure around the headquarters building, where he worked for Dr. Stuhlinger in the Research Projects Office. At noon, he opened his leather briefcase, which one would expect to be filled with esoteric calculations or designs, and removed his lunch, along with plate, knife, and fork. During the four years he remained in Huntsville, Oberth worked on a number of complex projects. Typical of them was a series of theoretical analyses of the stability of satellite orbits. However, he found time to contemplate further and elaborate some of his earlier space concepts. The finishing touches were put on his idea for a lunar vehicle that would permit astronauts to move over large areas of the Moon; unfortunately his 61-foot-tall, 11,900-pound design was impractical for a century or so to come. Additionally, he considered in greater detail his project for placing a very large mirror into orbit about Earth. It would be used for both military and civil purposes, providing light or intense heat at various spots on Earth below.

Oberth was almost archetypically the absent-minded professor. He looked the part, and he acted it. Motorists in his neighborhood and elsewhere in town soon learned to be aware of the little gray-haired man, with an equally gray beret and a briefcase, who seemed wholly unaware of the existence of traffic lights.

He also had little use for questions he considered frivolous. Once, when asked if the rockets being built at Redstone were not too complicated ever to operate properly, he answered in a parable: "I have an automobile with eight cylinders and a lawnmower with one cylinder. The automobile starts the first time I turn the key, but the lawnmower seldom starts at all the first time."

He was sharper with that ubiquitous civil service figure, the man from personnel who makes the desk audit to determine if the job is properly rated for the pay. The hapless individual asked Oberth how he spent his time—how many hours on which tasks, etc. With no outward display of annoyance, Oberth replied, "I spend ninety percent of my time thinking and ten percent answering foolish questions and making out forms."

In 1959, Oberth returned to Germany. At seventy years of age, he had reached mandatory retirement from the American civil service, but he had been employed by the government for only four years. Thus, his pension would not have been able to sustain him. Since he was a former schoolteacher in Germany, he qualified for a better pension in that country. However, he had to collect it in Germany.

As planning for Jupiter got under way, it became apparent that a new organization was needed for its development. On February 1, 1956, the US Army Ballistic Missile Agency (ABMA) was established at Redstone Arsenal with Brigadier General John B. Medaris as commander. Its technical core consisted largely of the Guided Missile Development

Division of Redstone Arsenal. The remainder of the arsenal was designated the US Army Rocket and Guided Missile Agency and remained under the command of Brigadier General Toftoy.

ABMA and the navy agreed that the new missile should be 58 feet in length and 8.7 feet in diameter. It would have an inertial guidance system, thus doing away with the need for expensive and cumbersome radars on the ground or aboard ship. Its radial dispersion at the target would be no more than 4,920 feet (quite sufficient for a missile with an atomic warhead). The navy would be responsible for developing the ship-borne launching equipment, while the army would design the missile and its land-based launching set.

Jupiter got under way with a top national priority; and millions of dollars and hundreds of jobs flowed into ABMA, where only a few years earlier Toftoy had pleaded for pennies.

The first Jupiter C, based on the Redstone, was launched on September 20, 1956, reached an altitude of 682 miles, and flew 3,335 miles, a record that stood until the advent of the first intercontinental ballistic missiles. A third missile later carried a $\frac{1}{3}$-scale model of the nose cone of the Jupiter missile then under development. The nose cone, made of a material that melted slowly during reentry, solved one of the most vexing technical problems facing engineers of the day: providing a means of protecting atomic warheads.

Just as everything was proceeding smoothly, near disaster struck on November 26, 1956. Wilson issued a "Roles and Missions Directive." In it, he directed that the army would henceforth be restricted to developing military missiles with a range of no more than 200 miles. Any weapon with a greater range would be the responsibility of the air force or navy.

Stunned by the decision, the people at ABMA were sent reeling by an event that took place less than a month later: The navy withdrew from the Jupiter project to develop its own fleet ballistic missile for use from submerged, nuclear-powered submarines. It had decided, quite logically, that the quantities of liquid oxygen and kerosene required for the Jupiter would present an unacceptable fire hazard aboard ship. Then, too, the navy decided that a submerged submarine would make a better and less vulnerable firing platform for launching such a rocket than a surface ship. For submarine launchings, liquid propellants were out of the question; only solid propellants could be used.

With their long experience in liquid propellants, the former Peenemünde team had never seriously considered the use of a solid propellant for very large rockets or missiles. They were fully aware that Oberth, as well as Tsiolkovsky and Goddard, had proved that solid propelled rockets could never be used to reach outer space. Their dedication to liquid propellants is exemplified by an anecdote concerning

Arthur Rudolph, one of the earliest designers of liquid propellant rocket engines and later manager of the Saturn 5 development.

While attending a dinner meeting in which the featured speaker extolled the merits of solid propellants for guided missiles, Rudolph, still unconvinced, approached the speaker after his address and raised his martini.

"That was a very interesting speech, but I must tell you that I am by birth a liquid propellants man," he said. Pointing to the olive in his cocktail, he continued, "This represents the ratio of my confidence in solids versus liquid."

Always one to hedge his bets technologically, Rudolph returned to the bar and modified his martini by adding four more olives. He then returned to the speaker.

"My friend, you see that my confidence in solids has increased; and the more I drink of this liquid propellant, the higher my level of confidence in solids rises."

Following the news of the navy's withdrawal from Jupiter, US Navy Captain Robert Freitag wrote to the team in Huntsville in December 1955. He informed them of a suggestion he had made to the Department of Defense concerning the future of the group. Freitag's message was:

> I sincerely regret that events took the turn they did and that Army and Navy requirements digressed to the point that our arrangement was forced to end. I sincerely believe that you and your team are the most outstanding group of missile people in this country . . . or in the world, and regret that the powers-that-be have decided that you will not be working on the Fleet Ballistic Missile. . . . I have proposed to Dr. Murphree's [Dr. E. V. Murphree was Special Assistant for Guided Missiles to the Secretary of Defense] office that he consider a rather drastic move. I have suggested that the ABMA be reconstituted as the "National Aeronautical Agency" and be given the responsibility for extraterrestrial projects such as satellites, sounding rockets, etc. . . .

After Murphree turned down Freitag's suggestion, the captain approached army officials with the same proposal and received a resounding no.

The first Jupiter flight test on March 1, 1957, followed in the familiar pattern of V-2 and Redstone. It was a failure. A second vehicle was launched on April 26, but sloshing of propellants within its tanks also produced a failure. A succession of successful flights then followed.

Jupiter was later modified to become a space carrier vehicle, called Juno 2, by placing on top a cluster of solid propellant rockets such as those on the earlier Jupiter C and Juno 1 under an aerodynamic shield. The

Juno 2 could place 90 pounds into Earth orbit and send some 15 pounds to the Moon. However, its record as a space booster was spotty at best. Out of ten launches only four were successful. (On March 3, 1959, however, a Juno 2 did launch the first US lunar probe, Pioneer 4, which missed the Moon but became the first man-made object to go into orbit about the Sun.)

On October 4, 1957, the new Secretary of Defense, Neil McElroy, visited ABMA. With him were Secretary of the Army Wilbur Brucker; General Lyman L. Lemnitzer, Chief of Staff of the Army; and Lieutenant General James Gavin, Chief of Research and Development of the Army. They were there for a briefing to the secretary on the Jupiter missile system.

Everyone was getting acquainted at an informal cocktail party at the Redstone officers' mess that evening when Gordon Harris, the public relations chief of ABMA, hurried in and announced to the group that the Soviets had a satellite in orbit.

Lieutenant Colonel Truman F. Cook, Brucker's miliary assistant, later described to the authors what then happened:

> I noticed that General Gavin was visibly shaken and understandably so since he had on several occasions recommended that the Army be allowed to rescue the faltering and expensive Vanguard . . . effort by a parallel effort using the Jupiter C. . . . [Von Braun] after several comments regarding the urgency of moving ahead with Jupiter C as a satellite vehicle said, "Today, man has taken his first step towards Mars."

The briefings on the following day were devoted more to Jupiter C than to Jupiter itself. McElroy was informed that two Jupiter Cs, leftovers from the Jupiter nose-cone reentry program, were in storage at Cape Canaveral. (By the addition of a single solid propellant rocket, attached to the payload, the Jupiter C became the Juno 1 space carrier.) McElroy wanted to know how long it would take to get them ready for launch, and von Braun impulsively replied 60 days. Medaris at once tempered the enthusiasm of his chief civilian manager and said that 90 days probably would be required.

Despite the pressure put on him for an immediate approval, McElroy decided to return to Washington and think things over.

"When you get back to Washington and all hell breaks loose, tell them we've got the hardware down here to put up a satellite any time," von Braun said to the departing secretary.

On November 3, the Russians orbited their 1,120-pound Sputnik 2, with the dog Laika aboard. Five days later, Medaris had his approval from

McElroy. As things turned out, the people at ABMA needed almost all of the ninety days forecast by Medaris. It took exactly eighty-five days before Explorer 1 was placed in orbit. Most of the time was needed to ready the satellite rather than the booster rocket, which had earlier been static tested. One of the stipulations made by Washington was that the satellite had to contain scientific instrumentation to provide data useful in IGY studies.

Dr. William H. Pickering, then director of California Institute of Technology's Jet Propulsion Laboratory, in part funded by the US Army, was in charge of the crash program to design, develop, test, and build the satellite. Pickering and his team worked around the clock. Another key man in the satellite project was Dr. James A. Van Allen, a distinguished radiation physicist of the State University of Iowa. He and his talented graduate student Wei Ching Lin provided cosmic-ray counters that proved to be the most valuable instrument aboard the 34-inch-long and 6-inch-diameter satellite that weighed only 18.13 pounds.

That Pickering's team could—and did—put together and test such a scientific instrument within the time specified was a much greater technological accomplishment than hauling out the Jupiter C vehicle at Cape Canaveral and preparing it for launch within the same time period.

On December 6, 1957, a Vanguard was launched, and less than one second later the first stage lost thrust and settled back onto the pad in flames.

Later in the month, the team at ABMA submitted to the Department of Defense a study that was ultimately to have a great impact on their future careers. It was entitled "Proposal for a National Integrated Missile and Space Vehicle Development Program." In it was discussed the need for a booster rocket with at least 1,500,000 pounds of thrust.

With the first US satellite still three weeks away, some members of the team at ABMA began planning a manned space flight. On January 10, 1958, a meeting was held in Huntsville to discuss Project Man Very High, the objective of which was "to carry a manned, instrumented capsule to a range of approximately 150 statute miles; to perform psycho-physiological experiments during the acceleration phase and the ensuing six minutes of weightlessness; and to effect a safe reentry and recovery of the manned capsule from the sea."

The project was being proposed "to improve the mobility and striking power of US Army forces through large-scale transportation by troop-carrier missiles." The need for such a delivery system and its expense were, of necessity, vaguely stated. At first, it was to be a tripartite program of the three armed forces; but after a month, the air force withdrew. Copies of the proposal were sent to the Ordnance Office of Chief of Research and Development, and the newly established Advanced

Research Project Agency (ARPA), hoping that funds would materialize from one of them. The project envisioned the use of an army rocket and an environmentally sealed capsule, the latter of which would be protected from heating during reentry by material developed to shield the Jupiter nose cone.

On May 16, the Secretary of the Army forwarded the proposal to ARPA, with the recommendations that it be approved and funded. Three days later, von Braun made a briefing on it to Roy Johnson, ARPA's director. In the following month, copies of Project Adam, as it was then called, were sent to the Central Intelligence Agency, for the missile's questionable utility in inserting agents into foreign countries had attracted the attention of the clandestine division of that agency. However, the CIA felt that the project was not feasible unless the development of an operational system could be speeded up significantly. By mid-September 1958, Project Adam was dead. With it went all hopes of the US Army's putting a man into space.

Meanwhile, things were getting sorted out on the launching pad at Cape Canaveral. On the night of January 29, 1958, the Explorer 1 satellite sat on top of its Juno 1 rocket booster, which was in the final countdown. In the blockhouse from Huntsville were Medaris and Drs. Kurt Debus and Hans Grüne, the latter two showing little emotion, after the many rocket launches they had witnessed, going back to the first ones at Peenemünde years before. Not so calm were their two fellow Peenemündians, Drs. Walter Häussermann and Ernst Stuhlinger, and Dr. Jack E. Froelich from the Jet Propulsion Laboratory. Not present were the three principals, von Braun, Pickering, and Van Allen. They were in Washington at a communications center of the Pentagon. In case of a successful orbiting, the army wanted them available immediately for the nation's press.

On that night Aeolus, god of winds, conspired against Juno. The winds at 40,000 feet were 165 miles per hour, clearly greater than those the rocket could withstand. The launch was scrubbed. The next night was worse; the winds reached 225 miles per hour. Again there was a scrub.

Major General Donald N. Yates, commander of Patrick Air Force Base, which supported ABMA's launching operations, gloomily forecast days more of the same high winds to Medaris, who had no reason to dispute him. Yates had been the meteorological officer who, with Royal Air Force Group Captain John Stagg, had brashly and correctly told Eisenhower, over a decade earlier, that the weather would break on the dawn of June 6, 1944, and that Operation Overlord, the invasion of France in World War II, should proceed. However, there was present at Cape Canaveral twenty-four-year-old First Lieutenant John L. Meisenheimer, a meteorologist from Yates' Patrick Air Force Base. He displayed the same brashness based on the advanced technology of a later generation of meteorologists. The high winds would abate on the late evening of January 31, Meisenheimer told Medaris, who took a gamble, and, lo, they did!

At 20.30 hours, the countdown got underway.

At 21.45 hours, trouble began. Someone noted a pool of liquid on the pad beneath the launcher.

A propellant leak?

A truly dedicated propulsion expert courageously ran onto the pad and thrust his head under the rocket to see what was wrong. Medical experts were of two minds on the dangers of unsymmetrical dimethylhydrazine, with which the Juno 1 was fueled. Some thought one whiff would be fatal; others thought that it could result only in baldness or impotency. The engineer quickly determined that the liquid was the result of a *spill* rather than a *leak*. A decade and a half later he was alive, hirsute, and otherwise functional.

The countdown resumed.

Things went smoothly until 22.35 hours. A caution light came on, and Robert Moser, who was later to become almost a permanent fixture in blockhouses and firing rooms at Cape Canaveral, told Debus that he had an indication one of the jet vanes on the Juno 1 was deflected. Debus glanced at his own panel, which indicated otherwise. No problem. Forget it. Resume countdown.

At 22.45 hours, Debus nodded affirmatively and almost casually to Moser, who pressed the switch that ignited the engine of the Juno 1.

The job of firing the upper stages fell to Stuhlinger, who followed their flight by radar tracking and a radio beacon. At the proper moment, he pressed a switch that sent the firing signal to them. All ignited on schedule.

At the communications center in the Pentagon, von Braun and Pickering waited impatiently. The tracking station on Antigua Island had reported the fourth stage had fired and the satellite had passed over it. But von Braun wanted confirmation from Pickering's tracking stations in California before he would say Explorer 1 was in orbit.

An hour and a half elapsed.

According to von Braun's calculation, they should have heard by then from California. Pickering got on the telephone to his tracking stations: "Why the hell don't you hear anything?"

Still there was silence.

Apparently, it had not attained orbit after all.

Then, within 30 seconds, all four stations reported they had Explorer 1's signals coming in. It was in a higher orbit than expected, which accounted for von Braun's prediction of a shorter period for the satellite.

From Peenemünde to Cape Canaveral had been a long voyage, but the team was in space at last.

Lieutenant Colonel Cook later described to the authors what then occurred:

> There was bedlam at all communications centers. Secretary Brucker personally congratulated General Medaris, Wernher von Braun and his team. A few minutes later, Secretary Brucker sent a message of congratulations to the Chief of Staff of the Army. In order to commemorate the event, the Secretary of the Army established the Concatenated Fraternity of Master Missileers, Circa 1958, Pentagon Chapter. Personalized and signed certificates of this order were presented to each person involved in that evening's vigil.

Following the announcement of the successful orbiting of Explorer 1, Huntsville found itself in the midst of a spontaneous celebration that was rivaled in enthusiasm only by those demonstrations that greeted the end of World War II, the end of World War I, and the appearance of General Nathan Bedford Forrest in town during the Civil War.

A crowd of several thousands gathered around the courthouse square, stopping traffic, blowing automobile horns, and thrusting hastily printed signs into the air. The sirens on fire engines and fire stations added to the general din. Firecrackers and, appropriately, skyrockets were fired all over town. Perhaps the most interesting display was the hanging in effigy of former Secretary of Defense Charles E. Wilson, who had earlier refused to let the rocket team at ABMA launch satellites.

From a vacation in Florida, Wilson professed puzzlement at being so singled out on the occasion. "I don't know why they're mad at me. I'm the one who put them in the Jupiter business [which rocket was not involved in the launching]. They must have me mixed up with the Ku Klux Klan," he said.

A few days later Pickering, von Braun, and Van Allen were back in Washington for dinner at the White House. In dressing at his hotel, von Braun discovered that he had come away from Huntsville without the white tie demanded by protocol. A hurried call was put through to James Hagerty, Eisenhower's press secretary.

"Tell Dr. von Braun not to worry," Hagerty said. "We'll have a white tie waiting for him at the White House tonight."

When von Braun arrived immaculate, except for tie, Hagerty handed him one, which he immediately placed around his neck and joined the guests in the reception room. A few minutes later the President entered wearing a *black* tie and clearly ill at ease. As he passed Hagerty, he said, "Jim, I can't seem to find my white tie anywhere."

* * * * *

On April 2, 1958, Eisenhower sent to the Congress a message requesting the establishment of a civilian space research agency. In it, Eisenhower said that the space agency should be civilian rather than military

> . . . because space exploration holds promise of adding importantly to our knowledge of the Earth, the Solar System, and the Universe, and because of its great importance to have the fullest cooperation of the scientific community at home and abroad in moving forward in the fields of space science and technology. Moreover, a civilian setting for the administration of the space function would emphasize the concern of our nation that outer space be devoted to peaceful and scientific purposes.

On July 29, the National Aeronautics and Space Administration (NASA) came into being, taking as its nucleus the personnel and facilities of the former National Advisory Committee for Aeronautics. Thus, national policy on space research again seemed to preclude participation by the Huntsville team at ABMA. Understandably, the former Germans and their fellow engineers began to wonder about their future.

Spirits revived on August 15, however, when ARPA gave them the task of developing a large space booster that could be used at some time in the future for lifting very heavy military payloads into orbit. It was to have a unique design—a central Jupiter propellant tank around which were clustered eight Redstone tanks. (The concept was later referred to by von Braun as "cluster's last stand.") The propellants were kerosene and liquid oxygen, and the booster's eight engines would produce 1,500,000 pounds of thrust. Eventually the vehicle received the name Saturn.

August also saw the departure from Redstone Arsenal of an old friend. Toftoy was transferred to Aberdeen Proving Ground, a traditional last post for Ordnance generals on the road to retirement. The US Army Ordnance Missile Command (AOMC) was formed with Medaris as commander, and he would remain there until retiring on January 31, 1960, being succeeded by Major General August Schomburg. A reorganized ABMA became a subordinate command under Brigadier General John A. Barclay.

The new year found von Braun, who already had two orders of the German *Kriegsverdienstkreuz*, in Washington for a special occasion. On January 17, 1959, he was presented with the Distinguished Federal Service Medal, which is the highest award made to civil servants. Never the one to miss an opportunity to do a bit of advertising when the press was around, von Braun said that ABMA could well spend an additional $50 million or $60 million "To do better and more thoroughly what we are doing now." Well aware that the man who best personified Uncle Sam was looking on, von Braun continued: "All our programs are a little undernourished. We need a rich uncle who would give us the money with no strings attached." However, the remarks fell on deaf ears so far as Uncle Ike was concerned.

By fall, morale at ABMA had plummeted to the point where more members of the team, ex-German and native American as well, were considering the desirability of better paying jobs in industry. For the past year, rumors, unchecked by the Department of the Army, indicated that ABMA or a large portion of it would be transferred to the newly established NASA. What remained would work only on short-range military missiles and rockets. The army, by Wilson fiat, had no role in space. The team, which von Braun by his optimism and enthusiasm had held together for more than two decades, was clearly in danger of disintegration. To add further to its woes, ARPA began cutting back funds for the Saturn booster because no clearly defined army payload could be identified for it. Some consideration was given to transferring Saturn to the air force; however, that branch of the service also could not define a need for it in the foreseeable future.

Gloomily, in a speech before the annual meeting of the National Association of Food Chains (in those days, everyone wanted to hear about space travel), von Braun made a prediction that was to prove unduly pessimistic: "I suspect we will have to pass Russian customs when *we* finally reach the moon."

On October 21, Eisenhower set minds at ease in ABMA by transferring its Development Operations Division (i.e., the team) to NASA as the Marshall Space Flight Center (MSFC). "The ... transfer provides a new opportunity for them to contribute their special capacities directly to the expanding civilian space program," he said. By the end of the year, a NASA Saturn Vehicle Evaluation Committee, under the chairmanship of Dr. Abe Silverstein, proposed in rocket technology an advance in magnitude that can well be compared to that vocalized by Neil Armstrong, as he set foot upon the Moon; "That's one giant step .." The Silverstein Committee recommended that all upper stages of Saturn vehicles be powered by the high-energy propellant combination of liquid hydrogen and liquid oxygen and that an engine to use such propellants be developed. For those members of the team used to such familiar propellants as kerosene, ethyl alcohol, and liquid oxygen, the use of liquid hydrogen presented a formidable engineering challenge.

On April 29, 1960, the city of Huntsville and nearby towns initially heard a sound to which they would later become accustomed. At ABMA the Saturn 1 first stage, with all eight engines firing, roared into life for the first time. The noise thundered through the city and country, rattling windows and shaking the earth. Later firings of Saturn stages and engines would bring claims against the Federal government from farmers who sought redress for cows that had gone dry, hens that refused to lay, cracks in plaster, and broken windows. On one occasion, a group of engineers, concerned about the effects of the sound on the environment, journeyed to a nearby town to set up instruments to measure the sound power level in it. At a very early hour in the morning, they were setting up their instruments on the sidewalk when a cruising police car became suspicious. The engineers were hauled off to jail by the officers, who were not convinced that the men were space engineers engaged in a scientific experiment rather than robbers who were attempting to blow open the doors of the local bank.

The changeover from ABMA to MSFC went smoothly enough on the surface, but it soon became apparent that there were some differences in operation. As part of ABMA, the engineers and scientists of the team had no need to worry about administrative matters since they were solely concerned with technical ones. Other elements of ABMA had swept the floors, paid for electricity and water, and made out the payroll. Suddenly, these mundane operations intruded and had to be dealt with. It took time, but they were sorted out. Also, as a part of ABMA, the team was concerned with one major project at a time, while future planning was a secondary effort. As MSFC, the team was to be the launch vehicle development center for all of NASA. The new organization became responsible for Redstone-Mercury, Juno 2, Centaur, and the Agena B stage for the air force's Atlas and Thor boosters, in addition to Saturn C-1, C-2, and C-3. If these projects were not enough, the responsibility for the development of the 1,500,000-pound thrust M1 rocket engine was assigned to the group as well.

On July 1, 1960, General Schomburg formally and officially bade farewell to 4,669 of his employees. The ceremony was held in front of Building 4488, the new joint headquarters of ABMA and MSFC. Returning to his office on the third floor as director of MSFC, von Braun got down to work in a job that superficially, at least, seemed to be the culmination of a lifetime dream. He was heading an organization whose sole purpose was to develop the large rockets needed to send satellites into orbit about Earth, instrumented probes into interplanetary space, and perhaps, eventually, place man on the Moon and on the neighboring planets of the solar system.

At the end of August, a rumor concerning von Braun was being circulated in a curious and unlikely quarter—Wall Street. After retiring from the army, Medaris had become president of the Lionel Corporation, a company whose major product was toy electric trains, which to the dissatisfaction of the stockholders were not selling well. The board of directors decided to diversify. The company wanted to enter the "Cinderella" field of electronics. Medaris had been hired to get it into the market. The story along Wall Street was that von Braun, tired of the squabbling between Department of Defense and NASA, had decided to join Medaris at Lionel. Predictably, some 74,000 shares changed hands at the highest quotation Lionel had reached for years. The rumor was false, of course, and some quarters of the street felt, correctly, that it had been floated solely to boost the company's sagging stock.

The Marshall Center was formally dedicated on September 8 by President Eisenhower, who journeyed to Huntsville to do so in memory of his deceased commander in World War II.

On September 28, von Braun was in Washington, but not for his usual reasons: conferences at NASA Headquarters or testifying before congressional committees. He was there to attend the premiere of a film, and one he would just as soon had never been made. *I Aim at the Stars* was supposed to be the biography of Wernher von Braun. Its production had been a tortuous one. The army insisted on reviewing, approving the script, and rewriting it at whim. The scriptwriters, in turn, took liberties with truth to add "suspense and drama" to his by then hopelessly distorted story. One such invention was that von Braun's secretary, at Peenemünde, was a British spy!

There were personality clashes as well. J. Lee-Thompson, the film's English director, said, "Von Braun and I disliked each other on sight. After a time, I grew almost to like him—almost, but not quite."

The picture was picketed in several cities, including London and New York. Even in Munich, it received boos and catcalls, directed more at a display of US Army missiles than at the film itself, from a group of leftists. The municipal authorities of Antwerp banned it.

The film won no awards, broke no box-office records, and years later it made only rare appearances on late-late-night television. *Time* magazine's review of the film was succinct: "Von Braun possibly has grounds for a libel suit, but then he might do better to ignore the picture. So might everybody else."

Von Braun was wholly of the same opinion.

On December 29, NASA's first administrator, Dr. T. Keith Glennan, submitted his pro forma resignation to become effective January 20, 1961, when President John F. Kennedy was to be inaugurated. The man who succeeded him was James E. Webb, who was sworn in on February 14. Webb, a former director of the Bureau of the Budget, was wise in the ways of Washington and friendly with many influential legislators, particularly Senator Robert Kerr, an oil millionaire from Oklahoma, who was chairman of the Committee on Aeronautical and Space Science, which approved the NASA budget. Webb was a voluble and very skilled manager who served NASA well in its formative days.

At the end of 1960, the team had launched its first satellite as a member of NASA. It was the Explorer 8, a 90-pound scientific satellite boosted into orbit by a Juno 2. With the arrival of 1961, the group in Huntsville was devoting much of its effort to the Redstone-Mercury vehicle. Originally, it had been designed to launch each of the seven astronauts in the Mercury program through a ballistic path that would produce some fifteen minutes of weightlessness and give them a chance to get the feel and control of their Mercury spacecraft. However, the schedule was changed to provide for only two manned launches. On January 31, just three years after the team had launched Explorer 1, Mercury-Redstone lifted off at Cape Canaveral, with what technically might be termed the first American. Its passenger was the chimpanzee Ham, who came through unscathed but clearly exhibited no enthusiasm for another space flight. Meanwhile, Astronaut Alan B. Shepard was training for his mission aboard another Redstone Mercury.

On April 12, however, there occurred an event that was to make Shepard's flight some three weeks later seem inconsequential. On that day Yuri Gagarin made a single orbit of the Earth and returned safely, becoming the first man to do so. The Russian triumph notwithstanding, Shepard's flight on May 5, which reached an altitude of 115.7 miles, was greeted with enthusiasm by the American public, which was warming to the incipient "space race" with the Soviet Union.

On May 25, President Kennedy said in his special message on urgent national needs to the Congress, "Now is the time to take longer strides, time for a greater new America, time for this nation to take a clearly leading role in space achievement which in many ways may hold the key to our future on earth. . . . I believe that this nation should commit itself to achieving the goal of landing a man on the Moon and returning him safely to Earth."

But Kennedy did not say *how*. It was the *how* that created dissension among various factions within NASA.

There were three ways of reaching the Moon. These were known as the direct, Earth-orbital, and lunar-orbital modes. Each had its pros and cons as well as its proponents and detractors. All three would accomplish the purpose. The arguments that arose within NASA were based on engineering determinations of which would be the optimum way to the Moon.

The direct mode was perhaps the simplest. Place the astronauts and their spaceship aboard a very powerful rocket and launch them to the surface of the Moon, from where they would return to Earth.

The Earth-orbital mode was more complex. It envisioned launching a rocket stage loaded with liquid oxygen into Earth orbit. It would later be joined by the astronauts with a third stage loaded with the requisite liquid hydrogen. The two spacecraft would rendezvous and dock, and the astronauts would effect the transfer of oxygen to their third stage, after which procedure they would take off for the Moon. The lunar-orbital mode was controversial, although it was not original. It had been proposed by such early pioneers as Hermann Oberth, in the 1920s, and several members of the British Interplanetary Society, in the 1940s. Fundamentally, it consisted of launching a two-part spacecraft from Earth directly into an orbit about the Moon. There, half of the craft with two astronauts aboard would detach itself and descend to the Moon. Having completed its mission on the Moon, the crew would then ascend to rendezvous and dock with the craft left in orbit. Having transferred to it, the lunar "taxi" would be detached; and the orbiting module would send the astronauts back to Earth.

At first, NASA Headquarters and the Space Task Group, the nucleus of which would form the Manned Spacecraft Center (later the Johnson Space Center) in Houston, Texas, favored the direct mode. The team at the Marshall Center, however, was partial to the Earth-orbital mode. The lunar-orbital mode was a latecomer in the competition and was espoused by a group of engineers at NASA's Langley Research Center, which was not directly involved in the manned lunar landing program.

Each of these modes had its good points and its bad points. The direct mode called for a super space booster known as Nova. The behemoth would have a clustered first stage with a thrust of 9 million pounds and could place into Earth orbit a payload of 290,000 pounds or send a payload of 100,000 pounds to the Moon. ("With Nova," Eberhard Rees said, with characteristic whimsy, "we could land a locomotive on the Moon—if anyone wanted one there.") But Nova was a vehicle that was well beyond technological realization within the time frame set by President Kennedy for a manned lunar landing.

The Earth-orbital mode also had some drawbacks. It called for the use of two Saturn C-5 (later called simply Saturn 5) rockets for each lunar mission. Since each Saturn would cost approximately $120 million apiece, there was an economic disadvantage. Additionally, the perfection of rendezvous, docking, and transfer of highly explosive propellants between spacecraft in orbit was a task not to be undertaken lightly.

The lunar-orbital mode had the disadvantage that it had not been completely worked out to the satisfaction of the proponents of the other two modes. Planners at NASA Headquarters, the Space Task Group, and MSFC simply did not trust the claims of its proponents. In short, it looked too good on paper to be true.

The most vocal advocate of the lunar-orbital mission was John C. Houbolt, an engineer at the Langley Research Center, who became an evangelist for several fellow engineers and began pushing their plan within NASA in late 1960. One of their most compelling arguments was that the lunar-orbital mode would permit a saving of 10,000 pounds in weight of the spacecraft. But even more important from the viewpoint of economy, the plan called for the use of only one Saturn 5 and there would be no need to undertake the costly Nova.

Throughout 1961, Houbolt became more and more vocal for the concept, addressing NASA management meetings and committee meetings as well as pushing it in speeches before professional societies. The more he talked, the more people listened: especially Max Faget and Dr. Joseph F. Shea, of the Manned Spacecraft Center. Engineers at the Marshall Center went back to their computers and did independent evaluations of the proposed lunar-orbital mode.

Despite the study being given the Langley plan, Houbolt failed to impress at least two NASA panels. An ad hoc committee chaired by William Fleming, of the agency's Office of Programs, was charged with looking into the problems involved in placing men on the Moon within six and a half to eight and a half years and returning them to Earth. The committee made its report on June 11, stating, wrongly as things turned out, that the critical item for such a project would be the launch vehicle rather than the manned spacecraft. However, the report endorsed the direct mode as the best way of achieving the goal.

Plans for reaching the Moon were temporarily laid aside as the fall approached. The team at MSFC turned its attention to the first launching of the Saturn 1 vehicle, the design and development of which it had begun earlier under the auspices of the US Army. The launching took place, with a dummy upper stage, from what is today known as the Kennedy Space Center, at Cape Canaveral, Florida, on October 27, 1961. The mission was to verify the structural integrity of the first stage, which it did. (Over the succeeding four years, nine additional Saturn 1 rockets, most with "live" second stages, were launched.) Producing a thrust of 1.5 million pounds, the vehicle was utilized to test the structural integrity of the Apollo spacecraft, to gain experience in the separation and ignition of stages in multistaged rockets, and to launch three Pegasus satellites which measured the micrometeoroid flux in near-Earth space.

By the end of the year, yet another NASA committee had a different idea on the best way to reach the Moon. On the day that Kennedy had asked Congress to commit the nation to landing a man on the Moon before the end of the decade, NASA had formed an ad hoc committee to study the nation's needs for large space boosters over the ensuing decade. It was a joint effort with the Department of Defense. Co-chaired by Dr.

Nicholas Golovin, NASA, and Dr. Laurence Kavanau, Department of Defense, the committee deliberated throughout the summer and submitted its finding as the year drew to a close. In effect, the report stated that the direct mode with Nova was not practical because of the time involved in producing such a launch vehicle and the tremendous complexities of it. The Golovin Committee recommended the Earth-orbital mode using Saturn 5.

On September 11, 1962, President Kennedy arrived at the Redstone Arsenal airfield, and strode down the steps of a ramp that fitted with precision to the door of the aircraft. Following him were Vice President Lyndon B. Johnson; Secretary of Defense Robert S. McNamara; Dr. Harold Brown, chief scientist of the US Air Force; the Right Honourable Peter Thorneycroft, British Minister of Defence; Dr. Jerome B. Wiesner, science adviser to Kennedy; and NASA Administrator and Deputy Administrator, James E. Webb and Dr. Robert C. Seamans.

Von Braun later recalled to the authors what happened as the party toured his facility:

> With the aid of a chart displayed in front of our immense first Saturn rocket, I explained the key maneuvers in LOR [lunar orbital rendezvous] when the President interrupted me and said, "I understand that Dr. Wiesner doesn't agree with this." Looking into a large crowd of dignitaries, newsmen, and radio commentators in the spacious assembly building, he demanded, "Where is Jerry?" Dr. Jerome Wiesner ... stepped forward and said, "Yes, that's right. I think the direct mode is better." Two minutes of lively argumentation furnished exciting raw data for the media, until JFK terminated the debate and moved on.

With the beginning of 1963, NASA formally authorized MSFC to initiate the Saturn 5. On January 25, the center undertook development of the large, liquid-propellant rocket that could place 120 tons in orbit about Earth or send 45 tons to the Moon.

By this time, too, it seemed to most people within NASA that Houbolt's group had the proper answer for the lunar mission. On July 5, the decision was made official: the lunar-orbital mode was the only safe and economical way to reach the Moon. Outside NASA some doubts still remained about the decision.

The development of the Saturn 5 continued apace, even though there was some residual controversy about the manned lunar landing. However, the efforts of the team at MSFC were not totally involved in that launch vehicle. Building on its earlier experience with the Saturn 1, the group at Huntsville had modified it into the Saturn 1B. The new rocket consisted of the first stage of the Saturn 1, lengthened and reduced in weight. Its eight engines were also reworked to give the vehicle a lift-off thrust of 1.6 million pounds. Thus, Saturn 1B was capable of placing approximately 39,000 pounds into orbit 100 miles above Earth. For its second stage, the rocket had what would become the third stage of the Saturn 5; and it also used essentially the same guidance and control unit that would later be employed by the Saturn 5.

The first Saturn 1B was launched from the Kennedy Space Center on February 26, 1966. Its mission was to test further the technology of the clustered design and multistage rocket as well as to flight-test components of the Apollo spacecraft. The Saturn 1B sent the Apollo command and service module as well as its launch escape system on a ballistic path that reached an altitude of 310 miles. The escape tower functioned as it should, and the service module rockets were tested before the command module, in which the men of Apollo would ride, plunged into the atmosphere at a velocity approximating what the craft would experience on returning from the Moon. (In addition to testing the third stage and guidance unit of the Saturn 5 and the Apollo spacecraft, the Saturn 1B was the first multistage rocket to be "man-rated" or certified for the transportation of men into space, as it was used in Projects Apollo, Skylab, and Apollo Soyuz Test Project. Eleven Saturn 1Bs were manufactured before the tenth one, built in 1967, was launched in 1975. Two rockets were left in storage, destined never to be launched because their pad at Cape Canaveral had been razed and sold for scrap metal. (One of them wound up ignominiously as a tourist attraction along Interstate Highway 65, near Huntsville.)

Dr. George E. Mueller, who had become associate administrator of the Office of Manned Space Flight in 1963, brought to NASA a great deal of systems engineering experience in large projects of the air force. His was native American engineering know-how, cost conscious of necessity, and included some ideas that were considered radical if not outright heretical by the cautious, step-by-step engineers of MSFC. Typical of these was Mueller's concept of "all up" testing. It simply meant that the first Saturn 5 would be launched with all three stages "live." There would be no dummy upper stage launches as there had been with Saturn 1.

The reaction to Mueller's concept of flight-testing at MSFC was predictable, immediate, and vocal. There was almost total and absolute opposition. Every old rocket hand knew that developmental programs simply were not run that way. In many cases the conservatism of the members of the team from Peenemünde and White Sands was shared by their younger American protégés. Some of the test engineers of Karl Heimburg's organization pointed out logically and soundly enough that if the first stage of Saturn 5 failed during an attempted launch, two very expensive upper stages would be lost and no performance data could be obtained from them.

Mueller's counterargument was that "if we announced that the purpose of an early launch was merely the flight-testing of the first stage, we'll get a pat on the back from Washington even in case any of the upper stages fail. On the other hand, if they don't fail, we'll be a quantum jump ahead in the game."

Von Braun listened to the pros and cons. His mind had been made up before the meeting at MSFC at which Mueller's concept was debated. He would support Mueller because he was the boss (as he had always supported Dornberger, who had been the boss on V-2), even though he had personal reservations. After all, von Braun figured that neither Mueller nor himself had any experience in building a monstrous three-stage rocket capable of carrying astronauts to the Moon. Von Braun put the concept of "all up" to his fellow team members in a ploy he frequently used. Let the fellows become outraged, thrash about, and give off steam. Sooner or later, they will begin to think about the problem rationally and in solid engineering terms.

That is exactly what happened.

The older hands began to point out that the experience with Saturn 1 and Saturn 1B indicated many of the fundamental concerns had been laid to rest. The success of the various Saturn 5 stages based on ground tests and static firings, as well as those on the Saturn 1B, gave no cause to doubt their performance in space. Several of the second-generation members of the team had a novel view: "If we have done as good a job as we have told Washington and Congress we have done, then we have nothing to worry about."

On November 9, 1967, the first Saturn 5 lifted off from Launch Complex 39 at the Kennedy Space Center. It was a "textbook" flight, with all three stages working perfectly. The vehicle carried an Apollo command and service module and a lunar module test article into space to check them out. In the control center, von Braun turned to Arthur Rudolph, the Saturn 5 project manager, and said, "I would never have believed it could happen." The second Saturn 5 was successfully launched on April 4, 1968, though longitudinal vibrations cropped up that produced a "pogo" effect. Fortunately, the problem was soon identified and solved. With two unmanned successes in a row, the third vehicle, now "man-rated," sent Astronauts Frank Borman, James Lovell, and William Anders on a trip around the Moon at Christmas. In March and May of 1969, two additional Saturn 5s were launched with astronauts to check out the Apollo spacecraft and its lunar landing module.

Even though the novelty of launching the huge Saturn had worn off and the team had no reason to doubt that Saturn 5 No. 506 would perform as had its predecessors, there was great anticipation and excitement in the control room and viewing stands at Kennedy Space Center on July 16, 1969. Aboard it were Astronauts Neil Armstrong, Edwin Aldrin, and Michael Collins, the first two of whom were to become the first men to walk on the Moon. No. 506's first two stages performed perfectly, and the vehicle's last task was completed when its third stage boosted the Apollo spacecraft out of Earth orbit and on the proper trajectory for a successful landing on the Moon.

Over the ensuing four years, six additional Saturn 5s sent an additional eighteen men to the Moon. During that period, the gross payload for the rocket grew from 85,000 pounds to 116,000 pounds, because of the flexibility built into it. This greater capability permitted the latter astronauts to lengthen their stay on the Moon from one to three days and allowed them to carry along Lunar Roving Vehicle (also developed by MSFC), a small electric car that extended their range of exploration from the landing site to a radius of 20 miles.

On May 14, 1973, the thirteenth and last Saturn 5 to fly launched Skylab, America's first space station, into orbit about Earth. It was subsequently visited by three crews flown to it by Saturn 1Bs. The team at the Marshall Center had developed not only the two launch vehicles but also the Skylab space station. It was made by refurbishing the second stage of what would have been Saturn 1B No. 12.

By the time that Skylab was circling Earth, few of the former Germans who had arrived with von Braun in 1950 were still employed at the Marshall Center. Death and retirement had taken their toll. While death was inevitable and fixed as a point in time, retirement was in many cases contrived and involuntary. For years, some people in NASA Headquarters had resented the former Germans in Huntsville. As their rank rose in Washington, that of the "Germans" fell in Huntsville. A final solution to the German problem began in January 1973; and the subtle pressures applied to them produced voluntary retirements among the older members of the team.

The situation is perhaps best described by reporter Peter Coburn, who wrote in the Huntsville *Times*, in 1976:

> Resentment began to fester within NASA. . . . Some were stunned at what they termed German arrogance. The Germans themselves became acutely aware of the detractors of the German way of operating. Criticism filtered to them of the tightness of their operation—filling key managerial and laboratory slots with their own; demanding full and unrestricted control of the operation; circumventing the system and its established policies; attempting to

> research, develop, and build a rocket system in-house in the labs and workshops of Marshall Center.
>
> "If they could have," grunts one such detractor, "the Germans would have launched the damned Saturn from Marshall."

A few of the indictments were true. The Germans did tend to be somewhat clannish and did occupy the top positions. However, at von Braun's urging, some of his top lieutenants adopted young American protégés and brought them up through the ranks. In a few cases, these younger men developed into competent managers who would serve the center capably during the next decade or so.

In commenting on this aspect of the operations at MSFC, James Webb said later, "They were unique in working together and knew whom they could trust, putting those men who could do best in the job. Those turned out to be the people they had known and worked with and trained in many cases." By the end of 1978, few former Peenemündians were left at the Marshall Center. Most were members of the team who had arrived late, such as Georg von Tiesenhausen and Willibald Prasthofer, a jovial Austrian and former VKN man who had spent ten years in France on missile programs.

Retirement at first proved to be a frustrating state for some of the former Peenemündians. After three decades of exciting and demanding work that taxed them intellectually and professionally, they simply were not prepared for what often and erroneously is called the "leisure life."

With an uncavalier lack of forethought considering his spouse, Bernhard Tessmann, who had been with von Braun at the Raketenflugplatz, Berlin, said, "Being in retirement is not easy. I always had the feeling of being needed. People to talk to. Only my wife is left. Now my days are filled with reading, walking, working in the garden. You have to do something. Otherwise life can become miserable."

Such gloomy feelings were not shared by all. The indefatigable Walter Wiesman retired at the first opportunity and plunged into a career of lecturing on the techniques of industrial communications for large corporations. Some of his clients were in his native Germany. Konrad Dannenberg, the propulsion engineer at Peenemünde and MSFC, channeled a portion of his energies into activities of the local chapter of the American Institute of Aeronautics and Astronautics and also became a part-time professor at the University of Tennessee Space Research Institute, in nearby Tullahoma, Tennessee.

Several team members remained active in the field of engineering and management. Hermann Weidner returned to Germany to manage an engineering company which he had inherited. Eberhard Rees, who had succeeded von Braun as director of MSFC, became a consultant to a

German manufacturer involved in developing Spacelab, a manned laboratory designed to orbit Earth aboard the space shuttle. Others turned to hobbies and talents unrelated to rocket engineering. William Mrazek devoted part of his abundant energies to building a swimming pool, a greenhouse in which he raised orchids, and assisting his second wife in managing two cafés.

The retirement of some members of the team proved a boon to the Johnson Environmental and Energy Center of the University of Alabama in Huntsville. Additionally, retirement provided a means by which those men could do something they had not been able to do in many years, return to the drawing board as engineers. Hans Paul, an expert in heat transfer, applied the skills he had early demonstrated on the V-2 and Saturn vehicles to solving problems relating to solar-energy collectors for heating, cooling, and providing hot water for homes and other buildings. Similarly, Tessmann and Heimburg, experts in rocket testing, transferred their abilities to the design of test stands for those collectors. Dr. Werner Sieber, whose speciality was materials engineering, likewise moved from the optimum materials for space vehicles to those for solar-energy collectors. Dr. Ernst Stuhlinger, world renowned for his studies in electric propulsion for the spacecraft of the future, put his knowledge to work designing an electric car for the urban driver. Wilhelm Angele developed for the university a device that can measure the roundness of a sphere to within one-tenth of one millionth of an inch (2.5 Angstrom). Earlier, at the Marshall Center, he had developed a machine to produce spheres that were round to less than one millionth of an inch. (One quartz sphere would be used in a space experiment to prove Einstein's theory of relativity.)

Inevitably, death began its subtraction of members of the team and their families. People who had survived World War II to settle, finally, in the small rural town in north Alabama chose to be buried in Huntsville's Maple Hill Cemetery, on a gentle slope to the south of the imposing mausoleum of Russel Erskine, a former president of the Studebaker Motor Company. Almost in its shadow is the lesser but still imposing mausoleum of John L. Robinson, who prepared his tomb in advance of his own death and saw his pet dog buried within before he entered it for the last time. Thus, Walter Burose, Hans Hüter, Mrs. William Mrazek, Josef Böhm, Gerhard Heller, and Count Friedrich von Saurma joined five former governors of the state, 120 unknown soldiers of the Confederate States of America, and a general of the Grand Army of the Republic who had fought against them.

Perhaps the final evaluation of the team could well be the words of one of them to reporter Peter Coburn. When asked what he considered that history would write of the men of Peenemünde, El Paso, and Huntsville,

Dr. Stuhlinger replied that the team would probably receive a few lines concerning their contribution to the landing of the first men on the Moon. Then, he added, "But that's enough. A footnote is enough."

Despite Dr. Stuhlinger's modest assessment, the history of technology is far more likely to afford the team a chapter rather than a footnote. In several areas, the team made significant advances in the state of the technology of rocketry. Typical of these are the development of the air-bearing gyroscope and the inertial guidance platform; supersonic wind tunnels for testing models of missiles; variable-thrust, liquid-propellant rocket engines; underwater launching of rockets; high-speed turbopumps for cryogenic propellants; industrial fabrication techniques for beryllium; and the electromechanical computer for testing missiles.

Among their accomplishments were participation in the successful realization of the first US ballistic missile with an inertial guidance system (Redstone), the first long-range missile flight (Jupiter C), the first demonstration of a US satellite launch capability (Juno 1), the first US satellite of Earth (Explorer 1), the first US lunar probe (Pioneer 4), the launching of the first US astronaut into space (Alan Shepard), the launching of the first men around the Moon (Apollo 8) and the first onto the Moon (Apollo 11), the development of the first vehicle to travel on the Moon (Lunar Roving Vehicle), and the first US space station (Skylab).

Epilogue

Wernher von Braun left Huntsville in February 1970, enthusiastically looking forward to the challenge of his new position at NASA Headquarters as Deputy Associate Administrator for Planning.

He did not leave his Alabama home of two decades, however, without serious consideration of what he was giving up. There were practical and possibly even compelling reasons for his remaining as head of the George C. Marshall Space Flight Center. For one thing, he knew that because he was German and other reasons, some at NASA Headquarters did not like him. Also, leaving Huntsville meant parting from the team he had led for almost four decades. Then there was the fact that within the NASA hierarchy, center directors generally carried more weight than the deputy, assistant, and associate administrator policy- and decision-making apparatus. In the end, it was only through the personal persuasion of NASA Administrator Dr. Thomas O. Paine that von Braun agreed to move to Washington.

Some members of the von Braun team and fellow Huntsvillians in general believed he was going to Washington to be groomed as an eventual NASA administrator. Von Braun had no such delusions. As Dr. Charles S. Sheldon, former White House staff member on the National Aeronautics and Space Council, later remarked, "There was always a lingering resentment at the Washington end toward von Braun and his team. There were always rumors that von Braun would someday be head of NASA. But there is a great sensitivity in Washington about racial and ethnic interests. . . . Von Braun would never be given a political position."

Von Braun's responsibilities at NASA Headquarters were focused on future planning, on staking out the programs for several decades of astronautical endeavor. Prior to his arrival on the Washington scene, the White House had indicated to Paine that President Nixon wanted a bold, new space project with which his name could be associated, much as President Kennedy's had been with Apollo. The only feasible space exploit that could top Apollo appeared to be the landing of man on Mars. While such a project could not be brought off until the late 1980s, at least Nixon would be credited as its initiator.

A full five months before the Apollo 11 landing on the Moon in July 1969 and a year before von Braun actually transferred to Washington, Nixon had established a Space Task Group, under the chairmanship of Vice President Spiro T. Agnew, that included Dr. Paine; Dr. Lee A. DuBridge, presidential science adviser; and Dr. Robert C. Seamans, Jr., secretary of the air force, as members. In September 1969, STG published its *The Post-Apollo Space Program: Directions for the Future* and, almost simultaneously, NASA released its *America's Next Decades in Space: A Report of the Space Task Group*. These reports, and another issued in March 1970 by the Space Science and Technology Panel of the President's Science Advisory Committee (*The Next Decade in Space*), all called for an aggressive post-Apollo space effort on the part of the United States.

First from his Huntsville base and later from Washington, von Braun argued for a program that would include a Mars landing before the end of the 1980 decade. Leading to that event would be the development of bases on the Moon and of a permanent manned space station supported by an Earth-to-orbit shuttle system. Other than the reusable shuttle, a nuclear Earth orbit–lunar orbit transfer stage and a highly maneuverable space "tug" for interorbital tasks were recommended, along with a lunar orbital station to support base activities below.

As the months went by, von Braun discovered that his arguments for an aggressive and well-conceived post-Apollo space program were being met with polite interest but no real enthusiasm or indication of support. Despite his unique combination of imagination, drive, practicality, and loquacious wit, so effective in the past, he and his NASA associates could not affect a changing tide. Nixon was losing interest, and even some of NASA's top administrators were beginning to show a general lack of enthusiasm.

The reasons why the United States failed to undertake an energetic space program based on the splendid Saturn-Apollo-Skylab foundation established in the 1960s and early 1970s are varied and complex. But one factor was dominant: the post-Apollo climate was not propitious for another great surge into space. America's priorities were shifting.

It was not surprising, then, that NASA Administrator Paine announced his resignation on July 28, 1970, to become a vice-president of the General Electric Company. His departure left von Braun adrift. Without Paine's presence and backing, he was vulnerable to those who disliked or disagreed with him, and he had no supporting base from which to operate. Without an aggressive space program to plan for, he became less and less effective.

To some within the NASA hierarchy, von Braun was on the road to becoming a "nonperson" at the agency, whose only alternative was to retire or resign. One manager behind the "green doors" (the portals of the sanctum sanctorum of NASA Headquarters) later reminisced to the authors, "He walked the halls with the smell of death upon him.... People had ceased speaking to him in the corridors, by and large." The picture is overdrawn, but it contains elements of truth. In preparing for 1971 budget hearings, for example, von Braun was seen to emerge visibly depressed from planning sessions with the new NASA administrator, Dr. James C. Fletcher. In post-Apollo NASA, von Braun was like the fleet admiral back from the glories of victory at sea who suddenly finds himself walking dazedly along the Pentagon corridors with nothing important to do. The trials and triumphs of Raketenflugplatz, Kummersdorf, Peenemünde, Fort Bliss, Huntsville, and Cape Canaveral were over. The space horizon had suddenly clouded.

Thus, when Wernher von Braun announced his retirement from NASA on June 10, 1972, no one was surprised. He simply could not work within what had become an essentially holding operation.

From NASA, he moved—for the first time in his life—into private industry, accepting a position as vice-president for engineering and development at Fairchild Industries in nearby Germantown, Maryland.

Once settled there, he undertook his tasks with characteristic energy and dedication. He was particularly intrigued with the potential of the Applications Technology Satellite (ATS) being developed by the company for NASA. One of its experiments included the transmission of educational television programs to isolated villages in various parts of India. Since the satellite could transmit on high power, a converter, a commercial TV set, and a small, homemade antenna were all that were needed for the villagers. Programs of an academic nature were sent to schoolchildren, while special programs were developed for their parents on such topics as crop fertilization, better practices in animal husbandry, and the importance of personal hygiene and family planning.

Von Braun's enthusiasm for and belief in the potential of the ATS for emerging nations led him to make a series of presentations on its benefits to the governments of Brazil, Iran, Spain, Venezuela, and Indonesia. He also became interested in the role his company could play in the field of alternate forms of energy, and was instrumental in getting a contract from the Energy Research and Development Administration in the field of coal gasification.

During his Fairchild years, von Braun was active in establishing and promoting the National Space Institute, the objective of which was to seek out a broad base of support for the space program from society at large. His message was simple: man's activities in space pay multiple dividends to man here on Earth.

At the peak of his Fairchild/National Space Institute activities, von Braun learned that he had cancer. Despite surgery, the disease progressed, forcing him to retire on December 31, 1976. Early in 1977, he was awarded the National Medal of Science by President Gerald R. Ford. Since illness prevented von Braun's meeting with the President, Fairchild Industries' Chairman Edward G. Uhl made the bedside presentation. In a letter to the President, Uhl reported that "Wernher was completely surprised and most appreciative. He reflected back to his coming to the United States with all his possessions in a single pasteboard carton, and he had tears in his eyes when he said, 'And now [said von Braun] President Ford is presenting me the highest honor any scientist can receive.'"

Wernher von Braun died on June 16, 1977 in Alexandria, Virginia, survived by his wife, three children, and two brothers. President Jimmy Carter said on that day: "To millions of Americans, Wernher von Braun's name was inextricably linked to our exploration of space and to the creative application of technology.... Not just the people of our nation, but all the people of the world have profited from his work. We will continue to profit from his example."

At a special memorial ceremony for von Braun at the Washington Cathedral, Dr. Ernst Stuhlinger, his colleague since Peenemünde, added this footnote to the life of a long-time friend:

> The world of technical accomplishment was only one part of his universe. "When my journey comes to an end," von Braun once remarked, "I hope that I can retain my clear mind and perceive not only those precious last minutes of my life, but also the transition to whatever will come then. A human being is so much more than a physical body that withers and vanishes after it has been around for a number of years. It is inconceivable to me that there should not be something else for us after we have finished our earthly voyage. I hope that I can observe and learn and finally know what comes after all those beautiful things we experience during our lives on earth."

APPENDIX A
Production Quantities of A4s

As noted in Chapter 5, Georg Johannes Rickhey's estimate of nearly seven thousand A4s produced during the war (Table 1) probably represents a composite of monthly target figures rather than actual missiles coming off the assembly line. In comparing his data with those extracted from the files of the Ministry of Armaments and War Production (Table 2), one finds a considerable difference. Starting with a composite number of 1,626 A4s manufactured through August 1944, the ministry shows a total of 5,953 (the 1,626 missiles compare with 1,400—Mittelwerk only—for the same period according to Rickhey's files).

At the very least, for the last four months Speer's Ministry figures are probably more accurate, since Rickhey's are almost assuredly estimates. But this alone does not explain the discrepancy of nearly a thousand missiles.

Part of the difference seems to be that the Ministry of Armaments and War Production was concerned only with Mittelwerk A4s, and not the experimental and pilot production output of Peenemünde. In an attempt to determine the truth, David Irving (the English author of *The Mare's Nest*) counted the actual invoices and shipping documents from the Mittelwerk records. His numbers are shown in Table 3.

Irving further records that 16 additional A4s were manufactured in July 1944, but were subsequently discarded as obsolete. Moreover, from November 1944 to March 1945 another 155 missiles were assembled that were later sent back to Mittelwerk from the field for refurbishment following damage en route to the firing point. As for production up north at Peenemünde, he estimates that from the year 1942 at least 314 A4s came off the lines, some 200 fewer than director Rickhey reckoned during the course of postwar interrogations.

Table 1. Production Estimates According to Mittelwerk Director Georg Rickhey

Period	Peenemünde Pilot Production	Mittelwerk	Total
Before 1944	300	–	300
January 1944	30	20	50
February	30	30	60
March	20	150	170
April	15	150	165
May	15	300	315
June	15	250	265
July	15	250	265
August	15	250	265
September	15	500	515
October	15	650	665
November	15	750	765
December	15	850	865
January 1945	—	750	750
February	—	800	800
March	—	700	700
Totals	515	6,400	6,915

Table 2. A4 Production According to Files from the Ministry of Armaments and War Production

Month	Number
September 1944	601
October	650
November	650
December	618
January 1945	700
February	618
March	490
Total (including the 1,626 through August 1944	5,953

Table 3. A4 Production According to Invoice Counts Made by David Irving

Month	No.	Month	No.
January 1944	50	September	629
February	86	October	628
March	170	November	662
April	253	December	613
May	437	January 1945	690
June	132	February	617
July	86	March (to 18th)	362
August	374	Total	5,789

Evidence available from Allied immediate postwar inspection team files supports Irving's count. The last invoice discovered was dated March 18, though missiles were produced nearly to April 1, 1945, according to Rickhey. Up to March 18, a total of 5,777 missiles were invoiced. Another 180 missiles are believed to have come out before the plant finally closed, making 5,957 units. Rickhey also estimated that another 260 to 300 A4s were built strictly as test units—they were moved immediately to test-firing ranges and launched.

Actual production and cost data are shown through March 31, 1945, in Table 4. Except for July 1944 and March 1945, these figures are identical to those compiled by Irving. The average cost of the A4 was RM 75,113 ($17,877), and the average monthly output of Mittelwerk was 397 units. It is believed that the unit cost would have dropped to RM 35,000 ($8,330) if production had continued to July 1945.

By December 1944, Allied intelligence had identified 43 A4 manufacturers' serial numbers, which allowed a reasonable guess to be made of monthly production figures. "Following the procedures used by the GAF [German Air Force], it is not unlikely [a report stated] that 700 rockets have been manufactured during the past month." During November, 662 missiles were logged out of the Mittelwerk.

Rudolph, after inspecting these various production figures, said that one cannot accept any of them as completely reliable. "Many of our missiles were not shipped on completion, but held until camouflaged railway cars had arrived; these would often be delayed because of the Allied bombing of railway lines and junctions. Damaged missiles were not returned to the Mittelwerk, but to a neighboring repair plant. Invoicing from this plant probably contributes to the discrepancies." He recalls that exactly 300 missiles were produced in April 1944, "for on the first of May (equivalent to the US Labor Day) we had a special celebration for Rickhey." Yet Irving came up with only 253 missiles for that month.

> Considering that the daily output of missiles was irregular [comments Rudolph], and that it varied between fifteen and thirty-six (the highest ever) in a twenty-four-hour period, the difference can easily be explained by delayed [April] shipping, which in turn contributed to the high shipment figure for May.

The 250 per month output figures for June, July, and August 1944 (according to the Mittelwerk information) have a simple explanation. The reentry airbursts discovered during the test flights at Heidelager cried out for solution. Production missile assembly at the underground plant was abruptly stopped and, instead, flight-test missiles were issued in lots of twenty. Each lot had a slightly different configuration, as innovations were introduced to determine if they would have any beneficial effect on flight-test performance, and hence on leading to the discovery of what caused the missiles to break up shortly after they entered the dense layers of the Earth's atmosphere.

Table 4. A4 Production and Cost Data According to Mittelwerk Documentation

Month/year	Production	Total Cost, RM	Unit Cost, RM	Total A4s Produced at Each Unit Cost
January 1944	50	5,000,000	100,000	1,000 units at RM 100,000 per unit from January 1 to June 2
February	86	8,600,000	100,000	
March	170	17,000,000	100,000	
April	253	25,300,000	100,000	
May	437	43,700,000	100,000	
June	132	11,920,000	90,000	From June 2, 900 units produced at RM 90,000 each
July	76	6,840,000	90,000	
August	374	33,660,000	90,000	
September	629	54,440,000	80,000	From September 23, 1,000 units produced at RM 80,000 each
October	628	50,240,000	80,000	
November	662	47,890,000	70,000	From November 8, 1,000 units produced at RM 70,000 each
December	613	41,710,000	60,000	From December 26, 1,000 units produced at RM 60,000 each
January 1945	690	41,400,000	60,000	
February	617	32,750,000	50,000	From February 9, precisely 967 units produced at RM 50,000 each, through to March 31, when plant closed
March	540	27,000,000	50,000	
15 months	5,957 units	447,450,000		

APPENDIX B

Addenda on Intelligence

A puzzling aspect of the secret weapon picture was the role played by German intelligence agents, who maintained more or less regular contact with the Allies throughout the course of World War II. Some were involved with conventional activities in Sweden, Turkey, Switzerland, Spain, and Portugal, while others were associated with anti-Hitler resistance figures who sought some kind of understanding with Allied political leaders in the event of a successful anti-Hitler coup.

Even before war broke out in September 1939, men like Major Ewald von Kleist-Schmenzin and Dr. Carl Goerdeler had been sent by Colonel-General Ludwig Beck and Admiral Wilhelm Canaris* to warn the British that unless a firm stance were taken against the Führer a war could not be averted. Later, as the war progressed, Canaris maintained numerous links with the Allies through neighboring neutral countries. He did this within the general framework of establishing an effective resistance movement against Hitler. Thus, Canaris and his Abwehr could have readily transferred information thought to be useful in bringing about the downfall of the German dictator.

The contact between German and Allied agents has intrigued many postwar writers. Certainly, Abwehr and other intelligence specialists would have had access to considerable knowledge of German political, industrial, technical, and military developments, attitudes, and policies. If highly placed persons with the German intelligence community were genuinely attempting to bring the war to a close short of an unconditional surrender, there is good reason to believe that information on promising and possibly decisive new weapons would have been passed on to the Allies. The *Oslo Report* seems to have been an early if not the first instance of the transfer of weapons information.

*Beck was Chief of the Army General Staff up to 1938 and an anti-Hitler leader; Canaris, Chief of German military intelligence—the Abwehr.

As one British expert pointed out to the authors, "It was during these 1942–45 transactions [e.g., the meeting between British and Abwehr personnel in neutral countries] that we got real information on Peenemünde, though we already had much of it through our wiretaps. But because Peenemünde was such a secret base, we couldn't bomb it right away on just this intelligence information. Otherwise, it would have betrayed the fact that we were in touch with their agents and were reading their ciphers—and they would change it all." In other words, the Germans would have taken precautionary measures, which of course the British wanted at all cost to avoid.

The "reading of ciphers" refers to the British technique—code-named *Ultra*—of intercepting and decoding enemy radio traffic sent through the supposedly uncrackable *Enigma* device. Developed by Polish and British mathematicians, cryptologists, and other specialists, Ultra was only revealed publicly in the spring of 1974. During the course of the war, Ultra permitted the Allies to monitor an amazing quantity of high-level military wireless messages. One knowledgeable source told the authors:

> Ultra gave us a description of the German Luftwaffe's involvement in the V-1 program in the greatest detail ... largely as a result of its carelessness in communicating with its various branches through tapable telegraph. In spite of their transportation systems and all sorts of weird things they did to complicate [their communications] we continued to read the Luftwaffe cipher. This gave us knowledge of the whereabouts of plants used for V-weapon [V-1] production.... When the data got into the aether, we read them in England at "Station X."

The use of Ultra intercepts in helping to unravel the secret weapon program is commented upon briefly by Group Captain Frederick W. Winterbotham in his book *The Ultra Secret* and more completely in R. V. Jones' *Most Secret War*.

What Jones called a "very long shot, in fact the longest that I made in the entire war," involved the radar tracking of German missiles along the Baltic coast. Jones correctly surmised that the Peenemündians would need to flight test their experimental weapons, and that they would do so in an east-northeasterly direction following the Baltic coastline. Intelligence had determined that the 14th and 15th companies of the German Air Signals Experimental Regiment were the best in the business, so Jones requested that they be carefully watched. Sure enough, one of the units—the 14th—was later found to have moved to the vicinity of Peenemünde and to be transmitting ranges and bearings on a "moving object." Jones explains:

> Starting with the station on Greifswalder Oie, the ranges and bearings made sense of something that was taking off from Peenemünde and was proceeding with a speed around 400 m.p.h. in an east-northeasterly direction ... We could locate the successive radar stations as each plotted a part of the track in turn. The stations were usually alerted before a firing took place, and there were security slips such as a reference to 'FZG 76' [an Fi103 designation] and so there was no doubt what they were plotting. It was a great moment, for my very long shot and landed us in a ring-side seat at all the trials of the flying bomb.

But what of the rocket? Only an occasional attempt to track one was observed from the "ring-side seat." Ultra intercepts soon revealed that A4 flight testing had been transferred to Blizna in Poland, leading in turn to photographic reconnaissance of the area and the discovery, on May 5, 1944, of an A4 there.

An interesting sidelight to message interceptions was the decoding by the United States of signals to Tokyo from the Japanese ambassador in Germany, Hiroski Oskima. In one message, for example he reported that "Germany's retaliation against England will come by rockets." And in another, he reported that "Talk in Germany centers more and more on a plan to use long-range rocket guns to shell London."

As important—indeed, as vital—as Ultra was, it was only a complement to other intelligence and not a substitute for it. The role of agents on the Continent was indispensable to the Allies, and is covered in a variety of sources. Among them are Jósef Garliński's *Poland, SOE and the Allies* (1969)—SOE is Special Operations Executive and *Hitler's Last Weapons: The Underground War against the V-1 and V-2* (1978). According to Garliński, the first "regularly repeated intelligence that the Germans were carrying out important experiments on the Baltic coast" came from a Danish unit under Major V. L. U. Glyth. Information was secured primarily by Danish coastal and fishing boat captains operating from the island of Bornholm.

A curious sidelight of intelligence and the V-1 weapons is described by Sefton Delmer, wartime head of Britain's so-called black radio. This operation prepared and broadcast programs from Soldatensender Calais (Soldiers' Radio Calais), which "was neither a soldier's radio, nor German, nor situated in Calais." Rather, it was a British counterfeit station charged with the mission of creating confusion among the Germans and providing instructions to foreign resistance organizations. Delmer also established the *Nachrichten*, a nightly news sheet printed in millions and dropped from Flying Fortress aircraft over Germany and occupied territories. It put into writing much that had been previously heard from "Calais" radio.

One of Delmer's colleagues, Robert Walmsley, made a specialty of analyzing the development of Dr. Joseph Goebbels' propaganda campaign insofar as the V-weapons were concerned.

> Not only my department found Walmsley's reports invaluable [writes Delmer]. The Joint Intelligence Committee and the Chiefs of Staff also paid attention to his analysis. They were right to do so, for Walmsley had discovered an important basic principle about the Goebbels propaganda. That was that, although the Germans did not mind how much they boasted in the output to foreign countries, they did their best not to promise victories to their own public unless they were fully confident of being able to bring them about.

An entry dated 17 June 1944 in the diary of Rudolf Semmler (the Propaganda Minister's press chief) is illuminating in this regard. It reads:

> Goebbels has personal reasons for being glad the V-weapon has appeared [referring to the V-1, first fired in combat on June 13]. On March 23, 1943, Hitler had told him for the first time of the plans for

> these weapons. At his request [Munitions Minister] Speer had sent him regular information about the progress being made with them.... For a year now Goebbels has been promising the coming retaliation in articles and speeches.... One month after another passed without the weapon appearing, and Goebbels' prestige fell lower and lower.... Now he feels he has been rehabilitated. He has turned out right. He has triumphed.

As soon as the V-1s started exploding on London and southeast England, Delmer put his Soldatensender and *Nachrichten* into action. Finding out that German launch crews often called the flying bomb "Der Dödel," it was decided to build up a campaign that would have the disquieting result of forcing the Germans to question the military decision to create the device in the first place. The Luftwaffe and the army have been unable to stop the invaders, Delmer's writers and broadcasters would trumpet. How, can the Dödel do it? Can it hit militarily useful targets? Shouldn't all the technology and the production capacity it represents be better employed by making aircraft to protect German cities and tanks to hurl the Allies back into the sea?

And what about the large quantities of fuel Dödel consumes at the very time when tank and fighter crews are sorely in need of it? The theme *Da fliegt der Spirit*—There goes our fuel—was repeated incessantly. One Soldatensender news bulletin had it that the Dödel consumed as much propellant in a single day as 100 Panther tanks, 50 bombers, or 100 fighters. "When there are no real weapons, then you must rely on the Führer's will to victory and his *Weltanschauung* [idealism] and to hell with the grenadier at the front...."

Still another aspect of the intelligence picture was the establishment of a network of double agents—spies whom the Germans believed were working for them when in reality they operated on behalf of the United Kingdom. Referred to as "double cross," they were controlled principally by Section B.1.A, Division B (counterespionage) of Britain's M.I.5 internal security organization in close cooperation with M.I.6 (external operations).

After D Day, the attention of the double agent system swung to the V-weapon program. Toward the end of June 1944, an agent known as Zigzag was parachuted into England armed with £6,000, two wireless sets, and some cameras. Among his several assignments were to report on damage caused by flying bombs and to obtain data on a suspected new Allied radio frequency that could upset the operations of V-2 missiles. After completing his work, he was to return to Germany. But before he did, he supplied the British with valuable information on the V-1 and the still-to-be committed rocket. "Quite early," notes Sir John Masterman in his *The Double-Cross System in the War of 1939 to 1945*, "we had become alive to the extent of the menace through the traffic of the agents."

In early September 1944, shortly before the V-2 offensive began, M.I.6 learned that agent Artist had warned agent Tricycle to leave London due to impending danger. Duncan Sandys was immediately informed. He also learned that agents operating in and near London had been instructed to report on impact locations of both V-1s and V-2s with the view of transmitting this information to launching crews across the Channel. Knowing where their first missiles were landing would enable the Germans to adjust their aim.

Masterman explains the role of the double agent in a carefully planned campaign of deception.

> It was soon realized [he wrote] that widely scattered though they were, the majority of the bombs [V-1s] were falling two or three miles short of Trafalgar Square, and the general plan of deception soon took shape. It was, in brief, to attempt to induce the Germans still further to shorten their range by exaggerating the number of those bombs which fell to the north and west of London and keeping silent, when possible, about those in the south and east.

The Germans would thus be led to believe that they were firing too deep and would hence shorten their aiming point. In reality, however, they were already undershooting. Masterman observes that the British could never be entirely certain that the Germans themselves had not devised a means of determining the actual V-1 impact locations, compelling, as he put it, "caution on our side."

When the V-2 offensive began in September 1944, similar tactics were employed. Masterman explains:

> There was, however, a technical difference. In the flying bomb attacks location had been the important factor because it was doubtful whether the enemy could tie the times we gave them to particular shots; in the rocket attacks timings were vital because the enemy could calculate accurately the time of arrival of any shot, and link this up with any information which we gave him.

In the light of this, it was decided to reveal real impact information that would show an aiming point in central London. But, the *times* for the impacts would be keyed to those of rockets falling 5 to 8 miles short of central London. "In this way," continues Masterman, "over a period of some months we contrived to encourage the enemy steadily to diminish his range." From mid-January to mid-February 1945, the mean point of impact edged eastward at the rate of a couple of miles a week. The result was that more and more V-2s fell well outside the central London area.

Bibliography

A selection of reference books, articles, letters, interviews, and other sources is given in the following pages. Space permits but a partial listing of the prewar, wartime, and postwar documentation examined by the authors. Many of the unpublished or limited distribution documents noted here have, over a period of years, been acquired or copied by the authors and are now in their files. Others are located at such places as the army's Redstone Scientific Information Center; the archives of the Adjutant General; the Office of Chief of Military History; the Ordnance Museum, US Army Ordnance Center and School, Aberdeen Proving Ground; the Research Studies Division, Air University, Maxwell Air Force Base; the Air Force Logistics Command, Wright-Patterson Air Force Base; the Naval History Division, Office of the Chief of Naval Operations; the National Air and Space Museum, Smithsonian Institution; the National Archives and Records Service; the Public Records Office; the Imperial War Museum; the Deutsches Museum; the Musée de l'Armée de l'Air; the Alabama Space and Rocket Center; the Willy Ley Special Collection at the University of Alabama in Huntsville; and elsewhere.

In order to avoid redundancy of listings, the bibliographies of some chapters have been grouped.

Chapter 1

BOOKS

Bergaust, Erik, *Reaching for the Stars*. New York: Doubleday, 1960.
Dornberger, Walter, *V-2*. New York: Viking Press, 1954.
Gartmann, Heinz, *The Men Behind the Space Rockets*. London: Weidenfeld & Nicolson, 1955.

Huzel, Dieter K., *Peenemünde to Canaveral.* Englewood Cliffs, N.Y.: Prentice-Hall, 1962.
Ley, Willy, *Rockets, Missiles, and Men in Space.* New York: Viking Press, 1968.
Rother, Rudolf, *Oberjoch, ein Buchlein vom Dorf und den Bergen.* Munich: Bergverlag Rudolf Rother, 1950.
Ruland, Bernd, *Wernher von Braun: Mein Leben für die Raumfahrt.* Offenburg: Burda Verlag, 1969.

ARTICLES AND REPORTS

Cahill, Jerry, "Wis. GI Held US Space Future Fate," Milwaukee *Sentinel*, February 12, 1958, 6.
Knebel, Fletcher, "Nazi Chemist Wined, Dined by US Army," Detroit *News*, February 3, 1958, 1, 10.
O'Hollaren, Bill, "V-2 Inventor Says He Could Have Won the War If He Had Been Given Two More Years," *The Beachhead News*, May 11, 1945, 3.
"World Needs Divine Guidance Along Dangerous Roads, Dr. von Braun," St. Louis *Globe-Democrat*, June 4, 1958, 10.

INTERVIEWS AND LETTERS

Interview on tape of Herbert Axster by Frederick I. Ordway III, Leutasch, Austria, September 11, 1971.
Interview on tape of Erich Neubert by Frederick I. Ordway III, Huntsville, Alabama, June 24, 1971.
Interview on tape of Theodor Poppel by David L. Christensen, Kennedy Space Center, Florida, August 9, 1971.
Interview on tape of Eberhard Rees by Frederick I. Ordway III, Huntsville, Alabama, June 21, 1971.
Interview on tape of Charles L. Stewart by Frederick I. Ordway III, New York, May 20, 1971.
Interview of Bernhard Tessmann on tape by Frederick I. Ordway III, Huntsville, Alabama, July 6, 1971.
Interview of Magnus von Braun by Mitchell R. Sharpe, London, November 2, 1972.
Letters to Mitchell R. Sharpe from Walter B. Dornberger, Chapala-Jalisco, Mexico, April 25, 1974, and undated, c. March 1974.
Letter to Mitchell R. Sharpe from Richard Fein, St. Louis, August 27, 1973.
Letters to Mitchell R. Sharpe from Dieter K. Huzel, Woodland Hills, California, May 8, 1974, and May 17, 1974.
Letter to Mitchell R. Sharpe from Walter E. Raack, Des Plaines, Illinois, February 17, 1974.
Letters to Mitchell R. Sharpe from Frederick P. Schneikert, Sheboygan, Wisconsin, February 7, 1972, and August 26, 1972.
Letter to Mitchell R. Sharpe from Charles L. Stewart, New York, February 4, 1974.

Chapter 2

BOOKS

Bergaust, Erik, *Reaching for the Stars.* New York: Doubleday, 1960.

Brandecker, Walter G., *Ein Leben für ein Idee, der Raketenpionier Max Valier.* Stuttgart: Union Verlag, 1961.

Clark, John D., *Ignition: an Informal History of Liquid Rocket Propellants.* New Brunswick, N.J.: Rutgers University Press, 1972.

Dornberger, Walter B., *V-2.* New York: Viking Press, 1954.

Essers, I., *Max Valier—a Pioneer of Space Travel.*, NASA TTF-664. Washington: National Aeronautics and Space Administration, Nov., 1976.

Gander, T. J., *Field Rocket Equipment of the German Army, 1939–1945.* London: Almark, 1972.

Gartmann, Heinz, *The Men Behind the Space Rockets.* London: Weidenfeld & Nicolson, 1955.

Heinkel, Ernst, *Stormy Life: Memoirs of a Pioneer of the Air Age.* New York: E. P. Dutton, 1956.

Ley, Willy, *Rockets, Missiles, and Men in Space.* New York: Viking Press, 1968.

———, *Events in Space.* New York: David McKay, 1969.

Nebel, Rudolf, *Die Narren von Tegeler.* Düsseldorf: Verlag Droste, 1972.

Ordway, Frederick I., and Ronald C. Wakeford, *International Missile and Spacecraft Guide.* New York: McGraw-Hill, 1960.

Philip, Charles G., *Stratosphere and Rocket Flight.* London: Sir Isaac Pitman & Sons, 1937.

Rynin, N. A., *Interplanetary Flight Communications*, Vol. 2, No. 4. Leningrad: P. P. Soikin, 1929. (NASA Technical Translation TT F-643, Washington, D.C., 1971.)

Von Braun, Wernher, and Frederick I. Ordway, *History of Rocketry & Space Travel.* New York: Thomas Y. Crowell, 1975.

Ward, Bob, *Wernher von Braun, Anekdotisch.* Esslingen: Bechtle Verlag, 1972.

Williams, Beryl, and Samuel Epstein, *The Rocket Pioneers on the Road to Space.* New York: Julian Messner, 1955.

ARTICLES AND REPORTS

Dornberger, Walter B., "European Rocketry After World War I," in L. J. Carter, ed., *Realities of Space Travel.* New York: McGraw-Hill, 1957.

———, *The Development of the Rocket Section of the Army Ordnance Office in the Years 1930–1943.* G.D. 600.5.R., September 1945. Aberdeen Proving Ground, Maryland: Ordnance Research and Development Center, Foreign Document Evaluation Branch.

———, *Life History.* Prepared for CIOS Team 183 at Garmisch-Partenkirchen, May 17, 1945.

Engel, Rolf, "Vom Raketenwagen zur Raketenforschung." Unpublished MS dated January 1, 1932, "planned for publication in *Umschau* had the HW-1 rocket been a success."

Ley, Willy, "Correspondence, Count von Braun," *The Bulletin of the British Interplanetary Society*, Vol. 6 August 1947, 154–156.

———, "The End of the Rocket Society," *Astounding Science Fiction*, Vol. 31, No. 6 and No. 7, August 1943, 64–78; September 1943, 58–75.

Nebel, Rudolf, *Rocket Flight.* (Translation of *Raketenflug*, Berlin, 1932.) (NASA Technical Translation TT F-11, 173, Washington, D.C., August 1967.)

Oberth, Hermann, "My Contributions to Astronautics," in Frederick C. Durant III and George S. James, eds., *First Steps Toward Space.* Washington; Smithsonian Institution Press, 1974.

Püllenberg, Albert, "Raketenpioneer Albert Püllenberg." Unpublished and undated article (c. March 1974), prepared by Püllenberg for Mitchell R. Sharpe.

Riedel, W. H. J., "A Chapter in Rocket History," *British Interplanetary Society Journal*, Vol. 14, No. 5 (September–October 1955), 208–212.

Sänger-Bredt, Irene, and Rolf Engel, "The Development of Regeneratively Cooled Liquid Rocket Engines in Austria and Germany, 1926–1942," in Frederick C. Durant III and George S. James, eds., *First Steps in Space.* Washington: Smithsonian Institution Press, 1974.

Thompson, G. V. E., "Oberth—Doyen of Spaceflight Today," *Spaceflight*, Vol. 1, No. 5 (October 1957), 170–171.

Von Braun, Wernher, "From Small Beginnings," in Kenneth Gatland, ed., *Project Satellite.* London: Allan Wingate, 1958.

———, "Reminiscences of German Rocketry," *Journal of the British Interplanetary Society*, Vol. 15, No. 3 (May–June 1956), 125–145.

———, "Das Geheimnis der Flüssigkeitsrakete," *Die Umschau*, No. 36, June 4, 1932, 449–452.

———, "Konstruktive, theoretische und experimentelle Beiträge zum den Problem der Flüssigkeitsrakete," (dissertation for PhD at Friederich-Wilhelm-Universität, April 16, 1934).

Von Opel, Fritz, "The Historical Development of Rockets and the Purpose and Limits of Technology." (NASA Technical Translation TT F-13,436. Washington, D.C., December 1970.)

Winter, Frank H., "Birth of the VfR: The Start of Modern Astronautics," *Spaceflight*, Vol. 19, Nos. 7–8, July–August, 1977, 243–256.

INTERVIEWS AND LETTERS

Interview on tape of Walter B. Dornberger by Frederick I. Ordway III and David L. Christensen, Huntsville, Alabama, November 4, 1971.

Interview on tape of Arthur Rudolph by Mitchell R. Sharpe, Huntsville, Alabama, June 21, 1978.

Interview on tape of Leo Zanssen by Frederick I. Ordway III and Krafft Ehricke, Bremen, September 4, 1971.

Letter from Wernher von Braun to the publisher, *Süddeutsche Zeitung* (Munich), undated (1958).

Letter to Frederick I. Ordway III from Walter B. Dornberger, Chapala-Jalisco, Mexico, undated (c. March 1974.)

Letters to G. Edward Pendray from P. E. Cleator, Liverpool, October 30, 1934; November 15, 1934.

Letter to G. Edward Pendray from Willy Ley, Berlin, January 30, 1935.
Letters to Mitchell R. Sharpe from Arthur Rudolph, San Jose, California, May 14, 1978; March 28, 1978; and April 6, 1978.

Chapter 3

BOOKS

Bar-Zohar, Michel, *La Chasse aux Savants Allemands.* Paris: Fayard, 1965.
Benecke, Theodor, and A. W. Quick, eds., *History of German Guided Missile Development: AGARD First Guided Missile Seminar, Munich, Germany, April, 1946.* Brunswick, FRG: Verlag E. Appelhaus, 1957.
Bergaust, Erik, *Reaching for the Stars.* New York: Doubleday, 1960.
Dannau, Wim, *Les Dossiers "Espace" de Wim Dannau.* Tournai: Casterman, 1966.
Dornberger, Walter B., *V-2.* New York: Viking Press, 1954.
Handbook on Guided Missiles of Germany and Japan. Washington, D.C.; War Department, Military Intelligence Division, February 1946.
Huzel, Dieter, *Peenemünde to Canaveral.* Englewood Cliffs, N.J.: Prentice-Hall, 1962.
Jones, R. V., *Most Secret War.* London: Hamish Hamilton, 1978. Also published as *The Wizard War.* New York: Coward, McCann and Geoghegan, 1978.
Klee, Ernst, and Otto Merk, *The Birth of the Missile: The Secrets of Peenemünde.* New York: E. P. Dutton; London: George G. Harrap, 1965.
Ordway, Frederick I., and Ronald C. Wakeford, *International Missile and Spacecraft Guide.* New York: McGraw-Hill, 1960.
Ruland, Bernd, *Wernher von Braun: Mein Leben für die Raumfahrt.* Offenburg: Burda Verlag, 1969.
Simon, Leslie J., *German Research in World War II.* New York: Wiley, 1947.
Speer, Albert, *Inside the Third Reich: Memoirs.* New York: Macmillan, 1970.
Von Braun, Wernher, and Frederick I. Ordway III, *History of Rocketry & Space Travel.* New York: Thomas Y. Crowell, 1975.
Ward, Bob, *Wernher von Braun, Anekdotisch.* Esslingen: Bechtle Verlag, 1972.

ARTICLES AND REPORTS

Angell, Joseph W., "Guided Missiles Could Have Won," *The Atlantic Monthly*, Vol. 121, No. 12 (December 1951), 29–34.
Dornberger, Walter B., "European Rocketry After World War I," *Journal of the British Interplanetary Society*, Vol. 13, No. 5 (September 1954), 245–268.
———, "The German V-2," in Eugene M. Emme, ed., *The History of Rocket Technology.* Detroit: Wayne State University Press, 1964.
———, *The Development of the Rocket Section of the Army Ordnance Office in the Years 1930–1943.* Aberdeen Proving Ground, Maryland; Ordnance Research and Development Center, Foreign Document Evaluation Branch, September 1945.
———, "The First V-2," in Arthur C. Clarke, ed., *The Coming of the Space Age*, New York: Meredith Press, 1967.

———, *Life History*. Prepared for CIOS Team 183 at Garmisch-Partenkirchen, May 17, 1945.

Ehricke, Krafft, "The Peenemünde Rocket Centre, Part 1," *Rocketscience*, Vol. 4, No. 1 (March 1950), 17–22.

History of German Guided Missiles; Hitler's Secret Weapons: Facts and Dreams, GD 630.0.7. Aberdeen Proving Ground, Maryland: Ordnance Research and Development Center, Foreign Document Evaluation Branch, April 1946.

Memorandum, "On the Practical Possibilities of Further Development of the Liquid Rockets and a Survey Taken of Tasks to Be Assigned to Research." Kummersdorf: Weapons Test Organisation, March 13, 1937. (A report translated by Dieter K. Huzel that shows liquid hydrogen was considered at that early date.)

Noeggerath, Wolfgang, *Development Work in Liquid Propellants in Germany*, Foreign Technical Seminar Report IRE/18. *1939–1945*. Wright Field, Dayton, Ohio, November 13, 1945.

Peabody, P. E., *Information on the History of German Military Rocket Development, Obtained Chiefly from 1st Lt. Schroeder of Nebellehrregiment (Smoke Training Regiment) 2*. Washington, D.C.: War Department, General Staff, Military Intelligence Service, September 4, 1945.

Püllenberg, Albert, "Raketenpioneer Albert Püllenberg," unpublished and undated (c. March 1974) article prepared for Mitchell R. Sharpe.

Schilling, Martin, "The Development of the V-2 Engine," in Theodor Benecke and A. W. Quick, eds., *History of German Guided Missiles Development*. Brunswick, FRG: Verlag Appelhaus, 1957.

Schulze, H. A., *Technical Data on the Development of the A4 (V-2)*. Marshall Space Flight Center, Huntsville, Alabama: Historical Office, February 25, 1965.

von Braun, Wernher, "From Small Beginnings," in Kenneth Gatland, ed., *Project Satellite*. London: Allan Wingate, 1958.

———, "Reminiscences of German Rocketry," *Journal of the British Interplanetary Society*, Vol. 15 No. 3, September 1954, 245–268.

Zwicky, F., "Report on Certain Phases of War Research in Germany," No. F-SU-3-RE, Wright Field, Dayton, Ohio, 1947.

INTERVIEWS AND LETTERS

Interview on tape of Walter B. Dornberger by Frederick I. Ordway III and David L. Christensen, Huntsville, Alabama, November 7, 1971.

Interview on tape of Krafft Ehricke by Frederick I. Ordway III, Bremen, September 5, 1971.

Interview on tape of Heinz Hilten by David L. Christensen and Ruth Heimburg, Huntsville, Alabama, August 12, 1971.

Interview on tape of Leo Zanssen by Frederick I. Ordway III and Krafft Ehricke, Bremen, September 4, 1971.

Letter to Robert L. Perry from Wernher von Braun, Huntsville, Alabama, February 9, 1966.

Letter to R. W. Reid from Wernher von Braun, Huntsville, Alabama, May 17, 1968.
Letters to Mitchell R. Sharpe from Albert Speer, Heidelberg, October 24, 1972; March 12, 1974; and April 6, 1974.

Chapter 4 and Chapter 13

BOOKS

Crankshaw, Edward, *Gestapo, Instrument of Tyranny.* New York: Viking Press, 1957.
Delaruc, Jacques, *The History of the Gestapo.* London: MacDonald & Jane's, 1964.
Dornberger, Walter, *V-2.* New York: Viking Press, 1954.
Handbook on Guided Missiles of Germany and Japan. Washington, D.C.: War Department, Military Intelligence Division, February 1946.
Huzel, Dieter, *Peenemünde to Canaveral.* Englewood Cliffs, N.J.: Prentice-Hall, 1962.
Irving, David, *The Mare's Nest.* London: William Kimber, 1964.
Lusar, Rudolf, *Die Deutschen Waffen und Geheimwaffen des 2 Weltkriegs und Ihr Wieterentwicklung.* Munich: J. F. Lehmann, 1962.

ARTICLES AND REPORTS

Dornberger, Walter B., "The Lessons of Peenemünde," *Astronautics*, Vol. 3, No. 2 (March 1958), 18–20, 58, 60.
Ehricke, Krafft, "The Peenemünde Rocket Centre," Parts 2 and 3, *Rocket-science*, Vol. 4, No. 2 (June 1950), 31–35; (September 1950), 57–88.
Hamill, James P., *Launching the A4 (V-2) Rocket Missile at Sea*, Tech. Rept. 16. Fort Bliss, Texas: Research and Development Service Sub-office (rocket), August 7, 1946.
Healy, Roy, "How Nazis' Walter Engine Pioneered Manned Rocket-craft," *Aviation*, Vol. 45, No. 1 (January 1946), 77–80.
Hermann, Rudolf, *The Supersonic Wind Tunnel of the War Ministry and Its Application in External Ballistics.* (Translation of a report prepared in 1939. It was made by H. A. Liebhafsky, Kochel, June 16, 1945.)
Mueller, F. R., "A History of Inertial Guidance." Redstone Arsenal, AL: Army Ballistic Missile Agency, n.d. (c. 1958).
Peabody, P. E., *Information on the History of German Military Rocket Development, obtained Chiefly from 1st Lt. Schroeder of Nebellehrregiment (Smoke Training Regiment) 2. British source. Received in Britain 28 July 1945, B-825.* War Department, Military Intelligence Division.
Porter, Richard W., *Control of the "Wasserfall," Ordnance Target Rept. 129, Elektromechanische Werke, Garmisch-Partenkirchen.* US Forces in Europe, Chief Ordnance Officer, Research and Intelligence Branch, n.d.
Sharpe, Mitchell R., and John M. Lowther, "Progress in Rocket, Missile, and Space Carrier Vehicle Testing, Launching, and Tracking Technology, Part II:

Facilities Outside the United States," in Frederick I. Ordway III, ed., *Advances in Space Science and Technology*. Vol. 7. New York: Academic Press, 1965.
Steinhoff, Ernst, "Early Developments in Rocket and Spacecraft Performance, Guidance, and Control," in Frederick C. Durant III and George S. James, eds., *First Steps in Space*. Washington: Smithsonian Institution Press, 1974.
The Story of Peenemünde, or What Might Have Been: Peenemünde East Through the Eyes of 500 Detained at Garmisch, n.d. (A document by various intelligence personnel interrogating Germans at Garmisch-Partenkirchen between May and September 1945.)
Stuhlinger, Ernst, "Vox Populi," *Space Journal*, Vol. 1, No. 2 (Spring 1958), 31–32.
Tsien, H. S., et al., *Technical Intelligence Supplement: A Report of the AAF Scientific Advisory Group*. Wright Field, Ohio: Air Materiél Command, Intelligence T-2, May 1946.
Zwicky, F., "Report on Certain Phases of War Research in Germany," No. F-SU-3-RE, Wright Field, Dayton, Ohio, 1947.

INTERVIEWS AND LETTERS

Interview on tape of Herbert Axster by Frederick I. Ordway III, Leutasch, Austria, September 11, 1971.
Interview on tape of Kurt Debus by Frederick I. Ordway III, Kennedy Space Center, Florida, April 27, 1972.
Interview on tape of Walter B. Dornberger by Frederick I. Ordway III and David L. Christensen, Huntsville, Alabama, November 7, 1971.
Interview on tape of Krafft Ehricke by Frederick I. Ordway III, Bremen, September 5, 1971.
Interview on tape of Dieter Grau by Frederick I. Ordway III and David L. Christensen, Huntsville, Alabama, June 17, 1971.
Interview on tape of Karl Heimburg by Frederick I. Ordway III and David L. Christensen, Huntsville, Alabama, June 26, 1971.
Interview on tape of Emil Hellebrand by Frederick I. Ordway III and David L. Christensen, Huntsville, Alabama, June 1, 1972.
Interview on tape of Rudolf Hermann by Frederick I. Ordway III, Huntsville, Alabama, May 23, 1971.
Interview on tape of Helmut Hölzer by Frederick I. Ordway III, Huntsville, Alabama, June 18, 1971.
Interview on tape of Walter Jacobi by David L. Christensen and Ruth Heimburg, Huntsville, Alabama, August 12, 1971.
Interview on tape of Hermann Kröger by Frederick I. Ordway III and Ruth Heimburg, Huntsville, Alabama, July 8, 1971.
Interview on tape of Hannes Lührsen by Frederick I. Ordway III, Heidelberg, September 9, 1971.
Interview on tape of Heinz Millinger by Frederick I. Ordway III, Frankfurt-Langen, September 8, 1971.
Interview on tape of Erich Neubert by Frederick I. Ordway III, Huntsville, Alabama, June 24, 1971.

Interview on tape of Theodor Poppel by David L. Christensen, Kennedy Space Center, Florida, August 9, 1971.
Interview of Rudolf and Dorette Schlidt by Frederick I. Ordway III, Bonn–Bad Godesberg, September 7, 1971.
Interview of Magnus von Braun by Mitchell R. Sharpe, London, November 2, 1972.
Interview of Leo Zanssen by Frederick I. Ordway III and Krafft Ehricke, Bremen, September 4, 1971.
Letter to Walter B. Dornberger from Bernhard Tessmann, Huntsville, Alabama, September 6, 1965.
Letter to Julius Klein from Wernher von Braun, Huntsville, Alabama, August 2, 1969.
Letter to Mitchell R. Sharpe from Walter B. Dornberger, Chapala-Jalisco, Mexico, April 25, 1974.
Letter to Mitchell R. Sharpe from Dieter K. Huzel, Woodland Hills, California, October 19, 1974.
Letters to Mitchell R. Sharpe from Albert Speer, Heidelberg, October 24, 1972; March 12, 1974; and April 6, 1974.
Disposition Form: "Submarine Launched Ballistic Missile," ORDAB-DLM. Redstone Arsenal, Alabama: Army Ballistic Missile Agency, April 21, 1960.

Chapter 5

BOOKS

Bornemann, Manfred, *Geheimprojekt Mittelbau, die Geschichte der Deutschen V-Waffen-Werke.* Munich: J. F. Lehmanns, 1971.
Dornberger, Walter, *V-2.* New York: Viking Press, 1954.
Höhne, Heinz, *The Order of the Death's Head: The Story of Hitler's SS.* New York: Coward-McCann, 1970.
Irving, David, *The Mare's Nest.* London: William Kimber, 1964.
Klee, Ernst, and Otto Merk, *The Birth of the Missile: The Secrets of Peenemünde.* New York: E. P. Dutton; London: George G. Harrap, 1965.
Ruland, Bernd, *Wernher von Braun: Mein Leben für die Raumfahrt.* Offenburg: Burda Verlag, 1969.
Speer, Albert, *Inside the Third Reich: Memoirs.* New York: Macmillan, 1970.

ARTICLES AND REPORTS

Bernards, Franz P. E. H., and Bernhard Kuhlbach, *Production Methods and Welding Techniques for the Combustion Chamber of V-2 (A4) Rockets.* Field Information Agency, Technical (FIAT) Final Report No. 776, March 5, 1946. (Office Military Government for Germany [US].)
Bilek, V. H., and J. D. McPhilimy, *Production and Disposition of German A4 (V-2) Rockets.* Staff Study No. A-55-2167-N1, DDC No. ATI-18315, March 1948. (Headquarters Air Materiel Command, Wright Field, Dayton, Ohio.)
Campbell, Colin, "Rocket Arsenal," *Royal Air Force Flying Review* (July 1948).
Dannenberg V-1 Bomb Plant. United States Strategic Air Forces in Europe,

Office of Assistant Chief of Staff A-2, Technical Intelligence Report No. N-24, May 18, 1945.

Foreign Logistical Organizations and Methods. A Report for the Secretary of the Army. Washington: War Department, 1946. (Contains wealth of information on German military organization, production, etc. Despite Allied bombardment, it records, "the highest output of war matériel in Germany was realized in September 1944." This was precisely the time that the V-2 campaign got under way.)

Hill, C. T., *German Production of V-2.* Chrysler Corporation Engineering Division, Tech Memo. No. A1-M-16, October 20, 1954.

Mittelwerk file 194/46 at Imperial War Museum, Foreign Documents Section, London. (Contains data on V-1 and V-2 missile invoices, serial numbers, dates of shipment, etc.)

Moureu, M. H., and P. Chovin, *Sur la Station Experimentale de Contrôle et de Reception des "V-2" de Ober-Raderach (Mission en Allemagne du 9 au 17 Mai 1945)*, Laboratoire Municipal de Paris, May 26, 1945.

Peenemünde Archives, Deutsches Museum, Munich, and Redstone Scientific Information Center, Huntsville.

Schade, H. A., *Investigation of Dannenberg V-1 Assembly Plant.* Washington, D.C.: Department of Commerce, June 11, 1945.

Target Identity Folders, CIOS teams (various interrogations, including Fleischer, Riedel III, Rees, and Gröttrup by E. H. Hull, May 21, 1945).

Target site documents for Allied intelligence personnel in Nordhausen-Niedersachswerfen vicinity (target and map reference numbers), 1944–45.

INTERVIEWS AND LETTERS

Interview on tape of Herbert Axster by Frederick I. Ordway III, Leutasch, Austria, September 11, 1971.

Interview on tape of Otto F. Cerny by Frederick I. Ordway III and Lily von Saurma, Huntsville, June 1, 1971.

Interview on tape of Dr. Walter Dornberger by David L. Christensen, Buffalo, New York, October 26, 1971, and by Frederick I. Ordway III and David L. Christensen, Huntsville, Alabama, November 7, 1971.

Interview on tape of Dr. Paul H. Figge by Frederick I. Ordway III and Boris Kit, Schwelm, September 5, 1971; Figge documentation.

Interview on tape of Alfred J. Finzel by Frederick I. Ordway III, Huntsville, Alabama, July 7, 1971.

Interview on tape of Werner K. Gengelbach by Frederick I. Ordway III and David L. Christensen, Huntsville, Alabama, June 15, 1971.

Interview on tape of Dieter Grau by Frederick I. Ordway III and David L. Christensen, Huntsville, Alabama, June 17, 1971.

Interview on tape of Karl Heimburg by Frederick I. Ordway III and David L. Christensen, Huntsville, Alabama, June 26, 1971.

Interview on tape of M. S. Hochmuth by David L. Christensen, Cambridge, Mass., October 27, 1971.

Interview of David Irving by Frederick I. Ordway III, London, September 29 and October 8, 1971; Irving documentation.

Interview on tape of Kurt Kettler by Frederick I. Ordway III and Boris Kit, Bremen-Lesum, September 5, 1971; Kettler documentation.
Interview of Hans Maus by Frederick I. Ordway III and Lily von Saurma, Huntsville, Alabama, April 26, 1971.
Interview on tape of Max E. Nowak by Frederick I. Ordway III and Ruth Heimburg, Huntsville, Alabama, July 13, 1971.
Interview on tape of Erich Neubert by Frederick I. Ordway III, Huntsville, Alabama, June 24, 1971; Neubert documentation.
Interview of Hermann Oberth by Frederick I. Ordway III, Nürnberg-Feucht, September 16, 1971.
Interview on tape of Robert Paetz by Frederick I. Ordway III and Ruth Heimburg, Huntsville, Alabama, July 6, 1971.
Interview on tape of Hannelore Bannasch Ranft by Frederick I. Ordway III, Schongau, September 15, 1971.
Interview on tape of Eberhard Rees by Frederick I. Ordway III and David L. Christensen, Huntsville, Alabama, June 21, 1971, and by Frederick I. Ordway III, February 23, 1974.
Interview on tape of Arthur Rudolph by Frederick I. Ordway III and Lily von Saurma, Huntsville, Alabama, April 7 and August 8, 1971.
Interview on tape of Colonel Terence R. B. Sanders by Frederick I. Ordway III and David L. Christensen, Huntsville, Alabama, March 6, 1972; Sanders' documentation.
Interview of Dr. Martin Schilling by David L. Christensen, Boston, Massachusetts, July 24, 1972.
Interview on tape of Bernhard Tessmann by Frederick I. Ordway III, David L. Christensen, and Ruth Heimburg, Huntsville, Alabama, July 6, 1971, and by Frederick I. Ordway III, February 16, 1974.
Interview on tape of Werner Voss by David L. Christensen and Ruth Heimburg, Huntsville, Alabama, August 3, 1971.
Interview on tape of Hermann Weidner by Frederick I. Ordway III, David L. Christensen, and Ruth Heimburg, Huntsville, Alabama, July 17, 1971.
Interview on tape of Albert Zeiler by Frederck I. Ordway III, Kennedy Space Center, Florida, April 27, 1972.
Letter to David L. Christensen from Hermann Weidner, Huntsville, Alabama, August 7, 1971.
Letter to Frederick I. Ordway III from Werner K. Gengelbach, Pasadena, California, November 7, 1972.

Chapter 6 and 9

BOOKS

Babington Smith, Constance, *Evidence in Camera: The Story of Photographic Intelligence in World War II*. London: Chatto & Windus, 1958.
Benecke, Theodor, and A. W. Quick, eds., *History of German Guided Missiles Development: AGARD First Guided Missile Seminar, Munich, Germany, April, 1946*. Brunswick, F.R.G.: Verlag E. Appelhaus, 1957.

Bergier, Jacques ("Verne"), *Secret Weapons—Secret Agents.* London: Hurst and Blackett, 1956.
Birkenhead, The Earl of, *The Prof in Two Worlds: The Official Life of Professor F. A. Lindemann, Viscount Cherwell.* London: Collins, 1961.
Cave-Brown, Anthony. *Bodyguard of Lies.* London: W. H. Allen, 1976.
Churchill, Winston S., *Closing the Ring.* Boston: Houghton Mifflin, 1951.
———, *Triumph and Tragedy.* Boston: Houghton Mifflin, 1953.
Delaruce, Jacques. *The History of the Gestapo.* London: Macdonald & Jane's, 1964.
Delmer, Sefton, *Black Boomerang.* New York: Viking Press, 1962.
Deutsch, Harold C., *The Conspiracy Against Hitler in the Twilight War.* Minneapolis: University of Minnesota Press, 1968.
Dornberger, Walter, *V-2.* New York: Viking Press, 1954.
Fest, Joachim L., *The Face of the Third Reich.* London: Weidenfeld & Nicolson, 1970.
Ford, Brian, *German Secret Weapons: Blueprint for Mars.* New York: Ballantine Books, 1969.
Grunberger, Richard, *Hitler's SS.* London: Weidenfeld & Nicolson, 1970.
Hahn, Fritz, *Deutsche Geheimwaffen 1939–1945.* Heidenheim: Erich Hoffman Verlag, 1963.
Handbook on Guided Missiles of Germany and Japan. Washington, D.C.: War Department, Military Intelligence Division, 1946.
Hartcup, Guy, *The Challenge of War: Scientific and Engineering Contributions to World War II.* Newton Abbot, Devon: David and Charles.
Hogg, I. W., *German Secret Weapons of World War II.* London: Arms and Armour Press, 1970.
Höhne, Heinz, *The Order of the Death's Head: The Story of Hitler's SS.* New York: Coward-McCann, 1970.
Huzel, Dieter, *Peenemünde to Canaveral.* Englewood Cliffs, N.J.: Prentice-Hall, 1962.
Irving, David, *The Mare's Nest.* London: William Kimber, 1964.
John, Otto, *Twice Through the Lines.* London: Macmillan, 1972.
Jones, R. V., *Most Secret War: British Scientific Intelligence 1939–1945.* London: Hamish Hamilton, 1978.
Joubert de la Ferté, Sir Philip, *Rocket.* London: Hutchinson, 1957.
Kahn, David, *The Codebreakers: The Story of Secret Writing.* New York: Macmillan, 1967.
Klee, Ernst, and Otto Merk, *The Birth of the Missile: The Secrets of Peenemünde.* New York: E. P. Dutton; London: George G. Harrap, 1965.
Krausnick, Helmut, Hans Buchheim, Martin Broszat, and Hans-Adolf Jacobsen, *Anatomy of the SS State.* New York: Walker and Co., 1965.
Leverkuehn, Paul, *German Military Intelligence.* London: Weidenfeld & Nicolson, 1954.
Lewin, Ronald, *Ultra Goes to War: The Secret Story.* London: Hutchinson, 1978.
Ley, Willy, *Rockets, Missiles, and Men in Space.* New York: Viking Press, 1968.
———, *Rockets, Missiles, and Space Travel.* New York: Viking Press; London: Chapman and Hall, 1951.

———, *Rockets: The Future of Travel Beyond the Stratosphere.* New York: Viking Press, 1944.
McGovern, James, *Crossbow and Overcast.* New York: William Morrow, 1964.
Manvell, Roger, and Heinrich Fraenkel, *The Canaris Conspiracy: The Secret Resistance to Hitler in the German Army.* New York: David McKay, 1969.
———, *The July Plot.* London: The Bodley Head, 1964.
Martelli, George, *Agent Extraordinary.* London: Collins, 1960.
Masterman, J. C., *The Double-Cross System in the War of 1939 to 1945.* New Haven: Yale University Press, 1972.
Newman, Bernard, *They Saved London.* London: Werner Laurie, 1952.
Ordway, Frederick I., III, and Ronald C. Wakeford, *International Missile and Spacecraft Guide.* New York: McGraw-Hill, 1960.
Pocock, R. F., *German Guided Missiles of the Second World War.* London: Ian Allen, 1967.
The Rise and Fall of the German Air Force. London: British Air Ministry, No. 248, 1948.
Ruland, Bernd, *Wernher von Braun: Mein Leben für die Raumfahrt.* Offenburg: Burda Verlag, 1969.
Schlabrendorff, Fabian von, *The Secret War Against Hitler.* New York: Pitman, 1965.
Speer, Albert, *Inside the Third Reich: Memoirs.* New York: Macmillan, 1970.
Thresmeyer, Lincoln R., and John E. Burchard, *Combat Scientists.* Boston: Little, Brown, 1947.
Trevor-Roper, Hugh R., ed., *Blitzkrieg to Defeat: Hitler's War Directives 1939–1945.* New York: Holt, Rinehart and Winston, 1965.
von Braun, Wernker, and Frederick I. Ordway III, *The Rockets' Red Glare.* New York: Doubleday, 1976.
Watson-Watt, Sir Robert, *Three Steps to Victory.* London: Odhams, 1957.
Winterbotham, F. W., *Secret and Personnel.* London: William Kimber, 1969.
———, *The Ultra Secret.* New York: Harper & Row, 1974.
Zeller, Eberhard, *The Flame of Freedom: The German Struggle against Hitler.* Coral Gables, Florida: University of Miami Press, 1969.

ARTICLES AND REPORTS

Big Ben Sub-Committee, *Minutes of First Meeting* (20 July 1944), with appendices (Bodyline Report, Fort Halstead, 24 November 1943); *The Ballistics of Long Range Rockets*, Branch for Theoretical Research, Fort Halstead, December 1943; *Sound Ranging, Hypothetical Long Range Rocket Mark I*, 18 July 1944; *Flying Bomb and Barrage Balloons*; R.A.E. *Flying Bomb dossier*, etc.).
Big Ben Sub-Committee, *Minutes of Second Meeting* (25 July 1944), with appendices.
Big Ben Sub-Committee, *Minutes of Third Meeting* (1 August 1944), with appendices (*The Design of Long Range Rockets*, CPD No. 11/44, April 1944).
Big Ben Sub-Committee, *Minutes of Fourth Meeting* (8 August 1944), with appendices [*Crossbow Committee Report* (44) 47, 3 August 1944 *Radio and Radar Aids to the Defeat of the Rocket and the Flying Bomb* by Sir Robert Watson-Watt; *The Organisation of Big Ben* (military organisation); Fraser, R. P.,

and J. M. Connor, *The Design of Long Range Rockets.* CPD Technical Report No. 11/44, April 1944; *Propulsive Liquids*, 1 August 1944 (report that concluded that the A4 was propelled by liquid oxygen and alcohol); memorandum by Isaac Lubbock dated 5 August 1944 entitled *Hypothetical Rocket "X.A.4, Mark I"*; appendix to memo: *Hypothetical Rocket "X.A.4 Mark I" Reconstruction and Analysis of Enemy Rocket on Basis of Swedish Parts*, B.B.S./P(44)16; *Interim Report on the Ballistic Control of B.B.*, Armament Research Dept., 31 July 1944; and *Extracts from Reports on Prisoners of War*].

Big Ben Sub-Committee, *Minutes of Fifth Meeting* (15 August 1944), with appendices [*Examination of Crossbow Sites*, A.I.2(g) Report No. 3/X; *Intelligence Report: German Documentary Information on A.4 Long Range Rocket Construction and Fuels*, 12 August 1944; *German Long Range Rocket Projectile Storage and Launching Sites*; *Radio and Radar Devices for the Defeat of "Big Ben,"* Sir Robert Watson-Watt, 11 August 1944; and *Radio Equipment of Big Ben: First Report on Items Received for Examination on 19th July*, by C. P. Edwards and G. F. Evans, R.A.E. Radio Department Report Radio/S.4840/CPE/34, 7 August 1944].

Big Ben Sub-Committee, *Minutes of Sixth Meeting* (22 August 1944), with appendices [*German Organisation: Interim Report to the Main Committee*, B.B.S./M(44)5: Conclusion (52)(b), 19 August 1944; *Intelligence: German Long Range Rocket Projectile Storage and Launching Sites*, 19 August 1944; *Radio and Radar Devices for the Defeat of the Rocket*, 19 August 1944; and *Warhead: Possible Use of Liquid High Explosives*, C.S.A.R., 19 August 1944].

Big Ben Sub-Committee, *Minutes of Seventh Meeting* (29 August 1944), with appendices [*The Swedish Rocket—Report of Reconstruction*, B.B.S./P(44)31, 25 August 1944; *Gyroscopic Control Possible Use in the Rocket; Rocket Trajectories; Probable Accuracy of the Rocket*; and *Warhead: Contribution of Unused Fuel to Blast Effect of Explosive*].

Big Ben Sub-Committee, *Minutes of Eighth Meeting* (12 September 1944), with appendices [*Examination of Storage and Launching Sites*, D/A.C.A.S. (I), 1 September 1944; *Radar and Radio Devices for the Defeat of the Rocket*, 1 September 1944; *Launching Sites for A4 Rocket*; *Memorandum for Director of Technical Services on Fall of First Two A4 Missiles on England*, 9 September 1944; and *Possible Fuzes for the A4 Rocket*].

Big Ben Sub-Committee, *Minutes of Ninth Meeting* (19 September 1944), with appendices [*Rocket Incidents Reports*; *Collection of Operational Data*; *Reconstruction of Remains of the Operational Rocket*; *Warhead: Summary of Information on Effect* (to mid-day, 15 September 1944); *Crater Formation Contribution of the Kinetic Energy of the Rocket to the Explosion*, 19 September 1944; *Vapour Trails Produced by the Rocket*; *Motion of the Rocket Through the Upper Atmosphere*, R.A.E. Aero Note 1514, 18 September 1944; *Radio Control: Enemy's Probable Technique*; and *Control and Accuracy of Big Ben,* Armament Research Department, Fort Halstead, 18 September 1944].

Big Ben, Ministry of Home Security Research and Development Report S.104/5230, July 19, 1944.

Bodyline Radio Subcommittee Minutes, 5 November 1943.

Crossbow memoranda; collection of high-level policy memoranda dealing with what became the German V-weapon threat and with Allied counter measures.

"Editorial," *Journal of the British Interplanetary Society*, Vol. 4, No. 2 (December 1937), 3.

Evidence in Camera, Vol. 7, No. 13 (July 24, 1944). (Flying bomb issue.)

German Long-Range Rocket Programme 1930–1945, War Office MI 4/14, October 30, 1945; Supplement, December 27, 1945.

Interim Report on the Projection of Long Range Rockets, A.D.D. Report No. 27/44, July 30, 1944.

Irving, David, "Unternehmen Armbrust," *Der Spiegel* Nos. 44, 45, 46, 47, and 48 (1965).

Jones, Professor R. V., *Scientific Intelligence.* (A lecture given to the Royal United Services Institution, London, February 19, 1949.)

Launching of the Flying Bomb, September 23, 1944 (based on POW interrogations).

Letter to the Editor concerning Wernher von Braun from Willy Ley, Correspondence section, *Journal of the British Interplanetary Society* Vol. 6, No. 5 (June 1947), 154.

Ley, Willy, "Evaluating the Vaunted V-2," *Aviation*, February 1945.

"News and Notes," *Astronautics* No. 44 (November 1939).

Note on the Long Range Rocket,, Assistant Director of Intelligence (Science), Air Ministry, July 16, 1944.

Putt, Colonel Donald L., "German Developments in the Field of Guided Missiles, *SAE Journal (Transactions)* Vol. 54, No. 8 (August 1946), 404.

"Rocket Experiments of 1934—News of German Experimenters," *Astronautics* No. 29 (September 1934), 9.

Status Memo, Air Ministry to SHAEF, August 12, 1944.

V-1 Development and Test Program. Interrogation of Prof. Karl Thalan, Kassel, July 10 and 21, 1945, United States Strategic Air Forces in Europe (Main), Office of the Assistant Chief of Staff, A-2, Exploitation Division.

V-2 Incident Tables (issued regularly from September 1944).

Woodruff, L. F., and G. H. Drewry, Jr., *Report on German Long Range Rocket*, August 16, 1944.

INTERVIEWS AND LETTERS

Interview of Constance Babington Smith by Frederick I. Ordway III, Cambridge, England, October 11, 1971.

Interview on tape of Sir William R. Cook by Frederick I. Ordway III, London, October 7, 1971.

Interview of Sir Alwyn Crow by Frederick I. Ordway III, Washington, D.C., June 29, 1964.

Interview of Captain Gerald O. C. Davies by Frederick I. Ordway III, Liphook, Hampshire, October 4, 1971.

Interview of Douglas N. Kendall by Anees Jung, Toronto, July 17, 1972.

Interview of Colonel A. D. Merriman by Frederick I. Ordway III, London, September 29, 1971.

Interview of H. Rowling by Frederick I. Ordway III, London, February 22, 1966.

Interviews of the Right Honourable Duncan Sandys by Frederick I. Ordway III, London, May 13, 1966 and October 12, 1971.
Letter to Frederick I. Ordway III from Colonel N. L. Falcon, Chiddingfold, Surrey, June 9, 1974.
Letters to Frederick I. Ordway III from Professor R. V. Jones, Aberdeen, December 9, 1971, and August 8, 1972.
Letter to Frederick I. Ordway III from Douglas N. Kendall, Toronto, May 25, 1972.
Letter to Frederick I. Ordway III from Alfred Price, Uppingham, Rutland, January 27, 1972.
Letter to Frederick I. Ordway III from Professor Walt W. Rostow, Austin, Texas, October 23, 1972.

Chapter 7

BOOKS

Andrews, Allen, *The Air Marshals.* London: Macdonald & Jane's, 1970.
Babington Smith, Constance, *Evidence in Camera: The Story of Photographic Intelligence in World War II.* London: Chatto & Windus, 1958.
Churchill, Winston S., *Closing the Ring.* Boston: Houghton Mifflin, 1951.
Craven, Wesley F. and James L. Cate, eds., *The Army Air Forces in World War II. Volume 3: Argument to V-E Day.* Chicago: University of Chicago Press, 1951. (See particularly Chapter 3, "Crossbow" and Chapter 15, "Crossbow—Second Phase" by Joseph W. Angell).
Dornberger, Walter, *V-2.* New York: Viking Press, 1954.
Foot, M. R. D., *SOE in France.* London: Her Majesty's Stationery Office, 1966.
Harris, Arthur, *Bomber Offensive.* London: Collins, 1947.
Huzel, Dieter, *Peenemünde to Canaveral.* Englewood Cliffs, N.J.: Prentice-Hall, 1962.
Irving, David, *The Mare's Nest.* London: William Kimber, 1964.
Joubert de la Ferté, Sir Philip, *Rocket.* London: Hutchinson, 1957.
Klee, Ernst, and Otto Merk, *The Birth of the Missile: The Secrets of Peenemünde.* New York: E. P. Dutton; London: George G. Harrap, 1965.
McGovern, James, *Crossbow and Overcast.* New York: William Morrow, 1964.
Olsen, Jack, *Aphrodite: Desperate Mission.* New York: G. P. Putnam's Sons, 1970.
Ruland, Bernd, *Wernher von Braun: Mein Leben für die Raumfahrt.* Offenburg: Burda Verlag, 1969.
Verrier, Anthony, *The Bomber Offensive.* London: Batsford, 1968.
Webster, Sir Charles, and Noble Frankland, *The Strategic Air Offensive Against Germany, 1939–1945.* 4 vols. London: Her Majesty's Stationery Office, 1961.

ARTICLES AND REPORTS

Angell, Joseph Warner, "Guided Missiles Could Have Won," *Atlantic Monthly*, 188 (December 1951), 29; and 190 (January 1952), 57.
Big Ben Rocket Files, United States Strategic Air Forces in Europe—Office of the

Director of Intelligence, December 1944 and January 1945. *Postscript* to above report, July 5, 1945.

Bomb Damage Folders, US Stategic Air Force No. 29, 1943–45. Bombing Reports, Vol. 29.

Case Histories of Vertical Bombs, Controlled Robot Aircraft and Glide Bombs, 1944.

Crossbow files: photographs, reviews, construction and test data, diagrams; analysis of Eglin Field tests, various dates, 1944.

Examination of Crossbow Sites: Storage, Launching and Motor Transport Parks and Servicing Depots, A.I.2(g) Report No. 7/2 X, August 25, 1944.

Hickman, C. N., *V-1 Launching Sites.* Report No. 321, 22, 24, 25. Washington, D.C., Department of Commerce, September 12, 1944.

Hickman, C. N., *Personal Notes on Trip to England and France, August 22–October 15, 1944*, November 4, 1944 (supplied by Dr. Hickman to authors).

Investigation of the Heavy Crossbow Installations in Northern France—Report by the Sanders Mission to the Chairman of the Crossbow Committee, February 21, 1945 (Volume I: general and history, conclusions, recommendations, appendices; Volume II: illustrations; and Volume III: Mimoyecques and Wizernes—two parts).

List of Noball Targets (ski and modified ski sites), August 28, 1944.

Marquise/Mimoyecques Bomb Damage Report (typed report), 1944.

Overacker, Colonel C. B., *Final Report on Crossbow Project.* Eglin Field, Florida, Proof Department, Army Air Forces Proving Ground Command, March 1, 1944.

Sanders, Colonel T. R. B., *A Summary of the Information to Date on the Crossbow "Heavy Sites" and a Consideration of Their Possible Use for the Projection of the Long Range Rockets*, A.D.D. Report No. 15/44, May 1, 1944.

Supplementary Test of Crossbow, Final Report, Proof Department, Army Air Forces Proving Ground Command, Eglin Field, Florida, April 17, 1944.

United States Strategic Bombing Survey: Overall Report (European War), September 30, 1945; *Statistical Appendix to Overall Report (European War)*, February 1947; *Report on the Crossbow Campaign—The Air Offensive Against the V-Weapons*, Military Analysis Division, January 1947.

INTERVIEWS AND LETTERS

Interview on tape of Herbert Axster by Frederick I. Ordway III, Leutasch, Austria, September 11, 1971.

Interview on tape of Konrad and Ingeborg Dannenberg by Frederick I. Ordway III and Lily von Saurma, Huntsville, Alabama, May 1, 1971.

Interview on tape of Kurt Debus by Frederick I. Ordway III, Kennedy Space Center, Florida, April 27, 1972.

Interview on tape of Friedrich Dohm by Frederick I. Ordway III, Huntsville, Alabama, June 16, 1971.

Interview on tape of Walter B. Dornberger by David L. Christensen, Buffalo, New York, October 26, 1971.

Interview of Walter B. Dornberger by David L. Christensen and Frederick I. Ordway III, Huntsville, Alabama, November 7, 1971.

Interview on tape of Werner Gengelbach by Frederick I. Ordway III and David L. Christensen, Huntsville, Alabama, June 15, 1971.
Interview on tape of Dieter Grau by Frederick I. Ordway III and David L. Christensen, Huntsville, Alabama, June 17, 1971.
Interview on tape of Rudolf Hermann by Frederick I. Ordway III, Huntsville, Alabama, May 22, 1971.
Interviews of Dr. C. N. Hickman by Frederick I. Ordway III, New York City, June 18, 1964, and May 5, 1970.
Interview on tape of Heinz Hilton by David L. Christensen and Ruth Heimburg, Huntsville, Alabama, August 12, 1971.
Interview on tape of Colonel Terence R. B. Sanders by Frederick I. Ordway III and David L. Christensen, Huntsville, March 6, 1972.
Interview on tape of Oscar Scholze by Frederick I. Ordway III, Ottobrun, September 14, 1971.
Letter to General H. H. Arnold from Brigadier General Grandison Gardner, Hq., Army Air Forces Proving Ground Command, Eglin Field, Florida, February 28, 1944 (re Crossbow Simulation Tests).
Letters to Frederick I. Ordway III from Colonel Terence R. B. Sanders, Buckland, Surrey, March 14, 1972 and October 16, 1972.
Letter to Frederick I. Ordway III from Wernher von Braun, Germantown, Maryland, April 15, 1974.
Letter to Mitchell R. Sharpe from Walter B. Dornberger, Chapala-Jalisco, Mexico, April 25, 1974.
"Crossbow." Memorandum for the Commanding General, Army Air Forces from Colonel Joe L. Loutzenheiser, Chief, Operational Plans Division, May 3, 1944, and other correspondence April and May 1944.
"Big Ben." Memorandum to Brigadier General George C. McDonald, Hq., US Strategic Air Forces in Europe, Office of the Director of Intelligence, November 7, 1944.

Chapter 8

POLAND

BOOKS

Bor-Komorowski, T., *The Secret Army.* London: Victor Gollancz, 1950.
Dornberger, Walter, *V-2.* New York: Viking Press, 1954.
Garlinski, Josef, *Poland, SOE, and the Allies.* London: George Allen and Unwin, 1969.
———, *Hitler's Last Weapons.* London: Julienn Friedmann, 1978.
Irving, David, *The Mare's Nest.* London: William Kimber, 1964.
Joubert de la Ferté, Sir Philip, *Rocket.* London: Hutchinson, 1957.
Piekalkiewicz, Janusz, *Spione Agenten: Geheime Kommandos in Zweiten Weltkrieg.* Munich: Südwest Verlag, 1969.
Polish Home Army Parachutist Association, *The Unseen and Silent.* New York: Sheed and Ward, 1954.

Werth, Alexander, *Russians at War 1941–1945*. New York: E. P. Dutton; London: Barric and Porkcliff, 1964.
Wojewodzki, Michal, *Akcja V-1, V-2*. Warsaw: Instytut Wydawniczy PAX, 1970.

ARTICLES AND REPORTS

Gollin, M. G. J., *The Bug Hunters of The Snark Was a Boojum* (manuscript on Blizna mission), July 8, 1957.
Iranek-Osmecki, Kazimierz, "Meldunek Specjalny 1/R Nr 242 Pociski Rakietowe," *Zeszyty Historyczne*. Paris: Instytyt Literacki, 1972.
Preliminary Report on the Visit of the Anglo-American Mission to the German Experimental Rocket Station in Poland, datelined Moscow,, September 22, 1944.
Unterlagen für die Vermessung der Feuerstellung und Nebenanlagen "Osteinzatz" (Blizna firing position survey), Deutsches Museum Peenemünde archives document E1777/43gK, September 14, 1943.
Von Braun, Wernher, "From Small Beginnings," in Kenneth W. Gatland, ed., *Project Satellite*. London: Allan Wingate, 1958.

INTERVIEWS AND LETTERS

Interview on tape of Herbert Axster by Frederick I. Ordway III, Leutasch, Austria, September 11, 1971.
Interviews on tape of Dr. S. G. Culliford by Frederick I. Ordway IV, February 11, 1973; and by Frederick I. Ordway III, November 2 and 10, 1973, Wellington, New Zealand.
Interviews on tape of Dr. Walter Dornberger by David L. Christensen, Buffalo, New York, October 26, 1971, and by Frederick I. Ordway III and David L. Christensen, Huntsville, Alabama, November 7, 1971.
Interview on tape of Friedrich Duerr by David L. Christensen and Ruth Heimburg, Huntsville, Alabama August 4, 1971.
Interviews of M. G. J. Gollin by Frederick I. Ordway III, London, February 22, 1966; August 4, 1971; and September 29, 1971.
Interviews of Colonel Kazimierz Iranek-Osmecki by Frederick I. Ordway III, London, April 19 and 21, 1973.
Interviews of Ernst Klee by Frederick I. Ordway III, Deutsches Museum—Peenemünde archives, Munich, September 13, 14, and 15, 1971.
Interview on tape of Carl H. Mandel by Frederick I. Ordway III and Ruth Heimburg, Huntsville Alabama, July 2, 1971.
Interview of Colonel A. D. Merriman by Frederick I. Ordway III, London, September 29, 1971.
Interview on tape of Heinz A. Millinger by Frederick I. Ordway III, Frankfurt-Langen, September 8, 1971.
Interview on tape of Theodor Poppel by David L. Christensen, Kennedy Space Center, Florida, August 9, 1971.
Interviews on tape of Eberhard Rees by Frederick I. Ordway III and David L. Christensen, Huntsville Alabama July 2, 1971, and by Frederick I. Ordway III, Huntsville, Alabama, February 23, 1974.

Interview on tape of Werner K. Rosinski by Lily von Saurma, Huntsville, April 13, 1971.
Interview on tape of Colonel Terence R. B. Sanders by Frederick I. Ordway III and David L. Christensen, Huntsville, Alabama, March 6, 1972.
Interviews on tape of Tadeus E. Schnitzer by Frederick I. Ordway III and David L. Christensen, Huntsville, Alabama, January 31 and February 10, 15, and 22, 1972.
Interview on tape of Albert Zeiler by Frederick I. Ordway III, Kennedy Space Center, Florida, April 27, 1972.
Letters to Frederick I. Ordway III from M. G. J. Gollin, Ashtead, Surrey, August 6, 1966, September 12, 1970, and July 31, 1971.
Letter to Frederick I. Ordway III from Colonel Kazimierz Iranek-Osmecki, London, February 18, 1972; April 12, 1972; and June 19, 1973.

SWEDEN

BOOKS

Dornberger, Walter, *V-2*. New York: Viking Press, 1954.
Irving, David, *The Mare's Nest*. London: William Kimber, 1964.
Jones, R. V., *The Wizard War*. London: Hamish Hamilton, 1978.

ARTICLES AND REPORTS

Description of Rocket Parts Found in Sweden, June 13, 1944. Air Intelligence Note, 20 June 1944.
German Long-Range Rocket Projectile. A.I.2 (g) D. of I. (R), Report No. 2254 (R), 22 July 1944.
Interim Report by the Director of Tactical Investigation Outlining the Organisation Likely to be Employed by the Germans for the Operation of Long Range Rockets . . . , War Office. T.I./1021, August 1944.
Drawings of Big Ben based on Swedish incident attached to a number of Big Ben Sub-Committee reports, Asiatic Petroleum Company, 1944.
The Swedish Rocket—Report on Reconstruction, B.B.S./P(44)31, 25 August 1944.
Dixon, Thomas F., "Solving the V-2 Mystery in 1944," *Air Power Historian* 10 (April 1963).
Edwards, C. P. and G. J. Evans, *Radio Equipment of Big Ben: First Report on Items Received for Examination on 19th July, 1944*. Radio Department, R.A.E., Radio/S.4840/CPE/34, 7 August 1944.

INTERVIEWS AND LETTERS

Interview on tape of Kurt Debus by Frederick I. Ordway III, Kennedy Space Center, Florida, April 27, 1972.
Interview on tape of Walter Dornberger by David L. Christensen, Buffalo, New York, October 26, 1971, and by Frederick I. Ordway III and David L. Christensen, Huntsville, Alabama, November 7, 1971.

Interview of Eugen Semitov by Frederick I. Ordway III, London, February 4, 1966.

Interview on tape of Ernst Steinhoff by David L. Christensen, West Carrollton, Ohio, August 19, 1971.

Letter from Colonel Bernt Balchen to Dr. Wernher von Braun, February 29, 1972.

Chapters 10, 11, and 12

BOOKS

Banger, Joan, *Norwich at War,* Norwich: Wensum Books.

Blumentritt, Günther, *Von Rundstedt; The Soldier and the Man.* London; Odhams, 1952.

Bradley, Omar, *A Soldier's Story.* New York: Henry Holt, 1951.

Braun, Wernher von, and Frederick I. Ordway III, *History of Rocketry & Space Travel.* New York: Thomas Y. Crowell, 1975.

Churchill, Winston S., *Triumph and Tragedy.* Boston: Houghton Mifflin, 1953.

Cole, Hugh M., *United States Army in World War II.* Washington, D.C.: Office of Chief of Military History, 1965.

Collier, Basil, *The Battle of the V-Weapons 1944–45.* London: Hodder and Stoughton, 1964.

———, *The Defence of The United Kingdom.* London: Her Majesty's Stationery Office, 1957.

Eisenhower, Dwight D., *Crusade in Europe.* Garden City, N.Y.: Doubleday, 1948.

Ellis, Major L. F., and others, *Victory in the West, 1944–45.* Volume 1: *The Battle of Normandy;* Volume 2: *The Defeat of Germany.* London: Her Majesty's Stationery Office, 1963.

Essame, H., *The Battle for Germany,* London: B. T. Batsford, 1969.

Gilbert, Felix, *Hitler Directs His War: The Secret Records of His Daily Military Conferences.* New York: Oxford University Press, 1950.

Guderian, Heinz, *Panzer Leader.* London: Weidenfeld & Nicolson, 1970.

Hill, Prudence, *To Know the Sky: The Life of Air Chief Marshal Sir Roderic Hill.* London: William Kimber, 1962.

Irving, David, *The Mare's Nest.* London: William Kimber, 1964.

———, *Hitler's War.* New York: Viking Press, 1977.

Joubert de la Ferté, Sir Philip. *Birds and Fishes: The Story of the Coastal Command.* London: Hutchinson, 1960.

———, *Rocket.* London: Hutchinson, 1957.

Kooy, J. M. J., and J. W. H. Uytenbogaart, *Ballistics of the Future.* New York: McGraw-Hill, 1946.

Liddell Hart, B. H., *History of the Second World War.* London: Cassell, 1970.

Longmate, Norman, *How We Lived Then: A History of Everyday Life during the Second World War.* London: Hutchinson, 1971.

Montgomery, Viscount, *The Memoires.* Cleveland: World, 1958.

Mosley, Leonard, *Backs to the Wall: London Under Fire 1940–1945*. London: Weidenfeld & Nicolson, 1971.
Patton, George S., Jr., *War As I Knew It.* Boston: Houghton Mifflin, 1947.
Pile, Sir Frederick, *Ack-Ack: Britain's Defence Against Air Attack During the Second World War.* London: George G. Harrap, 1949.
Pogue, Forrest C., *The Supreme Command.* Washington, D.C.: Office of the Chief of Military History, 1954.
Ramsey, Winston G., ed., *After the Battle*. No. 6. London: Algers Press, 1974 (V-Weapons issue).
Rawlings, John, *Fighter Squadrons of the RAF and Their Aircraft.* London: Macdonald & Jane's, 1969.
Saundby, Sir Robert, *Air Bombardment.* New York: Harper & Row, 1961.
Saunders, Hilary St. George, *Royal Air Force 1939–1945*. Volume 3: *The Fight Is Won*. London: H.M.S.O., 1954.
Shirer, William L., *The Rise and Fall of the Third Reich.* New York: Simon and Schuster, 1960.
Toland, John, *The Last 100 Days.* New York: Random House, 1965.
Turner, John Frayn, *Highly Explosive: The Exploits of Major "Bill" Hartley, MBE, GM of Bomb Disposal.* London: George G. Harrap, 1961.
Watson-Watt, Sir Robert, *Three Steps to Victory.* London: Odhams, 1957.
Wilmot, Chester, *The Struggle for Europe.* New York: Harper & Bros., 1952.

ARTICLES AND REPORTS

AAA Operating Against Flying Bomb. Headquarters, 12th Army Group, War Department Observers Board, ACF Report No. 165, August 17, 1944.
Accuracy of the A4, Special Projectile Operations Group Report SPOG/500/6, 1945.
Action. A4 Field Units. Ordnance Department, US Army (no date).
A4 Fibel. Published at Peenemünde in early 1944 to reflect situation of A4 missile as of January 7, 1944; launch manual used by troops in the field. English translation published in Huntsville: Army Ballistic Missile Agency, 1957 (translated and edited by John A. Bitzer and Ted A. Woerner).
Antwerp Under V-1 and V-2. Report in Dutch, French, and English; pictorial, 45 pages, 1945.
Armstrong, Brigadier General Clare H., et al., *Tactical Employment of Antiaircraft Artillery Units* (including defence against pilotless aircraft, V-1). The General Board, United States Forces, European Theater of Operations, AAA Section, November 20, 1945.
Army Radar Methods Used in Operation Crossbow. Army Operational Research Group Report No. 309, December 18, 1946.
Churchill, Winston S., Statement on V-2 Rocket. *New York Times,* November 11, 1944.
Dallmeyer, Captain A. R., Jr., "Antwerp X: "The Secret Command Which Saved the Allies' Number One Supply Port," *Coast Artillery Journal* 88 No. 5 (September–October 1945), 2.
Darnall, Colonel Joseph Rogers, "Buzz-Bomb Assaults on London," *Coast Artillery Journal* 90 No. 4 (July–August 1947), 50.

Delderfield, Flight Lieutenant R. F., "A Study in Passive Defence," *The Royal Air Force Quarterly,* Vol. 16, No. 3 (June 1945), 164 (deals with the Balloon Command).
Edgar, Captain William, "Ground Defense Plan: Antwerp X," *Coast Artillery Journal* 88 No. 5 (September–October 1945), 8.
Fall of Shot of the German A4 Long-Range Rocket used in Operations September 1944–March 1945, Army Operational Research Group Report No. 312, July 15, 1946.
Field Organisation and Establishments; and Deployment. Report on Operation "Back-fire," Special Projectile Operations Group SPOG/500/12, November 7, 1945.
"The Future of Long-Range Rockets," *Air Ministry Weekly Intelligence Summary* No. 280, January 13, 1945.
"German Propaganda and the Rocket," *Air Ministry Weekly Intelligence Summary* No. 281, January 20, 1945.
Harrison, Gordon A., *Cross-Channel Attack.* Washington; Office of the Chief of Military History, 1954.
Helfers, Lieutenant Colonel M. C. *The Employment of V-Weapons by the Germans During World War II.* Washington: Office of the Chief of Military History, Department of the Army, May 31, 1954.
Hill, Air Chief Marshal Sir Roderic, *Air Operations By Air Defence of Great Britain and Fighter Command in Connection with the German Flying Bomb and Rocket Offensives, 1944–1945,* Supplement to the London Gazette No. 38437, October 19, 1948.
Macmillan, Captain Norman, "German 'Reprisal Weapons' in Action.' Postwar, undated.
Marks, Lieutenant Colonel William S., Jr., Major Joseph P. D'Arezzo, Major R. A. Ranson, and Captain G. D. Bagley, *Detection and Plotting of the V-2 (Big Ben) Missile as Developed in the ETO* [European Theatre of Operations], 65 pages, July 4 1945.
Report on V-1. United States Strategic Air Forces in Europe, Office of Assistant Chief of Staff A-2, Technical Intelligence Report No. N-31, 29 June 1945.
Rüdel, Generaloberst Günther, *German Principles Covering Commitment and Control of Antiaircraft (Flak) Artillery, Project 5.* Translated by A. Rosenwald. Report MS No. P–009, datelined Munich, August 20, 1948 (deals in part with Fi103, later designated Flugzielgerät, or Fzg, 76).
Saunders, Hilary St. George, "The Flying Bomb," *Life,* November 20, 1944.
Thompson, Royce L. "Military Impact of the German V-Weapon" (unpublished ms.). Washington, D.C., Department of the Army, Office Chief of Military History, 1953.
V-2 Countermeasures in the ETO, Headquarters, Army Ground Forces, Army War College, 4 July 1945.
"V-Wapens;" *Bericht van de Tweede Wereld Oorlog,* Nr. 79, Band 5, August 7, 1971.
V-Weapons in London. United States Strategic Bombing Survey, Physical Damage Division, January 1947.
Walter, Eugen. *V-Weapons.* Translation of report MS B-689 prepared at Allen-

dorf, May 16, 1947, pencil edited, 95 pages. Office Chief of Military History.
"The War in the Air—1939–45," *RAF Quarterly,* XVI No. 3 (June 1945), 26.
Watkins, Major E. S., "Secret Phase of the Flying Bomb," *Coast Artillery Journal,* Vol. 88, No. 5 (September–October 1945), 43.
Wilmot, Chester, "Rocket Warfare: The Offensive Threat," *The Observer,* October 11, 1953; "The Defensive Counter," October 18, 1953.

INTERVIEWS AND LETTERS

Interview of Joan Banger by Frederick I. Ordway, Norwich, October 27, 1978.
Interview of Colonel Denis Ewart Evans by Frederick I. Ordway III, London, January 20, 1966.
Interviews on tape of Helmut Horn by Frederick I. Ordway III, Huntsville, Alabama, June 17, 1971, and by Frederick I. Ordway III and David L. Christensen, Huntsville, Alabama, June 26, 1971.
Interview on tape of Professor Dr. J. M. J. Kooy by Frederick I. Ordway III, Bremen, September 3, 1971.
Letter to Frederick I. Ordway III from General Sir Frederick A. Pile, Cottered near Buntingford, Hertfordshire, November 27, 1972.
Interviews of Colonel Kenneth Post by Frederick I. Ordway III, London, July 20, 1966, and October 11, 1971.
Interview on tape of Lieutenant Colonel Wilhelm Zippelius by Frederick I. Ordway III and Rudolf Schlidt, Bonn–Bad Godesberg, September 7, 1971.

Chapter 14 and Chapter 16

BOOKS

Bar-Zohar, Michel, *La Chasse aux Savants Allemands.* Paris: Fayard, 1965.
Bergaust, Erik, *Reaching for the Stars.* New York: Doubleday, 1960.
Bergwin, Clyde R., and William T. Coleman, *Animal Astronauts: They Opened the Way to the Stars.* Englewood Cliffs, N.J.: Prentice-Hall, 1963.
Dornberger, Walter B., *V-2.* New York: Viking Press, 1954.
Goudsmit, Samuel A., *Alsos.* New York: Henry Schuman, 1947.
Lasby, Clarence G., *Project Paperclip: German Scientists and the Cold War.* New York: Atheneum, 1971.
Ley, Willy, *Events in Space.* New York: David McKay, 1969.
McGovern, James, *Crossbow and Overcast.* New York: William Morrow, 1964.
Moore, John H., *The Faustball Tunnel, German POWs in America and Their Great Escape.* New York: Random House, 1978.
Ruland, Bernd, *Wernher von Braun; Mein Leben für die Raumfahrt.* Offenburg: Burda Verlag, 1969.
Scotland, Andrew P., *The London Cage.* London: Evans Bros., 1957.
Von Kármán, Theodore, with Lee Edson, *The Wind and Beyond, Theodore von Kármán, Pioneer in Aviation and Pathfinder in Space.* Boston: Little, Brown and Co., 1967.

ARTICLES AND REPORTS

Akens, Davis S., and Paul H. Satterfield, *Historical Monograph, Army Ordnance Satellite Programs.* Redstone Arsenal, Alabama: Army Ballistic Missile Agency, November 1, 1958.

Campbell, Colin, "Rocket Arsenal," *Royal Air Force Flying Review,* 1958, 31–33.

The Fedden Mission to Germany, Final Report. London: Ministry of Aircraft Production, June 1945.

History of AAF Participation in Project Paperclip, May 1945–March, 1947, 4 vols. Wright Field, Ohio: Air Matériel Command, 1948.

History of German Guided Missiles; Hitler's Secret Weapons, Facts and Dreams. GD 630.0.7 Aberdeen Proving Ground, Maryland: Ordnance Research & Development Center, Foreign Document Evaluation Branch, April 1946.

Lentzner, "Orders to Civilians Concerned." Boston, Massachusetts: Army Service Forces, First Service Command, Fort Strong, October 2, 1945.

"Paperclip, Part I," *Office of Naval Intelligence Review,* February 1949, 22–27; Part II is in the December 1949 issue, 15–21.

Staver, Robert B., "A Sequence of Events Related to the Initiation and Implementation of Operation Paperclip." Unpublished and undated material supplied to Mitchell R. Sharpe.

Tokady, G. A., "Foundation of Soviet Cosmonautics," *Spaceflight,* Vol. 10 No. 11, October, 1968, 340-346.

Zwicky, Fritz, *Report on Certain Phases of War Research in Germany,* Vol. 1. Pasadena, California: Aerojet Engineering Corp., October 1, 1945.

INTERVIEWS AND LETTERS

Interview on tape of Konrad and Ingeborg Dannenberg by Frederick I. Ordway III and Lily von Saurma, Huntsville, Alabama, May 1, 1971.

Interview on tape of Kurt Debus by Frederick I. Ordway III, Kennedy Space Center, Florida, April 27, 1972.

Interview on tape of Dieter Grau by Frederick I. Ordway III and David L. Christensen, Huntsville, Alabama, January 17, 1971.

Interview on tape of James P. Hamill by David L. Christensen, Washington, D.C., October 2–24, 1971.

Interview on tape of M. S. Hochmuth by David L. Christensen, Cambridge, Massachusetts, October 27, 1971.

Interview on tape of Walter Jacobi by David L. Christensen and Ruth Heimburg, Huntsville, Alabama, August 12, 1971.

Interview on tape of Hannes Lührsen by Frederick I. Ordway III, Heidelberg, September 9, 1971.

Interview on tape of Heinz Millinger by Frederick I. Ordway III, Frankfurt-Langen, September 8, 1971.

Interview on tape of Erich Neubert by Frederick I. Ordway III, Huntsville, Alabama, June 24, 1971.

Interview on tape of Theodor Poppel by David L. Christensen, Kennedy Space Center, Florida, August 9, 1971.

Interview on tape of Richard W. Porter by Frederick I. Ordway III, New York, May 4, 19–20, 1971.
Interview on tape of Hannelore Bannasch Ranft by Frederick I. Ordway III, Schongau, September 15, 1971.
Interview of Gervais Trichel by David L. Christensen, Birmingham, Michigan, April 28, 1972.
Interview of Magnus von Braun by Mitchell R. Sharpe, London, November 2, 1972.
Interview on tape of Werner Sieber by David L. Christensen and Ruth Heimburg, Huntsville, Alabama, July 22, 1971.
Letters to Mitchell R. Sharpe from Dieter K. Huzel, Woodland Hills, California, May 8, 1974; May 17, 1974; June 6, 1974; and August 20, 1974.
Letters to Mitchell R. Sharpe from Robert B. Staver, Los Altos, California, November 3, 1971; June 28, 1971; February 5, 1972; August 20, 1973; September 10, 1973; November 30, 1973; February 18, 1974; and March 25, 1974.
Letter to Robert B. Staver from William H. Peifer, Washington, April 2, 1963.
Letter to Robert B. Staver from Gervais W. Trichel, Detroit, Michigan, December 6, 1962.
Letter to Colonel H. N. Toftoy, "Enemy Equipment Intelligence Service Team," Washington: War Department, Office of the Chief of Ordnance, June 19, 1944, signed by Major General Gladeon M. Barnes.
Letter to Colonel J. L. Walker from Colonel R. D. Wentworth, Headquarters, US Forces European Theater, Assistant Chief of Staff G-2, January 24, 1946.
Letter, "Transfer of German Guided Missile Projects to the United States." Washington: War Department, Office of the Chief of Ordnance, June 30, 1945, signed by Major General Gladeon M. Barnes.
Teletype message to Chief of Ordnance, War Department, Washington, from Colonel Joel G. Holmes, Headquarters, Communications Zone, European Theater Operations, Chief Ordnance Officer. (This was written by Major Robert B. Staver and, in effect, initiated the Army's Operation Paperclip.)
Memorandum, "Ordnance Plan for Conduct of German Guided Missile Projects in the United States." Washington: War Department, General Staff, Assistant Chief of Staff for Intelligence, July 7, 1945, signed by Major General Clayton Bissell.
Cable from Eisenhower to Ordnance Office, Nordhausen, Germany, for R. B. Staver, City Military Government, Room 114, June 1, 1945.

Chapter 15

BOOKS

Bar-Zohar, Michel, *La Chasse aux Savants Allemands.* Paris: Fayard, 1965.
Huzel, Dieter K., *Peenemünde to Canaveral.* Englewood Cliffs, N.J.: Prentice-Hall, 1962.
Scotland, Andrew P., *The London Cage.* London: Evans Bros., 1957.

ARTICLES AND REPORTS

Erfahrungsbericht des Versuchskommandos Altenwalde, September 6, 1945, signed by Hans Lindenberg.

"The German A4 Rocket," Film, RSA No. 40. Huntsville, Alabama: Redstone Scientific Information Center, US Army Missile Command, Redstone Arsenal. (Produced by the British Ministry of Defence in December 1945.)

Interrogation of Major General Walter Dornberger by Flight Lieutenant H. W. Stokes and Captain W. N. Ismay of CIOS Team 183 at Garmisch-Partenkirchen, Germany, June 29, 1945.

Military Attaché Report, "Evaluation of British Report on Operation Backfire," No. R6133-45. Washington, D.C.: War Department, General Staff Intelligence Division, December 13, 1945, signed by Howard S. Seifert.

Military Attaché Report, "Operations Backfire and Clitterhouse (British Firing of V-2 Rockets)," No. R5499-45. Washington, D.C.: War Department, General Staff, Intelligence Division, October 19, 1945, signed by C. D. Martenson.

Report on Operation "Backfire." 5 vols. London: The War Office, January 1946.

Sharpe, Mitchell R., "Backfire and Clitterhouse: Britain Launches the V-2," *Aerospace Historian*, Vol. 24 No. 1, Spring/March, 1978, 36–44.

INTERVIEWS AND LETTERS

Interview on tape of Kurt Debus by Frederick I. Ordway III, Kennedy Space Center, Florida, April 27, 1972.

Interview on tape of Friedrich Dohm by Frederick I. Ordway III, Huntsville, Alabama, June 16, 1971.

Interview on tape of Walter B. Dornberger by Frederick I. Ordway III and David L. Christensen, Huntsville, Alabama, November 7, 1971.

Interview on tape of Karl Heimburg by David L. Christensen and Frederick I. Ordway III, Huntsville, Alabama, June 26, 1971.

Interview on tape of Gerhard Heller by David L. Christensen and Frederick I. Ordway III, Huntsville, Alabama, June 1, 1972.

Interview of Otto Hirschler by Frederick I. Ordway III and Lily von Saurma, Huntsville, Alabama, April 28, 1971.

Interview on tape of M. S. Hochmuth by David L. Christensen, Cambridge, Massachusetts, October 27, 1971.

Interview on tape of Helmut Hölzer by Frederick I. Ordway III, Huntsville, Alabama, June 18, 1971.

Interview on tape of Erich Kaschig by David L. Christensen and Ruth Heimburg, Huntsville, Alabama, July 29, 1971.

Interview on tape of Hans Palaoro by David L. Christensen and Ruth Heimburg, Huntsville, Alabama, August 4, 1971.

Interview on tape of Werner Rosinski by David L. Christensen and Lily von Saurma, Huntsville, Alabama, April 13, 1971.

Interview on tape of Arthur Rudolph by David L. Christensen and Ruth Heimburg, Huntsville, Alabama, August 8, 1971.

Interview on tape of Arthur Rudolph by Mitchell R. Sharpe, Huntsville, Alabama, June 21, 1978.

Interview of Walter Schuran by Frederick I. Ordway III and Lily von Saurma, Ottobrun, September 4, 1971.
Interview on tape of Albert Zeiler by Frederick I. Ordway III, Kennedy Space Center, Florida, August 27, 1972.
Letters to Mitchell R. Sharpe from W. S. J. Carter, Bristol, England, August 7, 1973; September 5, 1973; October 4, 1973; November 1, 1973; November 22, 1973; December 7, 1973; December 31, 1973; February 6, 1974; February 13, 1974; March 6, 1974; June 13, 1974; and July 31, 1974.
Letter to Mitchell R. Sharpe from Walter B. Dornberger, Chapala-Jalisco, Mexico, April 25, 1974.
Letter to Mitchell R. Sharpe from Dieter K. Huzel, Woodland Hills, California, October 19, 1974.
Letters to Mitchell R. Sharpe from Grayson Merrill, Syosset, New York, September 11, 1973, and October 7, 1973.
Letter to Mitchell R. Sharpe from William H. Pickering, Pasadena, California, May 22, 1974.
Letter to Mitchell R. Sharpe from Arthur Rudolph, San Jose, California, April 6, 1978.
Letter to Mitchell R. Sharpe from Howard S. Seifert, Stanford, California, June 10, 1974.
Letter, "Generalleutenant Dornberger," Headquarters US Air Forces in Europe, Office of the Assistant Chief of Staff, A-2, October 22, 1945, signed by George C. McDonald.
Letter, "Generalleutenant Dornberger," No. 17.9.1. War Office Control Commission for Germany (British Element), September 15, 1945, signed by J. S. Letheridge.
Memorandum, "Report of Travel in European Theater," Wright Field, Ohio: Army Air Forces, Air Technical Command, Engineering Division, November 9, 1945, signed by Charles W. Guy.

Chapter 17

BOOKS

Bar-Zohar, Michel, *La Chasse aux Savants Allemands*, Paris: Fayard, 1965.
Bornemann, Manfred, *Geheimprojekt Mittelbau, die Geschichte der Deutschen V-Waffen-Werke.* Munich: J. F. Lehmanns, 1971.
Gröttrup, Helmut, *Über Raketen.* Berlin: Ullstein, 1959.
Gröttrup, Irmgard, *Rocket Wife.* London: André Deutsch, 1959.
Kilmarx, Robert, *A History of Soviet Air Power.* New York: Frederick A. Praeger, 1962.
Lee, Asher, ed., *The Soviet Air and Rocket Forces.* New York: Frederick A. Praeger, 1959.
Parry, Albert, *Russia's Rockets and Missiles.* Garden City, N.Y.: Doubleday, 1960.
Tokady, G. A., *Comrade X.* London: Harvill Press, 1956.
Vladimirov, L., *The Russian Space Bluff.* London: Tom Stacey, 1971.

Zaehringer, Alfred, *Soviet Space Technology*. New York: Harper & Bros., 1961.

ARTICLES AND REPORTS

"A Summary of Soviet Guided Missiles Intelligence," US/UK-GM-4-52. Washington, D.C.: Central Intelligence Agency, July 20, 1953.
Bärwolf, Adelbert, "Captured Germans Gave Red Rockets a Boost," *Army, Navy, Air Force Register* (US), Vol. 82, February 25, 1961, 16–17.
"German Brains in Russia," *Aeronautics*, Vol. 34, No. 6, April 1952, 46–47.
Gröttrup, Helmut, "Aus den Arbeiten des deutschen Raketen-Kollektivs in der Sowjet-Union," *Raketentechnik und Raumfahrtforschung*, Heft 2, April 1958, 58–62.
Protsyuk, S., "Technical Chronicle: Who Runs the Program of Mastering Space in the USSR?" (NASA Technical Translation TT F-14,882, Washington, D.C., May 1973.)
The R-14 Project: A Design of a Long Range Missile at Gorodomlya Island. Central Intelligence Agency Information Report, CS-G-14851, August 26, 1953, Washington, D.C.
Ritchie, Donald, "Soviet Rockets Exploit German Technology," *Missiles & Rockets*, Vol. 5, No. 7, December 7, 1959, 17–19.
Schroeder, G. W., "How Russian Engineering Looked to a Captured German Scientist," *Aviation Week,* Vol. 62, No. 19, May 9, 1955, 27, 30, 32, 34.
"Simplicity Called Key to Red Success," *Missiles & Rockets*, Vol. 6, No. 4 (January 25, 1960), 59.
Sokolov, V. L., "Soviet Use of German Science and Technology, 1945–1946," Mimeographed Series No. 72. New York: Research Program of the USSR, 1955.
Sutton, George, "Evaluation of Russian Rocket Developments," *Journal of the British Interplanetary Society*, Vol. 13, No. 5 (September 1954), 262–268.
Tokady, G. A., "Soviet Rocket Technology," in Eugene M. Emme, ed., *The History of Rocket Technology*. Detroit: Wayne State University Press, 1964.
———, "Foundations of Soviet Cosmonautics," *Spaceflight*, Vol. 10, No. 11 (October 1968), 340–346.
Vladimirov, L., "From Sputnik to Apollo," *Posev* (Munich), September 1969, 47–51.

INTERVIEWS AND LETTERS

Interview on tape of Walter B. Dornberger by David L. Christensen and Frederick I. Ordway III, Huntsville, Alabama, November 7, 1971.
Letters to Mitchell R. Sharpe from Helmut Gröttrup, Munich, July 24, 1973; September 4, 1973; January 2, 1974.
Letter, LEM 1148-22, "Lecture by Dr. Tokady, Former USSR Rocket Specialist," April 26, 1961, North American Aviation Company, signed D. K. Huzel.
Letter, "Additional Information on Blaupunkt Werke, GmbH, Berlin-Wilmersdorf, Forckenbeckstr. 9/13," Headquarters, US Air Forces in Europe (Main), September 1, 1945, signed by Major Charles A. Crowley.
Letter to Mitchell R. Sharpe from Dieter K. Huzel, Woodland Hills, California, June 6, 1974.

Office memorandum, "Information on German Developments," Fort Bliss, Texas: Rocket Research and Development Service, Sub-office (rocket), June 24, 1944, signed by Lieutenant Colonel James G. Bain.

Chapter 18

BOOKS

Bergaust, Erik, *Reaching for the Stars.* New York: Doubleday, 1960.
Dannau, Wim, *Les Dossiers "Espace" de Wim Dannau.* Tournai: Castermann, 1966.
Huzel, Dieter K., *Peenemünde to Canaveral.* Englewood Cliffs, N.J.: Prentice-Hall, 1962.
Lang, Daniel, *From Hiroshima to the Moon.* New York: Simon and Schuster, 1959.
Lasby, Clarence G., *Project Paperclip: German Scientists and the Cold War.* New York: Atheneum, 1971.
McGovern, James, *Crossbow and Overcast.* New York: William Morrow, 1964.
Von Braun, Magnus, *Weg Durch Vier Zeitepochen.* Limburg: C. A. Starke Verlag, 1965.

ARTICLES AND REPORTS

Akens, David S., and Paul H. Satterfield, *Historical Monograph, Army Ordnance Satellite Program.* Redstone Arsenal, Alabama: US Army Ballistic Missile Agency, November 1, 1958.
"Bumper Shoot," *Douglas Progress* (McDonnell Douglas Aircraft Corp.), Third Quarter, 1965, 4–9.
Debus, Kurt, "From A4 to Explorer 1." Paper delivered to the 7th International History of Astronautics Symposium, 24th International Astronautical Federation Congress, Baku, USSR, October 8, 1973.
———, "High Altitude Research with V-2 Rockets," *Proceedings of the American Philosophical Society*, Vol. 91, No. 5 (December 1947), 430–446.
"German Scientists Plan Refueling Station in Sky on Route to Moon," El Paso, *Herald-Post*, December 4, 1946, 1.
Hail, Marshal, "White Sands Is 20 Years Old and Growing Stronger," El Paso *Herald-Post*, July 6, 1965, B-1.
Hargrave, Gene, "El Pasoan, Missile Sites 'Daddy,' Member of Space Range Pioneers," El Paso *Times Sunday Magazine*, November 26, 1972, 2.
"History of the AAF Participation in Project Paperclip, May 1945–March 1947." 4 vols. Wright Field, Ohio: Air Materiël Command, 1948.
"Klein Reveals Model in U.S. Had Moon Model in 1949," Chicago *Tribune*, October 30, 1957, 8.
Malina, Frank J., "America's First Long-range-missile and Space Exploration Program: The ORDCIT Project of the Jet Propulsion Laboratory, 1943–1946: a Memoir," in R. Cargill Hall, ed., *Essays on the History of Rocketry and Astronautics: Proceedings of the Third Through the Sixth History Symposia of the International Academy of Astronautics*, Vol. 2, NASA Conference Publication 2014.

Washington: National Aeronautics and Space Administration, 1977, 339–383.
Rivkins, Harvey, "A Report on the Ordnance Firings at White Sands," *Army Ordnance*, Vol. 27, No. 2 (March–April 1947), 429–431.
Strom, Virginia, "American Cooking 'Tasteless,' Says German Rocket Scientist; Dislikes 'Rubberized Chicken,'" El Paso *Herald-Post*, December 6, 1945, 1.
———, "German Scientists Tell How V-2 Rocket Was Developed from Experimental Mile-High Missile," El Paso *Herald-Post*, December 4, 1945, 1.
———, "Secrecy Is Lifted from Ft. Bliss V-2 Technicians," El Paso *Herald-Post*, December 4, 1946, 1.
"Daily Journal, Brigadier General H. N. Toftoy, 30 Dec. 1948–31 Mar. 1950" (unpublished).
Turner, Virginia, "Man's First Steps on Way to Moon Were Here," El Paso *Herald-Post*, July 15, 1969, 4.
Von Braun, Wernher, "*Rocket Launching from Ships. Special Report, Ordnance Research and Development Division, Sub-Office (Rocket), Fort Bliss, Texas, 1947.*"
———, "Technical Report No. 9, Investigation of the Upper Atmosphere with the A4 (V-2) Rocket Missile," Ordnance Research and Development Division, Sub-Office (Rocket), Fort Bliss, Texas, April 1946.
———, "The Firing of V-2 Missiles from an Aircraft Carrier," Research and Development Service, Sub-Office (Rocket), Fort Bliss, Texas, April 17, 1947.
White, L. D., "Final Report, Project Hermes V-2 Missile Program," No. R 52A0510. Schenectady, New York: General Electric Company, September 1952.
"White Sands Grows to $60 Million Plant in 12 years," El Paso *Herald-Post*, August 31, 1957, C-8.
Wiggens, Joseph W., "Hermes, Milestone in U.S. Aerospace Progress," *Aerospace Historian*, Vol. 21, No. 1 (Spring, March 1974), 34–38.
Zwicky, Fritz, "A Stone's Throw into the Universe: a Memoir," in R. Cargill Hall, ed., *Essays on the History of Rocketry and Astronautics: Proceedings of the Third Through the Sixth History Symposia of the International Academy of Astronautics*, Vol. 2, NASA Conference Report 2014. Washington: National Aeronautics and Space Administration, 1977, 325–337.

INTERVIEWS AND LETTERS

Interview on tape of Wilhelm Angele by David L. Christensen and Ruth Heimburg, Huntsville, Alabama, July 21, 1971.
Interview on tape of James Bramlet by David L. Christensen, Huntsville, Alabama, August 23, 1971.
Interview on tape of Kurt Debus by Frederick I. Ordway III, Kennedy Space Center, Florida, April 27, 1972.
Interview on tape of Dan Driscoll by David L. Christensen, Huntsville, Alabama, July 2, 1972.
Interview on tape of James J. Fagan by David L. Christensen and Ruth Heimburg, Huntsville, Alabama, July 30, 1971.
Interview on tape of Hans Grüne by David L. Christensen, Kennedy Space Center, Florida, August 9, 1971.

Interview of M. S. Hochmuth by David L. Christensen, Cambridge, Massachusetts, October 27, 1971.

Interview on tape of James C. Miller, Jr., by Mitchell R. Sharpe, Redstone Arsenal, Alabama, May 13, 1971.

Interview of Thomas Moore by Frederick I. Ordway III, Huntsville, Alabama, September 29, 1972.

Interview on tape of Hans Palaoro by David L. Christensen and Ruth Heimburg, Huntsville, Alabama, August 4, 1971.

Interview on tape of Theodor Poppel by David L. Christensen, Kennedy Space Center, Florida, August 9, 1971.

Interview on tape of Ramon Samaniego by David L. Christensen and Ruth Heimburg, Huntsville, Alabama, August 23, 1971.

Interview on tape of Klaus H. Scheufelen by Frederick I. Ordway III and Lily von Saurma, Überlingen, FRG, September 9, 1971.

Interview on tape of Ernst Steinhoff by Eugene Emme, Washington, D.C., September 14, 1973.

Interview of Holger N. Toftoy and James Hamill by Mr. Vince Hackett, Washington, D.C., February 10, 1959.

Interview of Magnus von Braun by Mitchell R. Sharpe, London, November 2, 1972.

Interview on tape of Albert Zeiler by Frederick I. Ordway III, Kennedy Space Center, Florida, August 27, 1972.

Letter to Mitchell R. Sharpe from Julius Klein, Chicago, Illinois, June 17, 1970.

Letter to Mitchell R. Sharpe from Horst W. Schneider, West Covina, California, July 11, 1976.

Letter to Mitchell R. Sharpe from Harold R. Turner, El Paso, Texas, September 27, 1973.

Letters to Mitchell R. Sharpe from Virginia S. Turner, El Paso, Texas, October 21, 1973; June 16, 1974.

Memorandum, "Pay and Allowances of German Scientists," February 21, 1946, Washington, D.C., War Department, Office of the Chief of Ordnance, Rocket Development Division, signed by Holger N. Toftoy.

Memorandum, "Restriction of Limits," February 2, 1948, Fort Bliss, Texas, Research and Development Division, Sub-office (rocket), signed by James P. Hamill.

Memorandum, "Employment of Dependents of Paperclip Specialists," June 23, 1948, Washington, D.C., Department of the Army, General Staff, signed by Colonel Lauren L. Williams.

Chapter 19

BOOKS

Anderson, Frank W., *Orders of Magnitude: A History of NACA and NASA, 1951–1976*, SP No. 4403. Washington, D.C.: National Aeronautics and Space Administration, 1976.

Belew, Leland F., ed., *Skylab, Our First Space Station*, SP-400. Washington, D.C.: National Aeronautics and Space Administration, 1977.

Bergaust, Erik, *Wernher von Braun*. Washington, D.C.: National Space Institute, 1976.

———, *Reaching for the Stars*. New York: Doubleday, 1960.

———, *Rocket City, U.S.A., from Huntsville, Alabama to the Moon.* New York: Macmillan, 1963.

Chapman, John L., *Atlas: The Story of a Missile.* New York: Harper & Bros., 1960.

Cox, Donald W., *The Space Race.* Philadelphia: Chilton Books, 1962.

Ezell, Edward C., and Linda N. Ezell, *The Partnership: A History of the Apollo-Soyuz Test Project*. Washington, D.C.: National Aeronautics and Space Administration, 1978.

Gavin, James, *War and Peace in the Space Age.* New York: Harper & Bros., 1958.

Green, Constance, and Milton Lomask, *Vanguard: A History.* Washington, D.C.: National Aeronautics & Space Administration, 1970.

Harris, Gordon, *A New Command: The Life of Bruce Medaris, Major General, U.S.A., Retired.* Plainfield, N.J.: Logos International, 1976.

Holder, William H., *Saturn 5, the Moon Rocket.* New York: Julian Messner, 1969.

———, *Skylab: Pioneer Space Station.* New York: Rand McNally, 1974.

Killian, James R., *Sputniks, Scientists, and Eisenhower: A Memoir of the First Special Assistant to the President for Science and Technology.* Cambridge, Mass.: MIT Press, 1977.

Kistiakowsky, George B., *A Scientist in the White House: The Private Diary of President Eisenhower's Special Assistant for Science and Technology.* Cambridge, Mass.: Harvard University Press, 1976.

Lange, Oswald, and Richard J. Stein, *Space Carrier Vehicle; Design, Development and Testing of Launching Rockets.* New York: Academic Press, 1963.

Lewis, Richard S., *Appointment on the Moon*, New York: Viking Press, 1968.

Logsdon, John M., *The Decision to Go to the Moon: Project Apollo and National Interest.* Cambridge, Mass.: Harvard University Press, 1970.

Medaris, John B., *Countdown for Decision.* New York: G. P. Putnam's Sons, 1960.

Odishaw, Hugh, ed., *The Challenge of Space.* Chicago: University of Chicago Press, 1962.

Raymond, Jack, *Power at the Pentagon.* New York: Harper & Row, 1964.

Rosholt, Robert T. *An Administrative History of NASA, 1958–1963.* Washington, D.C.: National Aeronautics & Space Administration, 1966.

Stuhlinger, Ernst, et al., *From Peenemünde to Outer Space: Commemorating the Fiftieth Birthday of Wernher von Braun, 23 March 1962.* Huntsville, Ala.: Marshall Space Flight Center, 1962.

Van Dyke, Vernon, *Pride and Power: The Rationale of the Space Program.* Urbana: University of Illinois Press, 1964.

Von Braun, Wernher, and Frederick I. Ordway III, *History of Rocketry & Space Travel.* New York: Thomas Y. Crowell, 1975.

Witkin, Richard E., *The Challenge of the Sputniks.* New York: Doubleday, 1958.

Young, Hugo, et al., *Journey to Tranquillity.* New York: Doubleday, 1970.

ARTICLES AND REPORTS

Akens, David S., *Historical Origins of the George C. Marshall Space Flight Center.* Huntsville, Alabama: Marshall Space Flight Center, December 1960.
———, *Saturn Illustrated Chronology*, MHR-5. Huntsville, Alabama: Marshall Space Flight Center, August 1, 1968.
———, and Paul H. Satterfield, *Historical Monograph: Army Ordnance Satellite Programs.* Redstone Arsenal, Alabama: Army Ballistic Missile Agency, November 1, 1958.
Bullard, John W., *History of the Redstone Missile System, Historical Monograph*, Proj. No. AMC 23M. Redstone Arsenal, Alabama: US Army Missile Command, October 15, 1965.
Bylinsky, Gene, "Dr. von Braun's All-purpose Space Machine," *Fortune*, Vol. 75, No. 5 (May 1967), 142–149, 214, 218–219.
Canright, Richard B., et al., *Report of the Joint ARPA-NASA Committee on Large Clustered Booster Capabilities.* Washington: National Aeronautics & Space Administration, January 8, 1959.
Daily Journals of Wernher von Braun, May 1958–March 1970.
"Guess You Could Call it a Success," *The Rocket* (US Army Missile Command) Vol. 22 No. 13, August 22, 1973, 9.
Debus, Kurt H. "From A4 to Explorer 1." Paper delivered to the 7th International History of Astronautics Symposium, 24th International Astronautical Federation Congress, Baku, USSR, October 8, 1973.
Eisenhower, Dwight D., "Are We Headed in the Wrong Direction?" *The Saturday Evening Post*, Vol. 235 (August 11–18, 1962), 24–28.
Furnas, Clifford C., "Birth Pangs of the First Satellite," *Research Times* (Cornell Aeronautical Laboratories), Vol. 18 (Spring 1970), 15–18.
Gibney, Frank, "The Missile Mess," *Harper's*, Vol. 220 (January 1960), 42–48.
Greever, Bill B., "General Description and Design of the Configuration of the Juno I and Juno II Launching Vehicles," *IRE Transactions on Military Electronics*, Vol. MIL-4, April, July, 1960, p. 60–64.
Hines, William, "Commando Rockets Bearing GIs Visioned," Washington *Star*, January 20, 1959, 21.
Houbolt, John C., "Lunar Rendezvous," *International Science and Technology*, Vol. 14, No. 2, February 1963, 62–65.
Joiner, Helen B., *The Redstone Arsenal Complex in the Pre-missile Era.* Redstone Arsenal, Alabama: US Army Missile Command, June 22, 1966.
Klawans, B., *The Vanguard Satellite Launching Vehicle: An Engineering Summary*, Engineering Rept. No. 11022. Baltimore, Maryland: The Martin Company, April 1960.
Medaris, John B., "The Explorer Satellites," *Army Information Digest*, Vol. 13, No. 10 (October 1958), 4–16.
NASA-DOD Relationship, Report of the Subcommittee of the Committee on Science and Astronautics, U.S. House of Representatives, Eighty-eighth Congress, Second Session, Serial 1. Washington, D.C.: US Government Printing Office, 1964.
O'Neill, Paul, "The Anachronistic Town of Huntsville," in *Editors of Fortune, The Space Industry, America's Newest Giant.* Englewood Cliffs, N.J.: Prentice-Hall, 1962.

"Saturn V Launch Vehicle Flight Evaluation Report AS-501, Apollo 4 Mission, MPR-SAT-FE-68-1." Huntsville, AL: Marshall Space Flight Center, January 15, 1968.

Sharpe, Mitchell R., "Lunar Roving Vehicle: Surface Transportation on the Moon," paper presented at the XVth International Congress for the History of Science, August 17, 1977, Edinburgh.

"Skylab Mission Report, Third Mission, JSC-08963." Houston TX: Johnson Space Center, July 1974.

Smith, Richard A., "Canaveral, Industry's Trial by Fire," *Fortune*, Vol. 65, No. 6 (June 1962), 200–208.

". . . They Didn't Think We'd Get Off the Ground," *The Rocket*, Vol. 22, No. 13, August 22, 1973, 8.

"Tight-knit Team Hand-made Each Missile," *The Rocket*, Vol. 22, No. 13, August 22, 1973, 7.

Toftoy, Holger N., "Army Missile Development," *Army Information Digest*, Vol. 11, No. 12 (December 1956), 28–37.

Von Braun, Wernher, "ABMA Presentation to the National Aeronautics and Space Administration, ABMA D-TN-1-59." Huntsville, Alabama: U.S. Army Ballistic Missile Agency, December 15, 1958.

———, *Concluding Remarks by Dr. Wernher von Braun About Mode Selection for the Lunar Landing Program Given to Dr. Joseph F. Shea, Deputy Director (Systems), Office of Manned Space Flight, June 7, 1962.*

———, "The Explorers," in F. Hecht, ed., *IXth International Astronautical Congress Proceedings, Amsterdam*, 1958, Vol. 2. Vienna: Springer Verlag, 1959, 914–931.

———, *Guided Missile (SSM) for Ranges of from 300 to 500 Nautical Miles.* Redstone Arsenal, Alabama: Ordnance Guided Missile Center, December 1950.

———, *A Minimum Satellite Vehicle Based on Components Available from Missiles Developments of the Army Ordnance Corps.* Redstone Arsenal, Alabama: Guided Missile Development Division, September 15, 1954.

———, "Rundown on Jupiter C," *Astronautics*, Vol. 3, No. 11 (October 1958), 34, 83.

———, "United States Space Carrier Vehicle Program," in C. W. P. Reutersward, ed., *XIth International Astronautical Congress, Stockholm, 1960*, Vol. 1. Vienna: Springer Verlag, 1961, 638–658.

Watson, Mark S., "Glennan Asks for Army Missile Unit," *The New York Times*, October 14, 1958, A-1.

———, "Council Headed by Ike Will Rule on Shift of Scientists & Engineers," *The New York Times*, October 15, 1958, A-1.

INTERVIEWS AND LETTERS

Interview on tape of James J. Fagan by David L. Christensen and Ruth Heimburg, Huntsville, Alabama, July 30, 1971.

Interview on tape of Hans Grüne by David L. Christensen, Kennedy Space Center, Florida, August 9, 1971.

Interview of Charles Hester by Frederick I. Ordway III and David L. Christensen, Huntsville, Alabama, June 21, 1971.

Interview of James C. Miller by Mitchell R. Sharpe, Huntsville, Alabama, May 13, 1971.

Interview on tape of Theodor Poppel by David L. Christensen, Kennedy Space Center, Florida, August 9, 1971.

Interview on tape of Arthur Rudolph by David S. Akens and David L. Christensen, Huntsville, Alabama, November 26, 1968.

Interview on tape of Hazel Toftoy by Mitchell R. Sharpe, Huntsville, Alabama, June 26, 1973.

Interview of Magnus von Braun by Mitchell R. Sharpe, London, November 2, 1972.

Interview on tape of Wernher von Braun by David L. Christensen, Roger E. Bilstein, and John Beltz, Huntsville, Alabama, November 17, 1971.

Letter to Eugene M. Emme from Herbert F. York, La Jolla, California, May 2, 1973.

Letter to Abraham Hyatt from Wernher von Braun, Huntsville, Alabama, November 14, 1960.

Letter to Lyndon B. Johnson from T. Keith Glennan, Washington, D.C., October 21, 1958.

Letter to John W. McCormack from T. Keith Glennan, Washington, D.C., October 21, 1958.

Letter to Donald A. Quarles from Hugh L. Dryden, Washington, D.C., November 12, 1958.

Letter to Mitchell R. Sharpe from Truman F. Cook, Dallas, Texas, August 24, 1973.

Letter to Mitchell R. Sharpe from Wernher von Braun, Germantown, Maryland, November 18, 1974.

Letter to Wernher von Braun from Robert Freitag, Washington, D.C., December 1, 1956.

Letter to Wernher von Braun from Abraham Hyatt, Washington, D.C., November 16, 1960.

Letter to Wernher von Braun from Mrs. John F. Kennedy, Washington, February 11, 1964.

Letter to Wernher von Braun from John G. Zierdt, Huntsville, Alabama, February 28, 1972.

Memorandum for Record, "Discussion with Secretary Brucker, General Hinrichs, and Colonel Gutherie," Washington, D.C., National Aeronautics & Space Administration, November 4, 1959, signed by T. Keith Glennan.

Memorandum, "Manned Lunar Landing and Return Attempt in 1967," Washington, D.C., National Aeronautics & Space Administration, May 8, 1961, signed by Abraham Hyatt.

Index

A1 rocket, 23
A2 rocket, 23
A3 rocket, 24, 27, 33–5
A4 rocket *see* V-2 rocket
A5 rocket, 27–8, 34, 39, 56
A6 rocket, 56
A7 rocket, 56
A8 rocket, 56
A9 rocket, 56–7, 322
A10 rocket, 56–7, 322, 351
A11 rocket, 57
A12 rocket, 57
Aberdeen Proving Ground, 352, 387
Abwehr, 409–10
Ackerman, E. G., 154
Adam project, 384
Advanced Research Project Agency (ARPA), 384, 387–8
Agena B rocket, 389
Agnew, Spiro T., 402
Akademgorodok, 31
Albert project, 356
Albring, Werner, 335, 339
Aldrin, Edwin, 397
Ambler, Air Commodore G. H., 212
American Interplanetary (*later* Rocket) Society, 91
Anders, William, 396
Angele, Wilhelm, 368, 399
Anklam, 50
Antwerp, 48, 183–4, 198, 200, 217*n*, 228–30, 234, 237, 239, 247–8, 251
Apollo spacecraft, 393, 395–6, 402
Applications Technology Satellite (ATS), 403
Arciszewski, Tomasz, 142, 145
Ardenne, Prof. Manfred von, 318
Armia Krajowa (AK: Polish Home Army), 132–45, 152–3, 164–5
Armstrong, Neil, 388, 397
Army Ballistic Missile Agency (ABMA), 379–83, 386–9
Army Ordnance Missile Command (AOMC), 387
Arnhem, 196–7
Arnold, Dr. (of Mittelwerk), 72
Arnold, Gen. H. H., 169
Astronautics, 92
Atlas rocket, 330, 350, 373, 377, 389
Attaya, Capt. Henry E., 371
Attlee, Clement, 275
Axster, Herbert, 7, 9, 11, 82, 188, 199, 264, 291
Axthelm, Gen. Walter von, 51

Baade, Brunolf, 340
Babington Smith, Constance, 101–3, 108, 160
Backfire operation, 295, 297–300, 303–9, 332
Bain, Col. James G., 314
Balchen, Col. Bernt, 150
Ball, Eberhard, 368
balloons (anti-aircraft), 217–19
Banger, Joan, 197

Bannasch, Hannelore, 70, 72, 115, 266
Barclay, Brig.-Gen. John, 387
Barnes, Maj.-Gen. Gladeon M., 287, 313, 344, 346, 349
Basse, Lt.-Col., 189
Baumpach, Gen. Werner, 180
Bechtle, Otto, 361
Beck, A., 36
Beck, Col.-Gen. Ludwig, 409
Becker, Col. Karl, 17–19, 25–6, 32, 58
Beek, Gerd de, 348, 373
Belhamelin (Normandy), 127
Berlin-Falckensee, 65
Bernard, Joan C. C., 294, 305
Bersch, Otto Karl, 67, 72
Beyer, Anton, 84
Big Ben sub-committee, 161–5, 170, 172–3, 224–6
Birkenhead, F. W. F. Smith, 2nd Earl of, 99
Birney, Hoffman H., 348
Bissell, Maj.-Gen. Clayton, 287
Blass, Josef, 335
Blizna *see* Heidelager
Blue Streak rocket, 330
Blumentritt, Gen. Günther, 201, 248
Böhm, Josef, 399
Bois Carré (Yvrench), 108
Bolster, Rear Adm. Calvin, 308
Bor-Komorowski, Tadeusz, 133, 140
Borkum island (Baltic), 23
Borman, Frank, 396
Bornemann, Manfred, 66
Börner (of Mittelwerk), 67
Boryna *see* Rządzki, J.
Bottomley, Air Marshal Norman H., 110
Boyes, Capt. (*later* Rear Adm.) Hector, 95
Bradley, Gen. Omar N., 295
Bradspies, Bob, 368
Brandner, Ferdinand, 327
Brauchitsch, Field Marshal Walther von, 36, 185*n*, 303
Braun, Eva, 305
Braun, Gerhard, 311–12
Braun, Iris von, 360
Braun, Magnus von, surrender to US troops, 2–3, 7–8, 10–11, 266, 275; at Peenemünde, 46; arrested by Gestapo, 46; and Mittelwerk production, 77; and bombing of Peenemünde, 114; in USA, 346, 348–9, 357, 369, 374
Braun, Magnus, Freiherr von, 4, 254
Braun, Margrit von, 369–70
Braun, Sigismund von, 6
Braun, Wernher von, surrender to US troops, 2, 4–11, 254–8, 261–2, 264–5, 268, 270–71, 275; early rocket interests, 12, 18, 92–4; early career, 19–20, 22–3; early experimental work, 22–6; tests rocket-propelled planes, 25; meets Hitler, 27, 44–5, 185; technical director at Peenemünde, 32, 35, 37, 40; and university aid, 35–6; German honours, 42, 45, 48; and V-2 programme, 45; Himmler courts, 45–6; arrested by Gestapo, 46–7; released, 48; on AA rocket, 51; under Storch reorganization, 53–4; and advanced rocket design, 55–7; and V-2 production, 61, 63, 250; and Schlier explosion, 86; and Ley, 92–4; and bombing of Peenemünde, 114–16; and V-2 in Sweden, 150; and 1945 evacuation from Peenemünde, 254–8, 261–2, 264–5; and Kammler, 265–6; interrogated by Americans, 271–3, 344; blacklisted, 278; and US Special Mission V-2, 281–2; Americans claim, 284–5; US contract, 289, 308, 310; Russians seek, 290–91; interrogated by British, 291; in USA, 310–13, 346–7, 349–52, 357; and US rocket design, 329, 337; marriage and family, 360; on space flight, 360–61; lectures, 362, 388; in Huntsville, 363–4, 366–9; music, 369; private life, 369–70; and US satellite, 374–7, 384–6, 389; on first Soviet satellite, 382; US honours, 387–8, 404; rumours of commercial interests, 389–90; and NASA, 389, 401–3; film on, 390; and Kennedy, 394; and Saturn testing, 396; and US protégés, 398; and post-Apollo plans, 402; retirement, 403–4; illness and death, 404
Bredt, Irene, 238, 327, 339
Brock, Col. B. H., 224–5
Bromley, Maj. William, 279, 285
Brown, Harold, 394
Brown, Lt.-Col. J. S., 302
Brucker, Wilbur, 382, 386
Brussels, 217*n*
Bucher (of AEG), 66
Buchhold, Prof. Theodor, 35, 372–3
Bühle, Gen. Walter, 44, 47, 191
Bull, Gen. Harold Roe, 289
Bullard, Prof. E. C., 231

Bumper missiles, 354, 356
Burden, Squadron Leader, 150
Burder, Flt.-Lt. C. H., 154
Burose, Walter, 367, 399
Bütting (of Mittelwerk), 67
Byrnes, James F., 276

C2 rocket *see* Wasserfall
Cameron, Maj.-Gen. Alexander M., 294–6, 302, 305–8
Canaris, Adm. Wilhelm, 95–6, 409
Carlson, Lt.-Col., 274
Carter, Hodding, 367
Carter, Jimmy, President of USA, 404
Carter, Col. W. S. J., 294–6, 301
Centaur rocket, 389
Central Interpretation Unit, Medmenham, 100–105, 107–9, 126–7, 129, 147, 160–61, 227
Cerny, Otto, 183
Chamberlain, Lt.-Gen. Stephen J., 358–9
Chciuk, Tadeusz, 145
Chemisch-Technische Reichsanstalt, 14
Chernikova, Katya, 332
Chertok, Major, 319
Cherwell, F. A. Lindemann, Viscount, 99–100, 105–7, 109, 126, 249
Chmielewski, Capt. Jerzy ("Rafel"), 134, 140–42, 145, 147–8, 152–5
Churchill, Winston S., 97, 99, 108, 162, 167–8, 248
Clark, Capt., 360
Clews, C. J. B., 224
Clitterhouse operation, 305, 309
Cobra project, 350
Coburn, Peter, 397, 399
Cockburn, Sir Robert, 95–6
Cole, Hugh M., 247
Collier, Basil, 244
Collin, Lt.-Col. G. J., 273
Collins, Michael, 397
Combined Intelligence Operations Subcommittee (CIOS), 273
Congreve, Sir William, 14
Coningham, Air Marshal Sir Arthur, 234
Cook, Lt.-Col. Truman F., 382, 386
Cook, Col. William J. R., 106, 283, 295
Cornelius (of Dörnten), 263
Crantz, Carl, 17
Cripps, Sir Stafford, 108–9
Crossbow operation (and Committee), 126–9, 140, 161–3, 167, 175, 232
Crow, Sir Alwyn D., 98–9, 106, 162, 291, 294
Culliford, Flt.-Lt. Stanley George, 142–6, 152–3

Dahm, Werner, 52, 261
Dannenberg, Konrad, 119, 398
Debus, Kurt, 114, 258–9, 299, 384–5
Degenkolb, Gerhard, 60–62, 64–7, 69, 73, 81, 86
Delderfield, Flt.-Lt. R. F., 219
Delmer, Sefton, 411–12
Dembiński, Marian, 141–2
Detko, George, 369
Diebner, Kurt, 58
Diesel-FT Rak. III, 16
Diver sub-committee, 163
Dixon, Thomas F., 151–2
Dobrick, Herbert, 350
Dohm, Friedrich, 117, 350
Dönitz, Grand-Admiral Karl, 7, 49
Dornberger, Frau Bunny, 40
Dornberger, Maj.-Gen. Walter, rescues von Braun, 5, 266; surrender to US troops, 5–9, 11; early rocket research, 17–19, 21–8; injured, 23; meets Hitler, 27, 44–5, 185; Peenemünde command, 32–4, 36, 40, 53, 188, 190; and V-2 guidance system, 40; and early V-2 launchings, 41–2; and V-2 programme, 42–3, 45, 185–6; and von Braun's arrest, 47–8; and Petersen, 49–50; and AA rocket project, 52; leaves Peenemünde, 53; AA responsibilities, 54; bans advanced rocket design, 55; on Degenkolb, 61; and V-2 production, 61, 63–4; and Mittelwerk, 66–7; Kammler and, 80–82; and bombing of Peenemünde, 114–17, 122; on Heidelager testing, 131; on Heidekraut targets, 153; and military use of V-2, 185–6; and mobile launching, 186; and field operations of V-2, 188–90, 192, 194, 201; promoted, 189; command change, 190, 193–4, 199, 255; and supply and storage, 201; on long-range missiles, 240; and V-2 effectiveness, 242–3, 250, 253*n*; and evacuation of Peenemünde, 255, 257; heads Working Staff, 255, 266; interrogated by Americans, 271–3, 282; documents recovered by Americans, 285, 314; interrogated by British, 291, 303; and Operation Backfire, 298–300, 303; detained by British, 303; in USA, 373

Dörnten (mine), 263–4, 283–4, 297, 314
Dorsch (of Todt Organization), 253*n*
Dreger, Alvin, 368
DuBridge, Lee A., 402
Durant, Frederick C. III, 374–5
Durkin, Flt.-Lt. T. J., 154

Easton, Air Commodore J. A., 171–2
Edse, Rudolf, 311–12
Ehricke, Lt. Krafft, 37, 58, 360, 373
Eisenhardt, Otto, 368
Eisenhower, Gen. Dwight D., 127, 198, 214, 248; and Peenemünde team, 282, 287; and Backfire operation, 297; and US V-2 launchings, 355; and Atlas rocket, 373; and US satellite programme, 377, 386–8
Ellis, Prof. Charles D., 98, 161, 165, 224, 294
Ellis, Clarence, 368
Elmi (engineer), 290
Engel, Rolf, 19
Engelbrunner, Gen. d'Aubigny von, 18
Enzian project, 80
Essame, Maj. G., 162
Explorer I, 383–6, 400
Explorer 8, 391

Faget, Max, 393
Falcon, Major (*later* Lt.-Col.) Norman L., 102, 108, 227
Falkenmayer, Karl, 339
Faulstich, Dr., 342
Fellersleben, 125
Ferrell, George, 368
Fi103 (F2G-76) *see* V-1 missile
Fichtner, Hans, 368
Fieseler, Gerhard, 50
Figge, Paul, 61–2, 65–6, 69, 72, 77–8, 81, 348
Finzel, Alfred, 356
Fischer, Prof. Hans, 4
Flakversuchstelle (FVS), 38
Flak E Flugabwehrkanonenentwicklung, 54
Fleischer, Karl Otto, 264, 281–2, 285
Fleming, William, 393
Fletcher, James C., 403
Fliet, C. van, 198
Floyd, Flt. Officer D., 154
Ford, Maj.-Gen. Elbert H., 371
Ford, Gerald R., 404
Förschner, Maj. Otto, 67–8, 72, 78
Fort Bliss (El Paso, Texas), 345–9, 352, 358–60
Frank, F. C., 148
Frankjel, Prof., 334
Fraser, Arthur, 369
Frau im Mond (film), 13–14
Freitag, Capt. Robert, 381
Friedland, 50
Friedrich, Hans R., 368, 373
Friedrichshafen, 60–61, 64–5, 83, 124–5
Fritsch, Maj.-Gen. Werner von, 24
Froelich, Jack E., 384
Fromm, Col.-Gen. Friedrich, 41, 49, 82, 188
Fuhrmann, Herbert, 368

Gagarin, Yuri, 391
Gaidukov, Gen., 306, 320–24
Galland, Maj.-Gen. Adolf, 124
Gardienenstangenraketen, 16
Garliński, Jòsef, 134, 137, 140–42, 144, 411
Gavin, Lt.-Gen. James, 382
Gedymin, Capt. Wiodzimierz, 145
Geiger, Prof. Johannes, 37
Geissler, Ernst, 259, 319, 347
Gengelbach, Werner, 88–9, 114, 255, 269–70
Genthe (purchasing officer, Peenemünde), 33
German Long Range Bombardment Commission, 174
Gesellschaft für Raketenforschung (Hanover), 16–17
Gesellschaft für Weltraumforschung (Breslau), 16
Gesman, George, 366
Gill, Air Vice Marshal W. C. C., 217–18
Ginsburg (electrical engineer), 331
Glennan, T. Keith, 390
Glushko, Col. Valentin P., 306, 321, 327, 334–5, 343
Glyth, Maj. V. L. U., 411
Goddard, Robert H., 28, 91, 151, 345, 378, 380
Goebbels, Josef, 63, 108, 248, 412
Goerdeler, Carl, 409
Gollin, Capt. G. J., 154, 157
Golovin, Nicholas, 394
Gonor, Gen. Leo, 325, 334–5, 338
Göring, Hermann, 27, 66, 255
Gorodomlya (Russia), 336–7, 339–41
Goslar, 263–4
Gothert, Bernard H., 316
Grau, Dieter, 89
Green, Charles F., 354
Greifswalder ('Oie') island, 33, 103

Gross, Lt. George, 286
Grosskowski, Prof. Janusz, 138
Groth, Wilhelm, 58
Gröttrup, Helmut, at Peenemünde, 46–8, 118; works for Russians, 56, 318–23, 326–7, 329, 331–42, 353; taken to USSR, 325–7
Gröttrup, Irmgard, 318, 323–6, 332–3, 335–6, 339, 342
Gröttrup, Peter, 335, 342
Gröttrup, Ulli, 335, 342
Grüne, Hans, 348, 372, 384
Grünow, Heinrich, 19, 21, 23
Gubbins, Maj.-Gen. Sir Colin, 140
Gunn, J. E., 369

Hackett, Vincent, 288
Hagerty, James, 386
Hamill, Maj. James P., 279–80, 306, 312–13, 315, 351–3, 357–60
Hamshaw Thomas, Wing Commander Hugh, 103, 107
Harbou, Thea von, 13
Harnish, Karl, 339
Harris, Air Chief Marshal Sir Arthur, 107, 111–13, 123
Harris, Gordon, 382
Hart, David, 368
Hart, Pastor George, 367–8
Harteck, Paul, 58
Hartmann, Maj.-Gen., 42
Haus Ingeburg, 2, 5, 6–8
Häussermann, Walter, 384
Hay, William, 369
Heeres Waffenamt, 32, 43
Heersversuchsstelle Peenemünde (HVP), 31–3, 36
Heidekraut (Tucheler Heide), 153
Heidelager (Blizna, Poland), as test site, 79–80, 82, 130–35, 147, 166, 192, 407, 411; evacuation and Russian occupation of, 152–3, 157; Allied mission to, 154–9; training at, 189, 191
Heigel, Maj., 36
Heimburg, Karl, 34–5, 85–7, 301, 363, 366, 372, 395, 399
Heinemann, Gen. Erich, 131, 175, 177, 179–82, 187, 190–91, 193, 199
Heinisch, Kurt, 14
Heisenberg, Prof. Werner, 58
Helfers, Lt.-Col. M. C., 175, 192, 240, 253*n*
Hellebrand, Bernd, 368
Heller, Gerhard, 4, 346, 351, 368, 399
Heller, Rainer, 368
Hermann, Rudolf, 33, 118, 122, 260
Hermes project and rockets, 277–8, 285, 315, 350, 370, 374
Hettlage, Prof. Karl Maria, 66–7, 87
Hill, Air Marshal Sir Roderic, 162, 206–13, 215–16, 220–22, 233–6
Himmler, Heinrich, courts von Braun, 45–6; and von Braun's arrest, 47–8; seeks control of rocket programme, 62–3, 80; and Kammler, 80–82, 194, 202; and Brauchitsch, 185*n*; and Hitler assassination attempt, 193; and rocket troops, 201; and Peenemünde evacuation, 258
Hitler, Adolf, death, 1, 7; visits Kummersdorf, 27; and V-2 programme, 28,40, 42–5, 60–62, 64, 66, 185–6, 193; and arrest of von Braun, 48; and labour difficulties, 63; and Mittelwerk production, 78; and Kammler, 81; and liquid oxygen production, 83; and secret weapons, 92, 94–5, 98, 174–5, 180, 240, 247, 412; Canaris and, 96; and V-1 targets, 181; assassination attempt on, 193; on V-2 targets, 198, 200; and V-weapons' effectiveness, 242, 249; scorched earth policy, 260; resistance to, 409
Hobbing, Lt. Enno, 316–17
Hochmuth, Lt. M. S., 280–83, 290, 306–7, 319, 351, 353
Hock, Johannes, 331, 335, 339–40
Hohmann, Lt.-Col., 187, 189, 194, 200
Höhne, Hans, 82
Holker, Rudolf, 268
Hollriegelskreuth, 125
Holmes, Col. Joel G., 283
Hölzer, Helmut, 40–41, 260, 301
Homer, Maj.-Gen. John L., 345–7
Hoover, Commander George W., 375
Hoover, J. Edgar, 358–9
Horstig, Col. Ritter von, 18, 25
Hosenthien, Hans, 118
Houbolt, John C., 393–4
Hubert (of Mittelwerk), 67
Hudson, Col. Carroll D., 371
Huntsville, Alabama, 363–9, 376–7, 386, 388
Hüter, Prof. Ernst, 35
Hüter, Hans, 11, 55, 399
Hüter, Uwe, 370
Huzel, Dieter, surrender to US troops, 6–8, 11, 257–8; and bombing of Peenemünde, 120, 122; hides secret rocket documents, 261–4, 281–2, 297, 314; and Operation Backfire, 297, 299–300; in USA, 357

Hydra operation, 112

Ilfeld, 69, 75
Iranek-Osmecki, Col. Kazimierz, 132, 134–5, 139, 141–2
Irving, David, 107, 405–7
Isayev, Aleksei M., 343
Isbell, Billy, 368
Ismay, Gen. Sir Hastings, 98, 107
Ismay, Capt. W. N., 298

Jaffke, Heinz, 338
Jannusch, Lt.-Col., 190
Jatos (Jet-assisted take-offs), 13, 32
Jodl, Col.-Gen. Alfred, 44, 46, 247
John, Lt.-Gen. Richard, 32
Johnson, Lyndon B., 394
Johnson, Roy, 384
Jones, Prof. R. V., on V weapons, 76, 94–7, 100, 104, 106–7, 113; on Polish intelligence, 133; on Heidelager, 147; on Polish and Swedish recoveries, 152; and Allied mission to Heidelager, 154, 157–8; on launch sites, 160; and Big Ben subcommittee, 162; on size of V-2 warhead, 167, 169–70; and anti-V-2 defences, 231; on economics of V-2 campaign, 247; on Ultra, 410
Journal of the British Interplanetary Society, 92, 93
Jungert, Wilhelm, 310–11
Juno rockets, 374, 377, 381–2, 384–5, 389, 391
Jupiter missile, 338, 374, 377–83, 387, 400

Kaiser, Hans K., 16
Kaltenbrunner, Gen. Hans, 48
Kame, Ingeborg, 119–20
Kammler, Obergruppenführer Hans, threats to Peenemünde scientists, 4–6, 256, 264–5; and labor programme, 63; and Mittelwerk, 66–7; ambitions and character, 80–83, 193, 195, 199–200; death, 81; and Lehesten plant, 87; and V-1 offensive, 184; and V-2 programme, 191; commands V-2 operations, 193–203, 222, 242; appointed Special Commissioner, 255, 328; and von Braun, 265–6; interrogated by Americans, 282; and Nuremberg trial, 303
Kapustin Yar, 332–4
Kármán, Theodore von, 238, 308
Karsch, Herbert, 349
Kaschig, Erich, 11, 353
Kavanau, Laurence, 394
Kegelduse rocket, 14
Keitel, Field Marshal Wilhelm, 44, 47, 175
Keller, K. T., 371–2
Kendall, Wing Commander Douglas, 101–2, 108–9, 161
Kennedy, John F., 390–94, 402
Kennedy Space Center, 354
Kenny, Flt.-Lt. André, 103, 108
Kerr, Senator Robert, 390
Kersten (*later* Schlidt), Dorette, 120, 255, 265
Kesselring, Maj.-Gen. Albert, 26, 200
Kettler, Kurt, 66–8, 71–2, 78
Kilmarx, Robert A., 326
Kirschstein, Dr., 301
Klein, Brig.-Gen. Julius, 361
Kleist-Schmenzin, Maj. Ewald von, 409
Klippel, Alfred, 331
Knacke, T. W., 316
Knerr, Maj.-Gen. Hugh J., 277, 281
Kochel (Bavaria), 50, 122
Kocjan, Antoni, 134–5, 138, 140
Kohn (master mechanic), 86
Kommission für Fernschiessen (Long-Range Bombardment Commission), 49–50, 174
Konev, Gen. I. S., 276
König, Capt., 38, 84
Kooy, J. M. J., 233
Korolev, Col. Sergei P., 306, 322–4, 333–6, 338, 343
Kostikov, A. G., 327
Krunichev, M. V., 327
Krupp, Alfred, 17
Küchen, Lt. Hartmut, 37, 40
Kühle, Capt. Dr., 69–70
Kummer, Maj., 266
Kummersdorf, 17–19, 21–2, 24, 27, 40, 92–3, 253
Kunze, Heinz, 66, 73
Kurilo, 331
Kurische Nehrung, Pillaru, 15
Kürs, Werner, 368–9
Kurzweg, Hermann, 118
Kutzevalov, Lt.-Gen. T. F., 327
Kuznetsov, Gen., 320

Lamplaugh, Maj.-Gen. S., 232
Lang, Alois, 265
Lang, Daniel, 354
Lang, Fritz, 13
Langer, Lt., 270
Langley Research Center, 392
Lasby, Clarence G., 276

Lee-Thompson, J., 390
Leeb, Gen. Emil, 32, 67–8, 78
Legut, Capt. Józef, 139
Lehesten (Thuringia), 86–7
Leigh-Mallory, Air Chief Marshal Sir Trafford, 206–8, 234
Lemnitzer, Gen. Lyman L., 382
Leningrad, 239*n*
Lennard-Jones, Prof. J. E., 161
Lewis, Richard S., 309
Ley, Robert, 120
Ley, Willy, 13, 15, 20, 91–3
Liebhafsky, Herman A., 272
Liège (Belgium), 87, 196, 198, 200, 239, 248, 251
Lindenberg, Hans, 8, 11, 74, 317
Lindenmayr, Hans, 38, 259
Lindner, Kurt, 266
Lippisch, Alexander, 316
Lloyd, Maj. C. W., 301
Lockhart, Brig. L. K., 295–6
Loki rocket, 351, 375
Long Range Bombardment Commission *see* Kommission für Fernschiessen
Long Range Proving Ground, 354
Love, Quincy, 368
Lovell, James, 396
Lovett, Robert A., 277
Löw, Albert, 16
Lubbock, Isaac, 106, 124, 162, 165, 172
Lührsen, Hannes, 288, 372
Lundquist, Charles A. & Mrs., 369

M1 rocket engine, 389
MSFC *see* Marshall Space Flight Center
MX 774 rocket, 350
McClelland, Brig.-Gen. H. M., 169
McClintic, Lt.-Col. S., 216
McDonald, Brig.-Gen. George C., 216
McElroy, Neil, 382–3
McNamara, Robert S., 394
Mäder, Dr., 33
Magnus, Kurt, 331, 342
Malenkov, G. M., 327
Man Very High project, 383
Manned Spacecraft Center (*later* Johnson Space Center), 392–3
Marienthal, 88
Marshall Space Flight Center, Huntsville, 389–90, 392–8, 401
Marshall, Gen. George C., 355
Massey, Capt. L. H., 154
Masterman, Sir John, 412–13
Masterman, Capt. Standish, 154
Mathes, Franz, 335
Maus, Hans, 118, 371
Mechtly, Eugene, 368–9
Medaris, Brig.-Gen. John B., 377, 379, 382–7, 389–90
Medmenham *see* Central Interpretation Unit
Meier, Otto, 331
Meisenheimer, Lt. John L., 384–5
Mercury spacecraft, 391; *see also* Redstone-Mercury
Merrill, Lt.-Commander Grayson P., 308
Merriman, Lt.-Col. A. D., 154
Merthin, Maj., 199
Merton, Sir Thomas, 162
Messerschmidt Me 163 aircraft, 104, 114
Metz, Gen. Richard, 190–94, 199, 241
Meyer, Capt. M. W. S., 299
Meyers, Dr., 311
Micinski, Czeslaw, 145, 147
Mikoš, Boleslaw, 135
Mikoyan, A. I., 327
Milch, Field Marshal Erhard, 41, 49–50, 64, 124
Millinger, Heinz, 265
Mimoyecques, 164, 166, 202
Mirak rockets, 13, 18–19, 92
Mittelbau, 66
Mittelwerk project, 65–77; labor, 70–71, 76; V-2 production, 73–5, 77–9, 89–90, 405–8; plant, 75; supplies, 76; Kammler and, 81; and US Special Mission V-2, 278–81; Russians and, 285, 320
Möller (engineer), 340
Montgomery, Field Marshal Sir Bernard Law, 183, 197, 198, 295
Morrison, Herbert, 167
Moser, Lt.-Col., 189
Moser, Robert, 385
Motyl (Poland) *see* Wildhorn III operation
Mrazek, William, 37, 267, 398
Mrazek, Mrs. W., 399
Mueller, George E., 395–6
Muenz, Willie, 40
Muller, Dr., 311
Müller, Gen. Heinrich, 48
Munnich, Karl, 331
Munro, A., 225
Murphree, E. V., 381
Musialek-Lowicki, Col. S. ("Miroslaw"), 142

Nadeau, Capt. Pershing J., 200
Närr, Anton, 338

National Aeronautics & Space Administration (NASA), 387–9, 391–4, 397, 401–3
National Space Institute, 404
Natter project, 80
Nebel, Rudolf, 13–16, 18–20, 26, 93
Nebelung (of Dörnten), 263–4, 282
Neu (production engineer), 65
Neubert, Erich ("Maxi"), 310–11, 314, 360
Neuhoffer, Kurt, 351
Newill, Prof. D. M., 161
Niedersachswerfen, 65, 67, 70, 76–7, 122, 278–81, 320; *see also* Mittelwerk
Niepokój, Zygmunt ("Norwid"), 137–8
Nike Hercules rocket, 350
Nimmegen, Erich, 256–8, 264–5
Nixon, Richard M., 402
Nöggerath, Wolfgang, 311–12
Nordhausen, V-2 production at, 26, 51, 59, 76–7; occupied by Russians, 278, 295, 306, 320; and US Special Mission V-2, 278–84, 294, 297
Nova booster, 392–4
Nye, Lt.-Gen. A. E., 98

Oak Ridge, Tennessee, 31
Oberammergau, 254–7, 270, 271
Oberjoch, 2, 6, 7
Oberth, Hermann, at Peenemünde, 12–14, 16, 26, 42, 56, 380; daughter killed, 85–6; Ley on, 93–4; and bombing of Peenemünde, 116–17; in USA, 378–9; and moon landing, 392
O'Hallaren, Bill, 10
Oie *see* Greifswalder
O'Mara, Lt.-Col. John A., 154
Opel, Fritz von, 13
Orbiter project, 376–8
Oskima, Hiroski, 411
Oslo report, 95–7, 409
Overcast operation *see* Paperclip operation

Paine, Thomas O., 401–3
Palaoro, Hans, 288, 301, 346
Paludin, Charles "Ted", 368
Paperclip (*formerly* Overcast) operation, 287, 292, 357, 362
Paris, 195, 196, 198, 200, 251
Parry, Albert E., 319
Patterson, Robert B., 286
Paul, Hans, 399
Pauley, Edward, 276
Paulus, Field Marshal Friedrich von, 42
Payne, Bob, 280
Peek, Flt.-Sgt., E. P. H., 103–4
Peenemünde, taken by Russians, 11, 59, 254–6, 261; as rocket research site, 26–31, 36, 185, 253; organization, 32–3; and university help, 35–6; recruiting for, 36–9; technical limitations, 39–40; bombed by RAF, 50–51, 62, 64, 106–7, 111–26, 128, 130–31, 168, 189; reorganized under Storch, 53–4; V-2 production at, 61–2; manpower, 62–3; British intelligence and aerial reconnaissance of, 95–7, 100–104, 107–8, 160–61, 165–6, 173, 410; training command, 186, 188; evacuation from, 254–7
Pegasus satellite, 393
Peiting, 11
Pendray, G. Edward, 91
Perren, W. G. A., 171–2
Petersen, Prof. Waldemar, 49, 255
Pflanz (master mechanic), 86
Pickering, William H., 308, 383–5
Pile, Gen. Sir Frederick, 162, 207–17, 231–2
Pioneer 4 space probe, 382, 400
Pobedonostsev, Col. Yuri A., 306, 325, 327, 334, 338
Poland, 130–48; V-2 recovered from, 139–48, 152–3; *see also* Heidelager
Polaris missile, 308
Poletz, von, 282
Poppel, Theodor, 310, 360
Portal, Air Chief Marshal Sir Charles, 109
Porter, Richard W., 272–4, 278, 285–6, 288, 315
Pose, Prof. H., 58
Post, Col. Kenneth G., 105, 162
Prufstand XII, 55, 58, 352
Pryor, Lt.-Col. R. M., 162
Püllenberg, Albert, 16–17
Putt, Lt.-Gen. Donald L., 362
Putze, Dr., 327

Quessel, W., 339
Quinn, Col. Horace B., 278, 283–5
Quistorp, Maria von (Frau Wernher von Braun), 360

R10 rocket, 329–31, 335, 337–9, 341
R14 rocket, 337–9, 341
R15 rocket, 339, 341
R134 rocket, 337

Raderach, 83–4, 87, 124–5
Raithel, William, 374
Raketenflugplatz, Berlin, 14–15, 18–20, 22; Hanover, 16
Raketenforschungsinstitut-Dessau, 19
Rashkov (engineer), 339
Rebitzki, Mrs. Engelhardt, 340
Rebstock plant, 88–9
Redl-Zipf plant, 84–7
Redpath, Maj. R. T. H., 299, 303
Redstone rocket, 350, 370–76, 378, 400
Redstone-Mercury rocket, 389, 391
Rees, Eberhard, at Peenemünde, 37, 53, 63, 123; in USA, 255, 281–2, 285, 310–11; interrogated by British, 291; on moon landing, 392; retirement, 398
Reinemund, Tech. Sgt. John A., 314
Reitsch, Hanna, 114
Repulsor rocket, 15
Retinger, Josef ("Brzoza"), 142, 145
Rheintochter project, 80
Richthofen, Maj. Wolfram von, 24–6
Rickhey, Georg Johannes, 71–2, 74, 77, 85–6, 124, 239*n*, 405–7
Riedel, Klaus, 14–15, 19–20, 26–7, 33, 35, 45–8, 55, 93
Riedel, Walter, 19, 21, 23, 32, 268–9, 281, 285, 351
Robertson, Howard P., 162, 281
Ronger, Dr., 324
Roosevelt, Franklin D., 168, 275
Rosen, Milton, 377
Rosenhead, Prof. L., 161
Rossberg, Maj., 190
Rossman, Maj.-Gen. Erwin, 259, 282
Rostow, Walter W., 169
Roth, Axel, 370
Roth, Ludwig, 33, 51, 54, 338, 351, 357, 371
Rother, Herr (innkeeper), 6, 8
Rowell, Capt. Robert, 108, 127, 161
Rowell, Maj.-Gen. S. F., 161
Rüdel, Gen. Günther, 241
Rudolph, Arthur, early career, 19, 22–3; at Peenemünde, 33, 37; on Saur, 62; and Mittelwerk, 67, 69–74, 77, 79; and Operation Backfire, 301; in USA, 352–3; on liquid propellants, 381; and Saturn rocket, 396; on V-2 production, 407
Rumschöttel, Oberstleutnant Richard, 261
Rundstedt, Field Marshal Gerd von, 181–2, 187, 197, 201, 248, 303
Russell, Bessie, 367
Rüsselsheim, 125
Rządzki, Józef ("Boryna"), 132

Sadelkow, 50
Safehaven project, 276
Sanders, Col. Terence R. B., 154, 156–8, 162
Sandys, Duncan, 98–103, 105–8, 110, 113, 157, 162, 413
Sänger, Eugen, 238, 327–9, 339
Satin, Alexander, 375
Saturn rockets, 376, 387–9, 392–7, 402
Sauckel, Fritz, 63
Saunders, Air Vice Marshal Robert H. W. L., 213
Saur, Karl Otto, 49, 60, 62, 66–7, 77–8, 81
Saurma, Count Friedrich von, 399
Sawatzki, 67, 69–73
Schade, Commodore H. A., 271
Scharnowski, Heinz, 299
Schattwald, 1–2, 7–8
Scheuffelen, Kurt, 350
Schilling, Harmuth, 368
Schilling, Martin, 34, 53, 86–7, 121, 353, 368, 373
Schlidt, Rudolf, 265
Schlier plant (Austria), 84–5, 87
Schmetterling (8–117) project, 80, 280
Schmidt, Helmut, 119, 347
Schmidt-Lossberg (of Rüstingkontor), 66–7
Schnarowski, Heinz, 11
Schneikert, Frederick P., 1–3, 7, 10
Schomburg, Maj.-Gen. August, 387, 389
Schroder, Paul, 38
Schuler, Albert, 11
Schüler, Max, 35
Schulze, August, 310, 360
Schulze, Sgt. Heinrich, 120, 259–60
Schüning, Dr. (of Mittelwerk), 72
Schwidetzky, Walter, 310, 312, 314, 373
Schwön, Dr. (of Mittelwerk), 72
Scotland, Lt.-Col. Andrew P., 303
Seamans, Robert C., jr., 394, 402
Searby, Group Capt. John, 113–14
Searcy, R. B. "Spec", 377
Seifert, Howard S., 308
Semmler, Rudolf, 412
Serov, Col.-Gen. Ivan A., 324, 328
Shawcross, Sir Hartley, 303
Shea, Joseph F., 393

Shedersky, Col., 306
Sheldon, Charles S., 401
Shepard, Alan B., 391, 400
Shershevsky, Alexander B., 13
Shirer, William L., 63
Shor, Lt.-Col. J. B., 154
Sieber, Werner, 399
Silnishnikov (engineer), 339
Silverstein, Abe, 388
Simon, Col. Leslie E., 313, 371, 374, 376
Simon, Capt. Neil, 108, 160
Singer, S. Fred, 375
Sinozersky, Lt.-Col. K., 154
Siracourt, 164, 166
Skylab space station, 395, 397, 400, 402
Smith, F. E., 162
Sokolov, Gen., 306
Solandt, Col. O. M., 162, 231
Somervell, Gen. Brehan Burke, 277
Sonderausschuss (Special Committee) A4, 61, 69–72, 81
Spaatz, Lt.-Gen. Carl, 277
Space Task Group, 392, 402
Sparkman, Senator John, 365
Special Committee A4 *see* Sonderausschuss A4
Special Operations Executive (SOE), 140–41
Speer, Albert, 174*n*; and building of Peenemünde, 31, 60; and forced labour, 38; sees first V-2 launchings, 41; and V-2 programme, 42–5, 185, 242, 249; and von Braun, 45, 48; decorates rocket men, 48–9; and long-range weapons, 49–50; and Dornberger's anti-aircraft responsibilities, 54; and rocket production, 60–62, 64, 405; and labour difficulties, 62–3; and Mittelwerk, 66; and Niedersachswerfen working conditions, 70; and Kammler, 81–2; on bombing of Peenemünde, 124; on effect of V weapons, 248–50, 253*n*; and V weapons, 412
Spiridonov, Col., 331, 336, 339
Spohn, Eberhard, 290
Spragins, M. B., 364
Sputnik 2, 382
Stacnowski, Mieczyslaw ("Sep"; "Macicj"), 132
Stagg, Group Capt. John, 384
Stahlknecht, Detmar, 61, 63
Stalin, Josef, 275, 328, 342
Stalin, Maj.-Gen. Vasily I., 328
Stalingrad, 42
Stauffenberg, Lt.-Col. Count Klause von, 193
Staver, Maj. Robert B., 265, 277–8, 280–88, 290, 315, 319
Stegmaier, Col. Gerhard, 63, 130–31, 188–91
Steinhoff, Ernst, at Peenemünde, 33, 37, 43, 318; after dispersal of facilities, 51; and bombing of Peenemünde, 114; and V-2 in Sweden, 148–9; surrender to Americans, 255, 266, 282, 285; US contract, 289; Russians seek, 290–91; interrogated by British, 291; in USA, 351, 355, 357
Steuding, Hermann, 33, 267–8
Stevenson, Flt.-Lt. D. W., 101–2
Stewart, Lt. Charles L., 2–5, 8–9, 11
Stewart, Homer J., 377
Stewart, Group Capt. Peter, 102–3
Stogner, Thomas, jr., 369
Stokes, Flt.-Lt. H. M., 273, 298
Storch, Paul, 53–4
Stradling, R. E., 225
Strohmayer, Anton, 84–5
Strom (*later* Turner), Virginia, 357, 361
Struszyński, Prof. Marceli, 138
Studien-Gesellschaft für Raketen, Frankfurt, 16
Stuhlinger, Ernst, in *VKN*, 37, 118; surrenders to Americans, 255, 285; in Huntsville, 365, 368; in USA, 379, 384–5; retirement designs, 399; on achievement, 399; tribute to von Braun, 404
Swanson, Conrad, 368
Sweden, V-2 recovered from, 140–41, 148–53, 163, 170–72
Szrajer, Kazimierz, 143–4

Taifun rocket, 54, 256, 258, 339, 351
Taylor, D., 231
Taylor, Sir Geoffrey, 162
Teeter, Ben, 368
Tessmann, Bernhard, surrenders to Americans, 6–8, 11; recruited by von Braun, 23; and submarine V-2, 55; and liquid oxygen, 84, 86–7; hides secret documents, 261–3, 281–2, 314; interrogated by Americans, 274; in USA, 357; retirement, 398–9
Thiel, Alfred K., 361, 373
Thiel, Walter, 28, 32, 35, 58, 63–4, 123
Thom, Lt.-Col. Georg, 188

Thomas, Gen. Georg, 28, 243
Thompson, Royce L., 252*n*
Thomson, Sir George, 99, 162, 169–70
Thor rocket, 378, 389
Thorneycroft, Peter, 394
"Those Fools at Tegel", 15
Tiesenhausen, Georg von, 55
Tikhonravov, Mikhail K., 334
Tizard, Sir Henry, 94
Todt, Fritz, 174, 186
Toftoy, Col. Holger N., responsibility for German scientists, 278–9, 283–5, 287–90, 313, 315, 318, 346, 352–3, 257–9, 371, 373–4; on space flight, 361; and space programme, 376; commands US Rocket & Guided Missile Agency, 380; retirement, 387
Tokady, Col. Grigori A. Tokaev, 320, 327–9
Trichel, Col. Gervais, 277–8, 285, 287, 346
Truman, Harry S., 275
Tschinkel, Johann G., 290, 368
Tsiolkovsky, Konstantin E., 12, 378, 380
Tumidajski, Col. K. ("Marcin"), 136
Turner, Lt.-Col. Harold R., 344–6, 354–5; marriage, 357
Tyulin, Col., 331

Ufa film company, 13, 15, 17
Uhl, Edward G., 404
Ultra, 410–11
Umpfenbach, Joachim, 321, 335–6, 339, 342
Unge, Theodor, 17
Usedom island *see* Peenemünde
Usher, Capt. E. M., 154
Ustinov, D. F., 327, 333–4, 336–7, 340
Utnik, Col. Marian, 140
Uytenbogaart, J. W. H., 233

V-1 missile (Fi 103; FZG 76), competition with V-2, 50, 64; production, 65, 73; Kammler and, 80; British intelligence and aerial photographs of, 109–10, 126–9, 410–13; bombing attacks on, 126–9; launch rates, 126, 210; operational use, 128, 147, 160, 166, 173–84, 209–20; and Polish intelligence, 135, 138, 147; and Diver sub-committee, 162; air launchings, 182, 184, 215, 244; British defensive action against, 206–20, 244; flight performance, 207; casualties from, 224, 245, 252; effectiveness and costs, 240–42, 244–7, 249–50, 253; numbers used, 244, 251
V-2 rocket (A4), development of, 2, 4, 10, 23, 26–8, 45, 60; technological roots, 12–13; production delays, 28–9, 36, 51; testing, 32, 39; Hitler and, 36, 42–5, 60, 185; guidance system, 40; first launchings, 41–2; support systems, 45; used against Antwerp, 48; submarine use, 55; later versions, 55–7; assembly and production, 59–90; contracts and production costs, 68, 74, 78, 239–50, 253; operational success rates, 79; Kammler and, 80–82; propellants, 82–9, 124, 163–5, 302–3; and British intelligence and technical analysis, 99–101, 133–4, 140–41, 147, 151–73, 410–13; effects of Allied bombing on, 124–6, 128; attacks on Britain, 128, 159, 167, 173, 192–4, 196–7, 222–38; launch platforms, 131, 158, 164–5, 167; and Polish intelligence, 132–8; fall and recovery in Sweden, 148–53, 163, 166, 170–72; and Big Ben sub-committee, 161–3; dimensions, 164, 166–9, 172, 221; and US intelligence, 168–70; operational use and deployment, 184–5, 189–90, 192–202; mobile launch sites, 186–7, 190, 192; training programme, 188–91; targets, 198–200; firing rate, 202, 234–5; firing record, 203–5, 251; British detection and defensive action against, 220–38; casualties from, 224, 236, 245, 252; warning systems, 227–30; performance, 236–7; effectiveness and cost, 239–50, 252; US Special Mission to recover, 278–85, 294; British demonstration firing (Backfire), 294–300, 302–9, 332; in USA, 313, 346, 349–50, 352–4; Russians and, 320–22, 327, 329–37; US launchings, 353–6; production quantities, 405–8
V-3 (*Hochdruckpumpe*), 80, 164, 202
V-4 rocket, 237–8
Valier, Max, 14
Van Allen, James A., 383–4, 386
Vanguard rocket, 377, 382–3
Vavilov, Maj. Anatoli, 261
Verein für Raumschiffahrt (*VfR*), 12–16, 18–20, 26, 92, 94, 98

Vershinin, Air Marshal K. A., 327
Versuchskommando Nord (*VKN*), 36–8, 52, 118, 122
Viebach, Fritz, 334, 342
Vöcklabruck, 84, 86
Voronov, Marshal Nikolai N., 331
Voskresensky, L. A., 334
Voznesensky, M. A., 327–8

Wa Prüf 10 & 11, 32, 40, 53
Wac Corporal rocket, 356
Wachtel, Col., 174, 180–82, 184
Waciorski, Stefan ("Funio"), 140
Waffenamt Prüfwesen, 32
Wagner, Prof. Carl, 35, 349, 357
Wagner, Prof. Herbert, 271, 275
Walmsley, Robert, 411
Walter, Col. Eugen, 177–8, 183, 187, 189, 193–4
Walther, Prof. Alwin, 35
Walther, Erich, 123
Wasserfall rocket project, 51–4, 58, 80, 123, 229, 249, 256, 258–9; recovered by Americans, 281, 283–4; Russians and, 339
Watson-Watt, Sir Robert, 162, 172, 225–6, 231
Watten (France), 106–7, 128, 164, 166, 168, 186, 193, 202
Webb, James E., 390–91, 394, 398
Weber, Lt.-Col. Wolfgang, 131, 194, 298, 301, 305
Weeks, Lt.-Gen. Sir Ronald, 294
Wehling (of Wifo), 66
Wei Ching Lin, 383
Weidner, Hermann, 86, 121, 268, 398
Weimar (Buchenwald), 125
Weise, Col.-Gen. Hubert, 113
Westphal, Gen. Siegfried, 200
Wheeler, W. H., 162
Whipple, Fred L., 375
White Sands, Texas, 306–7, 346, 349, 352–4, 356
Wiener-Neustadt (Austria), 61, 64–5, 124–5
Wiesman (Wiesemann), Walter, 38, 268–9, 348–9, 367, 398
Wiesner, Jerome B., 394
Wifo (*Wirtschaftliche Forschungsgesellschaft*), 65–6, 68
Wildhorn III operation ("Most III"), 140–47, 152
Wilkinson, Squadron Leader, 150
Wilkinson, Flt.-Lt. G., 154
Wilson, Charles E., 371, 377–8, 380, 386, 388
Wilson, Col. W. I., 296
Winkler, Johannes, 13, 15, 19
Winterbotham, Group Capt. Frederick W., 410
Wiśniewski, Col. Michal, 135
Witzell, Admiral Karl, 41
Wizernes (France), 26, 164, 166, 187, 193, 202
Wohlman, Prof. Walter, 35
Wolff, Waldemar, 335
Wolkowinski, Stanislaw ("Lubicz"), 141–2
Wormser, Eric M., 313–14, 352
Wright, T. P., 78

X-4 (8–344) project, 80

Yakolev, A. S., 327
Yates, Maj.-Gen. Donald N., 384
Young, David, 375

Zand, Lt.-Col. Steven J., 154
Zanker (of Mittelwerk), 67
Zanssen, Gerhard, 116
Zanssen, Col. Leo, 32, 43, 261
Zanssen, Frau Leo, 116, 261
Zawadski, Josef, 138
Zeiler, Albert, 42, 131–2
Zempin, 174–5, 178
Zhukov, Marshal Georgi K., 286
"Zigzag" (agent), 412
Zimmermann (engineer), 305
Zinnowitz, 125, 127
Zippelius, Lt.-Col. Wilhelm, 195–6, 199, 299–300, 302
Zobel, Theodor, 311–12
Zoike, Helmut and Frau, 119, 368
Zwicky, Fritz, 260, 273